Chemorheology of Polymers: From Fundamental Principles to Reactive Processing

Understanding the dynamics of reactive polymer processes allows scientists to create new, high value, high performance polymers. *Chemorheology of Polymers* provides an indispensable resource for researchers and practitioners working in this area, describing theoretical and industrial approaches to characterizing the flow and gelation of reactive polymers. Beginning with an in-depth treatment of the chemistry and physics of thermoplastics, thermosets and reactive polymers, the core of the book focuses on fundamental characterization of reactive polymers, rheological (flow characterization) techniques and the kinetic and chemorheological models of these systems. Uniquely, the coverage extends to a complete review of the practical industrial processes used for these polymers and provides an insight into the current chemorheological models and tools used to describe and control each process. This book will appeal to polymer scientists working on reactive polymers within materials science, chemistry and chemical engineering departments as well as polymer process engineers in industry.

Peter J. Halley is a Professor in the School of Engineering and a Group Leader in the Australian Institute for Bioengineering and Nanotechnology (AIBN) at the University of Queensland. He is a Fellow of the Institute of Chemical Engineering (FIChemE) and a Fellow of the Royal Australian Chemical Institute (FRACI).

Graeme A. George is Professor of Polymer Science in the School of Physical and Chemical Sciences, Queensland University of Technology. He is a Fellow and Past-president of the Royal Australian Chemical Institute and a Member of the Order of Australia. He has received several awards recognizing his contribution to international polymer science.

Chemorheology of Polymers

From Fundamental Principles to Reactive Processing

PETER J. HALLEY
University of Queensland

GRAEME A. GEORGE
Queensland University of Technology

CAMBRIDGE UNIVERSITY PRESS
Cambridge, New York, Melbourne, Madrid, Cape Town, Singapore, São Paulo, Delhi

Cambridge University Press
The Edinburgh Building, Cambridge CB2 8RU, UK

Published in the United States of America by Cambridge University Press, New York

www.cambridge.org
Information on this title: www.cambridge.org/9780521807197

First published 2009

Printed in the United Kingdom at the University Press, Cambridge

A catalogue record for this publication is available from the British Library

Library of Congress Cataloguing in Publication data
Halley, Peter J., 1966–
Chemorheology of polymers : from fundamental principles to reactive processing / Peter J. Halley, Graeme A. George.
p. cm.
Includes bibliographical references and index.
ISBN 978-0-521-80719-7 (hardback)
1. Reactive polymers. 2. Polymers–Rheology 3. Reactivity (Chemistry)
I. George, Graeme A. II. Title.
QD382.R43H35 2009
668.4′23–dc22 2009002702

ISBN 978-0-521-80719-7 hardback

Contents

Preface

Plastics are the most diverse materials in use in our society and the way that they are processed controls their structure and properties. The increasing reliance on plastics for high-value and high-performance applications necessitates the investment in new ways of manufacturing polymers. One way of achieving this is through reactive processing. However, the dynamics of reactive processes places new demands on characterization, monitoring the systems and controlling the complete manufacturing process.

This book provides an in-depth examination of reactive polymers and processing, firstly by examining the necessary fundamentals of polymer chemistry and physics. Polymer characterization tools related to reactive polymer systems are then presented in detail with emphasis on techniques that can be adapted to real-time process monitoring. The core of the book then focuses on understanding and modelling of the flow behaviour of reactive polymers (chemorheology). Chemorheology is complex because it involves the changing chemistry, rheology and physical properties of reactive polymers and the complex interplay among these properties. The final chapter then examines a range of industrial reactive polymer processes, and gives an insight into current chemorheological models and tools used to describe and control each process.

This book differs from many other texts on reactive polymers due to its

breadth across thermoset and reactive polymers
in-depth consideration of fundamentals of polymer chemistry and physics
focus on chemorheological characterization and modelling
extension to practical industrial processes

The book has been aimed at chemists, chemical engineers and polymer process engineers at the advanced-undergraduate, post-graduate coursework and research levels as well as industrial practitioners wishing to move into reactive polymer systems.

The authors are particularly indebted to students, researchers and colleagues both in the Polymer Materials Research Group at Queensland University of Technology (QUT) and at the Centre for High Performance Polymers (CHPP) at The University of Queensland (UQ). Special thanks are due to those former students who have kindly permitted us to use their original material. We would also like to thank Meir Bar for his countless hours of redrawing, editing and proof reading during his sabbatical at UQ. Thanks are extended also to Vicki Thompson and Amanda Lee from Chemical Engineering, UQ, for their tireless printing work. Thanks also go to the Australian Research Council, the Cooperative Research Centre scheme, UQ, QUT and individual industrial partners for their funding of reactive polymer research work.

1 Chemistry and structure of reactive polymers

The purpose of this chapter is to provide the background principles from polymer physics and chemistry which are essential to understanding the role which chemorheology plays in guiding the design and production of novel thermoplastic polymers as well as the complex changes which occur during processing. The focus is on high-molar-mass synthetic polymers and their modification through chemical reaction and blending, as well as degradation reactions. While some consideration is given to the chemistry of multifunctional systems, Chapter 2 focuses on the physical changes and time–temperature-transformation properties of network polymers and thermosets that are formed by reactions during processing.

The attention paid to the polymer solid state is minimized in favour of the melt and in this chapter the static properties of the polymer are considered, i.e. properties in the absence of an external stress as is required for a consideration of the rheological properties. This is addressed in detail in Chapter 3. The treatment of the melt as the basic system for processing introduces a simplification both in the physics and in the chemistry of the system. In the treatment of melts, the polymer chain experiences a mean field of other nearby chains. This is not the situation in dilute or semi-dilute solutions, where density fluctuations in expanded chains must be addressed. In a similar way the chemical reactions which occur on processing in the melt may be treated through a set of homogeneous reactions, unlike the highly heterogeneous and diffusion-controlled chemical reactions in the solid state.

Where detailed analyses of statistical mechanics and stochastic processes assist in the understanding of the underlying principles, reference is made to appropriate treatises, since the purpose here is to connect the chemistry with the processing physics and engineering of the system for a practical outcome rather than provide a rigorous discourse.

1.1 The physical structure of polymers

The theory of polymers has been developed from the concept of linear chains consisting of a single repeat unit, but it must be recognized that there are many different architectures that we will be discussing, viz. linear copolymers, cyclic polymers, branched polymers, rigid-rod polymers, spherical dendrimers, hyperbranched polymers, crosslinked networks etc., all of which have important chemorheological properties. Initially we will consider the theory for linear homopolymers (i.e. only a single repeat unit) in solution and the melt. This will then be extended to determine the factors controlling the formation of the polymer solid state.

The starting point for an analysis of the structure of linear polymers is the C–C backbone of an extended hydrocarbon chain, the simplest member of which is polyethylene. The

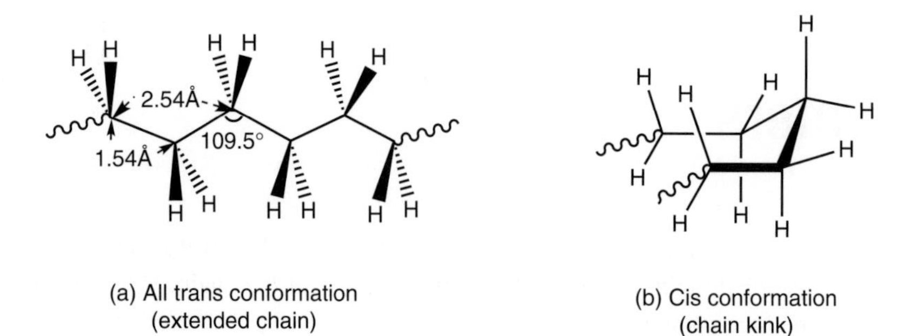

Figure 1.1. The carbon–carbon backbone of a polyethylene chain in its extended planar (all-*trans*) conformation (a) and its kinked, out-of-plane (*cis*) conformation (b).

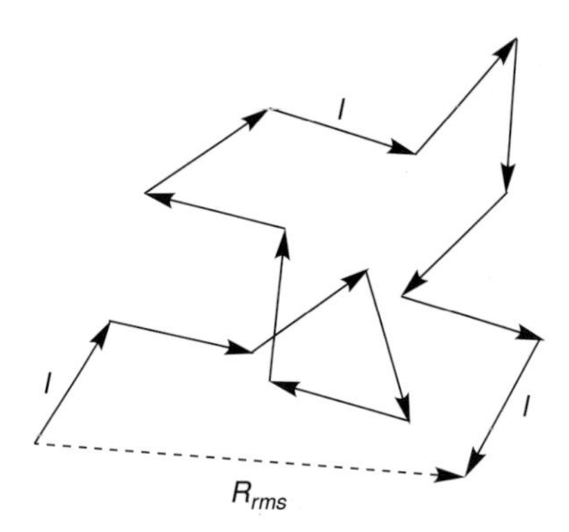

Figure 1.2. A schematic diagram of a freely jointed polymer chain with n segments of length l (in this case $n = 14$) showing the end-to-end distance, R_{rms}.

sp^3-hybridized tetravalent site of carbon that defines the angles and distances between the atoms along the backbone is shown in Figure 1.1 in

(a) an all-*trans* conformation with a planar C–C backbone and
(b) with the introduction of a *cis* conformation (as occurs in a cyclic six-membered hydrocarbon, cyclohexane), which allows the chain to kink out of the plane and change direction.

In the following we will initially consider the simpler concept of a freely jointed chain in which none of these constraints are present.

1.1.1 Linear polymers as freely jointed chains

The concept of polymer chains consisting of a freely jointed backbone which could occupy space as a random coil dated from 1933 when Kuhn defined a polymer chain as having n links of length l and the properties defined by a random flight in three-dimensions (Strobl, 1996). This is shown schematically in Figure 1.2.

This gave the coil the following properties: root mean separation of ends

$$R_{\text{rms}} = n^{1/2} l \tag{1.1}$$

and radius of gyration

$$R_g = (n/6)^{1/2} l. \tag{1.2}$$

Thus,

$$R_{rms}^2 = 6R_g^2. \tag{1.3}$$

From mechanics, the radius of gyration, R_g, is the average value of the first moment of all segments of the chain with respect to the centre of mass of the chain. If the chain is fully extended with no constraints regarding bond angles, i.e. a fully jointed chain as defined by Kuhn, then the maximum value of R_{rms} becomes

$$R_{max} = nl. \tag{1.4}$$

The ratio (R_{rms}^2/R_{max}) is a measure of the stiffness of the chain and is termed the Kuhn length. Thus, if there is a hypothetical freely jointed polyethylene that has 1000 carbon atoms separated by 1.54 Å then $R_{max} = 1540$ Å, $R_{rms} = 49$ Å and $R_g = 20$ Å.

The limitations of the random-flight model when applied to real polymer chains arise from

- the fixed bond angles
- steric interactions, which restrict the angles of rotation about the backbone.

This is apparent for polyethylene as shown in Figure 1.1. The restriction from a freely jointed chain to one with an angle of 109.5° between links increases R_{rms}^2 by a factor of two (namely the value of $(1 - \cos\theta)/(1 + \cos\theta)$). Other effects that must be taken into account are the restricted conformations of the chain due to hindered internal rotation and the excluded-volume effect, both of which may be theoretically analysed (Strobl, 1996). The excluded-volume effect was recognized by Kuhn as the limitation of real chains that the segments have a finite volume and also that each segment cannot occupy the same position in space as another segment. This effect increases with the number of segments in the chain as the power 1.2, again increasing the value of R_{rms} (Doi and Edwards, 1986).

When all of these effects are taken into account, a characteristic ratio C may be introduced as a measure of the expansion of the actual end-to-end distance of the polymer chain, R_0, from that calculated from a Kuhn model:

$$C = R_0^2/(nl^2). \tag{1.5}$$

Experimental values of this parameter are given in Table 1.1 and it may be seen that the actual end-to-end distance of a polyethylene molecule with 1000 carbon atoms (degree of polymerization DP of 500) is 126 Å from Equation (1.5) (i.e. $C^{1/2}n^{1/2}l$) rather than 49 Å from the Kuhn model, Equation (1.1). Data for several polymers in addition to polyethylene are given, including a rigid-rod aromatic nylon polymer, poly(*p*-phenylene terephthalamide) (Kevlar®), as well as the aliphatic nylon polymer poly(hexamethylene adipamide) (nylon-6,6).

Comparison of the values of C for the polymers with a flexible C–C or Si–O–Si backbone (as occurs in siloxane polymers) of about 6–10 with the value for the rigid-rod polymer of 125 demonstrates the fundamental difference in the solution properties of the latter polymer which has a highly extended conformation characteristic of liquid-crystal polymers. Equation (1.5) also shows that for a real chain the value of R_0 would be expected to increase as the half power of the number of repeat units, i.e. the degree of polymerization, $DP^{1/2}$.

Table 1.1. Experimental values of the characteristic ratio, C, for Equation (1.5)

Polymer	Characteristic ratio, C
Poly(ethylene)	6.7
Poly(styrene)	10.2
Poly(hexamethylene adipamide), Nylon-6,6	5.9
Poly(*p*-phenylene terephthalamide), Kevlar®	125

Conditions for observing the unperturbed chain

The data shown in Table 1.1 were experimentally determined from solutions under θ-temperature conditions. This involves measuring the properties when a solution has the characteristic properties which allow the polymer chain to approach ideality most closely. When a polymer chain is in solution the coil will expand due to polymer–solvent interactions and an expansion coefficient, α, is defined so that the actual mean square end-to-end distance $[R_{\rm rms}]_{\rm act}$ becomes

$$[R_{\rm rms}]_{\rm act} = \alpha R_0. \tag{1.6}$$

The magnitude of α depends on the forces of interaction between the solvent and the polymer chain. Thus, if the polymer is polar, when it dissolves in a polar 'good' solvent, it will expand and α is large. The converse is true for 'poor' (eg. non-polar) solvents and the chain will contract to lower than the unperturbed dimensions and, in the limit, the polymer may precipitate from solution. When a combination of solvent and temperature is found that is neither 'good' nor 'poor', i.e. $\alpha = 1$, then the chain–solvent and polymer–polymer interactions balance and R_0 is the unperturbed dimension of the chain. For a particular solvent, the temperature at which this occurs is the θ-temperature.

An interesting calculation is that of the volume occupied by the segments themselves compared with the total volume that the chain occupies. The diameter of a sphere within which the chain spends 95% of the time is about $5R_0$. Since the chain segments occupy only about 0.02% of this volume, the remaining space must be occupied by other chains of different molecules both when the polymer is under θ-conditions and in the presence of solvent molecules when it is expanded. Thus, except in very dilute solutions, polymer molecules interpenetrate one another's domains so that intermolecular forces between chains are significant.

Polymer chains in the melt

Polymer chains, in the melt, behave as if they are in the θ-condition, so the dimensions are those in the unperturbed state. This argument was put forward by Flory on energetic grounds and has been confirmed by neutron scattering (Strobl, 1996). The consideration begins with an analysis of the excluded-volume forces on an ideal chain. These arise from non-uniform density distributions in the system of an ideal chain in solution as shown in Figure 1.3.

This shows the way that the local monomer concentration, $c_{\rm m}$, varies from the centre of the chain ($x = 0$) to either end. The excluded-volume forces on the chain create a potential energy $\psi_{\rm m}$ sensed by each repeat unit, which depends on $c_{\rm m}$ and on a volume parameter $v_{\rm e}$ that controls their magnitude:

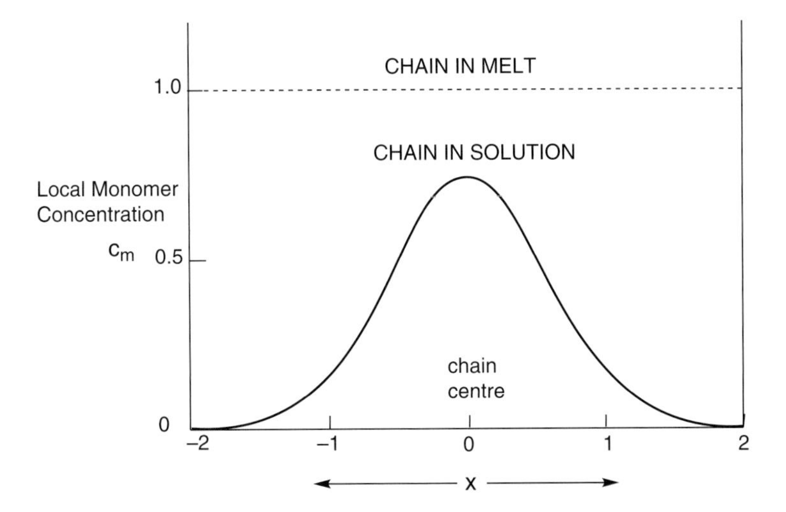

Figure 1.3. Comparison of the change in local monomer concentration with distance from the chain centre for a random chain in solution and in the melt. Adapted from Strobl (1996).

$$\psi_m = v_e c_m kT. \tag{1.7}$$

This produces a net force for all non-uniform density distributions so that for the bell-shaped distribution in Figure 1.3 there will be a net force of expansion of the chain. When the melt is considered, every chain is surrounded by a chain of the same type, so the concentration c_m is constant in all directions (the dotted line in Figure 1.3). No distinction is drawn between repeat units on the same or different chains. (As noted above, there will be interpenetration of chains in all but dilute solutions.) The result is that there is no gradient in potential and there are no forces of expansion. In effect, the polymer chain in the melt behaves as if the forces of expansion due to excluded volume were screened from each chain and the dimensions are those for the unperturbed chain.

This result may, by a similar argument, be extended to the interpenetration of chains as random coils in the amorphous solid state. These results will be of importance when the rheological properties of the melt through to the developing solid are considered in Chapters 2 and 3.

1.1.2 Conformations of linear hydrocarbon polymers

Figure 1.1 showed the planar zigzag (a) and the kinked chain (b) as two possible ways of viewing the chain of polyethylene. The conformation that the chain will adopt will be controlled by the energy of the possible conformers subject to the steric and energetic constraints dictated by the structure. The main feature of transforming from the stretched chain (a) to the coil through structures such as the *cis* conformation (b) depends on the rotation about the C–C backbone. The remaining degrees of translational and vibrational freedom will affect only the centre of mass and the bond angles and bond lengths, not the molecular architecture.

The possible rotational conformations possible for the chain can be envisioned by focussing on a sequence of four carbon atoms as shown in Figure 1.4(a).

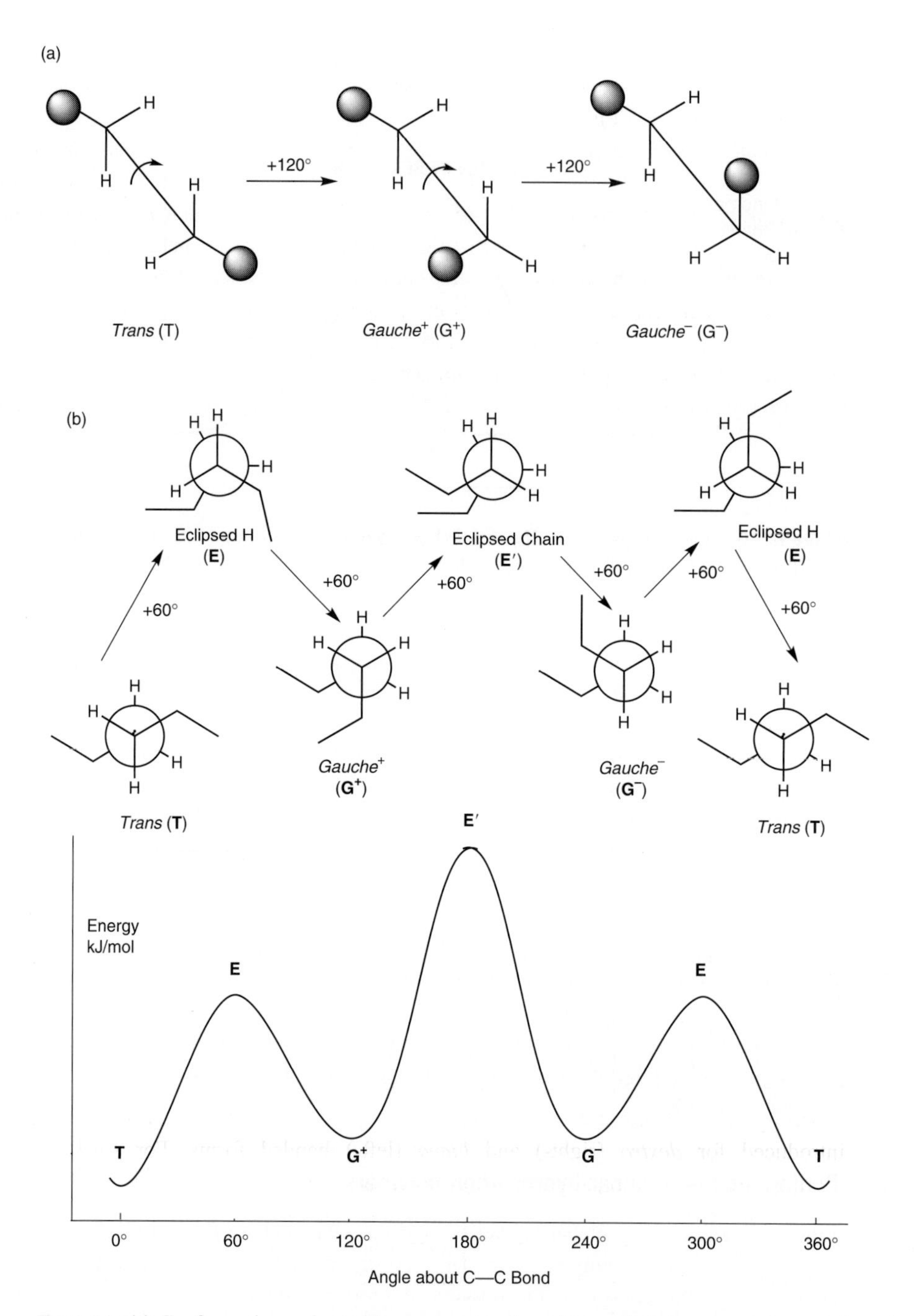

Figure 1.4. (a) Conformations adopted by a segment of a polymer chain by successive rotation about a C–C bond. The balls represent the carbon atoms from the continuing chain (initially in an all-*trans* extended-chain conformation). (b) Changes in conformational energy on successive rotation of an all-*trans* extended-chain conformation by 60° about a C–C axis.

This shows the successive rotation by 120° about the central C–C bond of the adjacent methylene group. Initially all carbon bonds lie in a plane and then after each rotation a hydrogen atom lies in the initial plane. A detailed analysis may be made of the rotational isomeric states of model compounds progressively from ethane, butane and pentane to determine the energy states of the conformers and, from the Boltzmann distribution, their populations (Boyd and Phillips, 1993).

The depiction of the conformers is facilitated by a simple schematic approach in which the atoms in Figure 1.4(a) are viewed along the central C–C bond initially in the *trans* (T) conformation and then rotation of the groups clockwise by 60° occurs in succession about this axis (Figure 1.4(b)). Analysis of these conformations identifies the energy maxima (eclipsed, E, conformations) and the energy minima (*trans* and *gauche* conformations) separated by up to 21 and 18 kJ/mol, respectively, as shown in the energy profile in Figure 1.4(b).

For polyethylene, the actual bond rotation from the *trans* (T) position to the other stable conformers (the *gauche* positions, G^+ and G^-, respectively) is slightly less than 120° due to unsymmetrical repulsions (Flory *et al.*, 1982). There are situations in which the repulsion due to steric crowding results in further deviations. For example, the sequence TG^+G^- will produce a structure with a sharp fold where the steric repulsion between methylene groups no longer allows an energy minimum. This is accommodated by a change in the angle of rotation giving an angle closer to that for the *trans* position (the so-called pentane effect) (Boyd and Phillips, 1993, Strobl, 1996).

When other groups are introduced into the polymer chain, such as oxygen in poly(oxy methylene) $[-CH_2-O-]_n$, the most stable conformation is no longer the all-*trans* chain but the all-*gauche* conformation $G^+G^+G^+$, etc. This means that the chain is no longer planar but instead is helical. The stability of the *gauche* conformation over *trans* is linked in part to the electrostatic interactions due to the polar oxygen atom in the chain (Boyd and Phillips, 1993). These conformations translate to the most stable structure expected at low temperatures. However, the low energy barriers between isomeric states mean that in the melt a large number of conformations is possible, as indicated in the previous section where the melt is seen to reproduce the properties of an ensemble of ideal random interpenetrating coils.

Asymmetric centres and tacticity

The structures considered above have been concerned with the behaviour of the backbone of the polymer. On proceeding from polyethylene to the next member in the series of olefin polymers, polypropylene, $[-CH_2-CH(CH_3)-]_n$, an asymmetric centre has been introduced into the backbone, in this case the carbon bearing the methyl group. An asymmetric centre is one where it is possible to recognize two isomeric forms that are mirror images and not superimposable. These are often described as optical isomers and the terms *d* and *l* are introduced for *dextro* (right-) and *laevo* (left-) handed forms. For small molecules these isomers may be resolved optically since they will rotate the plane of polarization in opposite directions.

For macromolecules it is useful to consider the structure of the polymer resulting from monomer sequences that contain the asymmetric centre. Figure 1.5 shows the two possibilities for the addition of the repeat unit as sequences of *d* units or *l* units to give meso (*m*) diads (*dd* or *ll*) when adjacent groups have the same configuration or racemic (*r*) diads (*dl* or *ld*) when they are opposite. If these sequences are repeated for a significant portion of the chain then we can define the tacticity of the polymer as being principally

isotactic if they are . . . mmmmmmmmm . . .
syndiotactic if they are . . . rrrrrrrrrrrrrr . . .
atactic if they are random . . . mmrmrrrmrmr . . .

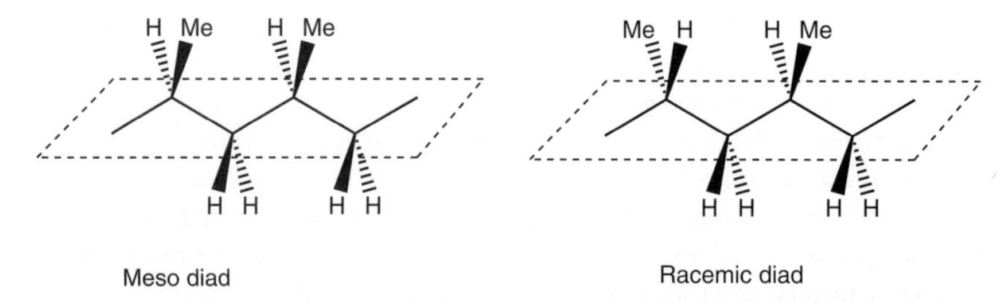

Figure 1.5. A schematic diagram illustrating meso (m) and racemic (r) diads.

As will be discussed later, special synthetic techniques are required to achieve isotactic and syndiotactic structures, and polypropylene, the example above, achieved commercial success only through the discovery of stereoregular polymerization to achieve the isotactic structure. The measurement of the degree of tacticity of a polymer is achieved through ^{13}C NMR studies of the polymer in solution (Koenig, 1999).

Isotactic polypropylene will adopt a conformation very different from the extended chain of polyethylene. In the early part of Section 1.1.2 it was noted that the minimum-energy conformations were considered to be attained by rotation about the C–C backbone and this introduced the possibility of *gauche* conformers as alternative energy minima. This can now be performed on the meso dyad in isotactic polypropylene by considering rotations about the two C–C bonds that will minimize the interactions between the pendant methyl groups. The starting point in this analysis is the nine near *trans* and *gauche* conformers since these define the local minima in energy of the backbone in the absence of the methyl groups. Introduction of the steric repulsion by the methyl groups in a TT conformation (Figure 1.5) suggests that this is not going to be a likely conformation and the conformers which are able to minimize the repulsion due to methyl groups in a meso dyad are limited to TG^- and G^+T. Just as a helix was generated when *gauche* conformers were accessible minima in poly(oxymethylene), so too we have two possible helices if the chain consists of m-dyads as in isotactic polypropylene. For TG^- it will be right-handed and for G^+T it will be left-handed. This helix will have three repeat units in one turn of the helix, i.e. a 3/1 helix, and this is the form which crystallizes.

In syndiotactic polypropylene, the methyl groups are well separated and the TT form is favoured, but there are other energy minima among the *gauche* conformations and TT/G^+G^+ and TT/G^-G^- sequences can generate left- and right-handed helices, respectively, where the repulsions are minimized (Boyd and Phillips, 1993). The chains may crystallize both in the TT and in the TTG^+G^+ form, so syndiotactic polypropylene is polymorphic.

1.1.3 Molar mass and molar-mass distribution

The length of the polymer chain or the degree of polymerization, DP, will have a major effect on the properties of the polymer since this will control the extent to which the polymer chain may entangle. The changes in this degree of polymerization that may occur on processing, resulting in either an increase (crosslinking) or a decrease (degradation) in DP, will have a profound effect on the properties both of the melt (e.g. viscosity) and of the resulting solid polymer (strength and stiffness). A formal definition of DP and thus the **molar mass** (or, less rigorously speaking, **molecular weight**) of a polymer is required in order to investigate the effect on properties as well as the changes on processing.

The addition polymerization reactions, discussed later in Section 1.2, result in the growth of polymer chains that consist of chemically identical repeat units arising from addition reactions of the original monomer, terminated by groups that will be chemically different from the repeat unit due to the chemistry of the reaction, the starting materials (e.g. initiators, catalyst residues), which may be attached to the chain, and impurities. Since these are generally only a very small fraction of the total polymer mass, the effect of the chemistry of the end groups can be ignored to a first approximation, although their quantitative analysis provides a method for estimating the number average molar mass as discussed below. Particular 'defects' such as chain branching, must be taken into account when the molar mass–property relationships are developed since the chain is no longer linear.

The mass of the linear polymer chain is thus related directly to the number of monomer units incorporated into the chain (DP) and will be $M_0 \times \mathrm{DP}$, where M_0 (g/mol) is the molar mass of the monomeric repeat unit. Thus, if all chains grew to exactly the same DP, then $M_0 \times \mathrm{DP}$, would be the molar mass of the polymer. If the end groups on the chain can be readily and uniquely analysed, then an average molar mass, M_n, or number-average molecular weight (as discussed in the next section) can be immediately determined since, if there are a mol/g of end group A and b mol/g of end group B then

$$M_n = 2/(a+b)\ \mathrm{g/mol}. \tag{1.8}$$

The conformation, end-to-end distance and radius of gyration of the polymer would be described by the simple considerations in Section 1.1.1. In the real polymer, the length of the polymer chain is controlled by the statistics of the chemical process of polymerization, so the distribution of chain lengths will depend on the reaction chemistry and conditions. The distribution is **discontinuous** since the simple linear chain can increase only in integral values of the molar mass of the repeat unit, M_0. The chain mass also includes that of the end groups M_e, so the first peak appears at $M_0 + M_e$, and then increments by $\mathrm{DP} \times M_0$ as shown in Figure 1.6(a). When the molar mass is low, as in oligomers, the individual polymer chains may be separated by chromatographic or mass-spectroscopic techniques and a distribution such as that shown in Figure 1.6(a) is obtained. For the large molar masses encountered in vinyl polymers ($>10^5$ g/mol) the increment in molar mass for each increase in DP is small and the end-group mass is negligible compared with the total mass of the chain. The distribution then appears to be continuous and sophisticated analytical methods such as MALDI-MS are required to resolve the individual chains (Scamporrino and Vitalini, 1999).

Size-exclusion chromatography (SEC) has become the technique of choice in measuring the molar-mass distributions of polymers that are soluble in easily handled solvents (Dawkins, 1989). The technique as widely practised is not an absolute method and a typical SEC system must be calibrated using chemically identical polymers of known molar mass with a narrow distribution unless a combined detector system (viscosity, light scattering and refractive index) is employed.

The effect of the chemical reactions during polymer synthesis on the molar-mass distribution is discussed in Section 1.2, but prior to this it is important to consider the various averages and the possible distributions of molar mass that may be encountered. It is then possible to examine the experimental methods available for measuring the distributions and the averages which are of value for rationalizing dependence of properties on the length of the polymer chain.

Molar-mass distributions and averages

The definitions of molar mass and its distribution follow the nomenclature recommended by the International Union of Pure and Applied Chemistry (IUPAC) (Jenkins, 1999).

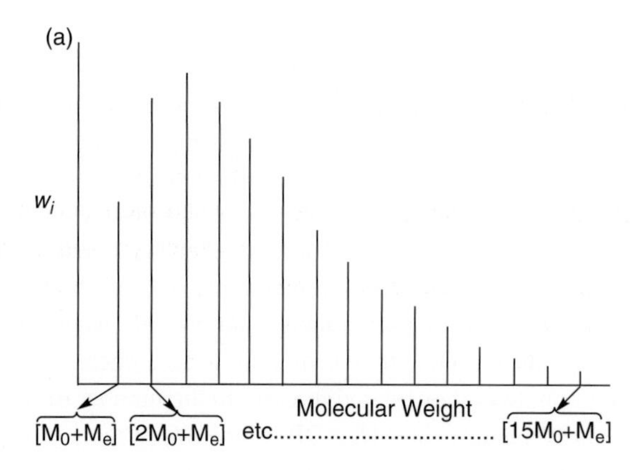

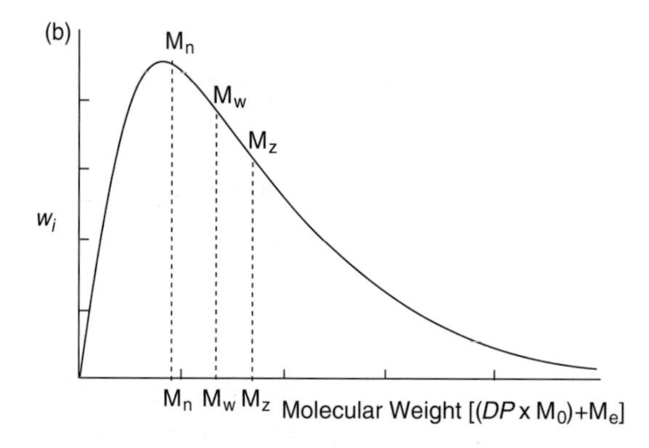

Figure 1.6. Illustration of the discontinuous distribution of polymer chain lengths (a) and the apparently continuous distribution and molar-mass averages (b) from a polymerization; w_i is the weight fraction of each chain length.

The simple averages that are used for property–molar-mass relations of importance in chemorheology as used in this book are the following:

(a) the **number average**, M_n:

$$M_n = 1 \Big/ \sum_i (w_i/M_i) = \sum_i N_i M_i \Big/ \sum_i N_i, \tag{1.9}$$

where w_i is the weight fraction of species i (i.e. $\sum_i w_i = 1$) and N_i is the number of molecules with molar mass M_i;

(b) the **weight average**, M_w:

$$M_w = \sum_i w_i M_i = \sum_i N_i M_i^2 \Big/ \sum_i N_i M_i; \tag{1.10}$$

(c) the z **average**, M_z:

$$M_z = \sum_i w_i M_i^2 \Big/ \sum_i w_i M_i = \sum_i N_i M_i^3 \Big/ \sum_i N_i M_i^2. \tag{1.11}$$

Thus, when considering what these different averages describe regarding the actual molar-mass distribution, it is useful to consider the continuous distribution as shown in Figure 1.6(b).

It is seen that the averages correspond to the first, second and third moments of the distribution and the ratio of any two is useful as a way of defining the breadth of the distribution. Thus the **polydispersity** is given by the ratio M_w/M_n. A normal or Gaussian distribution of chain lengths would lead to $M_w/M_n = 2$.

The experimental measurement of these averages has largely been performed on polymers in solution (Hunt and James, 1999). Since M_n depends on the measurement of the number of polymer chains present in a given mass, colligative properties such as vapour-pressure depression ΔP (measured by vapour-phase osmometry) and osmotic pressure (measured by membrane osmometry) relative to the pure solvent, can in principle provide the molar mass through an equation of the form

$$\Delta P = Kc/M_n, \tag{1.12}$$

where c is the concentration of the polymer in solution and K is a constant for the colligative property and the solvent. As noted before, end-group analysis also provides a measure of M_n. All of these techniques lose precision at high values of molar mass because the change in property becomes extremely small. As may be seen from Figure 1.6(b), the number-average molar mass is biased to low molar mass.

The weight-average molar mass, M_w, may be obtained by light scattering (Berry and Cotts, 1999). An analysis of the Rayleigh scattering of a dilute solution at various angles and concentrations as well as the difference in the refractive index between solution and solvent for these concentrations allows the measurement of the weight-average molar mass, as well as the mean square radius of gyration. The technique is extremely sensitive to any scattering impurity or particle and any aggregation that may occur.

Ultracentrifugation is less sensitive to these effects and also enables a value of M_w to be obtained, but because of the specialized nature of the equipment required is not as widely used for the study of synthetic commercial polymers as light scattering. It is possible to determine a value for M_z through measurements of sedimentation at various rotor speeds (Budd, 1989). As noted in Figure 1.6(b), the weight-average molar mass is biased to higher molar mass on the distribution.

Viscosity studies of dilute solutions provide a convenient relative measure of the molar mass and the resultant average, M_v, will lie closer to M_w than to M_n. The behaviour of polymer melts will be discussed in detail later, but it is noted that the melt viscosity is a strong function of the weight-average molar mass since the parameter m in the relation

$$\eta = kM_w^m \tag{1.13}$$

changes from 1.0 to 3.4 when the critical molar mass for entanglements is reached. This value varies with the chemical composition of the polymer.

1.1.4 Development of the solid state from the melt

Although this book is intended to address the chemorheology and reactive processing of polymers, the chemical reactions in the melt phase (e.g. branching reactions, degradation) may affect the subsequent solid-state and performance properties of the polymer. Furthermore, the end product of the reactive processing is the solid polymer and the transformation process from the liquid to the solid state of a polymer is fundamental to the success of the processing operation. It is therefore important to examine the way the polymer achieves its solid-state properties, and one of the most important properties

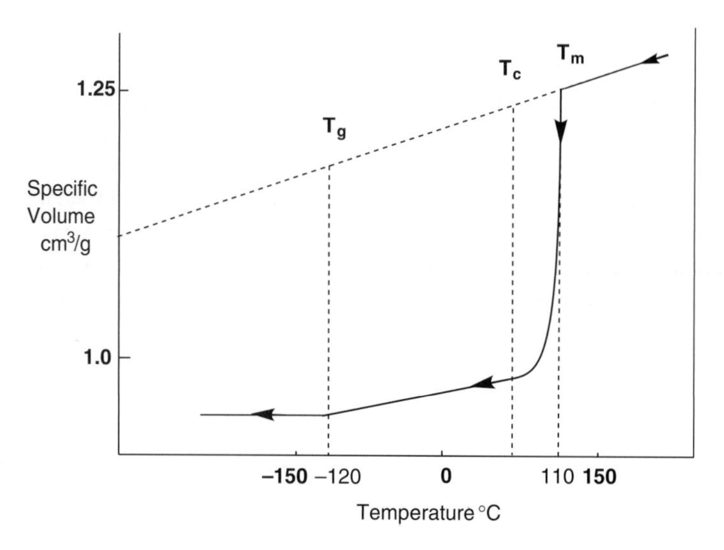

Figure 1.7. Changes in specific volume as a polymer melt (e.g. polyethylene) is cooled from above the melting temperature (T_m) to below the glass-transition temperature (T_g).

of the polyolefins is their semi-crystalline nature, i.e. the solid polymer contains both **amorphous** and **crystalline** material.

If we consider the process of cooling molten polyethylene, there will be a progressive decrease in the volume that the chains occupy. This specific volume, V_s, is the reciprocal of the density and this is shown in Figure 1.7 for the case on cooling the polymer from 150 °C to −150 °C. It is seen that there is a linear decrease in V_s with decreasing temperature, which is consistent with the coefficient of thermal expansion.

As the temperature of crystallization, T_c, is approached, there will be a sudden decrease in V_s and an exothermic process corresponding to a first-order phase transition. Thermodynamically this corresponds to a discontinuity in the first derivative of the free energy, G, of the system with respect to a state variable, i.e. in this case a discontinuity in volume:

$$[\partial G/\partial P]_T = V. \tag{1.14}$$

The extent to which this occurs in the polymer is controlled by a number of factors, and the volume fraction of crystalline material, φ_c, is related to the density, ρ, by

$$\varphi_c = (\rho - \rho_a)/(\rho_c - \rho_a), \tag{1.15}$$

where ρ_c and ρ_a are the densities of the crystalline and amorphous phases of the polymer, respectively. These values may be obtained from X-ray diffraction and scattering data. The actual structure of this crystalline material is considered in the following section.

The process of crystallization from the melt takes a considerable period of time, in contrast to the situation with a low-molecular-mass hydrocarbon (e.g. $C_{44}H_{90}$) (Mandelkern, 1989) that will crystallize over a temparature range of less than 0.25 °C. This results in the curvature in Figure 1.7, since significant undercooling, of up to 20 °C, is required in order for the crystallinity to develop. The detailed curve profile and the degree of crystallinity, φ_c, depend both on the degree of polymerization and on the molar-mass distribution (Mandelkern, 1989). These results highlight the reason for the undercooling, and this is the difficulty of extracting ordered sequences of the polymer chain and forming a thermodynamically stable structure. As discussed in the next section, this is even more difficult for polymers that have a more complex conformation than polyethylene. Thus a semi-crystalline structure will always result and the detailed morphology will depend on the cooling rate.

Further cooling of the polymer below T_c results in a further decrease in V_s, which again follows the coefficient of thermal expansion of the solid polymer, until there is a change in slope of the plot at the glass-transition temperature, T_g. This is a second-order transition (in contrast to melting, which is a first-order transition) since there is a discontinuity in the second derivative of the free energy with respect to temperature and pressure, i.e.

$$[\partial^2 G/\partial P^2]_T = [\partial V/\partial P]_T = -\kappa V, \tag{1.16}$$

where κ is the compressibility.

Similarly, there is a discontinuity in

$$-[\partial^2 G/\partial T^2]_P = [\partial S/\partial T]_P = C_p/T. \tag{1.17}$$

Thus there is a step change in heat capacity C_p at the glass transition, which is most conveniently studied by differential scanning calorimetry (Section 3.2).

It is also seen for the coefficient of thermal expansion,

$$(\partial/\partial T)[(\partial G/\partial P)_T]_P = [\partial V/\partial T]_P = a\,V. \tag{1.18}$$

As shown in Figure 1.7 there is a decrease in the coefficient of thermal expansion, α, to values lower than extrapolated from the polymer melt (dotted line). This results in the polymer having a lower density than would be predicted and there is thus a measurable free volume, V_f, which has an important bearing on the properties of the amorphous region of the polymer. This and the detailed analysis of the glass transition are considered after the molecular requirements for polymer crystallization and the structure of the crystalline region.

Polymer crystallinity

It was noted in Section 1.1.3 that, when one moves to polymers more complex than polyethylene, the likelihood of the polymer being able to crystallize depends on the chemical composition, in particular whether the repeat unit has an asymmetric centre. When it does, then the ability to crystallize rapidly diminishes with the amount of atactic material in the polymer. Fully atactic polymers will generally be amorphous and the properties of the glass resulting from the cooling of an atactic polymer from the melt are discussed in the following section.

As the simplest example of a linear polymer, the crystallinity of polyethylene has been most widely studied. A consideration of the thermodynamics of melting of a series of n-alkanes provides the starting point for extension to oligomers of differing chain length. The equilibrium melting temperature of a perfect polymer crystal composed of infinitely long chains cannot be determined, but it may be approached by extrapolation from calculations for chains of finite length. This enables the melting point to be calculated and compared with the experimental values, which reach an asymptotic value above $n = 300$ of 145 °C, compared with the observed value for high-density polyethylene of 138.5 °C. It was noted earlier that crystallization requires undercooling by 20 °C compared with T_m, so it is clearly a non-equilibrium process. Nucleation of crystallization is thus important and the process of creating a crystal analogous to an n-alkane from a melt that consists of highly entangled polymer

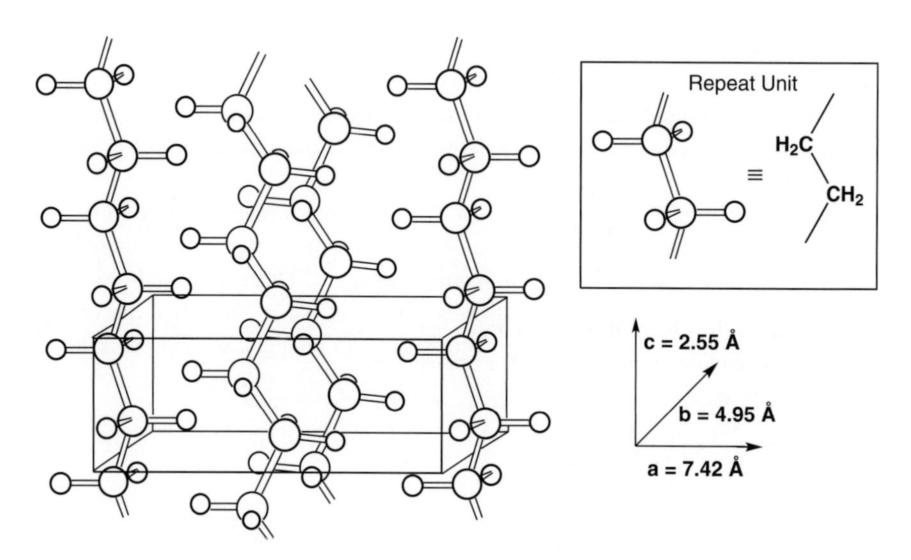

Figure 1.8. The orthorhombic crystal structure of polyethylene. Adapted from Strobl (1996).

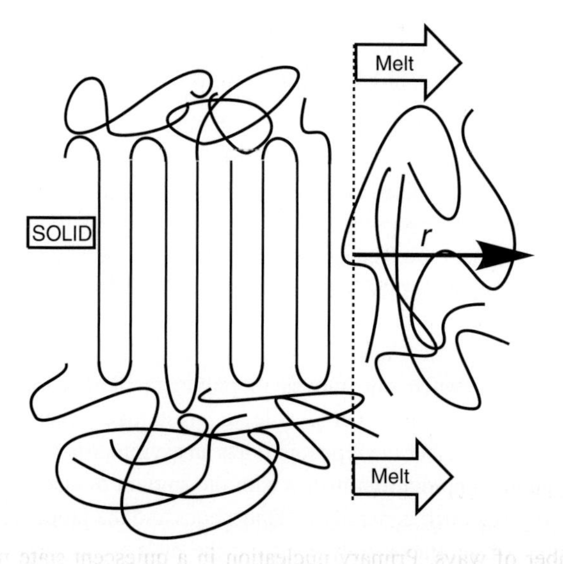

Figure 1.9. A schematic diagram of the development of the solid, semi-crystalline polymer by growth of crystalline lamellae in the direction r from the melt on cooling. Adapted from Strobl (1996).

chains is impossible. The process is believed to follow a two-step process whereby initially a partly crystalline phase separates and acts as the nucleation site. X-ray diffraction allows an orthorhombic unit cell to be associated with the growing polyethylene crystals as shown in Figure 1.8. The chain is in the all-*trans* planar zigzag, which, as discussed in Section 1.1.2, is the lowest-energy conformation of the chain.

The polymer crystal continues to grow by nucleation at the phase boundary as those sequences of the hydrocarbon chain that are untangled can add to the layer being formed. This results in a lamellar structure as shown in Figure 1.9 in which chains are attached over a length corresponding to the alkane chain which has formed the original crystallite face. It is this which determines the crystallite thickness, which is typically only 100 Å. Figure 1.9 shows many important feature of the crystal region of the polymer.

- The chain, being of fully extended length on the order of micrometres, must pass through the crystalline layer many times.
- The chain will not necessarily enter immediately at the end of the crystal, since this requires higher-energy conformations, and it may traverse the amorphous region before re-entry.
- Growth in the chain direction, rather than the lateral direction, is suppressed because of the folds and entanglements of the amorphous region at the upper and lower boundaries of the crystallite.

Pure single crystals are observed only by slow growth from solution, and these show crystalline lamellae of layer thickness 100 Å and have a higher incidence of tight folding (Keller, 2000) than is possible from the melt, with re-entry of chains occurring within three lattice sites of exit. Observation of the melt-cooled, partially-crystalline polymer under the polarizing microscope shows another level of structure, namely the formation of **spherulites**, the sizes of which are sensitive to the thermal history of the cooling melt. Diameters may range up to hundreds of micrometres and occasionally reach centimetres for slow cooling of polymers such as poly(ethylene oxide) melts. Spherulites are always depicted in two dimensions as in Figure 1.10 because of their appearance under polarized light, but are overlapping spheres formed by aggregations of lamellae each of thickness ~100 Å as they grow. This means that the chains forming as shown in Figure 1.9 are aggregating normal to the radius vector r. As shown in the micrograph in Figure 1.10(a) and schematically in Figure 1.10(b), the spherulites nucleate and grow radially, becoming distorted as they meet other growing centres. The amorphous material (the entangled chains in the solid in Figure 1.9) may be seen as that rejected by the growing lamellar fibrils and lying between the lamellae in the spherulite. This will be highly entangled, and there are **tie molecules** that traverse more than one lamella. These have a major role in the achievement of the mechanical properties of the solid polymer since they effectively couple the separate lamellae.

Nucleation and growth of polymer crystallites

The process of formation of the crystalline state is controlled by the kinetics of nucleation and this may arise in a number of ways. Primary nucleation in a quiescent state must be associated with foreign bodies such as deliberately added nucleating agents, such as fine talc particles, or residual impurities such as heterogeneous catalyst particles followed by spherulite growth. The plot of extent of crystallinity, φ_c, as a function of time is sigmoidal in nature and follows an Avrami equation of the form

$$\varphi_c = 1 - \exp[(-zt)^{\beta}], \tag{1.19}$$

where β is the Avrami exponent and z is the rate coefficient for crystallization (Strobl, 1996). It has been shown (Vaughan and Bassett, 1989) that nucleation may also occur at small segments of fully extended polymer chain that are sensitive to the entanglements and hence the degree of polymerization of the polymer. The sigmoidal relationship is a consequence of the rate of crystallization being proportional to the total area of free spherulite surface. As the spherulites touch, this decreases and the rate drops away. The

(a)

(b)

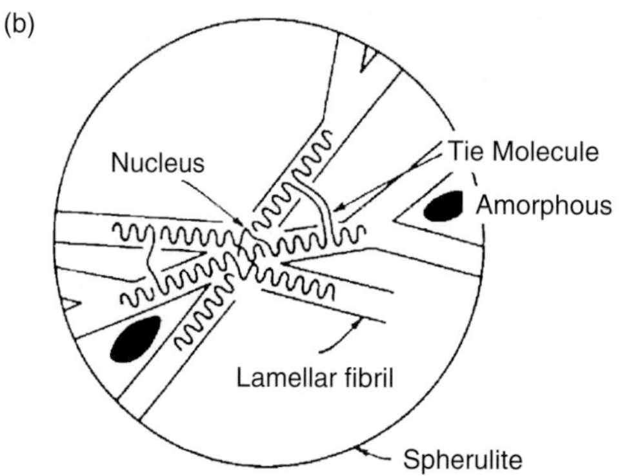

Figure 1.10. (a) The spherulitic habit of a semi-crystalline polymer on cooling from the melt. (A micrograph of polyethylene glycol viewed under crossed polarizers). (b) A schematic diagram of lamellar fibrils that have nucleated from the points shown and grow to the spherulite boundaries. Tie molecules connecting lamellae are shown.

growth rate is sensitive to temperature and is maximum between T_g and T_m. This results from a balance between the ease of nucleation at lower temperature (at which the polymer is more likely to be found in its lowest-energy conformation, which favours crystallization) and the increase in viscosity such that untangling and transport of chains to the growing crystal face becomes less probable at lower temperature and drops to zero at T_g. When the melt temperature is approached, chains are lost from the growing crystal surface faster than they can be attached. It has been proposed that the transition from the entangled melt to the partially crystalline state occurs through a state of lower order that forms the primary site, and these blocks are found, by atomic-force microscopy (AFM), to fuse into homogeneous lamellae (Heck *et al.*, 2000).

Crystallization can also be induced by the chain extension which occurs in an orientational flow. This is of obvious importance in rheology and processing, and it has been found (Somani *et al.*, 2002) that shearing of isotactic polypropylene resulted in oriented structures that did not relax even after 2 hours at temperatures 13 °C above the melt temperature. These structures were non-crystalline; their evolution was found to follow the Avrami equation; and they were proposed as the primary nucleation sites. It is known that in polyethylene the crystallization under extensional flow results in a 'shish-kebab' structure (Vaughan and Bassett, 1989). This consists of a chain-folded morphology that has nucleated on a core with an elongated structure due to chains with an extended all-*trans* conformation. Use of AFM has permitted this process to be followed at the molecular level, and it has been found that the rates of lamella growth vary widely, both spatially and with time (Hobbs *et al.*, 2001).

The thickness of the lamellae which form by secondary growth on the primary lattice sites may be understood by invoking the thermodynamics and kinetics of crystallization (Painter and Coleman, 1994). This analysis shows that

- the thermodynamics of the system allows the conclusion that the lowering of the melt temperature of the polymer relative to the equilibrium melting temperature, $T_m^o - T_m$, depends inversely on the thickness, l, of the polymer crystal, as given by the first term in the equation below;
- the crystal thickness is kinetically controlled and represents the thickness that allows the growing crystal to be stable,

$$l = (2\sigma_e/\Delta h)T_m^o/(T_m^o - T_m) + \delta l, \tag{1.20}$$

where σ_e is the free energy per unit area of the crystal face where growth occurs; Δh is the enthalpy of fusion and δl is the thickness of the primary crystallite, which is of the order of 10–40 Å.

The amorphous state and the glass transition T_g

The cooling of an amorphous polymer, such as atactic polystyrene, follows the specific-volume–time plot as shown in Figure 1.11(a). The effect on the specific volume, V_s, of passing through the glass-transition temperature, T_g, is to reach a glassy state with a density lower than for the ideal liquid. This second-order transition may be detected by observing the change in the coefficient of thermal expansion, α, and the heat capacity, C_P. Thus, unlike the crystallization of the polymer (as shown in Figure 1.7), there is no abrupt change in the volume or latent heat, as is characteristic of a first-order transition.

Figure 1.11(a) shows the specific-volume–temperature plot for an amorphous polymer and when it is extrapolated to 0 K the volume, V_{0g}, will be higher than if all of the atoms adopted the closest possible packing (V'). This difference represents the free volume, V_f, at absolute zero and arises because of the inability of the chains to reach their minimum energy and closest packing within a finite time frame during solidification. As the temperature increases above zero, V_f will increase due to thermal expansion, and the motions of the chain increase due to the newly available free volume. As the temperature increases, certain motions that had been inhibited due to the low free volume may now be unlocked. The modulus of the polymer decreases by four orders of magnitude on heating through T_g as it changes from an amorphous glass to a rubber because various motions of the polymer segments are now able to be accessed and dissipate energy. The molecular origin of relaxations below the glass-transition temperature may be seen in Figure 1.11(b), in which is plotted the change in available modes of energy dissipation in polystyrene as

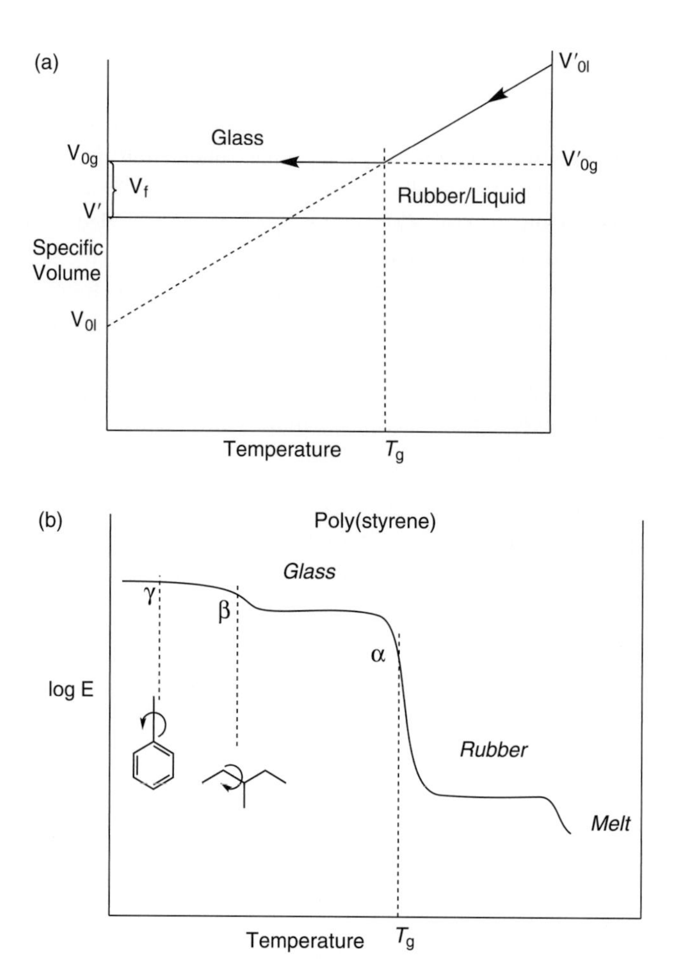

Figure 1.11. (a) The change in specific volume of an amorphous polymer as it is cooled from the melt through to the glassy state (contrast this with Figure 1.7). (b) Changes in the modulus (E) of amorphous poly(styrene) as it is heated from low temperature to the melt and the modes of energy dissipation accompanying the relaxations shown.

the temperature is decreased. At very low temperatures, the only modes available to polystyrene are rotations about side groups (γ-relaxation at ~–100 °C) and the backbone as a crankshaft (β-relaxation at ~0 °C). The effect of these motions on the modulus is modest compared with the change at T_g. Understanding the nature of the molecular relaxations at T_g requires an appreciation of the theories for the glass transition (McKenna, 1989). The glass transition is a kinetic effect and shifts with the frequency of observation. It is also sensitive to the molar mass of the polymer and the presence of crosslinks between chains. Any theory for the glass transition must accommodate these results.

In the rheology and processing of polymers the kinetic aspects of the glass transition are of particular interest since the achievement of thermodynamic equilibrium in the amorphous, high-molar-mass polymer is beyond the time frame of the dynamic environment of processing. One way of viewing the glass transition (Stachurski, 1987) is to consider the

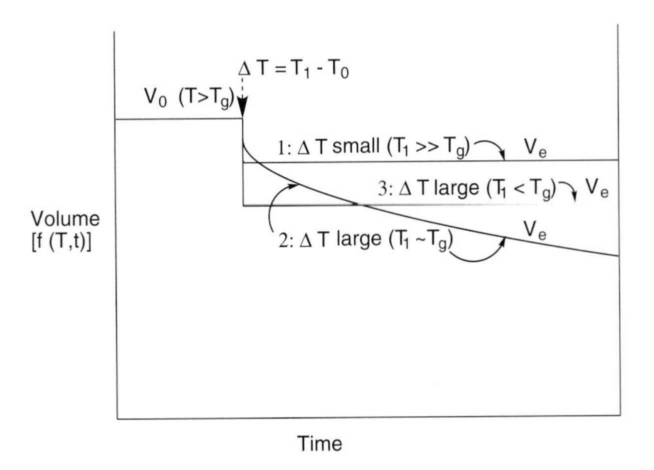

Figure 1.12. The change in free volume with time after applying a temperature drop, ΔT, to a polymer that is above, (1), close to, (2), or below, (3), T_g. Adapted from Stachurski (1987).

polymer at some point T_0 above T_g in Figure 1.12, where the volume is $V_e(T_0)$. From this point the temperature is suddenly decreased to T_1 at an infinitely fast rate. At the molecular level, the changes to the chain can be considered in the framework discussed in Sections 1.1.1 and 1.1.2 and consist of

- a decrease in thermal motion leading to a decrease in volume;
- an increase in the intermolecular forces as the chains adopt the conformations of lower energy; and
- a consequent decrease in the end-to-end distance, R_0, of the chain, which, due to entanglements, will be time-dependent.

The changes in conformation require co-operative motion to achieve the bond rotation needed to attain the *gauche*- and *trans*-conformers which are the lower-energy states of the chain. This will require motion of various segments of the chain and these transitions may be viewed as having particular relaxation times, τ_i. At a time after the application of the temperature change that is very much longer than the largest value of τ_i, the system will again be in equilibrium, so the volume is now $V_e(T_1)$.

The behaviour of the system's volume can be considered for the cases of T_1 being above, below and at the glass-transition temperature when $T_0 \gg T_g$.

(1) If $T_1 \gg T_g$ and $T_1 - T_0$ is small, then $V_e(T_1)$ will be attained instantaneously since there will be no change in the available conformations or relaxation times, so the new equilibrium volume is achieved within the time frame of the temperature drop.
(2) When $T_1 \approx T_g$, i.e. $T_1 - T_0$ is large, $V_e(T_1)$ is attained slowly because adoption of the new equilibrium conformation for that temperature is controlled by the relaxation times, τ_i.
(3) When $T_1 < T_g$, i.e. $T_1 - T_0$ is large, $V_e(T_1)$ is attained rapidly but corresponds to a non-equilibrium glass since at $T_1 < T_g$ the relaxation times are infinitely long compared with the time frame of the experiment.

This becomes important when available methods of measuring the glass-transition temperature of a polymer sample are considered. These include, among others, differential scanning calorimetry, DSC (which measures the change in heat capacity); thermomechanical analysis, TMA (which measures the change in coefficient of thermal expansion); and dynamic mechanical analysis, DMA (which measures the phase lag, $\tan\delta$, between a cyclically applied stress and the measured strain). In all cases, the sample of processed polymer is heated while the above properties are measured through the heat flow (DSC), change in volume (TMA) or the storage and loss moduli (DMA). As the system is heated, and the polymer passes through one of the relaxations described in Figure 1.11(b), the frequency of the motions becomes accessible to the method of measurement, i.e. they are both within the same time frame. Thus, if a technique such as DMA is used, whereby the cyclical frequency applied to measure the storage and loss moduli, and thus $\tan\delta$, may be varied, then the characteristic temperature (such as T_g) at which the frequency of measurement corresponds to the inverse of the relaxation time for the polymer will change with the applied frequency. Thus, T_g is sensitive to the method of measurement as well as the thermal history of the sample. The effect of thermal history may be seen by considering the DSC trace obtained when a sample is heated at a rate greater than or equal to the rate at which it was initially cooled from the melt or rubbery state to the glassy state (Chynoweth, 1989).

Consider Figure 1.13(a), which shows the change in volume of a polymer sample that was slowly cooled from point A, lying above the glass-transition temperature, to point X, where the material departs from the line A–B (the extrapolated super-cooled-liquid line) due to vitrification, and attains the volume given at point Y for the final sample temperature. If this sample is now measured in a DSC to determine the T_g, then two situations may be considered, as shown in Figure 1.13(b), depending on the rate of heating. In the first case the rate of heating corresponds to the rate of cooling and the T_g as given by the change in heat capacity ($\frac{1}{2}\,\Delta C_p$) corresponds to point X in Figure 1.13(a) since the volume change on heating follows the path Y–X–A. However, if the rate of heating in the DSC experiment *exceeds* that at which the sample was originally cooled, then the volume of the sample will follow the path Y–X–Z–A and the rate of conformational rearrangement lags behind the rate of heating. In this case a DSC trace such as the solid curve in Figure 1.13(b) is obtained with a maximum in C_p being observed as an artefact. The measurement of the 'true' T_g of polymers when the rate of heating in a DSC experiment exceeds the rate of cooling has been analysed in detail (Richardson, 1989) and a procedure described for determining the point of intersection of the enthalpy–temperature curves for the glass and liquid states from the DSC trace (equivalent to point X for the volume–temperature curve in Figure 1.13(a)).

This exemplifies the experimental difficulties inherent in determining the absolute value of T_g, which is considered in more detail when thermosets are discussed. Of particular interest is the value that a relaxation-dependent property may have when a system is in the vicinity of the glass transition. This is given by the empirical Williams, Landel and Ferry (WLF) equation:

$$\mathrm{Log}\ a_T = -17.4(T - T_g)/[51.6 + (T - T_g)], \qquad (1.21)$$

where a_T is the ratio of the value of the property at temperature, T, to that at the glass transition, T_g. From this it has been concluded that the value of the fractional free volume $V_f/(V_0 + V_f)$ at the glass transition is 0.025.

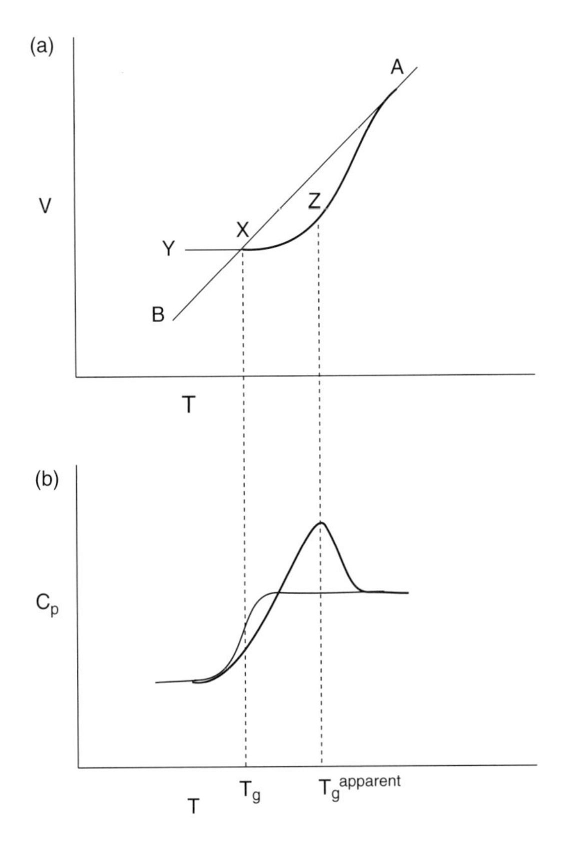

Figure 1.13. (a) A volume–temperature plot of a polymer sample on vitrification and (b) the effect of cooling history on the heat-capacity (C_p) anomaly observed as an apparent peak by DSC when the rate of heating exceeds the original rate of cooling for vitrification. Adapted from Chynoweth (1987).

Factors controlling the glass-transition temperature, T_g

It is possible to further understand the molecular basis of T_g by comparing the values for chemically different polymers as shown in Table 1.2 (Chynoweth, 1989). The comparison of the T_g values on proceeding from the simple, flexible backbone of polyethylene to the bulky side groups of polypropylene and polystyrene shows a progressive increase in T_g such that at room temperature (RT) polystyrene is stiff and rigid (RT $< T_g$) and fails in a brittle manner. Similarly, on increasing the forces between the chains (e.g. in poly(vinyl chloride)) the material stiffens compared with polyethylene, so T_g is higher (80 °C). This is most marked in Kevlar® (poly(*p*-phenylene terephthalamide)), where the strong intermolecular forces of hydrogen bonding between the amide groups on adjacent polymer chains as well as the stiffening effect of *p*-phenylene groups in the polymer chain result in the polymer behaving as a rigid rod with a T_g of 345 °C. Many of the correlations of T_g with macromolecular structure and molar mass can be reconciled through a consideration of the free volume. The strong intermolecular forces between the amide groups in the polyamides, nylon and Kevlar® result in a lowering of free volume. In contrast the flexible C–C backbone in polyethylene ($T_g = -90$ °C) discussed in Section 1.1.1 and the O–Si–O backbone of poly(dimethyl siloxane) ($T_g = -120$ °C)

Table 1.2. Glass-transition temperature (T_g) and correlation with polymer structural features

Polymer	T_g (°C)	Structural features
Poly(ethylene)	−100	Flexible C–C backbone
Poly(propylene)	0	Hindered C–C backbone due to pendant methyl groups
Poly(vinyl chloride)	80	Strong dipolar intermolecular forces
Poly(styrene)	100	Chain stiffening due to pendant phenyl groups
Poly(*p*-phenylene terephthalamide)	345	In-chain stiffening from *p*-phenylene groups together with amide hydrogen bonding

result in the chains sweeping out a large volume, so V_f is large and T_g is low. Similarly, if the side groups are flexible there will be a higher free volume and thus a lower T_g compared with a rigid and bulky side group (contrast poly(styrene), $T_g = 100\,°C$, with poly(1-butene), $T_g = -45\,°C$). Another example of the importance of free volume may be seen in the effect of the number-average molar mass of a homopolymer on T_g. The following Fox–Flory relation holds well for poly(styrene) polymer and oligomers with $B \approx 10^5$:

$$T_g = T_{g\infty} - B/M_n. \tag{1.22}$$

This reflects the effect of the greater number of chain ends at lower molar mass resulting in a larger local free volume and thus a lower T_g. While this relation hold well for linear polymers, there are exceptions linked to the nature of the end groups, e.g. when they are ionic or hydroxyl groups, and also when cyclics are studied. McKenna (1989) considers in more detail these and other factors which may affect T_g. The effect of crosslinking, which is important for reactive processing of both elastomers and three-dimensional networks, is considered in a later section.

The rubbery state

It was noted that, above the glass-transition temperature, the polymer is able to have considerable conformational freedom involving concerted motion over ≈ 50 atoms, which results in a decrease in modulus by a factor of about 10^4. The polymer is still far from the melt and the chain entanglements result in a material with viscoelastic properties. The rubbery nature may be seen from the tendency of the polymer to recover when a stress is applied. This recovery is a consequence of the higher order conferred on the chains when they are distorted so that when the stress is released there will be an entropic drive to return to the coiled state (Queslel and Mark, 1989). It is this entropic recovery that results in the shrinking of a loaded crosslinked elastomer (shown schematically in Figure 1.14) when it is heated.

If the stress is applied for a time much longer than the relaxation time and the chains have no physical or chemical crosslinks to prevent their viscous flow, they may disentangle and the deformation might not be recovered. The polymer is said to have undergone creep. Only light crosslinking is required to inhibit this permanent deformation without affecting the glass-transition temperature and so produce an elastomer with the unique properties of rubber elasticity.

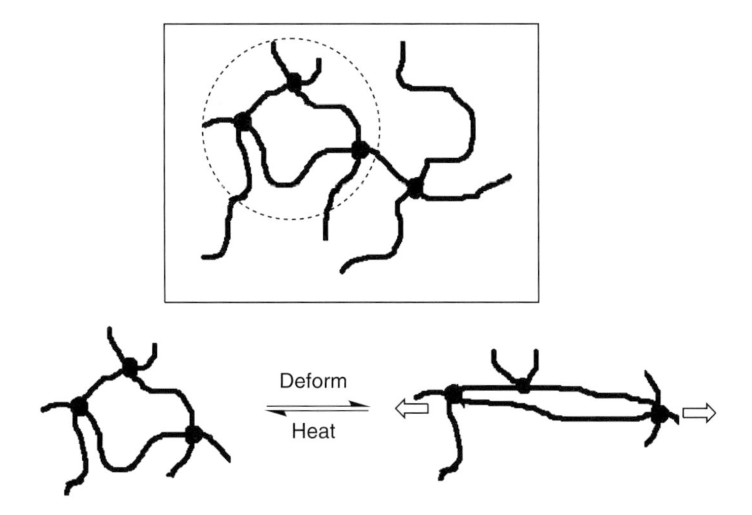

Figure 1.14. A schematic diagram of a crosslinked elastomer network and the changes in the (circled) section of the network when it is deformed in the direction shown by the arrows and then either is allowed to relax or is heated.

These crosslinks may be permanent covalent bonds, as in vulcanization of rubber (such as in Figure 1.14), or virtual, as in the crystalline microdomains linking the segments in a thermoplastic elastomer such as a polyurethane. It has been noted (Queslel and Mark, 1989) that the most important networks for rubber elasticity arise from functionalities of the crosslink points (or junctions) that are either four, as occurs in sulfur or peroxide vulcanization, or three, as occurs in end-group reaction of the polyol segment of a polyurethane with a trifunctional isocyanate. If the functionality is only two then end-group linkage may occur, leading to chain extension but not to crosslinking.

The crosslinking must be sufficiently infrequent (about one crosslink per hundred repeat units) as to allow the polymer to adopt a random coil configuration between crosslink sites and so exhibit entropic recovery when deformed. The chemistry of rubber crosslinking is discussed later.

1.2 Controlled molecular architecture

The discussion in the previous sections has focussed on the properties of a linear homopolymer chain. Attention has been paid to the way the conformation of the chain and the molar mass affect the properties in the melt and the development of the solid state on cooling the melt. The linear chain is an idealization of the real polymer and different architectures may be introduced by

- *cyclization* of all or part of the chain;
- the formation of short- or long-chain *branches*, which occur along the backbone;
- the formation of continuous branching so that the linear nature of the polymer is lost and an irregular *hyperbranched* architecture is formed, or a more regular *dendrimer*

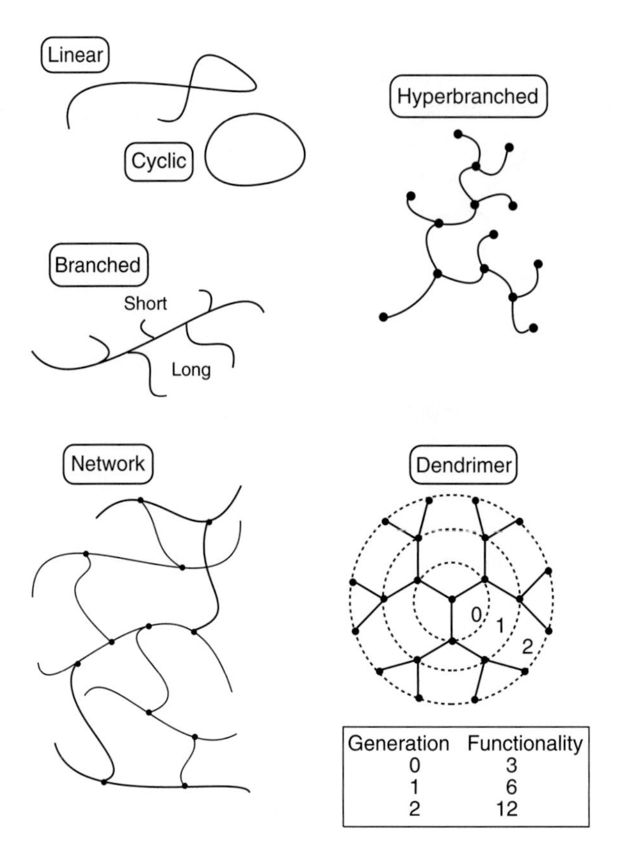

Generation	Functionality
0	3
1	6
2	12

Figure 1.15. A schematic diagram illustrating the various types of molecular architecture of polymers that can be obtained through synthesis.

architecture is obtained, in which uniform branching occurs at every branch point (Frechet and Tomalia, 2001); and

- crosslinking of the chains to a *network* so that the branches travel from chain to chain, resulting in an insoluble polymer with an infinite molar mass.

These possibilities are shown in Figure 1.15 and each will have a major effect on the chemorheological properties of the polymer compared with the linear parent. The detailed chemistry and mechanism of the reactions that lead both to linear polymers and to these different architectures are discussed in this section. The route to achieve these structures may involve stepwise polymerization; addition polymerization, or post-polymerization modification. Each of these polymerization reactions, with particular emphasis on the way they may be adapted to reactive processing and the chemorheological consequences, is considered separately. Further detailed architectures such as graft and block copolymers with several different chemical components are then considered.

1.2.1 Stepwise polymerization

This polymerization method is one of the first that was established commercially for polymer fabrication directly by reactive processing, i.e. the phenol–formaldehyde condensation reaction to form thermoset networks as developed by Baekeland and commercialized in 1910 as Bakelite. The chemistry of these reactions is complex and difficult to analyse because of the insolubility and intractability of the three-dimensional network of infinite molar mass. Simpler linear thermoplastics may be formed by using difunctional reagents, so the polymer is soluble to high extents of reaction and the molar mass is much lower.

A distinguishing feature of *stepwise polymerization* is that the reaction builds high-molar-mass polymer very slowly and does this throughout the reaction since dimerization, trimerization and higher oligomerization occur early in the reaction, followed by the further coupling of these low-molar-mass oligomers to form high-molar-mass polymer. The reaction then depends on the reactivity of the functional groups and their availability. This is to be contrasted with *chain polymerization* that results in high-molar-mass polymer almost instantaneously due to the rapid addition of the monomer species to the active centre, which may be an anion, cation or free radical. In some systems the same monomer may be polymerized by either addition or stepwise polymerization to give polymers that differ in properties due to molar-mass distribution, end groups etc.

The aromatic polyesters such as poly(ethylene terephthalate) (PET) were commercialized from about 1946 as fibres, but, because of the high processing temperatures, it was only some 20 years later that they appeared as engineering thermoplastics. The dominance of PET in beverage containers ensures the importance of the synthesis, processing and recycling of PET. Polyesterification is a suitable stepwise reaction to illustrate the principles of this industrially important polymerization. Applications in reactive processing will then be considered.

Polyesterification

Ester formation is a standard organic reaction between an alcohol and a carboxylic acid, which is an equilibrium reaction that has been shown to occur under catalysis by either acid or base. In polyesterification involving an organic acid, the substrate is itself the catalyst. It has been noted (Pilati, 1989) that, among the many reaction mechanisms, Scheme 1.1 is the most likely for acid-catalysed esterification, with the second reaction being the rate-determining step.

$$RCO_2H + H_3O^+ \longleftrightarrow [RC(OH)_2]^+ + H_2O$$

$$\mathbf{[RC(OH)_2]^+ + R'OH \longleftrightarrow [RC(OH)_2(R'OH)]^+}$$

$$[RC(OH)_2(R'OH)]^+ - H_3O^+ \longleftrightarrow [RC(OH)_2(OR')] + H_2O$$

$$[RC(OH)_2(OR')] + H_3O^+ \longleftrightarrow [RC(OH)_3]^+(OR') + H_2O$$

$$[RC(OH)_3]^+(OR') - H_2O \longleftrightarrow [RC(OH)(OR')]^+$$

$$[RC(OH)(OR')]^+ - H_3O^+ \longleftrightarrow RCO_2R' + H_2O$$

The net reaction is

$$\mathbf{RCO_2H + R'OH \longleftrightarrow RCO_2R' + H_2O}$$

Scheme 1.1. The reaction scheme for acid-catalysed esterification.

The reaction rate will depend on the factors which favour the rate-determining step moving to the right, i.e. the concentration of the free-ion species from the first reaction step. The removal of the water ensures that the reaction favours formation of the ester, RCO_2R'; otherwise the reverse reaction becomes significant.

In the absence of added catalyst, the reaction is formally of third order, since it will be second order in the acid RCO_2H which is both reagent and catalyst:

$$\text{Rate} = k[\text{RCO}_2\text{H}]^2[\text{ROH}]. \tag{1.23}$$

This may be extended to polyesterification by replacing the alcohol and acid with a diol and a diacid. Depending on the polarity of the medium, the reaction mechanism may involve different reactive intermediates, since the formation of charged species will be less probable in media of low dielectric constant as may occur in the polyesterification at high extents of reaction. The overall experimental kinetic order is the same as for simple esterification.

On considering the net reaction for esterification, it is seen that the reaction product will contain a terminal acid group and an alcohol:

$$\text{HOR}'\text{OH} + \text{HO}_2\text{CRCO}_2\text{H} \leftrightarrow \text{HOR}'\text{OCORCO}_2\text{H} + \text{H}_2\text{O} \qquad \text{dimer}$$

This new dimer species has end groups that may undergo three possible reactions, as shown in Scheme 1.2.

The formation of dimers, trimers and tetramers will consume the reagents early in the reaction, and the attainment of a high-molar-mass polymer requires the subsequent reactions of these oligomers, e.g. trimer + dimer → pentamer, or in general terms n-mer + m-mer to give $(n+m)$-mer. The degree of polymerization of the resulting oligomer, DP, is $(n+m)/2$. The analysis of the kinetics of this system is necessary to determine the molar mass of the polymer and also the factors that control the rate of polyesterification. This analysis is simplified by the observation from model compounds that the rate of esterification of a series of acids and alcohols is independent of the alkyl chain length, n, for both acid and alcohol after $n > 3$.

1. Reaction with another dimer to give a tetramer with the same end groups:

$$2\text{HOR}'\text{OCORCO}_2\text{H} \longleftrightarrow \text{HOR}'\text{OCORCOOR}'\text{OCORCO}_2\text{H} + \text{H}_2\text{O}$$

tetramer

2. Reaction of the dimer with further diol to give a dialcohol-terminated trimer:

$$\text{HOR}'\text{OCORCO}_2\text{H} + \text{HOR}'\text{OH} \longleftrightarrow \text{HOR}'\text{OCORCOOR}'\text{OH} + \text{H}_2\text{O}$$

trimer

3. Reaction of further diacid with the dimer to give a dicarboxyl-terminated trimer:

$$\text{HO}_2\text{CRCO}_2\text{H} + \text{HOR}'\text{OCORCO}_2\text{H} \longleftrightarrow \text{HO}_2\text{CRCOOR}'\text{OCORCO}_2\text{H} + \text{H}_2\text{O}$$

trimer

Overall:

$$n\text{HO}_2\text{CRCO}_2\text{H} + n\text{HOR}'\text{OH} \longleftrightarrow \text{H(OR}'\text{O}_2\text{CRCO})_n\text{OH} + (2n-1)\text{H}_2\text{O}$$

Scheme 1.2. Formation of oligomers during the early stage of polyesterification and the overall reaction.

The molar mass of the resultant polymer may be conveniently determined by the titration of the acid end groups of the separated polymer. Thus, if C is the number of moles of acid groups per gram of polymer, then $M_n = 1/C$ g/mol (since for a stoichiometric reaction the polymer would be expected to have an equal number of acid and alcohol end groups, so that each chain, on average, has one acid end group).

Kinetics of polyesterification and other stepwise reactions

The lack of dependence of the reaction rate for polycondensation on the extent of reaction (to a first approximation) allows a simple bimolecular reaction mechanism to be employed. Noting, from the previous section, that the reaction mechanism for simple esterification reactions (Scheme 1.1) had as the rate-determining step the second reaction, namely

$$[RC(OH)_2]^+ + R'OH \leftrightarrow [RC\,(OH)_2(R'OH)]^+,$$

then the rate of consumption of the alcohol, which will be the same as the rate of consumption of the acid (which can be followed by titration), is given by

$$-d[ROH]/dt = -d[RCO_2]/dt = k'[R'OH][RC\,(OH)_2]^+, \qquad (1.24)$$

and, since $[RC(OH)_2]^+$ will be given by the equilibrium value from the first reaction of Scheme 1.1,

$$[RC(OH)_2]^+ = K[RCO_2H][H_3O^+], \qquad (1.25)$$

where K is the equilibrium constant, then

$$\begin{aligned} -d[R'OH]/dt &= -d[RCO_2H]/dt \\ &= k'K[R'OH][RCO_2H][H_3O^+], \end{aligned} \qquad (1.26)$$

Thus, depending on whether the acid catalyst is an added strong acid, or is derived by ionization of the carboxylic acid reagent, the overall experimental order in the carboxyl may be second or first order, respectively.

Detailed studies of systems with no added acid and added strong acid have been performed, and lead to the following kinetic relationships (Manaresi and Munari, 1989).

Self-catalysed polymerization

The concentrations of the starting reagents in step polymerization are usually chosen to be the same, so that

$$-d[R'OH]/dt = k[RCO_2H]^2[R'OH] = k[R'OH]^3. \qquad (1.27)$$

It is convenient to introduce the extent of reaction, p, as the fraction of the reagent that has reacted at a time, t, so that

$$\begin{aligned} p &= ([R'OH]_0 - [R'OH])/[R'OH]_0, \\ \text{i.e. } [R'OH] &= (1-p)\,[R'OH]_0. \end{aligned} \qquad (1.28)$$

In an actual polycondensation, p may be measured by titration of the carboxylic acid end groups or by measurement of the amount of water evolved.

Solving for a simple third-order reaction, in terms of p,

$$1/(1-p)^2 = 2k[R'OH]^2 t + 1. \qquad (1.29)$$

This equation has been tested (Manaresi and Munari, 1989) for simple self-catalysed aliphatic polyester formation and marked deviations have been found at low and high conversions. This is attributed to the differences in polarity of the medium that may result in different reaction mechanisms since these are also seen for non-polymerizing systems. At high conversions there has been the removal of significant amounts of water and there has been a change in the reaction volume together with a large increase in viscosity. The ability for the reaction to proceed to completion depends on precise stoichiometry since side reactions and loss of volatiles may result in an apparent decrease in reaction rate.

The molar mass of the polymer is related to the extent of reaction, since this is linked to the end-group concentration. Then $M_n = M_0/(1-p)$, where M_0 is the molar mass of the polymer repeat unit. The development of the polymer chain length, the degree of polymerization, DP, is thus

$$\mathrm{DP} = 1/(1-p) = (2k[\mathrm{R'OH}]^2 t + 1)^{1/2}. \tag{1.30}$$

Thus, in an uncatalysed reaction the polymer chain grows with the square root of the time of reaction, a direct consequence of third-order kinetics. Consequently, in any practical application, the use of an external catalyst is necessary in order to achieve high-molar-mass polymer in the shortest possible time.

Externally catalysed polymerization: molar-mass distribution

The addition of a strong acid, such as *p*-toluenesulfonic acid, results in second-order kinetics since the $[H_3O^+]$ does not change significantly with time and is absorbed into the rate coefficient k'.

The kinetic equation then becomes

$$\begin{aligned}-\mathrm{d}[\mathrm{R'OH}]/\mathrm{d}t &= -\mathrm{d}[\mathrm{RCO_2H}]/\mathrm{d}t\\ &= k'[\mathrm{R'OH}][\mathrm{RCO_2H}]\\ &= k'[\mathrm{R'OH}]^2.\end{aligned} \tag{1.31}$$

Solving this second-order equation gives

$$1/[\mathrm{R'OH}] - 1/[\mathrm{R'OH}]_0 = k't, \tag{1.32}$$

or, in terms of the extent of reaction, p,

$$1/(1-p) = \mathrm{DP} = 1 + [\mathrm{R'OH}]_0 k't. \tag{1.33}$$

This shows the advantage of using the external catalyst, since the chain length of the polymer now increases linearly with time. A detailed experimental study of aliphatic polyesterification (Manaresi and Munari, 1989) shows that this relation is followed over a wide range of conversion, but with deviation at short time. This may again be linked to polarity effects, as for the uncatalysed reaction and simple ester formation.

Equation (1.33) does show that high-molar-mass polymer does not form until high extents of reaction at long time and shows the importance of the stoichiometry (and thus purity) of the starting materials, as given in Table 1.3.

Figure 1.16 shows the chain-length distribution of the polymer at increasing extents of conversion. At short times the polymer has a narrow molar-mass distribution since the reaction mixture contains only low-molar-mass oligomers. As the reaction proceeds it

Table 1.3. Dependence of the degree of polymerization, DP, on the extent of reaction, p, in polyesterification

Extent of reaction, p	$1-p$	DP	Relative time
0.9	0.1	10	1
0.99	0.01	100	11
0.999	0.001	1000	111

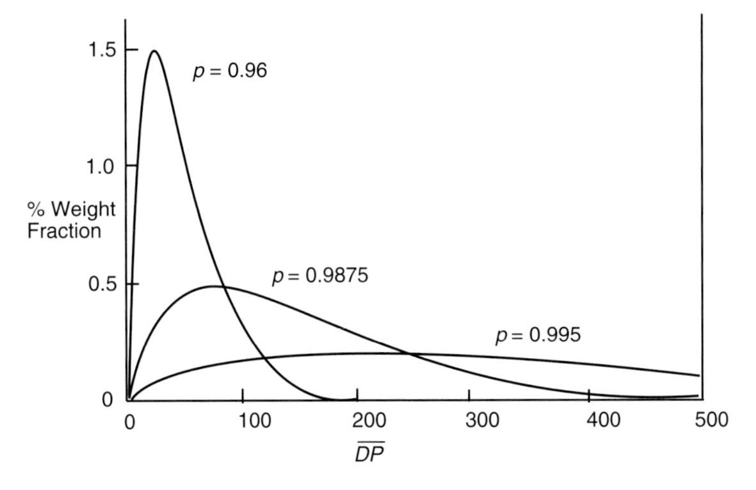

Figure 1.16. Changes in the chain-length distribution with conversion, p, for a typical stepwise polymerization. Adapted from Odian (1991).

broadens and shows a maximum at a degree of polymerization

$$DP_{max} = -\ln(1/p) \approx 1/(1-p). \quad (1.34)$$

The breadth of the molecular-weight distribution is given by the polydispersity:

$$M_w/M_n = 1 + p. \quad (1.35)$$

Thus at high extents of reaction it will approach the value 2.

If the stoichiometric ratio, r, deviates from unity, then a limiting polymer chain length is soon reached (Manaresi and Munari, 1989). In this case the extent of conversion of the two reagents at any time will not be the same, but will be in the ratio $r = p_A/p_B$, where A and B represent the diacid and diol, respectively. The degree of polymerization then becomes

$$DP = (1 + r)/(1 - r). \quad (1.36)$$

For example, if there is 3% deviation in the concentration of, say, the acid, then the limiting DP is 65. While high values of DP, e.g. over 200, may be required for high-strength polymers such as PET, for liquid polyesters (as used in unsaturated polyester composite fabrication where the resin impregnates the reinforcing fibre) a value of 10–20 is required and the stoichiometry may be used to control this. In the practical polyesterification reaction, the decision on the reagents is often based on the ease with which the starting material may be purified and the extent to which it will not participate in side reactions at the high temperature necessary to drive the reaction to completion. Thus, in the important reaction of formation of poly(ethylene terephthalate) (PET), because of the lack of high-purity terephthalic acid, alcoholysis of the methyl ester was used for many years. This reaction is related to the ester exchange reaction encountered in reactive processing of other thermoplastics (Lambla, 1992).

Polyesterification via alcoholysis and ester exchange

The replacement of the acid with an ester often allows the more facile purification of the starting material, so that stoichiometry is maintained to high extents of reaction. This facilitates a higher degree of polymerization. The reaction must be catalysed and the small molecule evolved is now an alcohol (Pilati, 1989). While *p*-toluenesulfonic acid has been employed, it is less effective than in the direct esterification, and metal catalysts, such as titanium alkoxides, are used instead, to avoid side reactions from the strong acid:

$$n\mathrm{R''O_2CRCO_2R''} + (n+1)\mathrm{HOROH} \leftrightarrow \mathrm{H(ORO_2CRCO)}_n\mathrm{OR'OH} + 2n\mathrm{R''OH}$$

According to this reaction scheme all of the chains should be –OH-terminated, but it has been noted that there are frequently side reactions that result in acidic by-products (Pilati, 1989). The kinetics of the reaction are of first order in both diol and ester, so the relationships described above should apply, although the mechanism is different. The role of the metal alkoxide is believed to be through co-ordination with the carbonyl group of the ester followed by reaction with an alkoxy group, but it is recognized as complex (Pilati, 1989).

Ester exchange reactions provide a related approach to polyesterification, but the practical application for *de novo* synthesis is limited.

$$\mathrm{RCO_2R'} + \mathrm{R''CO_2R'''} \leftrightarrow \mathrm{RCO_2R'''} + \mathrm{R''CO_2R'}.$$

The reaction is, however, important in the high-temperature melt blending of different polyesters because scrambling of the repeat units can occur. This opens up the possibility of creating novel copolymers through reactive extrusion. The reaction is very slow in the absence of catalysts, of which titanium alkoxides are again the most powerful. The reaction mechanism is expected to produce second-order kinetics and activation energies are of the order of 130–150 kJ/mol (Pilati, 1989).

In addition to the ester–alcohol and ester–ester exchange it is also feasible to have ester–acid exchange, or acidolysis.

$$n\mathrm{R''CO_2RO_2CR''} + n\mathrm{HO_2CR'CO_2H} \leftrightarrow \mathrm{R''CO_2(RO_2CR'CO_2)}_n\mathrm{H} + (2n-1)\mathrm{R''CO_2H}.$$

This reaction will be significant only if the acid product can be removed effectively from the reaction. This restricts the practical application to R″ being methyl (i.e. acetic acid is distilled off to generate high polymer). Of more importance is the role of acidolysis in the redistribution of functional groups and changes in molar mass when acidic compounds are present during melt blending. If the low-molar-mass monofunctional acid is scavenged during melt-blending reactive extrusion then the equilibrium will again be driven to the right.

Step polyesterification of bifunctional reagents

If a single reagent that has the end groups necessary for self-condensation is polymerized then the issues surrounding stoichiometry are no longer as significant. It is thus possible to purify the monomer and through appropriate catalysis generate the polyester:

$$n\mathrm{HO_2CROH} \leftrightarrow \mathrm{H(ORCO)}_n\mathrm{OH} + (n - 1)\mathrm{H_2O}.$$

Unfortunately, this simple approach is applicable only to the simple α-hydroxy acids, lactic acid and glycolic acid, because of competing side reactions. Other monomers, such as 3-hydroxy acids, undergo dehydration at the reaction temperature necessary to drive the reaction:

$$\mathrm{HOCH_2CH_2CO_2H} \rightarrow \mathrm{CH_2 = CHCO_2H + H_2O}.$$

If R has between five and seven methylene groups then there will be intramolecular dehydration, leading to lactones that will compete with polyesterification. This is an example of the cyclization reaction that competes with linear polymerization and is discussed in Section 1.2.2:

$$\text{H-(OR)-COOH} \xrightarrow{-\mathrm{H_2O}} \boxed{\text{(OR)COO}}$$

However, when the number of methylene groups in R is above seven, then linear polymers are again able to be formed due to the lower probability of cyclization when initially the ring becomes crowded and then, above 15 carbon atoms, the probability of the ends meeting tends to zero.

Ring opening of lactones in the presence of a suitable catalyst provides an alternative route to polyesters. The importance of ring-opening polymerization in reactive processing requires further discussion of this process for polymers such as polyamides and epoxides in addition to polyesters.

Other step-polymerization reactions important in reactive processing

The number of chemical systems for which stepwise polymerization of the polycondensation type, as discussed for polyesters, can be applied is large, but the systems of importance for reactive processing fall into two groups, depending on whether there is the evolution of a by-product necessary to drive the reaction forward.

Polycarbonates may be either aliphatic or, more commonly, aromatic with the generalised structure $\mathrm{R'O[ROCO_2]}_n\mathrm{R'}$. Lexan® is the most widely used aromatic polycarbonate engineering thermoplastic and may be formally regarded as the polyester from carbonic acid, $\mathrm{(HO)_2C = O}$, and the aromatic diol bisphenol-A. The high polymer may be formed by a number of routes, but the most widely used is the direct reaction of phosgene with bisphenol-A and the liberation of HCl (Clagett and Shafer, 1989). The polymerization is performed interfacially between an alkaline solution containing the bisphenol-A and methylene chloride containing dissolved phosgene. To avoid excessively high molar masses at the phase boundary, chain-transfer agents (phenols) are incorporated in the aqueous layer. The liberated HCl is neutralized by the alkaline aqueous layer:

$$n\mathrm{HO{-}C_6H_4{-}C(CH_3)_2{-}C_6H_4{-}OH} + n\mathrm{Cl{-}C(=O){-}Cl} \longrightarrow {*}{-}\left[\mathrm{O{-}C_6H_4{-}C(CH_3)_2{-}C_6H_4{-}OC(=O)}\right]_n{-}{*} + 2n\mathrm{HCl}$$

An alternative route is the transesterification reaction between diphenyl carbonate and bisphenol-A in the presence of a strong base catalyst with the liberation of phenol under reduced pressure. However, this method also requires the use of phosgene in the preparation of precursors, so there is no way that use of the toxic compound can be avoided (Clagett and Shafer, 1989). The use of copolymers such as the poly(ester carbonates) and blends with other thermoplastics has extended the range of properties

Polyamide formation is entirely analogous to polyesterification and all of the earlier considerations apply. For polymers of the type AABB, nylon-6,6 is the best example. Thus poly(hexamethylene adipamide) is formed by the condensation reaction between hexamethylene diamine and adipic acid:

$$n\mathrm{H_2N(CH_2)_6NH_2} + n\mathrm{HO_2C(CH_2)_4CO_2H} \leftrightarrow \mathrm{H(NH(CH_2)_6NHCO(CH_2)_4CO)}_n\mathrm{OH} + (2_n - 1)\mathrm{H_2O}.$$

The same considerations in achieving high molar mass of having well-controlled stoichiometry apply and this is achieved by forming a salt between the reagents, which may be recrystallized so that the stoichiometry is exactly 1 : 1. It is noted that the polymer should have on average one amine and one carboxylic acid terminal group, so the molar mass may be also obtained by end-group analysis of either or both end groups. It should be noted that an equilibrium is established in the same way as in polyesterification and the commercial production of polyamides is via a melt polycondensation at 200–300 °C so that water is driven off.

Polyamides may also be formed via the self-condensation of amino acids,

$$n\mathrm{H_2N(CH_2)}_x\mathrm{CO_2H} \rightarrow \mathrm{H[NH(CH_2)}_x\mathrm{CO]}_n\,\mathrm{OH} + n\mathrm{H_2O},$$

as well as the ring-opening reactions of the cyclic lactams (which are the intramolecular-dehydration products of the amino acids). The ring-opening polymerization reactions are important in reactive processing for the formation of nylon-6 and are considered separately (Section 1.2.4).

Amide interchange reactions will occur if a mixture of different polyamides is heated at the melt temperature and, in 120 minutes at 260 °C, a completely random copolymer with a single molar-mass distribution is formed (Gaymans and Sikkema, 1989). The properties of polyamides are affected by the strong hydrogen bonding that can occur between the amide groups in adjacent polymer chains. The polymers are semi-crystalline and high-melting (e.g. nylon-6,6 melts at 265 °C), but the amide group also results in a high moisture uptake (about 10%) for the polymer. This affects both the modulus of the material (by disrupting intermolecular hydrogen bonding) and also the processing, should all water not be removed prior to heating in an extruder.

Aromatic polyamides form an important class of high-performance liquid-crystalline polymers that have seen application as reinforcing fibres in composite materials (e.g. Kevlar®). Their superior properties are a result of the high aspect ratio of the rigid molecule and the strong hydrogen bonding, but their intractability means that they are not relevant in reactive processing under normal extruder conditions and most reactive processing applications are restricted to aliphatic polyamides.

Polyurethanes are an interesting hybrid of polyester and polyamide chemistry since they may be regarded as having an amide–ester repeat unit. The synthetic route is through the reaction of a diol with a di-isocyanate and, while this is a stepwise reaction, it must be distinguished from condensation reactions insofar as there is no small-molecule evolution and the reaction may occur at low temperatures:

$$n\mathrm{HOROH} + n\mathrm{OCNR'NCO} \rightarrow -\!\!(\mathrm{OROCONHR'NHCO})_n\!\!-$$

The versatility of urethane chemistry is seen in the wide range of properties that result from changes to the diol or the di-isocyanate. Thus, urethanes can range from soft elastomers and fibres to tough coatings and rigid foams. In one-shot formulations, the short-chain diol, polyol, isocyanates, chain extenders and additives (catalysts, plasticizers, fillers) are rapidly mixed and injected into the mould. These formulations are generally prepared as two separate parts, with the isocyanate being separate from the other components that are then rapidly mixed. The use of multifunctional diols or isocyanates allows crosslinking in the mould so that reactive processing such as reaction injection moulding may be employed (Frisch and Klempner, 1989). These are considered later in further detail.

Reaction with difunctional amines (see the next paragraph) allows the formation of poly (urethane-co-urea) and further extends the versatility of the segment architecture. For further control of the structure, a prepolymer is formed. The reaction, such as the polyurethane reaction shown above, is carried out with excess di-isocyanate so that an isocyanate-terminated prepolymer is obtained. The isocyanates used are typically aromatic, such as toluene di-isocyanate. This prepolymer is then reacted with a short-chain diol or diamine (for a polyurea) to form the final polymer.

Polyureas are analogues of the polyurethanes but with the diol being replaced by a diamine (Ryan and Stanford, 1989):

$$n\mathrm{H_2NRNH_2} + n\mathrm{OCNRNCO} \rightarrow -(\mathrm{HNRNHCONHR'NHCO})_{\overline{n}}.$$

While side reactions such as that of the urea with isocyanate are possible, they are generally limited by the high reaction rate for the isocyanate with the primary amine, provided that stoichiometric amounts of reagent are used. Polyureas form an industrially important class of polymers and, while they may be formed by the reaction of amines with carboxy compounds, the reaction route through the isocyanate has the advantage of there beiny no evolution of condensation by-products.

The properties of the polyurethane and polyurea homopolymers and copolymers depend on the chemical structures of the R and R′ groups. Thus, if R is a relatively long-chain polyester or a similarly flexible group then it will form a soft segment in the structure. Similarly, aromatic groups and other short-chain units will form a hard segment. The properties depend on the balance of these soft and hard segments as well as the degree of crosslinking.

Step polymerization of cyclic monomers

Ring-opening of cyclic monomers to give the high polymer may be undertaken using either anionic or cationic initiators. In this case the reactions are formally addition polymerizations and are considered later. However, the ring opening of **ε-caprolactam** to form nylon-6 may, in addition to the ionic reactions, occur as a step reaction involving first hydrolytic ring opening (Gaymans and Sikkema, 1989):

$$\mathrm{HN{-}[(CH_2)_n]{-}C(=O)} + \mathrm{H_2O} \longrightarrow \mathrm{H_2N{-}(CH_2)_n{-}COOH}$$

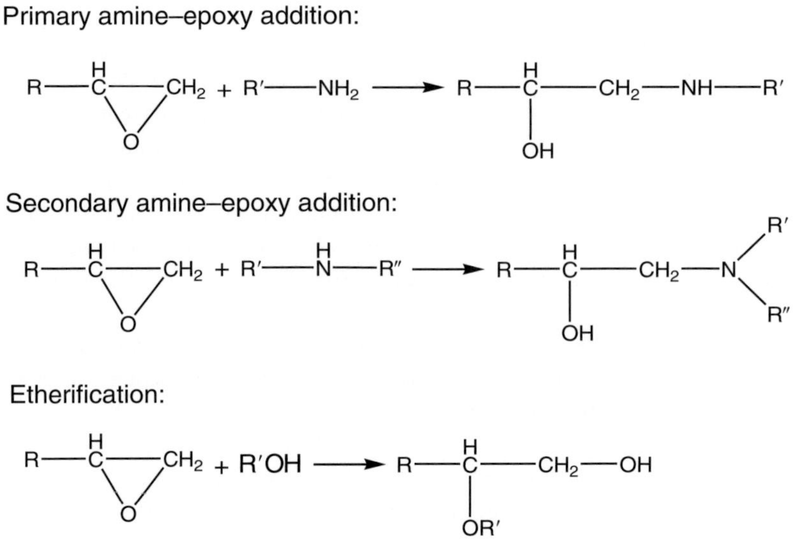

Primary amine–epoxy addition:

$$R-\overset{H}{C}(-O-)CH_2 + R'-NH_2 \longrightarrow R-\overset{H}{\underset{OH}{C}}-CH_2-NH-R'$$

Secondary amine–epoxy addition:

$$R-\overset{H}{C}(-O-)CH_2 + R'-\overset{H}{N}-R'' \longrightarrow R-\overset{H}{\underset{OH}{C}}-CH_2-N(R')(R'')$$

Etherification:

$$R-\overset{H}{C}(-O-)CH_2 + R'OH \longrightarrow R-\overset{H}{\underset{OR'}{C}}-CH_2-OH$$

Scheme 1.3. The principal reactions between an epoxy resin and an amine.

This reaction is industrially important since it is used in batch and continuous industrial processes but, because of the slow reaction rate, is inappropriate for carrying out in an extruder. (The anionic polymerization has, however, been demonstrated in a twin-screw extruder, and is discussed later.) The reaction takes place in the presence of 5%–10% water and, to increase the rate of initiation, a small of amount of ε-amino caproic acid (the product of ring-opening of the monomer) is added to provide the starting material without the need for hydrolytic ring opening, which may be slow. Since the final DP depends on the equilibrium water concentration, this is driven off at high conversion and the resulting polymer will contain about 8% residual monomer and 2% cyclic oligomers. The oligomers up to $n = 9$ are seen , but only in small amounts (Odian, 1991).

Other step-polymerization reactions of large cyclic monomers by entropically driven ring-opening polymerization have been reviewed (Hall and Hodge, 1999). An important example is the heating of cyclic bisphenol-A carbonates over a titanium isopropoxide transesterification catalyst to yield a polycarbonate with a relative molar mass of about 250 000. This has also been performed under conditions of reaction injection moulding (Hall and Hodge, 1999).

Perhaps the most widely exploited cyclic monomer in reactive processing of composite materials via a stepwise reaction is the **oxirane** or **epoxy group** (Hodd, 1989). Epoxy resins are principally used to form three-dimensional networks, but linear polymerization is possible. The main linear polyaddition reactions involve catalysed ring-opening in an ionic chain reaction. However, it is appropriate to consider the chemistry of the oxirane group in its reaction with nucleophilic reagents, principally amines, at this point so that the range of possible reactions may be introduced.

Epoxy–amine reactions

The epoxy group is characterized by high reactivity to both nucleophilic and electrophilic reagents, and reactions with a primary amine are shown in Scheme 1.3. These reactions

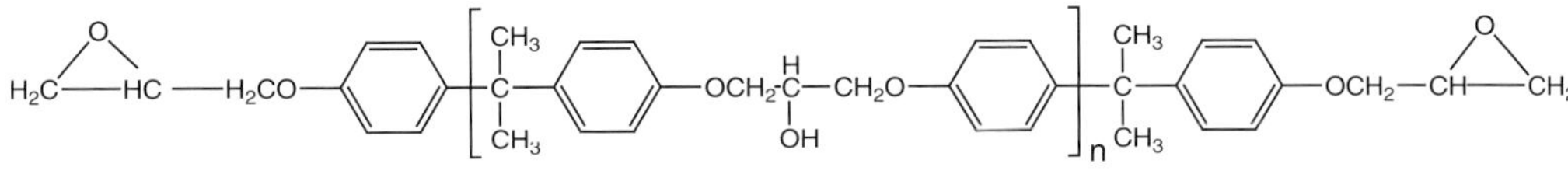

DGEBA

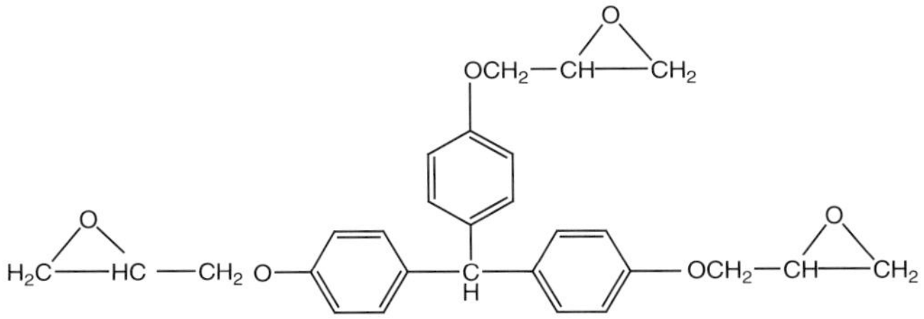

TACTIX

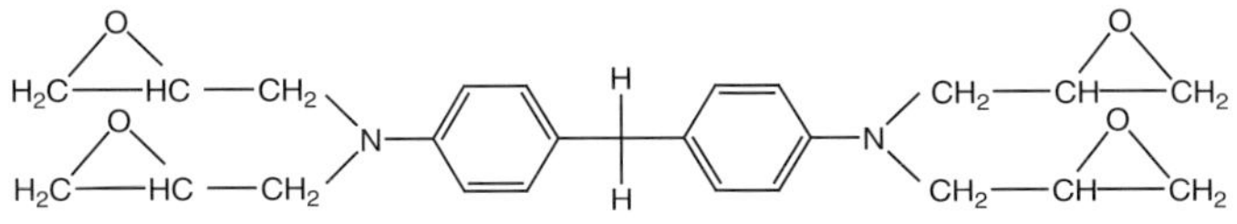

TGDDM

Scheme 1.4. Structures of typical difunctional, trifunctional and tetrafunctional epoxy resins.

have been elucidated by the use of model compounds such as phenyl glycidyl ether and aniline for which R and R′ are the phenyl group (Mijovic and Andjelic, 1995).

The first reaction results in the formation of a secondary amine and an alcohol. Each of these may react with another mole of epoxide, so further chain extension may occur. If the epoxide is difunctional as in the common commercial epoxy resin DGEBA (diglycidyl ether of bisphenol-A) then a three-dimensional network will rapidly form (Dusek, 1986). This is discussed later. The viscosity of the resin depends on the extent of oligomerization (n) that occurs during synthesis and, in the fabrication of fibre-reinforced epoxy-resin composites, this has an important effect on the rheology of the system during fibre impregnation and in the cure kinetics of the system and thus the chemorheology during cure as the network forms, since the –OH groups participate in the chemistry of the cure.

It has been noted that, when dealing with bulky epoxy resins such as DGEBA or the even bulkier TGDDM, rather than model systems, the rate coefficients for each of the three reactions are different (St John and George, 1994). Thus the rate of primary amine reaction is much greater than that for the secondary amine (reaction 2 in Scheme 1.3), so the first stage of the reaction is to form a linear polymer by a stepwise addition. It has also been found that the reaction between primary amine and epoxide will be catalysed by hydroxyl groups either present on the backbone of the polymer or present as impurities in the commercial resin (St John *et al.*, 1993). The effect of the –OH is believed to arise by virtue of hydrogen bonding to the oxirane as shown in Scheme 1.5.

There are several hydrogen-bonding configurations that may be drawn that also involve another molecule of amine, so the system may exhibit self-catalysis. A detailed study of the

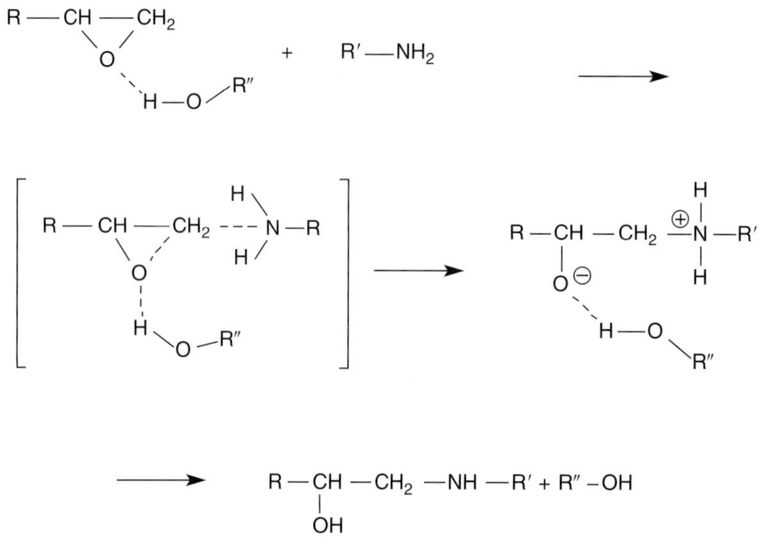

Scheme 1.5. Alcohol catalysis of the epoxy-ring-opening reaction by amine groups (the –OH may arise from impurities in the resin, the resin itself, e.g. DGEBA, Scheme 1.4, or following ring-opening of another epoxy group, Scheme 1.3). Reproduced with permission from St John (1993).

kinetics of model epoxy–amine systems has shown that self-catalysis is important, and it is often difficult to separate this from impurity effects in actual epoxy systems (Kozielski *et al.*, 1994). The etherification reaction (i.e. epoxy–hydroxyl reaction) has a high activation energy and is important at high temperatures.

It is also important to recognise that the structure of the amine nucleophile will have a major effect on the reactivity of the system. Thus aromatic amines such as 4,4′-diamino-diphenyl sulfone (DDS), shown in Scheme 1.6, will be much weaker bases and less reactive than aliphatic amines, so high temperatures are required for reaction to occur in the absence of added catalysts.

In many catalysed systems the reaction mechanism will be a competition between stepwise reactions and addition, so the resulting network may be very complex. This is addressed in more detail when considering the chemorheological development of the three-dimensional network.

1.2.2 Different polymer architectures achieved by step polymerization

In Section 1.2 it was noted that the linear homopolymer was only one possible chain conformation and it is appropriate to examine the types of different molecular architecture that have been demonstrated to be formed through simple step-polymerization reactions.

Cyclization

Cyclization may occur in a number of types of polymerization reaction where there is a finite probability that the reactive chain end may react with the other end of the same chain rather than with a different molecule. The result is a polymer with no chain ends and with a set of conformation-dependent properties different from those of the linear precursor. For

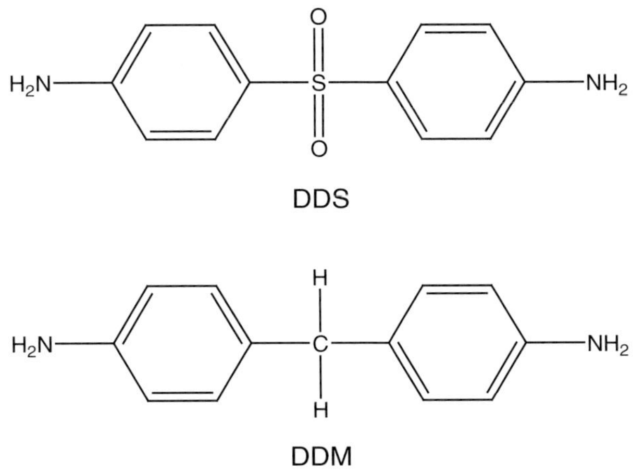

DDS

DDM

Scheme 1.6. Aromatic amine curing agents: DDS, 4,4′-diamino diphenyl sulfone; and DDM, 4,4′-diamino diphenyl methane.

example, a cyclic polymer with the same molar mass as a linear polymer will have one half the radius of gyration (R_g) (Richards, 1989).

As discussed earlier regarding polyesterification, simple condensation polymerization reactions operate by reactions of the type $AA + BB \rightarrow ABAB$ and, as this continues, there will be successive formation of dimers, tetramers etc. as the chain builds in a stepwise fashion. The rule governing these reactions is that the end group A may react only with end group B and, while at low extents of reaction the probability of an A group and a B group from the *same* chain reacting will be low, this probability will increase as the chain length increases. This will be governed in part by the thermodynamic stability of the resulting ring. Thus it is well known that for a simple cycle of methylene groups $(–CH_2–)_n$ there is a minimum in ring strain energy at $n = 6$, after which it increases again, until beyond $n = 14$ it is vanishingly small. The components that make up this strain are angle strain and conformational strain (the most important forms of these being torsional strain, due to eclipsed conformations, and transannular strain, due to ring crowding). Similar concepts apply for other cyclic systems such as polydimethyl siloxanes (PDMSs), $(–Si(CH_3)_2O–)_n$ and it has been reported that, at $n = 22$, the ring is planar (Richards, 1989). Siloxanes are of interest also since they are synthesized by the ring-opening reaction of the cyclic precursor with $n = 4$, commonly known as D_4, and the polymerization is an equilibrium reaction between the linear and cyclic forms.

In addition to the above thermodynamic conditions for ring formation, the kinetics of the reactions must be considered. Thus, for a reaction to take place the two ends of the polymer chain must be in the correct conformation for sufficient time for the new bond to form. The kinetic factor for cyclization is proportional to $R_g^{-5/2}$, so the net effect of the thermodynamic and kinetic factors is that rings are not favoured between $n = 8$ and $n = 11$. Suter (1989) has considered the theoretical approaches of Jacobsen and Stockmayer and compared theoretical and experimental values for macrocyclization equilibrium constants. This has also been performed for Monte Carlo as well as rotational-isomeric-state calculations for the statistical conformations of cyclic esters (decamethylene fumarates and maleates) and agreement with experimental molar cyclization equilibrium constants found (Heath *et al.*, 2000).

In polycondensation such as polyester and polyamide synthesis the fraction of cyclics is up to 3%, whereas in PDMS the cyclic content may be around 10%. In the case of polyesters the extent of cyclization is reduced by operating at high reagent concentration so that the bimolecular reaction of linear polymer formation is favoured over the unimolecular cyclization reaction of the end groups. It should be noted that there is a class of rigid-rod polymers, such as certain liquid-crystal-forming aromatic polyesters, for which the stiffness of the backbone results in a high aspect ratio for the chain so that there is a low probability of cyclization because the end groups cannot undergo any intramolecular reaction.

It was noted above that the reason for the low yield of cyclic oligomers in polycondensation reactions is the high concentration of the reagents. If the reaction occurs in dilute solution, the proportion of cyclics increases markedly (Hall and Hodge, 1999). For example, nylon-11, when synthesized at high dilution, yielded 57% cyclic polyamide (Hodge and Peng, 1998). A second approach to achieving a high concentration of cyclics is the cyclo-depolymerization reaction, and a yield of 50%–80% was possible using a solution concentration of 2% for the depolymerization. The use of cyclic oligomers in reactive processing has been discussed (Hall and Hodge, 1999), and the low viscosity of the oligomers was seen as attractive for reinforced reaction injection-moulding applications. By operating at high concentrations of oligomer and at elevated temperature one ensures that the equilibrium favours the formation of the linear polymer. While this enables interesting polymer reaction chemistry, the viscosity soon increases rapidly and, at the high temperatures necessary for reactive processing, side reactions can become important.

Step copolymers

The earlier example of the formation of a poly(urethane-co-urea) showed the change in properties made possible by including a comonomer (in that case a difunctional amine) together with the usual diol for reaction with a di-isocyanate. This can be extended to a wide range of step polymerizations where an additional reactant is added. Examples could be the use of two AB-type monomers (e.g. amino acids) or two AA (e.g. diacids) to react with one BB (a diamine) to form co-polyamides. Several features of step polymerization help in understanding the resultant copolymer. For example, since high-molar-mass polymer is formed only late in the reaction, the composition of the copolymer will be that of the feed ratio of the monomers.

Random step copolymers: unsaturated polyesters

An important example of a copolymerization of an industrially-important polyester is the formation of unsaturated polyesters that are the starting material for styrene crosslinked networks. In this case the number of unsaturated groups that are sites for crosslinking must be controlled by having spacers of saturated groups between the unsaturated groups. This is achieved by using maleic anhydride and phthalic anhydride as the acids, with a diol such as neopentyl glycol or ethylene glycol (Scheme 1.7).

Upon ring-opening, the maleic acid may isomerize from the *cis* form to the *trans* form, giving fumaric acid. This is an equilibrium reaction that depends on the acid catalyst as well as on the nature of the diol with which it reacts (Scheme 1.8).

The synthesis of the unsaturated polyester is analogous to the formation of a saturated polyester, and the side reactions discussed earlier, such as ester exchange, can occur at high temperature. The presence of the unsaturated group does mean that there are further possible side reactions:

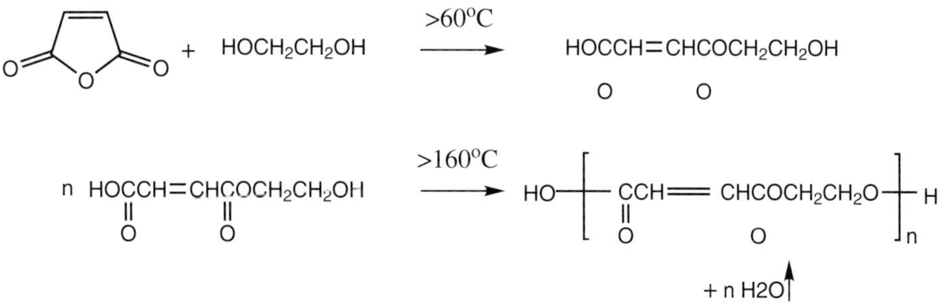

$$\text{maleic anhydride} + HOCH_2CH_2OH \xrightarrow{>60^{\circ}C} HOCCH{=}CHCOCH_2CH_2OH$$

$$n\ HOCCH{=}CHCOCH_2CH_2OH \xrightarrow{>160^{\circ}C} HO{-}\left[CCH{=}CHCOCH_2CH_2O \right]_n{-}H + n\ H_2O\uparrow$$

Scheme 1.7. The reaction scheme for maleic anhydride condensation with ethylene glycol, showing fumarate group formation in unsaturated polyester resin. Adapted from Hepburn (1982).

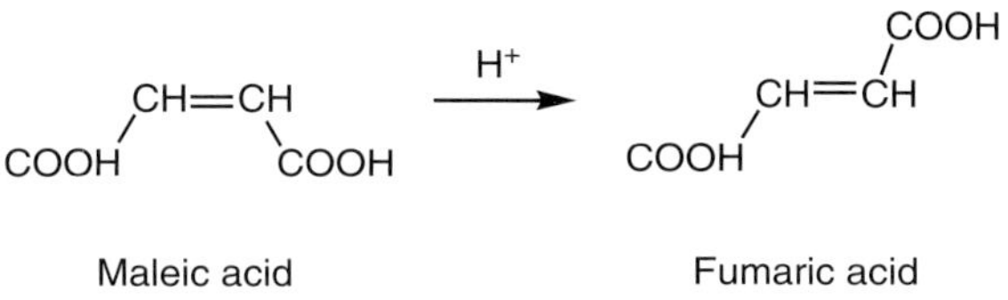

$$\text{COOH-CH=CH-COOH} \xrightarrow{H^+} \text{COOH-CH=CH-COOH}$$

Maleic acid → Fumaric acid

Scheme 1.8. Maleic acid–fumaric acid isomerization.

Scheme 1.9. Chain branching in unsaturated polyester formation due to addition of diol to a double bond.

- there is a need for rigorous exclusion of oxygen in order to prevent auto-oxidation through peroxide formation at the unsaturation and thus crosslinking;
- the hydroxyl groups from the diol may add to the double bond, resulting in a side chain, a loss of unsaturation and loss of stoichiometric balance in the synthesis (Fradet and Arlaud, 1989).

The latter reaction (Scheme 1.9) is a reversible, acid-catalysed reaction and may occur to the extent of 10%–20% loss of potential unsaturation.

The resulting copolymer will be a statistical copolymer with a random distribution of the saturated (aromatic) and unsaturated (fumarate) groups. The addition of monomers might not always be simultaneous, for practical reasons. For example, the tendency of phthalic anhydride to sublime after reforming from the half esters during polyesterification is overcome by firstly reacting the phthalic anhydride with the diol and then adding the maleic anhydride. This would be expected to form a block copolyester, depending on the extent of oligomerization that had occurred. However, the facile ester exchange reactions that occur in these systems mean that a statistical distribution is soon achieved, so there will be random occurrence of unsaturation along the backbone, depending on the original feed ratio of the components.

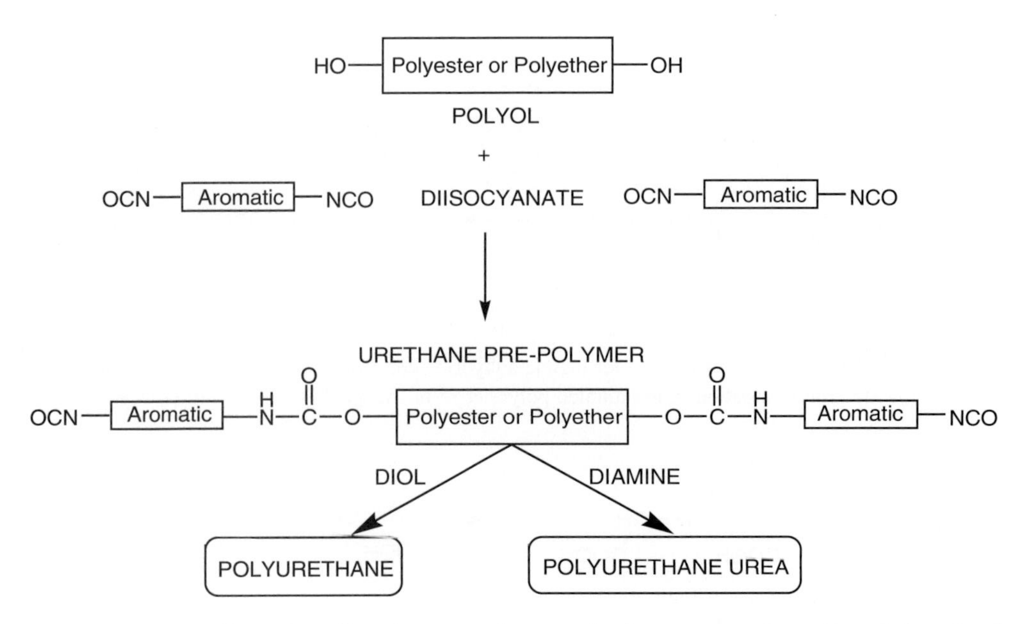

Scheme 1.10. Formation of urethane prepolymer and subsequent reaction with a diol or diamine extender to give a polyurethane or a polyurethane–polyurea. Adapted from Hepburn (1982).

The reaction of the double bond to form crosslinked networks and the chemorheology of these systems are considered later. Different acid and diol starting materials result in polyester resins with superior properties. For example, the use of *iso*-phthalic acid instead of phthalic anhydride (that produces *ortho*-phthalate units) results in improved water resistance of the resins, but at increased cost.

Block copolymers

The formation of step copolymers that are forming blocks is best illustrated through considering polyurethanes that are formed by prepolymers. This was discussed earlier in outline for the polyurethane co-polyureas. In the following example, a polyester-urethane is formed by reacting an –OH-terminated polyester with an –NCO-terminated polyurethane. Each of these is formed by the appropriate stepwise polymerization and then reacted to give a larger –NCO-terminated prepolymer:

$$\text{HO}-[\text{polyester}]-\text{OH} + \text{NCO}-[\text{polyurethane}]-\text{OCN}$$
$$\rightarrow \text{OCN–}\{[\text{polyurethane}]\text{NHCOO–}[\text{polyester}]\text{–OCONH–}[\text{polyurethane}]\}_{\overline{n}}\text{NCO}.$$

This prepolymer (Scheme 1.10) is then reacted with a chain-extender diol or diamine to give the final polyurethane or polyurethane–polyurea block copolymer, respectively. The diversity in properties that the final urethane block copolymers exhibit is governed by the nature of the three building blocks: the diol-terminated polyester, the di-isocyanate-terminated polyurethane and the chain extender (Hepburn, 1982). As shown in Figure 1.17, these form flexible blocks (from the polyester) that undergo phase segregation as structures of size 1000–2000 nm.

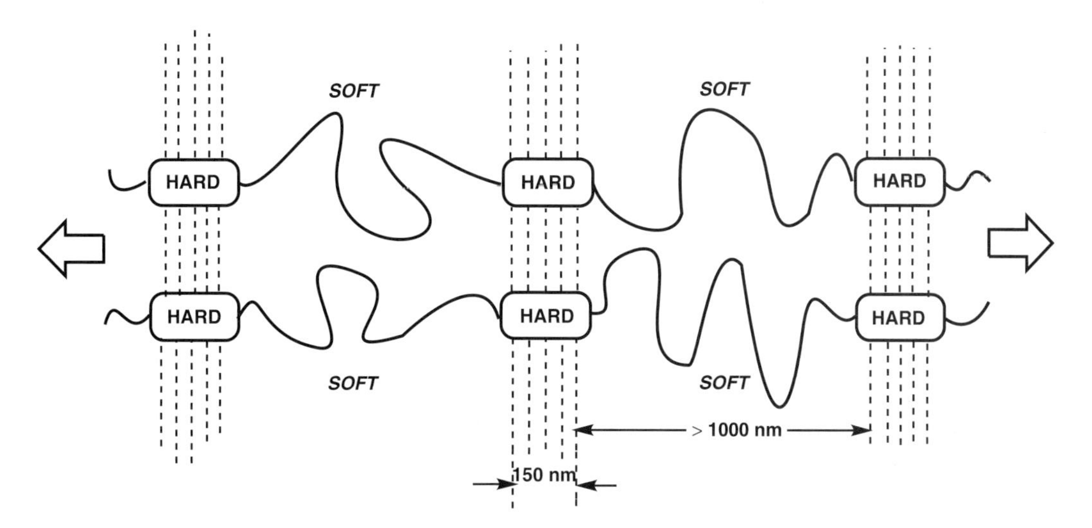

Figure 1.17. A schematic diagram of a polyurethane thermoplastic elastomer consisting of short, hard, aromatic segments that may crystallize and long, soft and extensible, polyol segments. The soft segments allow extension when stressed, as shown. Adapted from Hepburn (1982).

As an alternative to polyester diols, polyethers (such as polypropylene glycol) that are even more flexible are often used. These are bonded to smaller (150-nm) rigid domains from the (aromatic) isocyanate. The chain extender in the formulation may be either flexible or rigid.

If the diol chain extender is used in exact molar proportion to the unreacted isocyanate, then a linear polyurethane elastomer is obtained. The resulting thermoplastic elastomer may be extruded or injection-moulded, and the properties arise from the ability of the hard and soft segments to form semi-crystalline domains that act as virtual crosslinks in the polymer and give it elastomeric properties, as shown in Figure 1.17.

On elongation in the direction of the large arrows, the hard blocks resist deformation and the soft segments extend. On release of the stress, the original dimensions are recovered.

Catalysts are generally not employed for linear block copolymer systems such as this, but, where a higher reaction rate is required, as in reaction injection moulding, or the components are of widely differing reactivity (e.g. secondary hydroxyl groups), then catalysts are necessary. The most widely employed catalysts for isocyanate reactions are tertiary amine bases. Foam formation also requires a careful choice of catalyst since there needs to be fast polymer formation coupled with gas generation and a high extent of reaction in order to form the final product with high strength. Triethylene diamine (1,4-diazo-[2,2,2]-bicyclo-octane) (DABCO) is widely used as the catalyst for foams (Hepburn, 1982).

Branching in step-growth polymerization

Chain branches in stepwise polymerization may occur due to side reactions such as the addition of an –OH to a double bond in unsaturated polyester synthesis. Alternatively, chain branches can be a deliberate addition such as in polyurethane synthesis where a functionality greater than 2 is brought to bear (e.g. by using a triol chain extender) so that chain branching can occur. In systems of stepwise polymerization in which the reagents are polyfunctional these branches will lead to the occurrence of crosslinks and gelation. This is

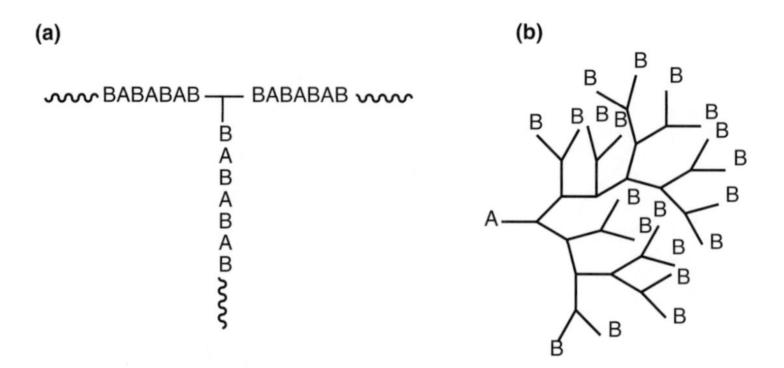

Figure 1.18. Branching during step polymerization. (a) Addition of a trifunctional agent (e.g. a triol, BBB) to a self-condensing difunctional system (e.g. a hydroxy acid, BA). (b) Hyperbranching from a system AB_2. In both cases A may react only with B.

in contrast to the situation in addition polymerization of linear polymers, where controlled branching is a widely used strategy for achieving particular rheological and mechanical properties.

In certain cases it is possible to have branches introduced without crosslinking. If we consider the self-condensation reaction of, say, a hydroxy acid, AB, then, if a triol, BBB is added at low concentrations, there will be a possibility that the acid in the monomer will react with one of the B groups from the triol rather than with the B group on another molecule of monomer. If the remaining B groups from the triol now react with acid groups that then continue to undergo self-condensation, there will now be two chains growing where previously there was one, i.e. a branch point has been introduced, as shown below. The reason why there is no crosslinking is that the B groups cannot react with each other, so the branch cannot terminate on another chain. This is shown in Figure 1.18(a).

The molar-mass distributions for such multichain polymerizations have been calculated and shown to be narrower than for a single-chain polymerization since the probability of formation for very long chains is reduced. The polydispersity for chain growth in the presence of a multifunctional reagent A_f with functionality f then becomes, as the conversion, p, approaches unity,

$$M_w/M_n = 1 + 1/f, \tag{1.37}$$

i.e. it should approach 1.33 for the system discussed. This is to be contrasted with the system without branching, where M_w/M_n approaches 2 as the reaction reaches completion (Equation (1.35)).

Crosslinking will occur when a multifunctional reagent, B_f, that can react with the second group is added, since then the branches may connect the chains. Alternatively, it can occur when either A_f or B_f is added to a difunctional system AA plus BB. The importance of this in network formation and chemorheology is considered in more detail in Section 1.2.4. At this stage it is appropriate to note the conditions under which gelation may occur in order to contrast network formation with hyperbranching.

The condition for gelation was demonstrated by Flory (1953) to occur, for a system consisting of a triol (B_3) added to a difunctional monomer (A_2), at the critical conversion, α_c, of 0.5 (i.e. $\alpha_c = 1/(f-1)$, where f is 3 for the triol).

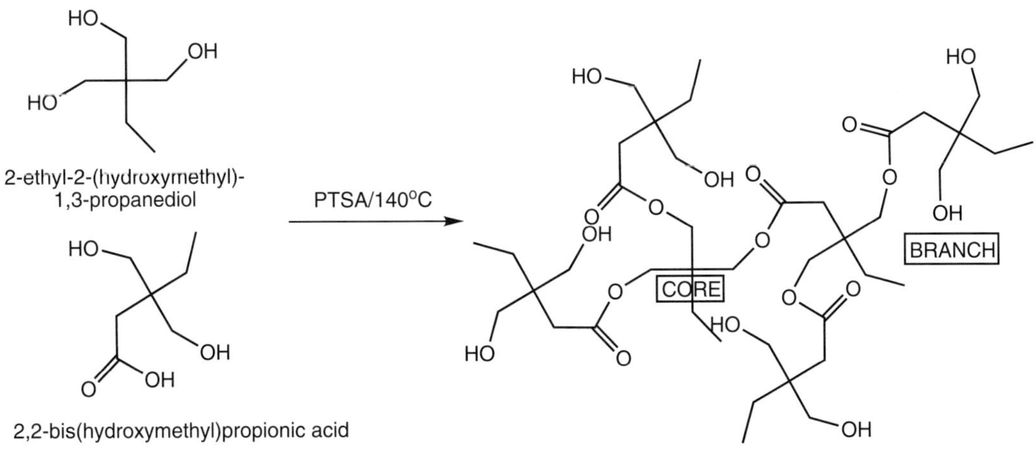

2-ethyl-2-(hydroxymethyl)-1,3-propanediol

PTSA/140°C

CORE

BRANCH

2,2-bis(hydroxymethyl)propionic acid

Scheme 1.11. Early stages of the synthesis of a hyperbranched polyester. After Jikei and Kakimoto (2001).

Since the conversion, α, at any time is related to the conversion of A_2, p_A, for a mixture with stoichiometric ratio ($r = [B_3]/[A_2]$) by the relation

$$a = p_A^2/r, \tag{1.38}$$

a calculation of the extent of conversion for the stoichiometric amounts of the polyol and polyurethane ($r = 3/2$) considered before shows that this will occur at conversions of the isocyanate groups of 0.87. Thus gelation will result before full conversion of isocyanate.

In the special case of *hyperbranched* polymers, as shown in Figure 1.18(b), the branching is controlled so that novel molecular architectures are achieved without gelation. If, instead of a system consisting of difunctional and trifunctional reagents, A_2 and B_3, a single reactant of structure AB_2 is employed, the conversion with time is given by

$$a = p_A/(f-1) \leq a_c. \tag{1.39}$$

Thus a step-polymerization system synthesized from an AB_2 monomer should be highly branched but never reach gelation even at full conversion of the available functional groups. This is the basis of the formation of hyperbranched polymers by step-growth polymerization (Jikei and Kakimoto, 2001) and a reaction scheme for AB_2 hyperbranching is shown in Scheme 1.11.

Dendrimers and hyperbranched polymers

The controlled branching that diverges at each branch point in a stepwise polymerization results in a novel architecture, which in three dimensions becomes globular. The term 'dendrimer' reflects the tree-like hyperbranching that results when such a system is constructed so that the branches contain no linear components but diverge at each branch point (Frechet and Hawker, 1996, Frechet and Tomalia, 2001). This is shown in Figure 1.19 (Frechet and Hawker, 1996).

It is seen that the globular dendrimer (**1**) may be considered as four separate *dendrons* (**2**) that branch from the core, **C** (**3**). In the nomenclature of Frechet, **R** is a reactive site, **F** is a focal point and **B** is a building block and branching point. At each branching point of the dendron there are two new arms formed, which, prior to the next generation, *G*, of reaction will have 2^G reactive sites. Thus, at the completion of the reaction, the number of terminal

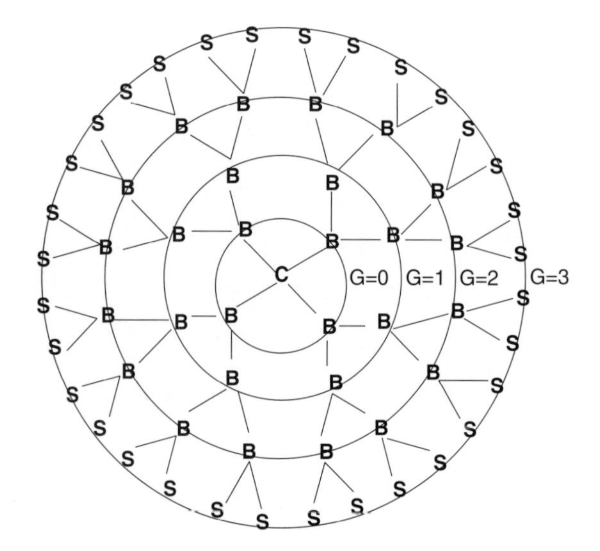

1: DENDRIMER (Generation 3)

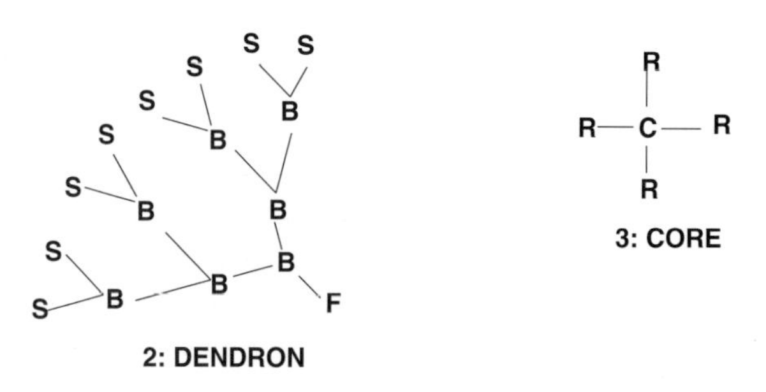

Figure 1.19. Schematic of a Dendrimer with generations (G) indicated and the building blocks B. At generation 3 this dendrimer has 32 active sites (S), which may continue reaction to generation 4. Nomenclature is explained in the text. After Frechet and Hawker (1996).

groups (labelled **S**) on the total dendrimer of generation 4 will be $4\times2^4=64$. The number of building blocks **B** is $4\times(2^3+2^2+2^1+2^0)=60$. Examination of the dendrimer structure shown in Figure 1.19 reveals the unique feature of the large number of end groups compared with a linear polymer of the same molar mass. The globular structure is also achieved without the chain entanglements that would be required of a linear polymer. These features result in unusual rheological and reaction properties that are still being elucidated.

The feature that distinguishes dendrimers from hyperbranched polymers (Figure 1.18(b)) is the much more well-controlled architecture in which there are no linear components in the structure. Two broad strategies have been employed to achieve dendrimers, namely divergent and convergent growth, which differ in terms of the direction that the synthesis follows with respect to the core (Frechet and Hawker, 1996). In both cases it is important that the sample be purified after each generation to remove any by-products of side reactions (e.g. cyclization) that may affect the growth of the next generation. The divergent approach has allowed very large amino-functionalized dendrimers based on a trifunctional core and up to generation 10 to be prepared in commercial quantities, such as poly(amidoamine)

(PAMAM) or Starburst dendrimer. This has $3 \times 2^{10} = 3072$ end groups. It has been noted (Frechet and Hawker, 1996) that the difficulty in achieving a perfect structure increases with each generation. The convergent approach which builds individual dendrons commencing from the reactive sites, **S**, and proceeding inwards has fewer unwanted side reactions. However, it involves the challenge of coupling the large and bulky dendron fragments to the core, **C**, before the dendrimer structure is achieved.

The divergent approach has been shown (Bosman *et al.*, 1999) in examples of a poly (propylene imine) dendrimer to produce only 23% dendritic purity by the fifth generation, which has 64 end groups. This is a simple consequence of the statistics of the process used, since incomplete reaction of amino groups or intramolecular cyclization between arms of the growing dendrimer will produce occasional defect groups. Since it requires 248 reactions to reach generation 5 commencing from a tetrafunctional core, an incidence of defects per step of only 0.5% will result in the above level of dendritic purity. If it is possible to purify each generation (as happens in the convergent approach) then low levels of impurities may be achieved and polymers can appear with a single mass, as demonstrated by MALDI-MS (Bosman *et al.*, 1999). The actual conformation that the dendrimer may adopt, the hydrodynamic radius and the molar-mass dependence of the viscosity are discussed later when these systems are compared with linear polymers. The main point noted here is that, for the same molar mass, dendrimers differ from linear polymers by having low entanglements and a large dependence of properties on the end groups. This is most marked in the glass-transition temperature (T_g) (Bosman *et al.*, 1999).

Hyperbranched polymers are synthesized by a 'one-pot' strategy, in contrast to the sequential approach for dendrimers, and are characterized by high levels of 'impurity' structures, due to the large number of possible configurations that may arise from the statistics of reaction of an AB_2 system (Kim, 1998). Also the polydispersities of hyperbranched polymers are very large since, unless special care is taken, the larger molecules grow at the expense of the smaller molecules since they have a large number of available reactive sites and statistically are more likely to react. The degree of branching, DB, is a convenient measure in terms of which to differentiate hyperbranched polymers. An ideal degree of branching is that seen for the pure dendron, where every reaction point is a branch, i.e. DB = 1. This obviously reduces for each reaction site that results in a linear unit L, rather than a dendritic unit, D. There will be terminal units, T, both for dendrons and for hyperbranched polymers, and the definition is

$$\mathrm{DB} = (N_{\mathrm{D}} + N_{\mathrm{T}})/(N_{\mathrm{D}} + N_{\mathrm{T}} + N_{\mathrm{L}}), \tag{1.40}$$

where N_{D} is the number of dendritic units, and so on. The other extreme is the linear polymer, where DB = 0, so the hyperbranched systems' DB will always fall between these extremes and, for a one-pot synthesis of AB_2, will statistically approach 0.5. Ways of achieving a higher degree of branching (Jikei and Kakimoto, 2001) are

- use preformed dendron fragments, e.g. by pre-reacting three AB_2 units to give AB_4; the theoretical DB then becomes 0.75;
- polymerize in the presence of a core unit (if a core unit of functionality B_3 is used then a DB of 0.8 is possible);
- enhance the further reactivity of any linear units that are formed

These techniques will also result in a decrease in the polydispersity of the system.

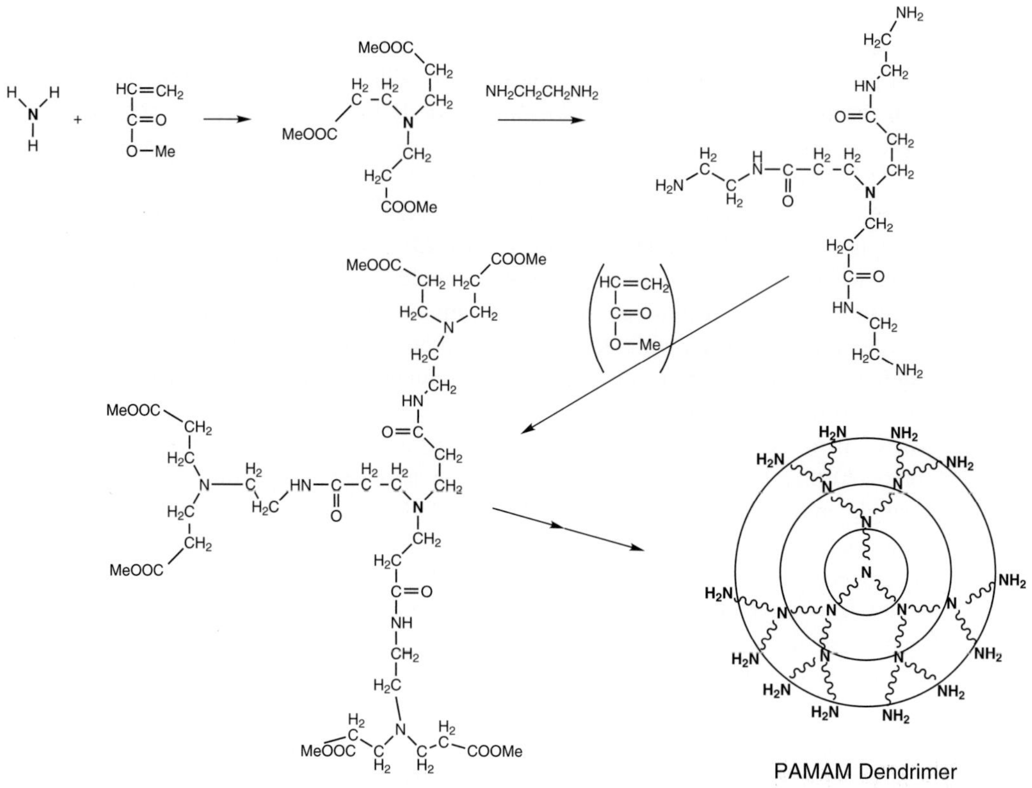

Figure 1.20. Synthesis of an amine-terminated polyamide dendrimer by a divergent route (the Starburst PAMAM system). Adapted from Frechet and Hawker (1996).

Dendrimer step-growth synthesis

A range of divergent and convergent step-growth polymerization strategies has been employed to create dendrimers. The choice of a divergent synthetic route depends on the use of high-yielding reactions and symmetrical building blocks as well as excess reagents to drive the reaction. One of the first successful dendrimers was the Starburst PAMAM system (Frechet and Tomalia, 2001) which had ammonia as the trifunctional core, which was reacted with methyl acrylate in a Michael addition to give the first product, a trifunctional ester. Excess ethylene diamine was then added in the second reaction to give a standard amidation reaction and produce three arms that are primary-amine-terminated. The reaction sequence of Michael addition on the amine with methyl acrylate followed by reaction with ethylene diamine may then be repeated to create the subsequent generations, as shown in Figure 1.20.

The resulting dendrimer is thus a giant amino-terminated aliphatic polyamide with a globular conformation, quite unlike the amine- and acid-terminated polyamides described earlier and synthesized by condensation polymerization. Many other divergent dendritic synthetic routes have been described that lead to terminal groups such as –OH, –Br and –OPh (Frechet and Hawker, 1996), and the limiting feature is often the steric crowding that makes higher generations unachievable.

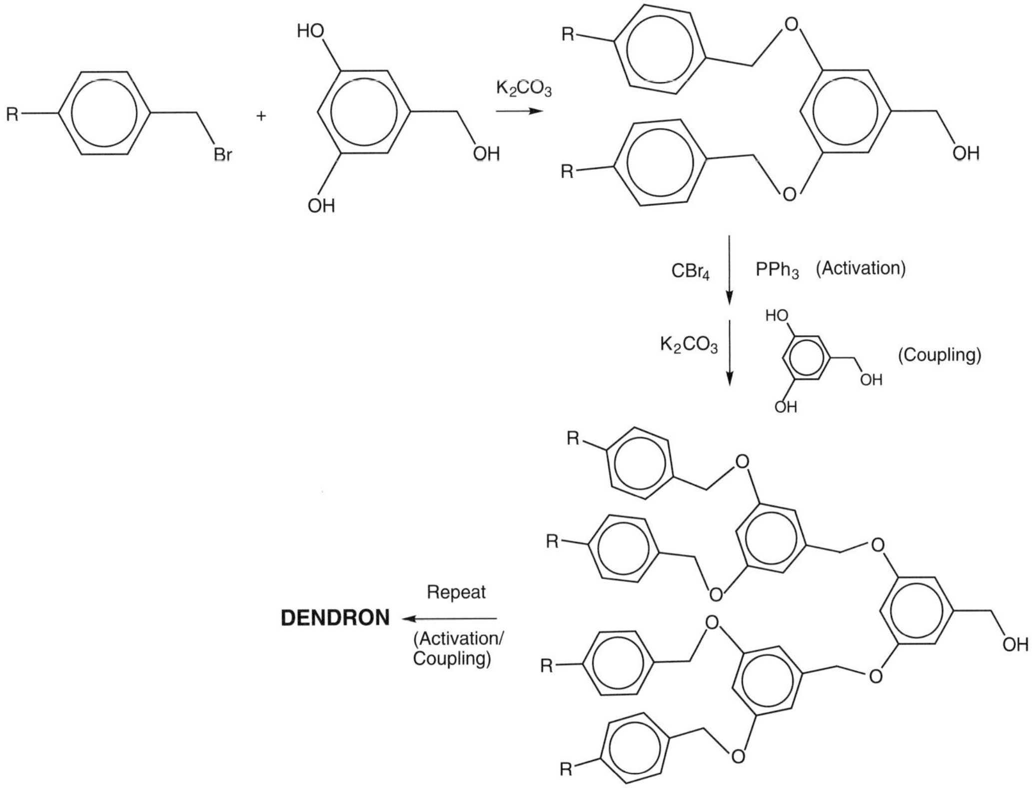

Figure 1.21. Synthesis of a dendron, precursor to a dendrimer, by a convergent route. Adapted from Frechet and Hawker (1996)

Convergent synthesis has been illustrated through the formation of polyethers based on 3,5-dihydroxybenzyl alcohol built from the surface functional group by reaction with benzyl bromide (Figure 1.21). Alkylation occurs at the two phenyl –OH groups so that the hydroxymethyl group is left available for the next step, which is regeneration of a benzyl bromide group that is able to repeat the reaction with 3,5-dihydroxybenzyl alcohol. This sequence was chosen because it proceeds to high yield under mild conditions and has no significant side reactions. This two-step procedure may then be repeated to build dendrons of the desired generation prior to coupling to a core molecule. This is shown for a tri-functional core of 1,1,1-*tris*(4′-hydroxyphenyl)ethane. Because of the purification procedure at each step, the most structurally pure dendrimers are produced by the convergent approach, but the number of generations achievable will be limited by steric considerations in the final coupling reaction to the core. Results from MALDI-MS confirm that the convergent dendrimers are monodisperse, with the theoretical and experimental molar masses in agreement.

Synthesis of hyperbranched polymers

In contrast to the careful development of dendrimer structures generation by generation in a sequence of reactions, it was noted above that the one-pot synthesis of hyperbranched polymers leads to complex and often ill-defined globular structures. This is the consequence of the often unequal reactivity of the groups in the AB_2 monomer, such that, when one group A is added to B, the reactivity of the second B group will be reduced for steric reasons. One of the commercially available hyperbranched polymers used in applications such as thermoset toughening is an aliphatic polyester (Boogh *et al.*, 1999), and this is an appropriate example to choose to illustrate the polymerization reactions.

Polycondensation of 2,2-*bis*(hydroxymethyl)propionic acid has been carried out in the melt at 140 °C using *p*-toluenesulfonic acid (PTSA) as the catalyst. A core molecule, 2-ethyl-2-(hydroxymethyl)-1,3-propanediol was used, as shown below, to achieve DB values in the range 0.83–0.96 (Jikei and Kakimoto, 2001). The early stage of the reaction is shown in Scheme 1.11, where the core molecule has all –OH groups reacted and one branch has extended by reaction of the two –OH groups.

Self-polycondensation would be expected to compete with the reaction with the core, but it has been found that it is possible to drive the reaction to achieve the high degree of branching required. The resulting polymer has predominantly –OH terminal groups.

Similar considerations apply for the aromatic polyesters such as those prepared from 3,5-dihydroxy benzoic acid or the trimethylsililated adduct (Jikei and Kakimoto, 2001). In the use of the alternative monomer 3,5-*bis*(trimethylsiloxy)benzoyl chloride the monomer was required to be of high purity in order to avoid the occurrence of side reactions and formation of an insoluble product. The terminal hydroxy groups are regenerated by hydrolysis of the trimethylsiloxy (TMS) groups in methanol on completion of the reaction and the DB was 0.55–0.60. It is noted that the T_g value for the aromatic ester was 197 °C, compared with 40 °C for the aliphatic polyester, although both have thermal stabilities of over 300 °C. If the aromatic –OH groups were esterified with adipic acid, T_g fell to 6 °C, showing that the properties of hyperbranched polymers, like those of dendrimers, are very dependent on the end-group composition.

Network formation by step growth

The formation of crosslinked networks has been the basis for polymer technology since the development of phenol–formaldehyde resins by Baekeland in 1910. The changes to the rheology of the phenolic system, as it develops from a liquid to a rubber and then to a glass, arise from the formation of a three-dimensional network as the step-growth chemical reactions occur. Water is evolved and the end product is a solid infusible mass.

It is thus apparent that chemorheology has been at the foundation of polymer science, but the intractability of the three-dimensional network has greatly limited the development of a precise understanding of all the chemical changes that underlie the rheological properties. Statistical analysis has enabled a model for the progressive development of the network to be built on the basis of the functionality of the species involved and criteria for the rheological changes leading up to the formation of an infinite, gelled network. Modern spectroscopic and analytical methods have enabled a greater understanding of the chemical changes occurring and have enabled the augmentation of the statistical models with chemical kinetic data. The detailed analysis of the models for network polymers is presented in Chapter 2 and the chemorheological analysis of network-forming polymers as the transition from a liquid through a rubber to a glass is presented in Chapter 3. In this section we will look at the underlying chemistry for various network-forming polymers and then examine the kinetics of these reactions.

Phenolic resins

Phenolic resins are generally prepared commercially by the crosslinking reaction of prepolymers. The crosslinking is normally produced by the elimination of water at high

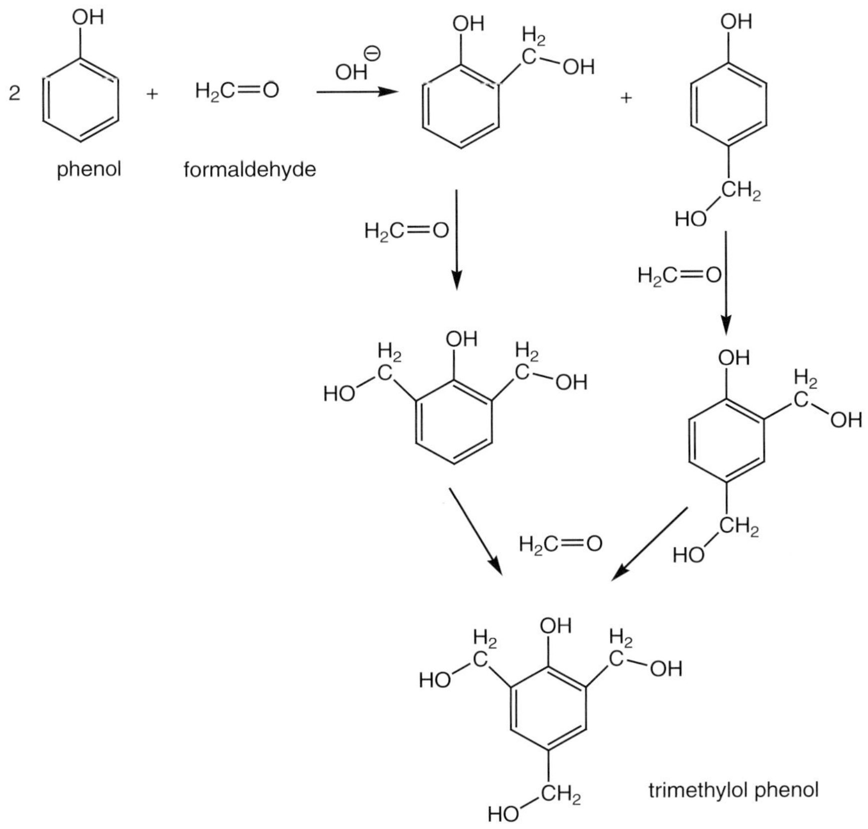

Scheme 1.12. The reaction scheme for trimethylol phenol formation.

temperature (>160 °C), but can also be made to occur at lower temperature (60 °C) by the use of strong acid catalysts. In this case the water phase separates and remains in the resin. There are many forms of phenolic resins, but the general categories are resoles and novolacs. The chemistry of resole formation is illustrated here (Knop *et al.*, 1989).

Commencing from phenol and formaldehyde, the products of base-catalysed condensation reaction at $T < 60\,°C$ are a series of mono-, di- and tri-methylol-substituted phenols (Scheme 1.12).

The reaction mechanism probably involves the formation in base of phenolate anion (PhO^-), followed by the electrophilic attack of the formaldehyde preferentially in the *ortho* and *para* positions. Depending on the molar ratio of formaldehyde to phenol, the further methylolation reactions may occur. This reaction is of second order, i.e.

$$\text{Rate} = k[\text{PhO}^-][\text{CH}_2\text{O}]. \tag{1.41}$$

Resole prepolymer formation is then made to occur by heating above 100 °C to evolve water. Above 130–150 °C a methylene bridge (to give a novolac by the evolution of formaldehydc, as shown in Scheme 1.13) is formed, rather than an ether link of the resole.

The prepolymers are soluble in common solvents (e.g. ethanol), to give varnishes, or may be emulsified to impregnate cellulosic reinforcement, to give a composite. The crosslinking

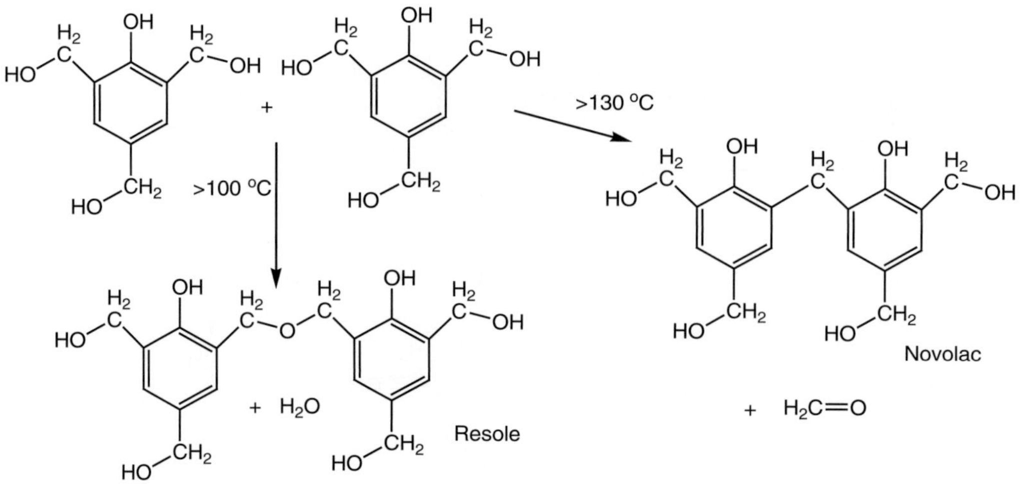

Scheme 1.13. The reaction scheme for resole and novolac formation.

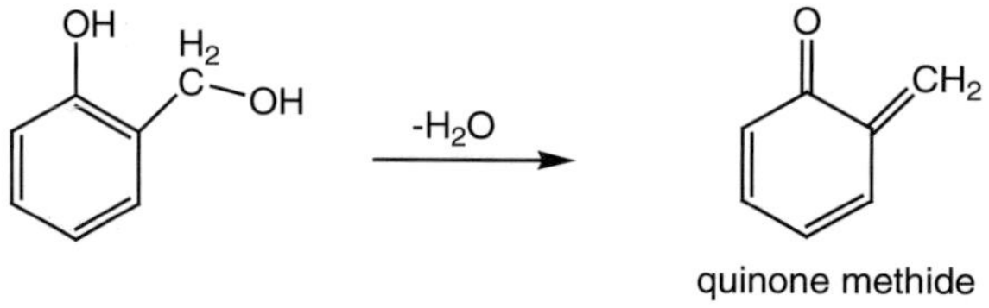

Scheme 1.14. The reaction scheme for quinone methide formation.

reaction at 200 °C drives off water and results in a network with very high crosslink density and T_g, so without reinforcement the fully cured resins are brittle.

A particular feature of the high-temperature-crosslinked resin is the darkening due to the formation of quinone methide (Scheme 1.14). Since this reaction is important only above 160 °C, by using acid catalysts at $T < 80$ °C darkening can be avoided. In principle the acid-catalysed reaction will occur at room temperature, but for the best properties a post-cure reaction up to 80 °C is required.

Cyclic products from the reaction of *p*-substituted phenols with formaldehyde may be formed during the alkaline condensation reaction. These calix(4)arenes contain both methylene and dimethylene ether links and form even-numbered cycles ranging from $n = 4$ to $n = 8$ (Knop *et al.*, 1989).

While novolac links may form at elevated temperature during resole synthesis (Scheme 1.12), the preferred synthesis of novolac resins takes place under acid conditions (pH from 1 to 4) and with an excess of phenol. In this case the methylol intermediates cannot be isolated since they react rapidly to give methyene-bridged structures with relative molar mass ≤ 2000, as shown in Scheme 1.15.

The absence of the methylol groups means that the systems cannot crosslink on heating but require an added curing agent such as hexamethylene tetramine.

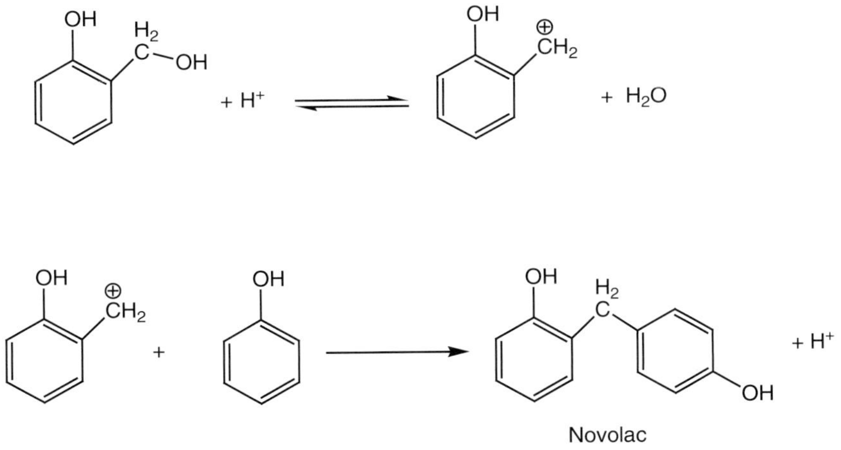

Novolac

Scheme 1.15. The reaction scheme for formation of novolac.

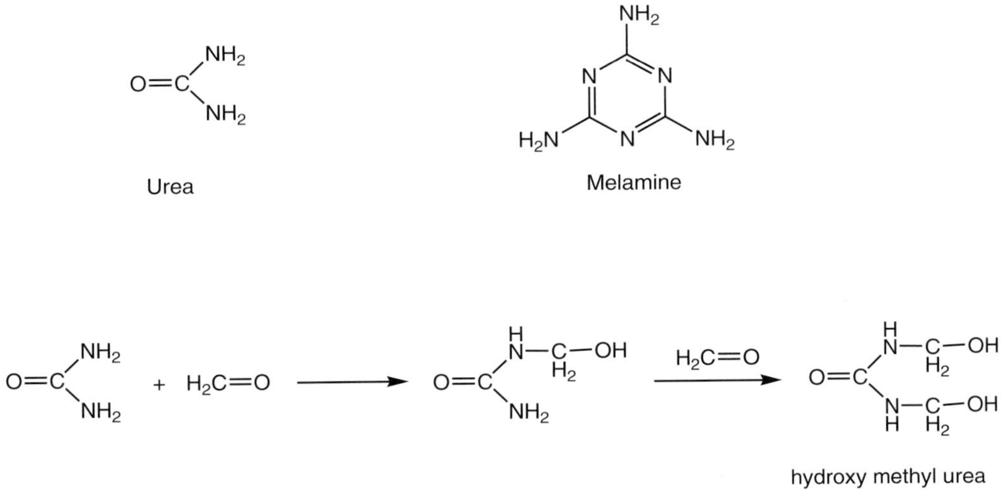

Urea

Melamine

hydroxy methyl urea

Scheme 1.16. Structures of urea and melamine, and the reaction scheme for the condensation reaction of urea and formaldehyde.

Urea– and melamine– formaldehyde resins

The reaction of urea and melamine with formaldehyde is a condensation reaction that allows firstly the formation of methylol compounds, then the reaction of these to form oligomers and finally the formation of a crosslinked network (Braun and Ritzert, 1989). The first stage of the reaction may be either acid- or base-catalysed, and the subsequent oligomerization is acid-catalysed if condensation to give methylene bridges rather than dimethylene ether linkages is desired (Scheme 1.16).

These intermediates are water-soluble and may be used to impregnate cellulosic fillers prior to the subsequent crosslinking reaction. The crosslinking reaction (Scheme 1.17) involves firstly the reaction of methylol groups and urea to give a methylene bridge (1).

$$\text{-CONHCH}_2\text{OH} + \text{H}_2\text{NCO-} \rightarrow \text{-CONHCH}_2\text{NHCO-} + \text{H}_2\text{O} \qquad (1)$$

$$2\text{-CONHCH}_2\text{OH} \rightarrow \text{-CONHCH}_2\text{OCH}_2\text{NHCO-} + \text{H}_2\text{O} \qquad (2)$$

$$\text{-CONHCH}_2\text{OCH}_2\text{NHCO-} \rightarrow \text{-CONHCH}_2\text{NHCO-} + \text{H}_2\text{CO} \qquad (3)$$

$$2\text{-CONHCH}_2\text{OH} \rightarrow \text{-CONHCH}_2\text{N(R)CH}_2\text{OH} + \text{H}_2\text{O} \qquad (4)$$

where R = $\text{-CH}_2\text{NHCO-}$.

Scheme 1.17. Possible crosslinking reactions of urea–formaldehyde resins.

Self-condensation of the methylol groups to give a dimethylene ether linkage (reaction (2)) may, at high temperatures, also lose formaldehyde to generate the methylene bridge (reaction (3)). Tertiary-amine sites may also be produced by the reaction of a methylol group with the amido nitrogen (reaction (4)).

The methylol group may then undergo further crosslinking reactions. While the completion of crosslinking should in principle give methylene linkages and tertiary-amine sites, there will also be dimethylene ether linkages and free methylol groups as well as imino groups from incomplete reaction with formaldehyde.

The reactions of melamine (the structure of which is shown in Scheme 1.16) are similar to those of urea except that there are three amino groups per molecule, so it is hexafunctional with respect to formaldehyde and it is possible to separate the different reaction products with formaldehyde up to hexamethylol melamine. The crosslinking reactions are identical to those of urea–formaldehyde resins.

The lower curing temperature of these resins (100 °C under acidic conditions) avoids the discolouration of phenolics but there is still the evolution of water (and formaldehyde) on crosslinking that requires special procedures during autoclave curing of composites containing these resins. These problems do not arise on crosslinking of epoxy resins, in which there is no small molecule evolved, but this is still a step-growth polymerization with the appropriate choice of curing agent.

Epoxy resins cured with amines and other nucleophiles

The class of thermosetting resins for which polymerization and crosslinking are based on the ring-opening of a three-membered cyclic ether (the oxirane, glycidyl ether or epoxy group) is one of the most industrially important groups of materials. The resins were commercialized in 1946 and are based on the reaction of aromatic phenols (e.g. bisphenol-A) or aromatic amines (e.g. 4,4′-diaminodiphenyl methane) with epichlorohydrin under basic conditions. The sequence of reactions employed to produce the most common epoxy resin, the diglycidyl ether of bisphenol-A (DGEBA), is given in Scheme 1.18 since it shows the origin of the possible oligomers and by-products of the reaction that can affect the subsequent curing reactions.

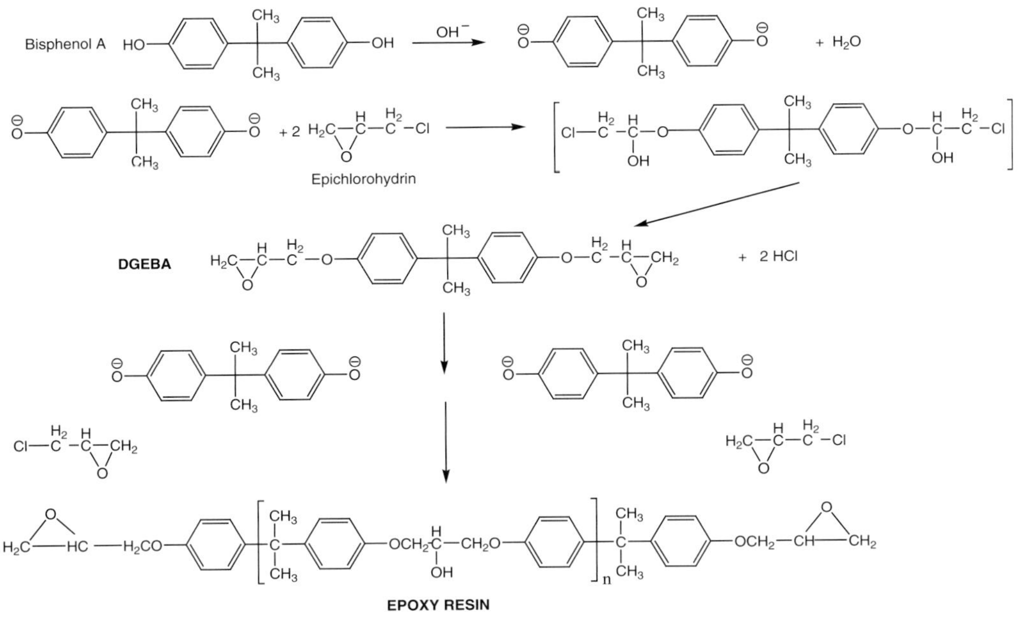

Bisphenol A

OH^-

+ H_2O

Epichlorohydrin

DGEBA

+ 2 HCl

EPOXY RESIN

Scheme 1.18. The reaction sequence for DGEBA-based epoxy-resin synthesis.

It is noted that the value of the degree of polymerization in the final product may range from $n = 0.15$, giving a molar mass of 380 g/mol (as in the commercial resins Shell Epon 828 or Ciba Araldite 6005 and Dow DER 331 (Bauer)), up to $n = 13$, giving a molar mass of 3750 g/mol in Shell Epon 1009. The actual DGEBA resin as shown ($n = 0$) is not commercially used. The viscosity also increases with the value of n, with the resin becoming semi-solid at $n \approx 1$ and the T_g progressively increasing from 40 to 90 °C. Normally the resins are difunctional, but, if ring closure of the chlorohydrin group, –CH(OH)CH_2Cl, is incomplete, then the group will remain as an impurity and the resin will have a small amount of a monofunctional epoxy. The value of n depends on the extent of reaction of the formed epoxy resin (with $n = 0$) with more epichlorohydrin to give the oligomers $n = 1, 2, \ldots$ and thus is linked to the original stoichiometry. Chromatographic (HPLC and SEC) analysis of the commercial Shell Epon 828 shows that the majority of the resin (83%) is $n = 0$, with the amounts of other oligomers decreasing to be vanishingly small by $n = 3$. The total difunctional resin content is about 94%. As n increases the molar mass between crosslinks will also increase, so the resulting networks will have a lower crosslink density (as discussed in more detail later).

Other high-performance epoxy resins have been synthesized with the goal of achieving higher-temperature performance. Initially epoxy novolacs, which have a more rigid aromatic backbone and a higher epoxide functionality, were prepared.

More recently, trifunctional and tetrafunctional resins (Tactix 742, namely the triglycidyl ether of *tris*(hydroxyphenyl) methane, and Shell 1153, namely a diaryl-methylene-bridged DGEBA) have been trialled, but the alternative approach of curing glycidyl amine resins with aromatic amines has proven to be more widely employed to achieve high T_g. These resins were shown earlier in Scheme 1.4.

$NH_2(CH_2)_2NH(CH_2)_2NH_2$ Diethylene triamine (DETA)

$NH_2(CH_2)_2NH(CH_2)_2NH(CH_2)_2NH_2$ Triethylene tetramine (TETA)

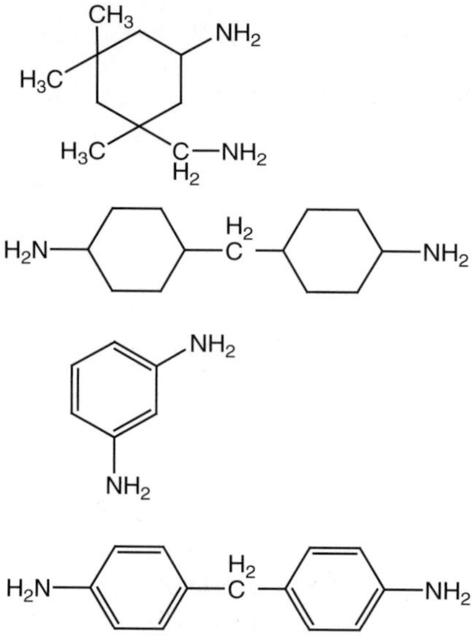

CH_3, NH_2, H_3C, H_3C, $\underset{H_2}{C}-NH_2$ Isophorone diamine (IPD)

H_2N-, $\overset{H_2}{C}$, $-NH_2$ 4,4′-diamino dicyclohexyl methane (PACM)

NH_2, NH_2 m-phenylene diamine (mPDA)

H_2N-, $\overset{H_2}{C}$, $-NH_2$ 4,4′- diamino diphenylmethane (DDM)

Scheme 1.19. Selected aliphatic, cycloaliphatic and aromatic curing agents for epoxy resins.

The reaction mechanism of DGEBA with a range of aliphatic amines and polyamines has widely been studied by thermal analysis as well as chemical analysis of the extent of reaction. Some of the many aliphatic and aromatic difunctional and polyfunctional amines that have been employed are shown in Scheme 1.19 (Pascault *et al.*, 2002).

Instrumental analytical methods including HPLC, NMR and FT-IR have enabled the course of the reaction to be delineated by analysing the sol and gel fractions over time. In Section 1.2.1 the individual amine–epoxy reactions were presented, since the first stage of the reaction with a primary amine involves chain extension. This reaction competes with crosslinking since the reaction of the primary amine with epoxide is much faster than the reaction of the secondary amine. It is the latter reaction that results in branching of the chain and thus the formation of the first crosslinks.

These reactions are of importance in the chemorheology of the network formation since it is possible to link the chemical changes with the viscosity changes that signal the development of an infinite network. The statistical analysis of the network and the relation to rheological properties are discussed in detail in Chapters 2 and 3. At this stage it is useful to consider the chemistry of the formation of an infinite network and the role that the changes in the viscosity of the system play in the chemistry. There is a feedback process between the physical changes occurring to the system and the rate of the reactions as they become diffusion-controlled.

As an example of the crosslinking reaction of multifunctional epoxy resins with amines through a step-growth polymerization, we will examine the details of the reaction of TGDDM (*N,N,N′,N′*-tetraglycidyl-4,4′-diamino diphenyl methane) with aromatic amines (St John and George, 1994). The kinetic approach developed has also been employed for glycidyl ether resins (Kozielski *et al.*, 1994). The main reason for the development of the current level of understanding of these reactions is the wide use of the resins in aerospace systems and the importance of ensuring reliable cure to achieve the desired balance of thermal and mechanical properties.

The understanding of the reaction mechanisms and structures of likely products of the chain extension and crosslinking reactions of TGDDM with DDS has been based initially on the reaction of model compounds such as diglycidyl aniline (representing the epoxy reactive sites) and various secondary amines representing the structure formed after one mole of epoxide has reacted. The aim has been to elucidate the feature that distinguishes TGDDM systems from DGEBA systems: the steric crowding about the epoxy groups versus the likely catalytic effects of hydroxyl groups adjacent to the second epoxide group after the first has reacted (Matejka and Dusek, 1989, Matejka *et al.*, 1991). St John (1993) has summarized the results from these and other model-compound studies regarding the features of the reaction as follows.

1. Reaction of a primary amine with epoxide is catalysed by species that may undergo hydrogen bonding to it, including primary amines and the hydroxyl groups produced following other epoxy-ring-opening events.
2. Reactions of secondary amines with epoxide occur in a similar way to those of primary amines, though often at a significantly slower rate due to steric hindrance.
3. Etherification (hydroxyl reaction with epoxide) has a high activation energy, so it occurs only at elevated temperatures and is catalysed by teriary amines.
4. N-glycidylanilino compounds, on reaction with aromatic amines, give rise to several cyclic structures, including an eight-membered ring, through an intramolecular secondary-amine–epoxy addition reaction, morpholine and seven-membered ether rings as well as a 1,2,3,4-tetrahydro-3-hydroxyquinoline ring.
5. *N*-diglycidylanilino compounds, after one epoxy group has reacted, exhibit enhanced reactivity of the remaining epoxy group, presumably through catalysis by the hydroxyl group formed in the first reaction.
6. In *N*-glycidylanilino compounds there occurs little chain growth due to etherification or an anionic growth mechanism, especially when hydroxyl groups are present.

When these results are applied to the TGDDM/DDS system the likely chain extension, crosslinking and competing reactions (e.g. cyclization from the intramolecular reaction) are those shown in Scheme 1.20 (St John, 1993).

These reactions assume a TGDDM molecule of perfect composition. A detailed analysis of the synthesis of TGDDM shows that a range of impurities can result from incomplete reaction, of the type shown earlier for the synthesis of DGEBA but exacerbated by the steric crowding once a single epoxide ring has closed (St John and George, 1994). Many of these have been shown to have an important bearing on the shelf life of the epoxy resin (Podzimek *et al.*, 1992). The resin in its pure form (>90% TGDDM) is very stable and should remain constant for over one year at room temperature. However the impurities **B**, **C**, **D** and **E** in Scheme 1.21 have been determined to play a role in reducing the shelf life; all impurities reduce the functionality of the resin and lower the crosslink density.

The effect that these resin impurities have on the chemorheology is pronounced insofar as it has been shown that the gel time of the resin is extremely sensitive to the impurities, particularly the hydroxyl-containing species (St John *et al.*, 1993). This is discussed in more detail later.

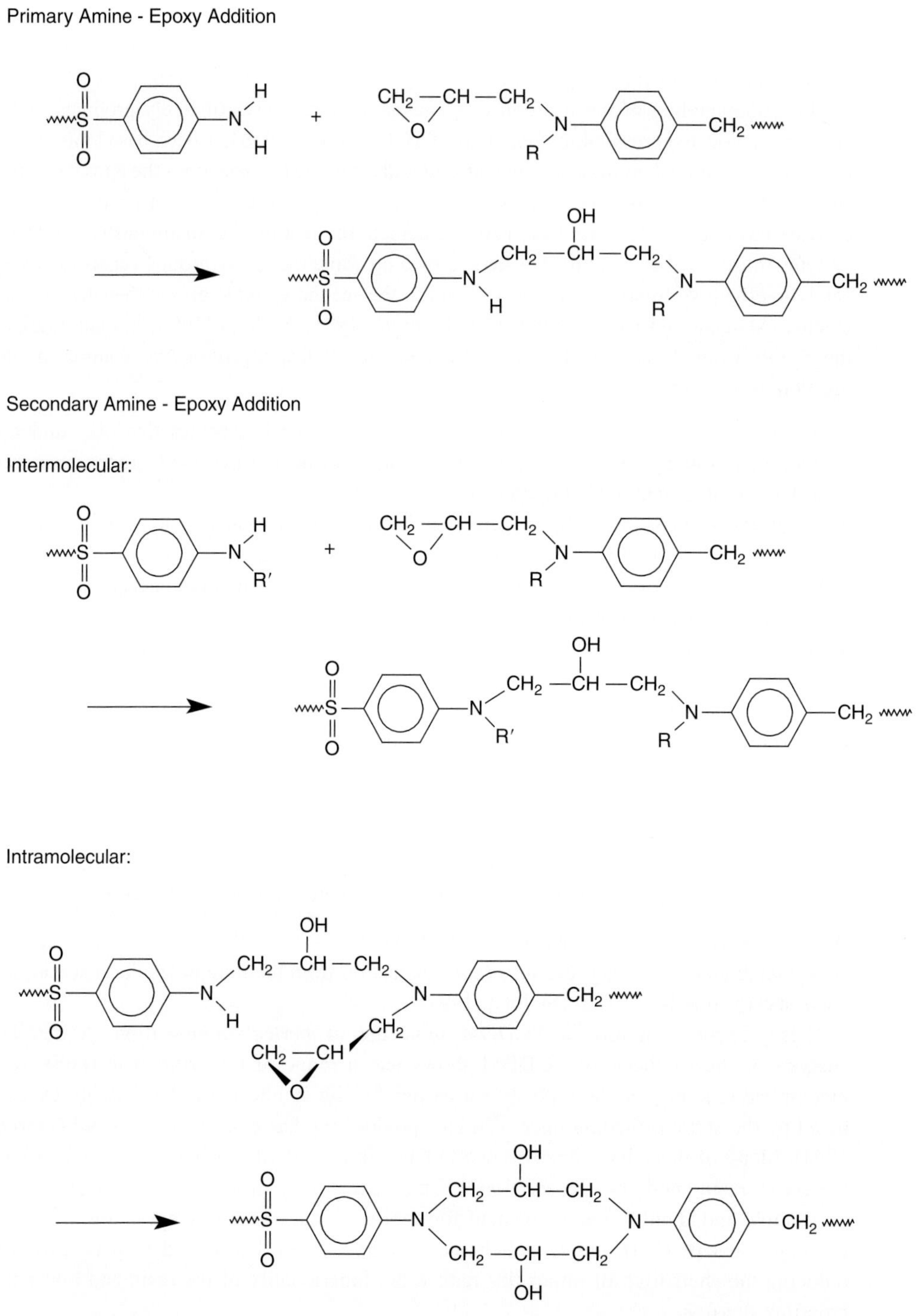

Primary Amine - Epoxy Addition

Secondary Amine - Epoxy Addition

Intermolecular:

Intramolecular:

Scheme 1.20. The possible reaction of TGDDM with DDS, showing the possible intramolecular reaction leading to cyclization instead of crosslinking. Reproduced with permission from St John (1993).

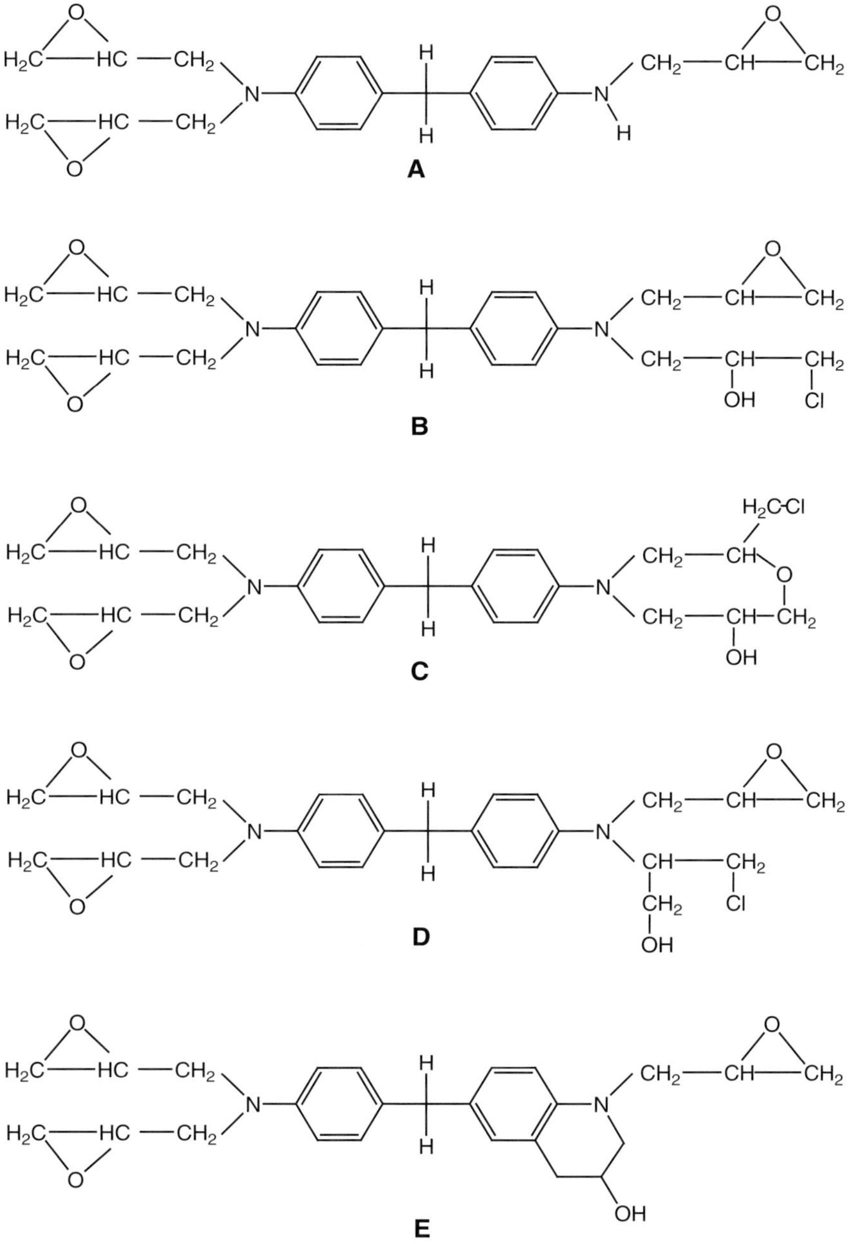

Scheme 1.21. Possible impurities in commercial TGDDM resins. Reproduced with permission from St John (1993).

Kinetics of amine–epoxy cure: TGDDM/DDS

The kinetics of the amine–epoxy crosslinking reaction have been studied in detail by a wide range of methods, and of particular importance is the correlation of the chemical changes, as the network develops, with the changing viscosity of the system. This has been facilitated by spectroscopic methods such as FT-NIR (Section 3.3.6), and the results of this analysis together with thermal analysis (Section 3.2.2) have allowed a quantitative kinetic model to be developed. For the application of this in real-time process control, fibre-optic techniques (Section 3.4.2) as well as a suitable processing model (Chapter 6) are required.

In kinetic models of the development of rheology during the step-growth crosslinking of an epoxy resin with a nucleophile (e.g. an aromatic primary amine) the equations derived must be not only empirically useful for predicting reaction behaviour but also mechanistically sensible. In many empirical equations derived, for example, from DSC results, the mechanistic information is necessarily limited and certain assumptions must be made (e.g. the relative rate coefficients of reactions of primary and secondary amines) (Cole *et al.*, 1991). This results in a set of equations that fit the thermal-analysis data and enable reactive-group profiles to be constructed. However, these profiles predict total consumption of secondary amine when, in fact, residual groups are present on completion of cure. This is important because it affects the subsequent moisture uptake of the resin (since water can undergo hydrogen bonding with the secondary-amine groups). Similarly (but more seriously), in the development of a thermochemical model for the cure of a commercial epoxy–amine formulation under autoclave conditions (Ciriscioli and Springer, 1990) an empirical set of equations was developed from DSC results. These included a negative pre-exponential factor for part of the conversion, but no comment was made regarding the physical meaning of this result.

In the following kinetic analysis of the cure of TGDDM with DDS at elevated temperatures, the equations are developed from both DSC and NIR data (in some circumstances collected simultaneously (de Bakker *et al.*, 1993)), and enable quantitative estimation of the reactive-group profile at any time during isothermal (and in some cases temperature-ramped) cure.

1. *Primary-amine reaction*

$$-\mathrm{d}[\mathrm{PA}]/\mathrm{d}t = k_1[\mathrm{PA}]^2[\mathrm{EP}] + k_2[\mathrm{E}_0][\mathrm{E}_1][\mathrm{OH}] + k_3[\mathrm{PA}][\mathrm{E}_1][\mathrm{OH}]. \qquad (1.42)$$

The first term represents the self-catalysis of the reaction by primary amine that may undergo hydrogen bonding with the epoxide. The presence of the [OH] term recognizes both the auto-catalytic nature of the reaction, since –OH groups are produced by epoxy-ring-opening, and the catalytic effect that pre-existing hydroxyl-group-containing impurities have on the reaction rate. (The chemical structures of these were shown at the end of the previous section.) This rate equation also recognizes that the reaction rate of an epoxy group (E_1) after an adjacent epoxide has reacted differs from that before one has reacted (E_0) (Scheme 1.22). This requires a statistical approach in determining the relative concentration of the two species (St John, 1993).

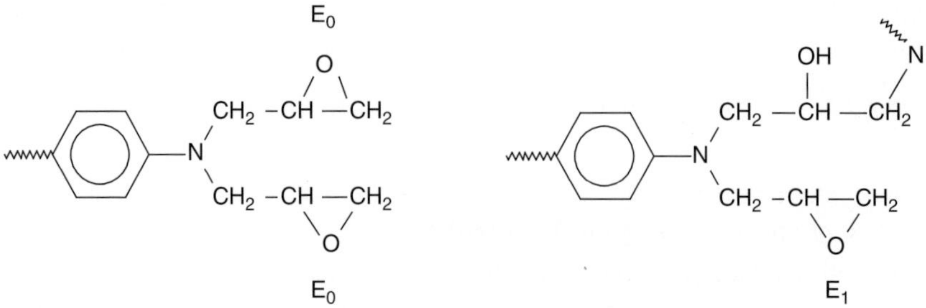

Scheme 1.22. The naming convention for epoxy groups of TGDDM before (E_0) and after (E_1) reaction with amine has occurred. Reproduced with permission from St John (1993).

2. Secondary amine reactions

$$-\mathrm{d[SA]}/\mathrm{d}t = \mathrm{d[TA]}/\mathrm{d}t = k_4[\mathrm{SA}][\mathrm{E_0}][\mathrm{OH}] + k_5[\mathrm{SA}][\mathrm{E_1}][\mathrm{OH}]. \qquad (1.43)$$

The rate coefficients for the secondary-amine reactions were found to be only 17% of those for the primary-amine reaction, thus explaining the residual secondary amine found at the end of cure. This equation was found to explain the development of the main crosslinking site, namely the tertiary-amine site formed on the DDS, corresponding to network interconnection. However, overlaid with the rate equation for chemical conversion that implies that all reagents are accessible to one another is the effect of the development of the network so that the reactions become diffusion-controlled. This is of interest since this means that the rate coefficients now reflect the chemorheology of the system, not just the chemistry. Thus, if k_d and k_c represent the rate coefficients for diffusion and chemical control, the measured rate coefficient, k_e, will be given by (Cole *et al.*, 1991)

$$1/k_e = 1/k_c + 1/k_d. \qquad (1.44)$$

When this analysis is applied to this system it is found that the two rate coefficients are sensitive to different phases formed during cure. Thus k_4 is sensitive to gelation whereas k_5 is not. This follows since the intramolecular reaction to form a cyclic structure is of higher probability for k_5, so this reaction should not be affected by the diffusion of the reagents together as is k_4.

3. Etherification reactions

$$\mathrm{d[ET]}/\mathrm{d}t = k_6[\mathrm{EP}][\mathrm{OH}] + k_7[\mathrm{EP}][\mathrm{OH}][\mathrm{TA}]. \qquad (1.45)$$

This reaction is important at elevated temperatures because of the high activation energy of the reaction. It is important to consider the products of the reaction since it is also a route to formation of cyclic species that lowers the total crosslink density. The possible etherification reactions are shown in Scheme 1.23, and, in view of the different possibilities, the rate equation is relatively straightforward. The inclusion of the term for catalysis by tertiary amine reflects the importance of catalysis in etherification (Cole *et al.*, 1991).

The tertiary-amine catalysis will favour the intramolecular cyclization reactions which will lower the crosslink density of the network in comparison with the situation if only intermolecular reactions took place.

Kinetic analysis of other epoxy-resin systems

A detailed analysis of the kinetics of cure has also been performed for polyfunctional diglycidyl ether resins (Kozielski *et al.*, 1994). Interestingly, the systems did not exhibit diffusion control of the secondary-amine–epoxy reactions until vitrification. Furthermore, etherification reactions (i.e. those of –OH and epoxy) were insignificant under the conditions of reaction, which suggests that the amine group present in TGDDM must have an activating effect. These studies have shown that complementary techniques are required in order to determine the factors that are important in achieving a three-dimensional network.

1.2.3 Addition polymerization

Addition or chain-growth polymerization is the most important industrial process for the production of polymers. Polyethylene, polypropylene and polystyrene are all formed

Etherification

Intermolecular:

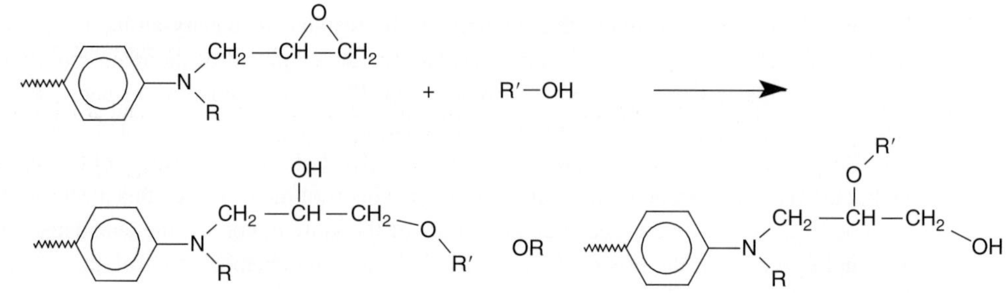

Intramolecular:

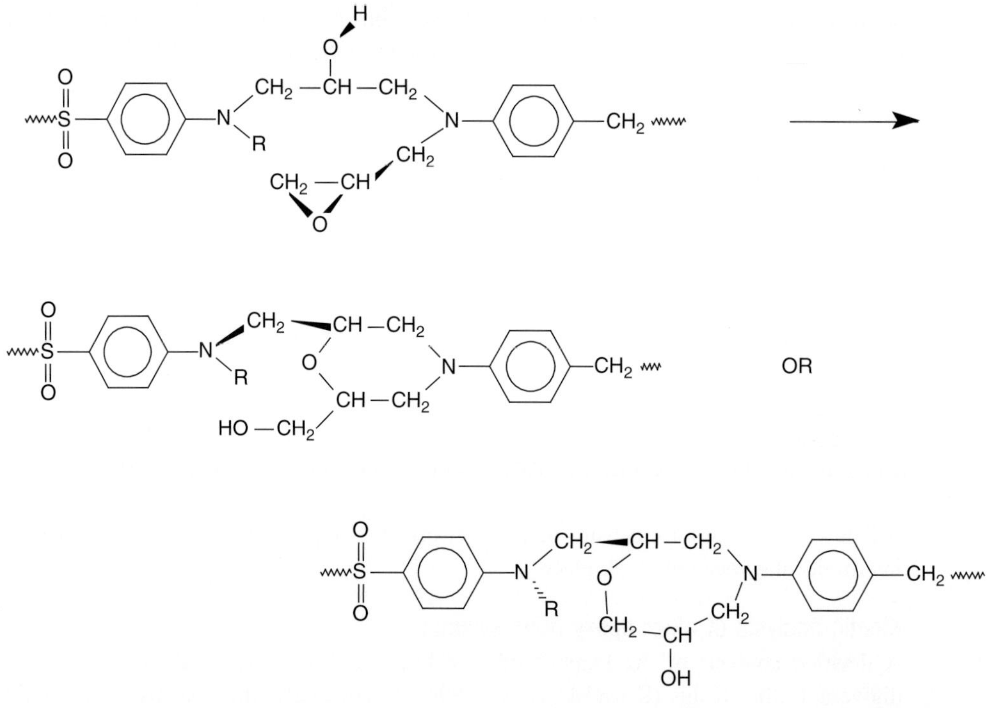

Scheme 1.23. Possible etherification reactions during the cure of TGDDM with DDS. Reproduced with permission from St John (1993).

through chain-growth polymerization. The polymer is formed by the addition of a monomer to a reactive centre to extend the chain length by one repeat unit at a time. The reactive centre is generally created by the use of an initiator that is decomposed to the reactive species (generally by heat or radiation). The addition of monomer occurs rapidly to give high-molar-mass polymer in a propagation reaction that preserves the reactive centre until a termination event destroys it. The chain length or degree of polymerization, DP, is then defined.

Table 1.4. Differences between chain-growth and step-growth polymerization

Chain growth	Step growth
Repeating units add one at a time to active site at the end of the polymer chain	Reactive monomer end groups add together to give oligomers with reactive end groups that then react further
Monomer disappears steadily throughout the reaction as it is converted to polymer	Original reactants disappear early and at DP < 10 only 1% monomer remains
High polymer forms at low extents of conversion	High polymer is formed only at high extents of reaction (>99.9%)
Only short reaction times are needed to give high DP	Long reaction times are needed for high DP

The nature of the reactive centre defines the chemistry of the polymerization, the rate and conditions under which high polymer may form, and particular features of the polymer architecture (such as the tacticity; see Section 1.1.2). The nature of the reactive centre and the monomer may also control the side reactions (such as branching) and defect groups that may be introduced, which may affect the subsequent performance of the polymer. In the following, we will consider the most common types of addition polymerization since this may define the properties of the polymer that then control the chemorheology. Certain of these reactions are more important than others in reactive processing, and the particular examples of reactions that occur in forming networks as well as modification of polymers will be considered in more detail.

Some of the types of polymerization that come into this category include

- free-radical polymerization
- anionic polymerization
- cationic polymerization
- co-ordination polymerization
- ring-opening polymerization
- living polymerization

In some cases, the monomer may be polymerized by more than one method, and in ring-opening polymerization the reaction may also occur through step-growth polymerization (Section 1.2.1) in addition to the initiated chain-growth polymerization.

The chemistry of chain-growth polymerization may be readily distinguished from step-growth polymerization by the features in Table 1.4.

Free-radical polymerization

The starting point for the free-radical polymerization is to choose a monomer (M) that will react with a free radical $R'\cdot$ to add to the radical and also create another radical centre $R'M\cdot$ of high reactivity that may add another monomer molecule. The source of the free radical to initiate the process may be the monomer itself or an added initiator (I) chosen to produce two free radicals, $R'\cdot$, per molecule cleanly and efficiently at a temperature suitable for the addition of monomer to occur. In the case of self-initiation, heat or radiation must be supplied in order to fragment the monomer or otherwise create the radical to **initiate** the chain polymerization. Table 1.5 gives examples of typical vinyl monomers that undergo free-radical polymerization, and Table 1.6 shows some initiators. Provided that certain thermodynamic

Table 1.5. Chemical structures of vinyl monomers and their polymers

Monomer	Structure	Polymer
Ethylene (ethene)	$H_2C{=}CH_2$	Poly(ethylene), PE $-\!(CH_2-CH_2)_n-$
Styrene (phenyl ethene)	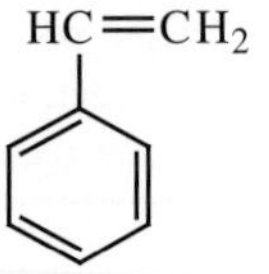 $HC{=}CH_2$ with phenyl ring on HC	Poly(styrene), PS 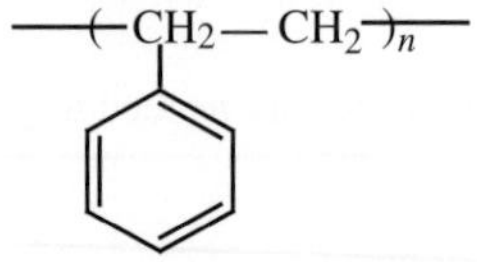 $-\!(CH_2-CH_2)_n-$ with phenyl ring
Vinyl chloride (chloro ethene)	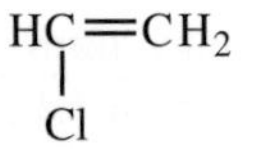 $HC{=}CH_2$ with Cl on HC	Poly(vinyl chloride), PVC 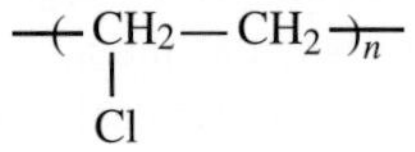 $-\!(CH_2-CH_2)_n-$ with Cl
Methyl methacrylate (2-(methoxycarbonyl) -1-propene) or (2-methyl 2-propenoic acid, methyl ester)	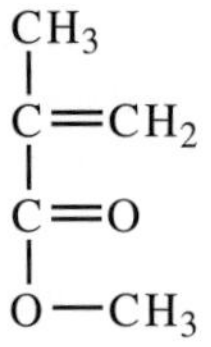 CH_3–C(=CH_2)–C(=O)–O–CH_3	Poly(methyl methacrylate), PMMA 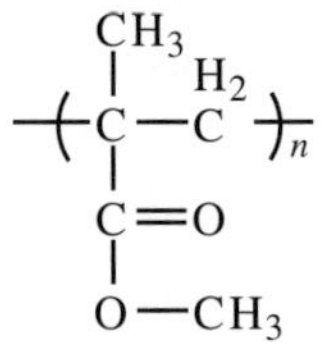 $-\!(C(CH_3)(C(=O)O{-}CH_3)-CH_2)_n-$

considerations are met (in particular that, at the temperature of reaction, the free-energy difference, ΔG, between monomer and polymer is negative), the reaction of monomer, M, with the radical $R'M\cdot$ may **propagate** through m steps, and high polymer rapidly forms. At each step of monomer addition, the radical centre at the chain end is recreated: $R'(M)_{m-1}M\cdot$. This reaction will continue, consuming monomer steadily, until the radical centre undergoes a reaction to **terminate** the chain polymerization. One way for this to occur is by two chains colliding to give bimolecular termination of the radicals, so creating inert polymer: $R'(M)_{2m}R'$. The degree of polymerization, DP, is thus $2m$. These processes and other competing reactions such as chain transfer are considered in detail below.

Kinetics of free-radical polymerization

In order to determine the factors that control the **rate**, r_p, and the **chain length**, ν_p, of the polymerization, it is important to determine the rate of the separate reactions involved, namely initiation, propagation and termination, as well as the rates of any other processes that may compete with them. These rate equations will contain reactive intermediates (i.e. radical concentrations $[R\cdot]$) that cannot be explicitly determined. The process of solution requires the steady-state approximation, which states that there is no net change in radical population with time during the steady-state polymerization, i.e. $d[R\cdot]/dt = 0$, in order to

Table 1.6. Chemical structures of initiators and temperatures at which each initiator has a half-life of one hour

Initiator	Structure	T(°C) for $t_{1/2} = 1$ h
Dilauryl peroxide	$CH_3(CH_2)_{10}C(=O){-}O{-}O{-}C(=O)(CH_2)_{10}CH_3$	79
Dibenzoyl peroxide	$C_6H_5{-}C(=O){-}O{-}O{-}C(=O){-}C_6H_5$	91
Dicumyl peroxide	$C_6H_5{-}C(CH_3)_2{-}O{-}O{-}C(CH_3)_2{-}C_6H_5$	137
2,2′-Azo *bis*-(isobutyronitrile)	$H_3C{-}C(CH_3)(C{\equiv}N){-}N{=}N{-}C(CH_3)(C{\equiv}N){-}CH_3$	82
2,2′-Azo *bis*-(2,4-dimethyl valeronitrile)	$(CH_3)_2CH{-}CH_2{-}C(CH_3)(C{\equiv}N){-}N{=}N{-}C(CH_3)(C{\equiv}N){-}CH_2{-}CH(CH_3)_2$	64

obtain an expression for r_p and v_p in terms of the rate coefficients for the elementary reactions and the concentration of monomer and any other reagents (e.g. chain-transfer agents) added to modify the polymerization.
In Table 1.7, the simplest situation is considered.

- The initiator decomposes thermally to give the initiating radicals $R'\cdot$ with an efficiency f. This is typically between 0.1 and 0.8, depending on the type of initiator and the viscosity of the polymerizing medium, since it is reduced by radical recombination. The initiators such as AIBN shown have a higher efficiency than, say, dialkyl peroxides, since they liberate nitrogen to produce two radicals that are more widely separated. The rate coefficient, k_d, will be described by an Arrhenius temperature dependence.
- The propagation reaction is described in terms of an average rate coefficient, k_p, and the polymer radical concentration, $[R\cdot]$. The value of k_p is very large, $\approx 10^2$–$10^4\ \mathrm{dm^3\,mol^{-1}\,s^{-1}}$, and is independent of chain length. This is the reaction that consumes monomer and grows the polymer chain.
- The termination reactions shown are bimolecular; these may occur either by recombination to give one chain, or, depending on the monomer, disproportionation to give two polymer chains (where DP is half that for recombination). The elementary reaction rate is

Table 1.7. Rate coefficients and reaction rates for elementary reactions for free-radical addition polymerization

Process	Elementary reaction	Rate coefficient	Rate of reaction
Initiation	$I \rightarrow 2R'\cdot$	k_d	$r_i = 2f\,k_d[I]$
Propagation	$R'\cdot + M \rightarrow R'M\cdot$	k_{p1}	$r_p = k_p[M][R\cdot]$
	$R'M\cdot + M \rightarrow R'MM\cdot$	k_{p2}	(setting:
	$R'M\cdot_{(m-1)} + M \rightarrow R'M\cdot_m$	k_{pm}	$\sum[R'M\cdot_m] = [R\cdot]$)
		$(k_{p1} \approx k_{p2} \approx k_{pm} = k_p)$	
Termination	$R'M\cdot_m + R'M\cdot_m \rightarrow R'M_{2m}R'$	k_t	$r_t = 2k_t[R\cdot]^2$
	$R'M\cdot_m + R'M\cdot_m \rightarrow 2R'M_m$		

the same for both cases. A simplification has been made, namely that all chains are of the same length, m, whereas in fact there will be a statistical distribution of chain lengths, leading to a distribution of molar mass.

The features of a linear chain reaction may be recognized in the above table. The initiation step creates free radicals and the termination step destroys them. The rate-determining step will be the propagation reaction, which is the product-forming step and has the characteristic feature of a linear chain reaction, viz. one radical is formed for every radical that reacts. Noting these features, we have

$$\text{Rate of polymerization } r_p = -d[M]/dt = k_p[M][R\cdot]. \tag{1.46}$$

The value of $[R\cdot]$ is obtained by applying the steady-state approximation that $d[R\cdot]/dt = 0$, i.e. the rates of initiation, r_i, and termination, r_t, must be equal since these are the radical-creation and -destruction steps:

$$r_i = r_t. \tag{1.47}$$

Thus

$$2fk_d[I] = 2k_t[R\cdot]^2, \tag{1.48}$$

so then

$$[R\cdot] = (fk_d[I]/k_t)^{1/2}. \tag{1.49}$$

Substituting for $[R\cdot]$ gives

$$r_p = k_p(fk_d[I]/k_t)^{1/2}[M]. \tag{1.50}$$

A more general expression that allows initiation by other means (e.g. radiation) in addition to thermal initiation by an added initiator is

$$r_p = -d[M]/dt = k_p[r_i/(2k_t)]^{1/2}[M] = k_{eff}[M]. \tag{1.51}$$

From this relation it can be seen that the free-radical polymerization will be of first order in monomer, M, and the effective rate coefficient, k_{eff}, will have a temperature dependence that will depend on the activation energies E_i, E_p and E_t of the elementary reactions:

$$k_{eff} = A\exp[-E_{eff}/(RT)], \tag{1.52}$$

where

$$E_{\text{eff}} = E_{\text{p}} + \frac{1}{2}E_{\text{i}} - \frac{1}{2}E_{\text{t}}. \tag{1.53}$$

Since the termination reaction is a radical-recombination reaction the activation energy, E_t, is very low, ≈10 kJ/mol. A typical activation energy for a peroxide initiator is $E_i = 130$ kJ/mol, and for a vinyl monomer E_p is in the range 20–40 kJ/mol. Thus E_{eff}, a typical activation energy for a thermally initiated vinyl polymerization, is between 80 and 90 kJ/mol. The dominant factor in this calculation is the activation energy for initiation, and it is seen that, if this can be lowered, then the reaction rate will be higher at the same temperature. This may be achieved in a number of ways.

- Initiating polymerization by absorption of radiation by a photoinitiator that produces radicals with a quantum efficiency of φ. If I_a is the absorbed intensity in quanta $dm^{-3}\,s^{-1}$ then $r_i = 2\varphi I_a$, $E_i \approx 0$ and $E_{\text{eff}} \approx 15$–35 kJ/mol.
- Adding a catalyst that will lower the activation energy for initiation. An example is the use of a transition-metal ion to act as a redox agent to decompose peroxides and hydroperoxides. Typical values are $E_i \approx 40$ kJ/mol and $E_{\text{eff}} \approx 35$–55 kJ/mol, so that the activation energy is approximately half that of the uncatalysed reaction.

Polymer chain length

The molar mass of the polymer depends on the effective number of addition steps that occur per initiation step. In the description of free-radical chain reactions, this is the kinetic chain length, ν_p:

$$\nu_{\text{p}} = r_{\text{p}}/r_{\text{i}} = k_p[r_{\text{i}}/(2k_t)]^{1/2}[\text{M}]/r_{\text{i}}, \tag{1.54}$$

$$= k_{\text{p}}[\text{M}]/(2k_{\text{t}}r_i)^{1/2}. \tag{1.55}$$

This relation is of fundamental importance in free-radical polymerization since the kinetic chain length decreases with an increase in the rate of initiation. Thus an attempt to accelerate polymerization by adding more initiator will produce a faster reaction but the polymer will have shorter chains. This can also be seen as a consequence of the steady-state approximation in a linear chain reaction since the rate of termination is equal to the rate of initiation and, if the rate of termination increases to match the rate of initiation, the chains must necessarily be shorter.

The effect of temperature on the polymer chain length may be seen by determining E_ν:

$$E_\nu = E_{\text{p}} - \frac{1}{2}E_{\text{t}} - \frac{1}{2}E_{\text{i}}. \tag{1.56}$$

Thus, since E_p is less than $\frac{1}{2}(E_i + E_t)$, a typical value being −45 kJ/mol, the chain length will decrease with temperature for thermally initiated polymerization. In contrast, if photochemical initiation is used, for which $E_i \approx 0$, then E_ν is positive (a typical value being 20 kJ/mol) and the molar mass will then increase with increasing temperature. For the redox-catalysed, peroxide-initiated polymerization discussed above, E_ν will be close to zero, so the chain length will not be sensitive to temperature.

The relationship between the chain length, ν_p, and the degree of polymerization, DP, will depend on the mechanism of termination. If recombination occurs, DP $= 2\nu_p$, whereas if disproportionation is dominant, DP $= \nu_p$. Values for DP can range up to many thousands depending on the monomer.

The lifetime of the growing polymer chain: the Trommsdorff effect

It was noted earlier for step-growth polymerization that the polymer chain length increased with the time of reaction, whereas in free-radical and other addition polymerizations the chain grew rapidly and over a moderate range of conversions the chain length was unchanged. The time taken to polymerize a chain may be determined by a simple calculation of the lifetime of a radical since this is reflecting the balance of propagation and termination. The average lifetime of a growing radical, τ, is given by the steady-state radical concentration divided by the steady-state rate of disappearance (Odian, 1991):

$$\tau = [R\cdot]/(2k_t[R\cdot]^2) = 1/(2k_t[R\cdot]). \qquad (1.57)$$

Since, in the steady state, from (Equ 1.48),

$$[R\cdot] = [r_i/(2k_t)]^{1/2}, \qquad (1.58)$$

we have

$$\tau = 1/(2k_t r_i)^{1/2}. \qquad (1.59)$$

It is noted that the kinetic chain length may be written in terms of the radical lifetime as

$$\nu_p = k_p[M]/(2k_t r_i)^{1/2} = k_p[M]\tau. \qquad (1.60)$$

Thus, if for any reason the average lifetime of the growing radical increases, then there will be an increase in the polymer chain length. An often-encountered example of this is the Trommsdorff or gel effect that occurs in the polymerization of solutions of high monomer concentration when the viscosity, η, increases and, after a certain extent of conversion, there is a rapid acceleration in the rate of polymerization. This is interpreted as an indication of the decrease in the rate of termination as this reaction becomes diffusion-controlled. A feature of diffusion-controlled reactions is that the rate coefficient k_t is not chemically controlled but depends on the rate at which the terminating radicals can collide. This is most simply given by the diffusion-controlled rate coefficient, k_d, in the Debye equation:

$$k_t = k_d = 8RT/(3\eta). \qquad (1.61)$$

A practical example of this is seen (Odian, 1991) in the polymerization of methyl methacrylate, in which the term $[k_p/(2k_t)]^{1/2}$ (that reflects the rate of polymerization) increases sixfold between 10% and 40% conversion. There is some compensation for the decline in translational diffusion by an increase in segmental diffusion, but inevitably auto-acceleration occurs. Over the conversion range from 10% to 80% the radical lifetime increases from $\approx$1 s to over 200 s. The full understanding of kinetics at high conversion is further complicated by the effect of the viscosity on the initiator efficiency, f, as well as a decline in k_p. The final stage may be reached when the glass-transition temperature, T_g, exceeds the reaction temperature and the reaction ceases because the system vitrifies. Radicals may be trapped in the glassy state and have very long lifetimes because there are no termination reactions.

Table 1.8. Effects of addition of an inhibitor, S, or chain-transfer agent, XA, on the elementary reactions and their rates in free-radical polymerization

Process	Elementary reaction	Rate coefficient	Rate of reaction
Inhibition	$R'M\cdot + S \rightarrow R'MS$	k_{st} ($\gg k_p$)	$k_{st}[R'M\cdot][S]$
Chain transfer	$R'M\cdot + XA \rightarrow R'MX + A\cdot$	k_{tr}	$k_{tr}[R'M\cdot][XA]$
Re-initiation	$A\cdot + M \rightarrow AM\cdot$	k_a	k_a $[A\cdot][M]$

Reactions competing with propagation: inhibition and chain transfer

The uncontrolled free-radical polymerization of monomers will produce polymers of molar-mass distributions that depend on the rate parameters, the temperature and the type of initiator, as described in the above equations. Further control of the molar mass can be achieved by controlling the chain-propagation process. This is generally achieved by the addition of a molecule that is able to compete with the monomer for the growing radical chain. At one extreme, all the radicals could be captured by having a molecule that had a much higher rate of reaction than monomer and also killed off the chain reaction. An example is a hindered phenol that is able to donate the phenolic hydrogen to terminate a growing chain and form a phenoxy radical that is sterically hindered and thus unable to re-initiate polymerization. The kinetic chain length ν_p is reduced to ≤ 1. This is the requirement for an inhibitor or stabilizer, as discussed later in Section 1.4.3, and is technologically important for the storage stability of monomers. This reaction becomes the dominant termination reaction and the time for which it is able to stabilize the monomer depends on the concentration of radicals, which under storage conditions is low, thereby enabling lengthy storage stability for low concentrations of stabilizer.

It should be noted that oxygen is an efficient scavenger of free radicals and the reaction rate of a monomer free radical, such as the styryl radical, with a molecule of oxygen is much greater than that for its reaction with a molecule of monomer. Thus oxygen is an efficient inhibitor of polymerization and must be rigorously removed for efficient chain growth. In the particular case of scavenging of alkyl radicals by oxygen, the product is the peroxy radical, which can then abstract a hydrogen atom to form hydroperoxides. These are stable at low temperatures but initiators of polymerization at elevated temperature. Alternatively, the peroxy radical can react with monomer to produce a peroxide copolymer that is potentially explosive.

If, instead of killing off the radicals, the molecule terminates the growing chain and then provides another initiating radical, then there will be control of molar mass of the polymer without stopping the linear chain reaction. This is the purpose of adding a chain-transfer agent. The effect on the polymerization kinetics is the addition of another term in the propagation reaction sequence. Thus in the following scheme (Table 1.8), the radical A· produced by the chain-transfer reaction is able to initiate another chain with the monomer radical M· at the chain end that will propagate with the same rate as before the addition of the transfer agent XA. The only difference will be the end group, which is a fragment of the transfer agent rather than an initiator fragment. The effect that the additive has on the polymerization depends on the reactivity of the agent, its concentration and whether the rate of re-initiation is comparable to the rate of propagation. Chain transfer can in principle occur to all species that are present in the polymerization solution: the polymer, monomer, solvent and initiator. The effect that this has on the polymerization depends on the reactivity of the radical that is formed. Certain solvents such as chlorinated hydrocarbons are powerful chain-transfer agents. A chain-transfer coefficient can be defined for all species as the ratio of the rate coefficient for the transfer reaction to that for propagation: k_{tr}/k_p. Chain transfer to polymer will result in the formation of branches and broadening of the molar-mass distribution. The role of chain transfer to polymer that results in branching cannot be realized by examination of DP since it does not change the number of polymer molecules in the system (Barson, 1989).

The case of chain transfer to initiator may be exploited in special forms of controlled free-radical polymerization, e.g. with sulfur-containing initiators that are discussed later within the topic of 'living' free-radical polymerizations.

The practical case industrially is when radical transfer to the added chain-transfer agent, XA, is the dominant reaction. The effect on DP may be determined by noting that in the presence of XA the rate of formation of polymer has increased above that produced by the normal termination reaction. The kinetic chain length, ν_p, is the ratio of the rate of propagation to the rate of formation of chain ends:

$$\mathrm{DP} = 2\nu_p = 2r_p/(r_t + k_{tr}[\mathrm{R'M\cdot}][\mathrm{XA}]). \tag{1.62}$$

If the degree of polymerization in the absence of the chain transfer agent is DP_0, then, using Equations (1.46) and (1.47), the following form of the Mayo equation is obtained:

$$1/\mathrm{DP} = 1/\mathrm{DP}_0 + k_{tr}[\mathrm{XA}]/(k_p[\mathrm{M}]). \tag{1.63}$$

The ratio k_{tr}/k_p is the chain-transfer coefficient, C_A. Similar coefficients, C_S, C_M and C_I, may be defined for chain transfer to solvent, monomer and initiator, respectively. In commercial systems, the value of C_A is chosen to be >1 so that only small concentrations of XA are required in order to have a marked effect on the value of DP. For example, if 0.1% of XA is added with respect to monomer and DP_0 is 1000, then, if C_A is 1, the DP is reduced to 500. Also, since $k_{tr} \approx k_p$, the agent is consumed slowly and there is no effect on the rate of polymerization.

Conversely it is possible to produce low-molar-mass oligomers or telomers by deliberately choosing an agent with a large value of C_A (e.g. methyl mercaptan, $C_A \approx 2 \times 10^5$ in styrene), so that DP is reduced to 5 for a concentration of 0.001%. Further particular examples of chain transfer (e.g. to polymer to form branches) will be discussed later, together with the use of reversible-addition fragmentation transfer (RAFT) and other radical-mediated synthetic strategies.

Equilibria in free-radical polymerization

The discussion of the kinetics of polymerization was based on the premise that the overall change in free energy, ΔG, upon polymerization of a monomer was negative at the temperature at which the reaction would take place. To determine whether this is indeed so, it is necessary to consider the changes in enthalpy ΔH and entropy ΔS that are responsible for ΔG at a particular temperature:

$$\Delta G = \Delta H - T\,\Delta S. \tag{1.64}$$

These changes are for converting one mole of monomer to one mole of polymer repeat unit. There is an obvious loss of entropy on taking a monomer molecule with a high

Table 1.9. Enthalpy and entropy changes for conversion of monomer to polymer, and the corresponding ceiling temperature

Monomer	ΔH^0 (kJ/mol)	ΔS^0 ($J\,K^{-1}$ mol)	T_c (°C)
Styrene	−70	−105	394
α-Methyl styrene	−35	−104	66
Methyl methacrylate	−55	−117	198
Ethylene	−93	−155	327
Propylene	−84	−116	451

degree of freedom and constraining it on a polymer chain, which is countered by the highly exothermic nature of most vinyl polymerizations. Table 1.9 gives some typical values so it is possible to calculate the temperature at which ΔG may change sign, so that the polymerization equilibrium at that temperature will favour monomer rather than polymer, i.e. $T_c = \Delta H^0/\Delta S^0$.

This can give a comparative measure of the ceiling temperature, T_c, in polymerization (which is important for the onset of thermal depolymerization as discussed in Section 1.4.1). The thermodynamic ceiling temperature is rarely achieved in practice, both because of the requirement for a closed system and also owing to the onset of other degradation reactions such as crosslinking.

For the situation of polymerization in a closed system, it is necessary to consider the position of the equilibrium in the propagation reaction as given by the value of K, the equilibrium constant. This shows (Odian, 1991) that the ceiling temperature depends on the monomer concentration:

$$T_c = \Delta H^0/(\Delta S^0 + R\ln[\mathrm{M}]_c). \tag{1.65}$$

Thus the values shown in Table 1.8 are for standard conditions and represent just one of a series of ceiling temperatures for various monomer concentrations above which polymer formation is not favoured. Thus, in a bulk polymerization reaction the ceiling temperature may change with conversion in such a way that complete conversion is not achieved. For example, if methyl methacrylate is polymerized at 110 °C the value of $[\mathrm{M}]_c$ calculated from the above equation is 0.139 M and this will be the monomer concentration in equilibrium with the polymer. The polymer, when removed from the monomer, will have the expected ceiling temperature as given in Table 1.8 and will depolymerize only if there is a source of free radicals to initiate the depolymerization (Section 1.4.1)

Anionic polymerization

Polymerization by ionic initiation is much more limited than that by free-radical initiation with vinyl monomers, but there are monomers such as carbonyl compounds that may be polymerized ionically but not through free radicals because of the high polarity. The polymerization is much more sensitive to trace impurities, especially water, and proceeds rapidly at low temperature to give polymers of narrow molar-mass distribution. The chain grows in a 'living' way and, unlike in the case of free-radical polymerization, is generally terminated not by recombination but rather by trace impurities, solvent or, rarely, the initiator's counter-ion (Fontanille, 1989).

Initiation:

$$n\text{-BuLi} + CH_2{=}CHR \rightarrow n\text{-Bu-}CH_2\text{-(HR)}C^- Li^+$$

Propagation:

$$n\text{-Bu-}CH_2\text{-(H)(R)}C^- Li^+ + CH_2{=}CHR \rightarrow n\text{-Bu-}(CH_2CHR)_nCH_2\text{-(H)(R)}C^- Li^+$$

Termination by chain transfer to water (or alcohol):

$$\text{---}CH_2\text{-(H)(R)}C^- + H_2O \rightarrow \text{---}CH_2\text{-(R)}CH_2 + OH^-$$

Scheme 1.24. The sequence of reactions for anionic polymerization of an electron-withdrawing monomer (e.g. acrylonitrile) initiated by n-butyl lithium.

The range of vinyl monomers that may be polymerized via an anionic route is restricted to those with strong electron-withdrawing groups attached to the vinyl group so that attack by the negatively charged (nucleophilic) anionic initiator is favoured. They also then stabilize the propagating anionic species, as shown in Scheme 1.24 for the polymerization by n-butyl lithium of a monomer such as acrylonitrile. The propagation step is insertion of monomer at the anion and the initiator is recreated so that the chain grows until the monomer is exhausted or there is chain transfer to the solvent. The living chain end may spontaneously decay through an elimination reaction (e.g. hydride elimination from the polystyryl carbanion (Odian, 1991)). Usually the chain is terminated by the addition of a chain-transfer agent such as water or alcohol.

The effectiveness of water as a chain-transfer agent is seen from its value of C_s of 10 at 25 °C in the polymerization of styrene. The reactions are often carried out at low temperature in order to limit unwanted chain transfer and side reactions. The long life of the carbanion is seen by virtue of the maintenance of its bright colour (e.g. red for the styryl carbanion).

One feature of anionic polymerization is that it is possible to have polymerization occurring at both ends of an initiating dimer formed by electron transfer from a metal aryl compound such as sodium naphthalenide. This electron transfer results, in the case of styrene monomer, in the formation of a dianion:

$$Na^+[^-C(H)(\varphi)CH_2CH_2(\varphi)(H)C^-]Na^+$$

Propagation can then occur at both carbanion centres. Polymers synthesized by an anionic route have characteristic features of a narrow molar-mass distribution and controlled degree of polymerization, and these follow from the mechanism and kinetics of polymerization.

Kinetics of anionic polymerization

The kinetics of polymerization with anionic initiation are less well understood than free-radical polymerization, but a simple treatment allows the important results to be drawn from a consideration of the initiation, propagation and termination reactions (Muller, 1989).

The initiation reaction often involves the ionization of the initiating electron-transfer agent, MtA, to liberate the anion, A^-, followed by reaction with the monomer, M, to produce the initiator, AM^- as shown in Table 1.10. If $[M^-]$ is the average concentration of

Table 1.10. Rate coefficients and reaction rates for elementary reactions for anionic addition polymerization

Process	Elementary reaction	Rate coefficient	Rate of reaction
Initiation	$MtA \overset{K}{\longleftrightarrow} Mt^+ + A^-$ $A^- + M \overset{k_i}{\longrightarrow} AM^-$	k_i	$r_i = k_i[A^-][M]$ $= k_i K[MtA][M]/[Mt^+]$
Propagation	$AM^- + M \overset{k_p}{\longrightarrow} AMM^-$ $AM_{n-1}^- + M \overset{k_p}{\longrightarrow} AM_n^-$	k_p	$r_p = k_p[M^-][M]$
Termination[a]	$AM_n^- + H_2O \overset{k_{tr}}{\longrightarrow} AMH + OH^-$	k_{tr}	$r_{tr} = k_{tr}[M^-][H_2O]$

[a] Termination occurs by chain transfer to water.

all propagating species and termination is taken to be chain transfer to water present as an impurity, then solving using the steady-state approximation applied to $[M^-]$ gives

$$r_p = k_p k_i K[M]^2[MtA]/(k_{tr}[H_2O][Mt^+]). \tag{1.66}$$

The degree of polymerization (r_p/r_i) is given, from Table 1.10, by

$$DP = k_p[M]/(k_{tr}[H_2O]) \tag{1.67}$$

$$= [M]/(C_{trW}[H_2O]), \tag{1.68}$$

noting that the coefficient for chain transfer to water $C_{trW} = k_{tr}/k_p$.

In many systems where the initiator decomposes completely (so the anion concentration is that of the initiator, MtA) and there is no chain transfer, i.e. the chain length is governed by the consumption of monomer, on complete reaction

$$\begin{aligned} DP &= \text{Monomer consumed/number of chain ends} \\ &= [M]/[MtA]. \end{aligned} \tag{1.69}$$

The striking feature of anionic polymerization of having a narrow molar-mass distribution results from the fact that initiation occurs rapidly and totally before propagation, so every anion represents a growing polymer chain and the termination processes that may lead to a distribution of chain lengths are limited. A feature of such **living polymerization** is that at the end of the reaction, with complete consumption of monomer, the reactive centres are still available for further reaction and, if monomer is added, the reaction proceeds with the molar mass continuing to increase linearly with conversion. General aspects of living polymerization, particularly those which may occur at higher temperatures and with other types of initiation (including the use of free-radical initiators) are discussed later.

Anionic polymerization of carbonyl compounds

The polymerization of aldehydes to give polyacetal is readily undertaken anionically in the presence of base, due to the susceptibility of formaldehyde to nucleophilic attack (Odian, 1991) as shown in Scheme 1.25. It should be noted that the resulting polymer is thermally unstable and stabilization is achieved by an esterification reaction of the unstable hemiacetal end groups after polymerization.

Initiation: $Mt^+ A^- + CH_2{=}O \rightarrow A\text{-}CH_2\text{-}O^- + Mt^+$

Propagation: $A\text{-}CH_2\text{-}O^- + nCH_2{=}O \rightarrow A\text{-}[CH_2\text{-}O\text{-}]_n\text{-}CH_2\text{-}O^-$

Termination by chain transfer:

$$A\text{-}[CH_2\text{-}O\text{-}]_n\text{-}CH_2\text{-}O^- + H_2O \rightarrow A\text{-}[CH_2\text{-}O\text{-}]_n\text{-}CH_2\text{-}OH + OH^-$$

Scheme 1.25. Elementary reactions in the anionic polymerization of formaldehyde to form polyacetal.

The polymer has a relatively low ceiling temperature (119 °C), but this is the highest of all the formaldehyde polymers. The lack of success at polymerizing other monomers was due to the low ceiling temperatures (e.g. −39 °C for acetaldehyde) (Odian, 1991). Aldol condensation can be a side reaction competing with polymerization for these monomers. There are other routes, such as cationic ring-opening polymerization of cyclic acetals, to achieve the same polymer (Penczek and Kubisa, 1989).

Cationic polymerization

Cationic polymerization shares several of the features discussed above for anionic polymerization, but the requirements in terms of the monomer structure will be opposite, i.e. the vinyl monomers should have an **electron-donating** side group since the propagating centre carries a positive charge. Alternatively, the presence of an aromatic ring can stabilize the cation by making available resonance structures. Examples of these are alkyl vinyl ethers, isobutylene, isoprene, styrene and α-methyl styrene (Sauvet and Sigwalt, 1989).

Initiation of cationic polymerization may be achieved by any proton-donating species, but the limitation in, for example, protonic acids is the nucleophilic nature of many of the counter-ions, e.g. halide, such that hydrogen halides are ineffective initiators since they produce only the addition product with vinyl compounds. Lewis acids such as BF_3 and $AlCl_3$ are more widely used but require a proton donor, e.g. water, and it is this proton donor that is the actual initiator and the counter-ion is a product of the resulting reaction: e.g. BF_3 initiation of the polymerization of isobutylene (Odian, 1991) shown in Scheme 1.26.

In recognition of the fact that the initiating species is not BF_3 (since under anhydrous conditions this polymerization could not occur) but rather the proton donor (or *protogen*), the BF_3 is described as the *co-initiator*. The initiating system in Scheme 1.26 is thus the initiator–co-initiator complex. The donor may be a carbocation rather than a proton, and the initiator is then a *cationogen*, with an example being $AlCl_3$ and t-alkyl chlorides. Self-initiation by a self-ionization process may also occur, but the reaction rate is much lower than that for the initiator–co-initiator system. The details of carbocation generation leading to different initiation routes are discussed in detail by Sauvet and Sigwald (1989).

Propagation occurs through the addition of monomer at the cation by insertion between it and the counter-ion. A rapid succession of additions to the cation then produces high polymer. The temperature must be kept sufficiently low to enable this reaction to compete with chain transfer to monomer (shown below) that terminates the chain and results in re-initiation, but prevents the formation of high-molar-mass polymer.

Isomerization may occur to compete with propagation, particularly in the 1-alkenes, other than isobutylene (the illustration used here). This is linked to the possible rearrangements of the propagating cation to more stable cations. Monomers such as styrene and vinyl ethers are not affected since they do not have structures available for rearrangement. An example

Initiation:

$$BF_3 + H_2O \leftrightarrow BF_3{\cdot}OH_2$$

$$BF_3{\cdot}OH_2 + (CH_3)_2C{=}CH_2 \rightarrow (CH_3)_3C^+ \ldots (BF_3OH)^-$$

Propagation:

$$(CH_3)_3C^+ \ldots (BF_3OH)^- + n(CH_3)_2C{=}CH_2 \rightarrow$$

$$H\text{-}[CH_2C(CH_3)_2]_n\text{—}CH_2C^+(CH_3)_2 \ldots (BF_3OH)^-$$

Chain transfer to monomer:

$$H\text{—}[CH_2C(CH_3)_2]_n\text{—}CH_2C^+(CH_3)_2 \ldots (BF_3OH)^- + (CH_3)_2C{=}CH_2 \rightarrow$$

$$H\text{—}[CH_2C(CH_3)_2]_n\text{—}CH_2C(CH_3){=}CH_2 + (CH_3)_3C^+ \ldots (BF_3OH)^-$$

Termination:

$$H\text{—}[CH_2C(CH_3)_2]_n\text{—}CH_2C^+(CH_3)_2 \ldots (BF_3OH)^- \rightarrow$$

$$H\text{—}[CH_2C(CH_3)_2]_n\text{—}CH_2C(CH_3)_2OH + BF_3$$

Scheme 1.26. Elementary reactions in the cationic polymerization of isobutylene initiated by boron trifluoride.

of an isomerizing monomer is 3-methyl-1-butene, in which the rearrangement to a tertiary cation can occur from the (initially propagating) secondary cation, so the increased stability of the tertiary cation results in the polymer containing a mixture of these two isomers at the end of reaction. Depending on temperature, the tertiary isomer may dominate. The complexity of the carbocationic polymerization of these systems and dienes has been reviewed (Nuyken and Pask, 1989).

Termination will occur when the carbocation undergoes reaction with nucleophilic species other than monomer to produce a dead chain and no re-initiation. Since cationic polymerizations are carried out with high-purity reagents and under rigorous conditions this reaction is much less likely than chain transfer to monomer. The mutual repulsion of the charged polymerization sites ensures that bimolecular termination cannot occur (unlike in free-radical polymerization, where this is the most probable termination route). Recombination of the cation with the counter-ion will occur, and these termination reactions are often very specific to the chemistry of the initiator.

Deliberate termination of the reaction on completion of polymerization is generally performed with alcohols or water in the presence of base in order to deactivate the co-initiator.

Kinetics of cationic polymerization

The specificity of the reaction mechanism to the chemistry of the initiator, co-initiator and monomer as well as to the termination mechanism means that a totally general kinetic scheme as has been possible for free-radical addition polymerization is inappropriate. However, the general principles of the steady-state approximation to the reactive intermediate may still be applied (with some limitations) to obtain the rate of polymerization and the kinetic chain length for this living polymerization. Using a simplified set of reactions (Allcock and Lampe, 1981) for a system consisting of the initiator, I, and co-initiator, RX, added to the monomer, M, the following elementary reactions and their rates may be

Table 1.11. Elementary reactions, rate coefficients and reaction rates for a simple cationic polymerization

Process	Elementary reaction	Rate coefficient	Rate of reaction
Initiation	$I + RX \overset{K}{\longleftrightarrow} R^+IX^-$	k_i	$r_i = k_iK[I][RX][M]$
	$R^+IX^- \leftrightarrow R^+ + IX^-$		
	$R^+IX^- + M \overset{k_i}{\longrightarrow} RM^+\ IX^-$		
Propagation	$RM^+IX^- + nM \overset{k_p}{\longrightarrow} RM_nM^+IX^-$	k_p	$r_p = k_p[RM^+IX^-][M]$
Termination	$RM_nM^+IX^- \overset{k_t}{\longrightarrow} RM_{n+1}X + I$	k_t	$r_t = k_t[RM_nM^+IX^-]$

summarised (Table 1.11). In the above relations, the average concentration of all reactive species is assumed. In the steady state, the rates of initiation and termination are equal, so the concentration of the carbocation reactive intermediate may be determined and substituted into the propagation rate equation:

$$r_p = k_p[RM^+IX^-][M] = (k_pk_iK/k_t)[I][RX][M]^2. \tag{1.70}$$

The degree of polymerization, DP, is the ratio r_p/r_i as before, so

$$DP = (k_p/k_t)[M]. \tag{1.71}$$

It was noted above that chain transfer to monomer was a more frequent chain-breaking reaction than termination, but the chain-transfer reaction resulted in re-initiation of the chain. The degree of polymerization in this case is reduced from DP_0 to DP depending on the value of the chain-transfer coefficient, $C_M = k_{tr}/k_p$, hence

$$DP = r_p/(r_t + r_{tr}). \tag{1.72}$$

Thus

$$1/DP = 1/DP_0 + C_M. \tag{1.73}$$

It has been noted (Odian, 1991) that the steady-state approximation cannot be applied in many cationic polymerizations because of the extreme rate of reaction preventing the attainment of a steady-state concentration of the reactive intermediates. This places limitations on the usefulness of the rate expressions but those for the degree of polymerization rely on ratios of reaction rates and should be generally applicable. The molar-mass distribution would be expected to be very narrow and approach that for a living polymerization (with a polydispersity index of unity), but this is rarely achieved, due to the chain-transfer and termination reactions discussed above. Values closer to 2 are more likely.

The temperature of reaction for cationic polymerization is usually kept very low. The rate of initiation is largely insensitive to temperature, so the rates of propagation and termination alone determine the temperature dependence of the polymerization. Thus the observed activation energy, E_{obs}, will be just the difference between propagation and termination activation energies:

$$E_{obs} = E_p - E_t. \tag{1.74}$$

Generally E_t is greater than E_p, so the observed activation energy is **negative** (so that high reaction rates occur at low temperature). The same conclusion is reached for the degree of polymerization, DP, although, as noted above, the temperature dependence of the chain-transfer rate constant k_{tr}, (through the activation energy E_{tr}) must be considered. Since this may be the dominant form of termination, the difference $E_p - E_{tr}$ controls the temperature dependence of the degree of polymerization. This difference is also usually negative, so formation of higher-molar-mass polymer as well as high polymerization rate are favoured by low temperatures.

Co-ordination polymerization

The use of co-ordination polymerization is best known in the Ziegler–Natta polymerization of olefins containing an asymmetric centre (e.g. propylene) to give stereoregular polymers of industrial significance (Tait, 1989). The importance of the polymer tacticity in achieving conformations that allow crystallization and thus useful mechanical properties was discussed earlier (Section 1.1.2). For monomers without an asymmetric centre, such as ethylene, Ziegler–Natta catalysts allow polymerization at low pressures and at room temperature to give polymers with higher density and lower degrees of branching than those achievable by free-radical polymerization.

Modern, heterogeneous Ziegler–Natta catalysts are prepared from a (restricted) combination of the following species:

(1) a group IV–VIII transition-metal compound, e.g. $TiCl_3$
(2) an organometallic compound of group I–IV, e.g. $Al(C_2H_5)_3$ or $Al(C_2H_5)_2Cl$
(3) a donor such as a Lewis base or an ester, e.g. butyl benzoate
(4) a support such as $MgCl_2$

The goal in catalyst development is to achieve a high activity (defined as kilograms of polymer produced per gram of initiator) together with high stereospecificity. Thus a modern catalyst for isotactic polypropylene may have an activity of 2000 kg/g Ti and an iso-specificity of 98% (Odian, 1991). The support has an important role to play in achieving high activity since the number of available sites on Ti may be less than 1% of the theoretical maximum if no support is used. In some cases the catalyst may covalently bond to the support. The catalyst remains in the polymer after polymerization, and this may become an issue affecting the long-term stability of the polymer (Section 1.4). Many catalysts based on soluble metallocenes have been developed in the attempt to achieve higher iso-specificity with high yield.

Mechanism of stereopolymerization

There is no single mechanism for Ziegler–Natta polymerization because of the variety of catalyst and co-catalyst systems as well as the different phases in which the reaction may take place. The process of stereoregular polymerization can be understood from the mechanism of initiation and propagation as the monomer is inserted at the polymerization site on the catalyst surface. The detailed mechanism for achieving stereospecificity is an active area of research (Corradini and Busico, 1989, Tait and Watkins, 1989), but some general principles may be learned from the simple Ziegler–Natta catalysts (Allcock and Lampe, 1981).

The mechanism involves the formation of a transient co-ordination complex between the olefin and the transition metal (Ti) because of a vacant orbital on the metal as shown in Scheme 1.27. The aluminium complex (not shown) also participates by constraining the direction of addition of the monomer. This may also involve alkylation of the titanium as

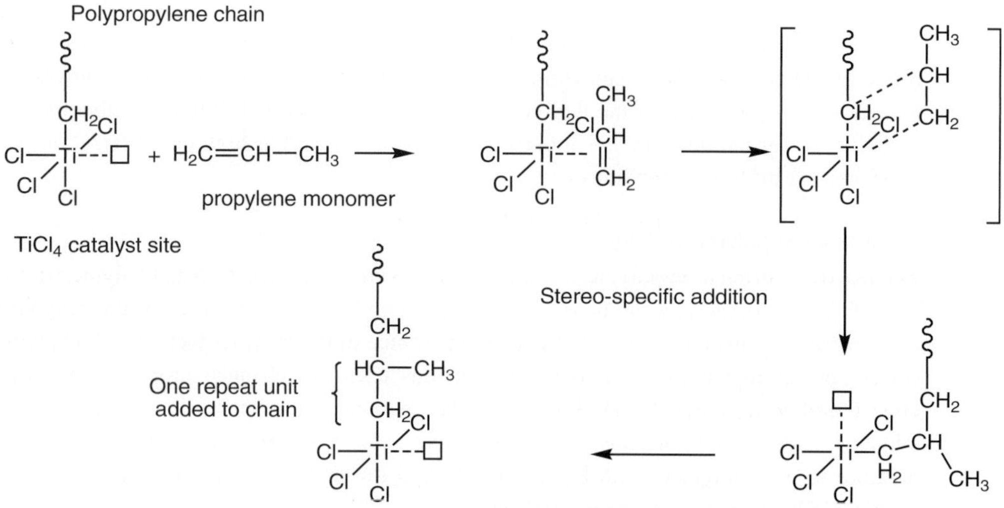

Scheme 1.27. A simple reaction scheme for polymerization of isotactic polypropylene using a $TiCl_4$ Ziegler–Natta catalyst. After Boor (1979).

well as stabilizing the reaction site. The reaction will occur without the aluminium complex, but the stereospecificity and yield are lower. The above references provide the detail of the mechanisms that have been postulated, for the purposes relevant here we note only that reaction takes place via an insertion followed by migration of an alkyl group through a four-centred transition state (Boor, 1979).

The reason for stereospecific polymerization of the olefin is the combined effect of the steric and electrostatic repulsions between the substituent (e.g. methyl group on propylene) and the ligands of the transition metal. These sites occur on the edges of the titanium crystal, so the efficiency is often related to the physical structure of the catalyst. This can change on proceeding through the polymerization because the pressure from the growing polymer can shatter larger crystals, and the rate progressively increases until a steady rate corresponding to the smallest achievable particle size has been reached.

The rate of polymerization is thus

$$r_{\mathrm{p}} = k_{\mathrm{p}}[Ti^{*}][M], \tag{1.75}$$

where $[\mathrm{Ti}^{*}]$ is the concentration of active sites, which may be between $10^{-2}\%$ and 10% of the actual [Ti]; [M] is governed by the adsorption isotherm for the monomer on the catalyst surface under the conditions of the polymerization. To determine the expected molar mass of the polymer the termination reactions need to be known. The lifetime of the growing chain is governed by the chain-transfer reactions that may occur. It is well known that polyethylene polymerized through a Ziegler–Natta route contains a range of unsaturated end groups. These may arise by the mechanisms shown in Scheme 1.28.

These reactions account for terminal vinyl-group formation, but there are other species, viz. *trans*-vinylene and vinylidene, that cannot be accounted for in this way. It is noted that these do not terminate the chains since re-initiation at the Ti site may occur.

The polydispersity with heterogeneous Ziegler–Natta polymerization may range from 5 to 30. High-density polyethylene (HDPE) produced by co-ordination polymerization has a degree of branching of about 1–5 per 1000 monomer units, compared with 25–50 for

Chain transfer to monomer:

$$\mathbf{Ti}\text{—}CH_2CH(CH_3)\text{—} + CH_3CH{=}CH_2 \rightarrow CH_2{=}C(CH_3)\text{—} + Ti\text{—}CH_2CH_2CH_3$$

Intramolecular β-hydride transfer:

$$\mathbf{Ti}\text{—}CH_2CH(CH_3)\text{—} \rightarrow CH_2{=}C(CH_3)\text{—} + Ti\text{—}H$$

Scheme 1.28. Possible termination mechanisms of the growing chain in Ziegler–Natta polymerization of isotactic polypropylene.

low-density (free-radical, high-pressure polymerized) polyethylene (LDPE). It is consequently higher melting (135 °C versus 110 °C) and has a degree of crystallinity of >70%, compared with ≈50% for LDPE. The molar mass is typically 50 000–250 000 g/mol, but there are applications (e.g. in biomedical orthopaedic implants) where ultra-high molar masses of 1.5×10^6 and above are needed. In this case the chain entanglements persist well above the melt temperature and the materials have very high wear resistance.

Ring-opening polymerization

Ring-opening polymerization of a cyclic monomer to yield a linear polymer is considered a separate class of polymerization even though it may involve an initiation system that is either anionic or cationic and thus have some features in common with those systems discussed before. Not all ring-opening polymerizations are addition polymerizations; an example of a stepwise reaction is the industrially-important ring-opening of ε-caprolactam to give nylon-6.

The main difference arises in the thermodynamic requirements for polymerization. In the case of the polymerization of a vinyl monomer, there is a large enthalpic difference between the monomer and polymer that overcomes the loss in entropy that accompanies the constraining of the monomer to form a linear chain. In the case of conversion of a cyclic molecule to a polymer, there is little change in the enthalpy per repeat unit between the monomer and the polymer, just a loss of the ring strain energy.

Considering a hypothetical system such as the polymerization of cyclic alkanes to give polymethylene, if the ring is unstrained (as in six-membered cyclics) then polymer formation is not favoured thermodynamically since ΔG is positive for this system. The actual polymerization of the smaller and larger rings, for which ΔG is negative, is generally not achieved since there is no reactive site for ring-opening and thus no mechanistic pathway to make the reaction kinetically feasible. Introduction of a hetero-atom or a functional group such as an ester or amide immediately makes the reaction feasible by offering a reaction pathway.

These reactions are all commercially significant and the reactions may in some cases take place in a reactive extruder. The mechanism of the reaction depends on whether the reaction is anionically or cationically initiated, and the effects of reaction conditions are those as discussed before.

Nylon-6

The example in Scheme 1.29 of the ring-opening polymerization of ε-caprolactam is given since it has also been exploited in reactive processing (Brown, 1992). In the example given the initiator is 0.2%–0.3% sodium lactamate. The anionic polymerization (such as that in Scheme 1.29) is faster than cationic or hydrolytic polymerization.

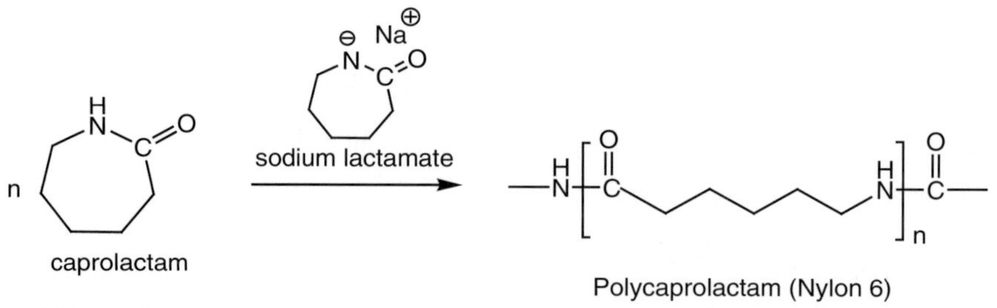

n caprolactam —sodium lactamate→ Polycaprolactam (Nylon 6)

Scheme 1.29. Polycaprolactam polymerization from ring-opening of caprolactam. After Brown (1992).

$$BF_3 + H_2O \rightarrow BF_3OH^- + H^+ \xrightarrow{epoxy} BF_3OH^-\ {}^+CH_2CH[OH]$$

Scheme 1.30. Ring-opening polymerization of epoxides initiated by BF_3.

Epoxides

It was noted earlier that, among the reactions of diglycidyl epoxide resins, the linear polymerization with an initiator such as BF_3 was feasible. This is often exploited in reactive processing of carbon fibre pre-impregnated with cpoxy resin that contains a latent curing agent that is a complex that does not initiate at room temperature such as BF_3·MEA. Whether the polymerization occurs through a cationic or an anionic mechanism depends on the detailed composition of the catalyst (Hodd, 1989). For Lewis acids and other protonic reagents the mechanism involves a co-catalyst such as water (Scheme 1.30).

The bulky anion then stabilizes the intermediate adduct from protonation of the epoxy group and then facilitates insertion of epoxide at the cationic propagation site. Rapid polymerization can then occur. Cationic photopolymerization of epoxides often involves the photo-generation of acid from an initiator such as diaryliodonium or triaryl sulfonium salts (Crivello, 1999). The anions are important in controlling the addition at the cationic site and are typically BF_4^- and PF_6^-. The reactivity of the system depends also on the structure of the epoxide.

Polydimethyl siloxanes

There are several inorganic polymers that are formed by a ring-opening polymerization, among which polydimethylsiloxanes (PDMSs) are most relevant to studies in reactive processing. The starting material is the cyclic monomer, D_4, $[-Si(CH_3)_2-O-]_4$, as shown in Figure 1.22. The anionic polymerization is initiated by base such as KOH to give the anionic centre shown in Scheme 1.31.

Propagation occurs by addition and ring-opening of D_4 at the anionic site. Water may act as a chain-transfer agent or be used to terminate the polymerization, as shown above. Alternatively, to create the more familiar trimethyl-terminated PDMS, hexamethyl disiloxane is added to end-cap the chain, as shown in the last reaction in Scheme 1.31. This may then initiate further chains at the new anion site. Consequently PDMS oligomers may have a range of trimethyl and hydroxyl terminating groups. Thus, while PDMS is the most common silicone polymer, a range of other groups, such

ethylene oxide → $-\!(\!-OCH_2CH_2-\!)_n-$ Poly (ethylene oxide) or Poly (oxy ethylene)

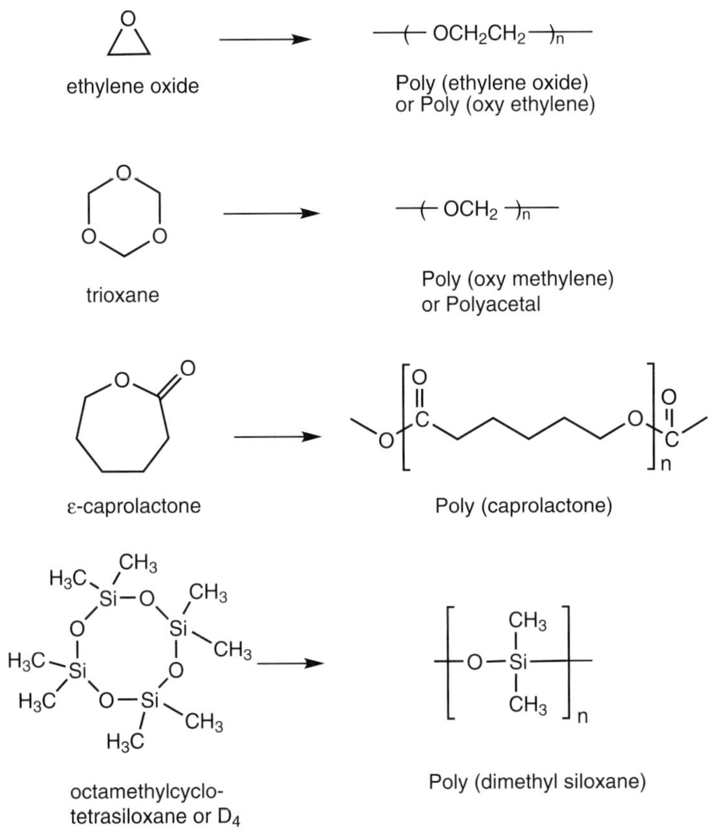

trioxane → $-\!(\!-OCH_2-\!)_n-$ Poly (oxy methylene) or Polyacetal

ε-caprolactone → Poly (caprolactone)

octamethylcyclo-tetrasiloxane or D_4 → Poly (dimethyl siloxane)

Figure 1.22. Examples of cyclic monomers and their polymers from ring-opening polymerization.

Initiation:

$KOH + D_4 \rightarrow HO\text{-}Si(CH_3)_2\text{-}O\text{----}Si(CH_3)_2O^- K^+$

Propagation:

$HO\text{-}Si(CH_3)_2\text{-}O\text{----}Si(CH_3)_2O^- K^+ + nD_4 \rightarrow HO\text{-}[Si(CH_3)_2\text{-}O]_{4n+4}^- K^+$

Termination:

$HO\text{-}[Si(CH_3)_2\text{-}O]_{4n+4}^- K^+ + H_2O \rightarrow HO\text{-}[Si(CH_3)_2\text{-}O]_{4n+4}H + KOH$

Chain Transfer:

$\text{----}Si(CH_3)_2O^- K^+ + (CH_3)_3Si\text{-}O\text{-}Si(CH_3)_3 \rightarrow$

$\text{----}Si(CH_3)_2O\ Si(CH_3)_3 + Si(CH_3)_3O^- K^+$

Scheme 1.31. The reaction mechanism for ring-opening polymerization of the cyclic siloxane D_4 to give poly(dimethylsiloxane) or PDMS.

as vinyl and phenyl, may be introduced to achieve different physical and chemical properties.

It has been noted that this polymerization is thermodynamically unusual in that the driving force for the polymerization is the increase in entropy, ΔS, of 6.7 J mol^{-1} K^{-1} on moving from the constrained cycle of D_4 to the extremely flexible polydimethylsiloxane backbone, which has many available conformations (Odian, 1991). Since ΔH is close to zero the only reason why the polymerization proceeds is the increase in entropy on polymerization.

Cyclic oligomers

The above example of the entropically driven ring-opening polymerization of D_4 to yield PDMS may be extended to other cyclic oligomers (Hall and Hodge, 1999). It was noted earlier that cyclic oligomers offered a route to high-molar-mass polycarbonates by the ring-opening reaction using a transesterification catalyst. Similar reactions have been observed under conditions for addition polymerization. The equilibrium between the cyclic monomer and polymer is driven in the direction of polymer at the high concentration used in the polymerization. Among the attractive features of the use of cyclic oligomers are the absence of heat of reaction and the control of molar mass achieved by controlling the number of end groups.

It is noted that these reactions need not always proceed in the same way as for the non-cyclic monomer. For example, in the thermal polymerization of a macrocyclic ethylene terephthalate dimer, it has been noted (Nagahata *et al.*, 2001) that, from the molar-mass distribution, the polymerization cannot be occurring through a stepwise polymerization but must rather go through an ionic addition reaction. It was suggested that an alkoxide ion generated through thermal scission of an ester linkage initiated the reaction by ring-opening at the ester carbonyl of other macrocyclics.

Living polymerization – ionic and free-radical

The particular features of anionic polymerization that made the polymer chains 'living' were discussed above. The main requirement for a living polymerization is the absence of any process for spontaneous termination so that the degree of polymerization is controlled by the ratio of monomer to initiator concentrations. The molar-mass of the polymer therefore increases linearly with monomer conversion. On exhaustion of the monomer, the initiation centres remain, so chains may be re-initiated by addition of further monomer. Termination or chain transfer is controlled by the deliberate addition of a reagent to remove the living end. The resulting polymers will also have very narrow molar-mass distributions since rapid initiation ensures that all chains are initiated at the same time.

This behaviour is very different from that in cationic or free-radical polymerization, where initiation is slow and the conditions are chosen to be such that the concentration of the reactive intermediates is low. This ensures that the termination reactions (which for free-radical reactions are usually bimolecular) are kept under some control. As shown in Figure 1.23, the molar mass of the polymer (provided that conditions for the Trommsdorff effect are not reached) does not change with the extent of conversion, i.e. high-molar-mass polymer is formed from the onset of the reaction. In the case of cationic polymerization, bimolecular termination is not possible due to repulsion of the positively charged growing chain ends, and the high rate of reaction is the main reason for keeping the intermediate concentration low. Both reactions yield controlled molar masses only through the control of chain transfer. In the case of cationic polymerization, low temperature is required in order to

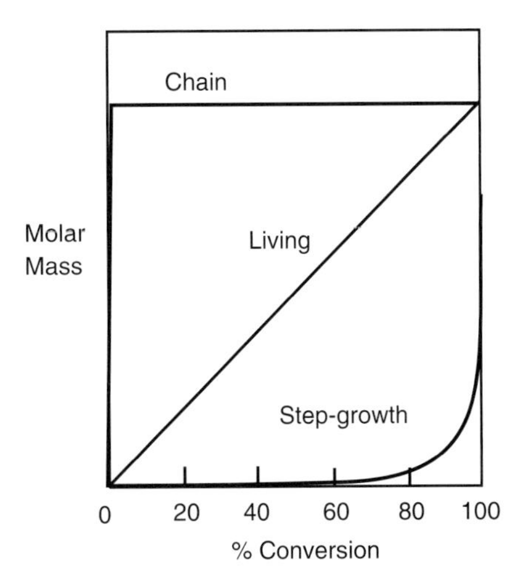

Figure 1.23. The relation between polymer molar mass (molecular weight) and conversion for chain, step-growth and living polymerization. After Odian (1991).

achieve high-molar-mass polymer since the activation energy for propagation is much lower than that for chain transfer (often to monomer) or termination by adventitious impurities. Unlike with anionic polymerization, the presence of these reactions results in a broader molar-mass distribution.

Since high-molar-mass polymer is produced from the onset of polymerization, neither of these reactions, when performed with the initiator system and conditions described so far, can be regarded as a 'living' polymerization. However, it has been noted that, if the necessary low concentration of the reactive intermediate could be achieved by having a 'dormant' species in equilibrium with the propagating species (Matyjaszewski, 1999), then it is possible to control the extent of the polymerization more precisely, so that it increases linearly with conversion. Whether this can be truly classed as a living polymerization has been debated (Darling *et al.*, 2000), but there is agreement that features of living polymerization occur when propagation and *reversible* termination are faster than any *irreversible* termination. A further feature is that living polymers can be re-initiated by supplying further monomer. In the following, just a few of the many examples of approaches that have been employed for living free-radical and cationic polymerizations are considered, which may be added to the anionic living polymerizations discussed earlier. Some of these systems have been applied in the area of reactive processing and, since these are principally based on free-radical systems, only these are considered here.

Nitroxide-mediated polymerization

The key feature of the use of a 'dormant species' may be seen in the following general scheme (Scheme 1.32) that involves complexation of the propagating species by means of a stable nitroxide radical (Hawker *et al.*, 2001). The P_n–O bond of the alkoxy amine P_n–O–NR is thermally labile at the polymerization temperature, so this becomes the site for the insertion of monomer. Propagation then occurs at a rate that is much slower than for a simple free-radical addition reaction since the propagating radical concentration (which is governed by the position of the equilibrium with the alkoxy amine

P_n-O-NR ↔ P_n· + ·O-NR

+M

↓

P_{n+1}-O-NR ↔ P_{n+1}· + ·O-NR

etc.

Scheme 1.32. Nitroxide-mediated polymerization showing formation of the 'dormant' polymeric alkoxy amine species.

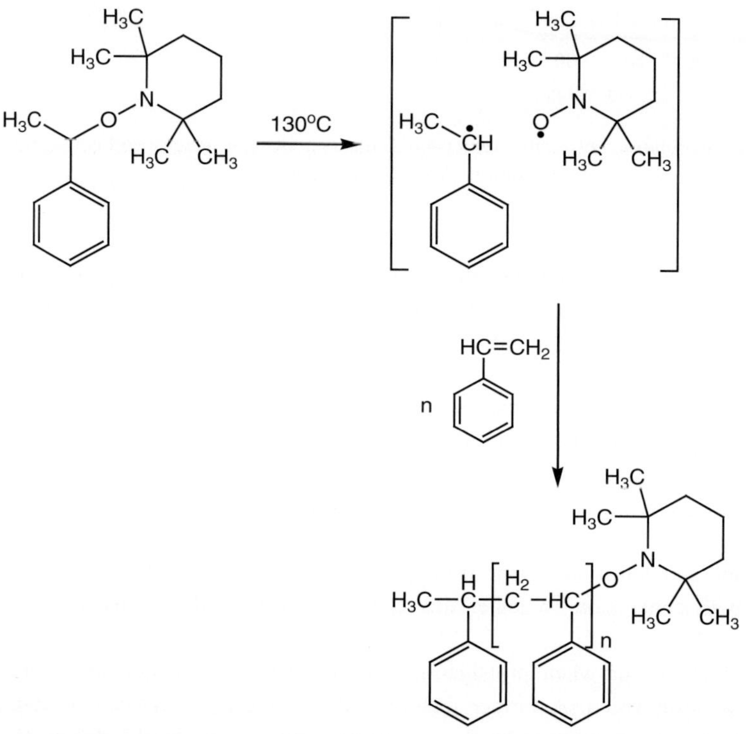

130°C

$HC{=}CH_2$

n

Scheme 1.33. Nitroxide-mediated living radical polymerization of styrene using TEMPO. After Hawker *et al.* (2001).

at the reaction temperature) is very low. It is this low concentration of reactive chain ends that minimizes irreversible termination reactions that occur in the normal free-radical polymerization. A further 'living' feature is that DP increases linearly with conversion.

The initiation of the polymerization shown above may take place from a preformed alkoxy amine with an end group commensurate with the monomer, M. This is shown in Scheme 1.33 for styrene polymerization (Hawker *et al.*, 2001).

While the initial alkoxy amines were based on the nitroxyl radical TEMPO (2,2,5,6-tetramethyl piperidinyloxy) as shown above, this system is of limited applicability because of the high temperatures (>125 °C) and long polymerization times (>24 hours)

$$P_n\text{-}X + Mt^{m+}\text{–Ligand} \leftrightarrow P_n\cdot + Mt^{(m+1)+}X\ (\text{Ligand})$$

$$+M$$

$$\downarrow$$

$$P_{n+1}\text{-}X + Mt^{m+}\text{–Ligand} \leftrightarrow P_{n+1}\cdot + Mt^{(m+1)+}X\ (\text{Ligand})$$

etc.

Scheme 1.34. Control of polymerization by formation of the metal–ligand dormant species in ATRP.

as well as its unsuitability for living polymerization of important monomers such as acrylates and methacrylates. Changes to the structure of the mediating radical and in particular the move away from the highly hindered structure of TEMPO to give a more unstable nitroxide resulted in a more flexible initiator that achieves a lower polydispersity. This arises from the lowering of chain-end degradation by virtue of the lower temperatures and reaction times. The side reaction leading to termination and thus broadening of the MWD involves the loss of nitroxyl mediating radical by reduction to form a free alkoxy amine, which can then terminate another growing chain as the nitroxyl radical is regenerated. This behaviour is very monomer-specific and, while it is insignificant in styrene polymerization, reduction of the nitroxyl in methacrylates is dominant and the living character is not achieved. In contrast, for butyl acrylate with a second-generation nitroxide there is a linear increase in molar mass (up to 22 000 g/mol) with conversion (up to 90%) during polymerization for 16 hours at 123 °C. Of particular importance is the insensitivity of the method to impurities and the presence of air, so this method approaches the versatility of techniques such as atom-transfer radical polymerization (ATRP) and reversible addition–fragmentation transfer (RAFT). These techniques are discussed below.

Atom-transfer radical polymerization (ATRP)

The 'dormant species' in ATRP arises from the polymer chain being capped with a halogen atom (P_n–X), while in the active state the halogen is chelated to a metal complex, thus allowing monomer to add. This takes advantage of the Kharasch reaction in which halogenated alkanes add to vinyl monomers by a free-radical reaction that is catalysed by transition-metal ions in their lower-valent state (Fischer, 2001).

The transition-metal undergoes a one-electron oxidation as the halogen atom is transferred to give the active species, consisting of a polymer radical, $P_n\cdot$ and the complex, as shown below. This then allows monomer to add, before it is reversibly deactivated to give the halogen-terminated chain that is longer by one repeat unit (Scheme 1.34).

In this Scheme the ligand is often bipyridyl if the metal is Cu I and, in early ATRP, ruthenium compounds were used. The nature of the ligand controls the effectiveness of the catalyst. The system is 'living' since the concentration of the transition metal in the higher oxidation state is sufficient to ensure that deactivation to the dormant state is much faster than any other polymer radical termination steps, including bimolecular recombination. Because of the absence of a Trommsdorff effect the reaction may be carried out in the bulk (Matyjaszewski, 1999). The detailed mechanism is complex since the reaction rate depends on the metal, the organic halide, the ligand and the counter-ion (Fischer, 2001).

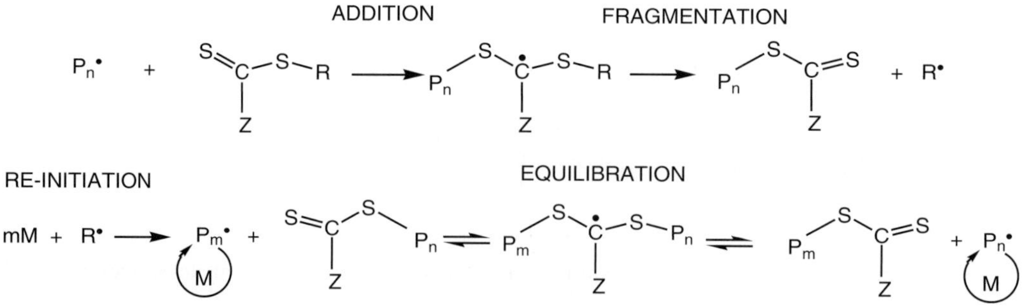

Scheme 1.35. The principle of living radical polymerization using a RAFT agent. After Mayadunne *et al.* (1999).

Reversible addition–fragmentation chain transfer (RAFT)

Polymerization by RAFT differs from nitroxide-mediated polymerization and ATRP by the use of a chain-transfer agent, usually a thiocarbonyl thio compound of the general structure S=C(Z)S–R, to capture the propagating radicals and then fragment to produce a dormant species and a new initiating radical, R·, as shown in Scheme 1.35. The requirement for an effective RAFT agent is that it is rapidly consumed to produce equilibration between the dormant and active species. This formally requires the rate of propagation to be slower than the rates of addition and fragmentation (Mayadunne *et al.*, 1999).

In studies of the effect of the leaving group (R) and the activating group (Z), general conclusions regarding the effectiveness of various agents have been drawn (Chiefari *et al.*, 2003, Chong *et al.*, 2003). The most effective compounds have high transfer coefficients ($C_s > 20$ at a concentration of 3×10^{-3} mol/dm^3 in methyl methacrylate (MMA)), and the range of effective compounds is much smaller for MMA than for styrene and methyl acrylate, for which the values of C_s are much higher. This is found to depend not on the rate constant for addition to the thiocarbonyl group (the first reaction, above) but rather on the properties of R and P_n· as leaving groups as well as on the reverse reaction of R· with the polymeric RAFT agent to reform the initial RAFT agent. The features of R· that control the effectiveness of the RAFT agent are the stability of the radical, the polarity and steric factors. In the case of Z, the effectiveness of the RAFT agent is linked to the interaction of Z with the C=S bond to activate or deactivate the group with respect to free-radical addition.

The successful operation of a RAFT agent may be seen from the low polydispersity of the polymer, the slow formation of a high molar mass and the close agreement between the measured M_n and the value calculated from

$$M_n = ([\text{Monomer}]/[\text{CTA}])\times\text{conversion}\times M_0. \tag{1.76}$$

This is seen in Figure 1.24 (Mayadunne *et al.*, 1999), in which two potential RAFT agents, at a concentration of 2.97×10^{-2} M, are assessed in the bulk polymerization of styrene at 100 °C. It is found that the first agent, *S*-benzyl *N*, *N*-diethyldithiocarbamate; (i.e. Z is *N*, *N*-diethyl-) produces polymer after 6 hours that has $M_n = 317\,100$ g/mol and $M_w/M_n = 1.86$ (GPC trace A). The theoretical value after this time is 4 700 g/mol and it is apparent that RAFT polymerization is *not* occurring for this compound and normal free-radical polymerization of the bulk monomer is occurring, with a high polydispersity. In contrast, if *S*-benzyl *N*-pyrrolocarbodithioate (Z is *N*-pyrrolo-) is used, then trace B is obtained after 6 hours ($M_n = 6480$ g/mol) and trace C ($M_n = 15\,600$ g/mol) after 30 hours. The

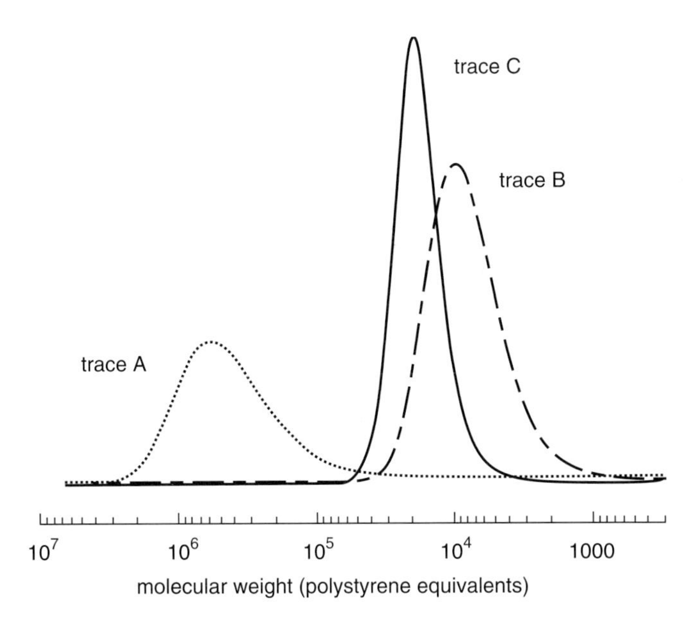

Figure 1.24. Assessment of the effectiveness of RAFT agents from the low polydispersity of the polymer. Trace A corresponds to the use of *S*-benzyl-*N*,*N*-diethyldithiocarbamate, which is not an effective RAFT agent, whereas traces B and C correspond to use of the effective agent, *S*-benzyl-*N*-pyrrolocarbodithioate, after 6 and 30 hours of polymerization. Reprinted with permission from Mayadunne (1999). Copyright 1999 American Chemical Society.

calculated value of M_n in the latter case is 18200 g/mol and the measured polydispersity is $M_w/M_n = 1.20$. It is clear that this compound is an effective RAFT agent.

The kinetics and mechanisms of these systems are being elucidated from studies of model polymer–RAFT-agent adducts. It has been shown that a high value of the exchange rate coefficient is required in order to ensure that low-polydispersity polymer is formed early in the polymerization (Goto *et al.*, 2001). The usefulness of RAFT, and indeed all living polymerization systems such as those discussed above, lies in the existence of a 'living' site that is available, after the exhaustion of monomer, for re-activation by the addition of a second monomer to give a block copolymer of low polydispersity. The range of molecular architecture that may readily be formed in living polymer systems is discussed in the next section.

1.2.4 Obtaining different polymer architectures by addition polymerization

Addition polymerization offers a wide range of options for modifying the linear polymer chain to produce molecular architecture ranging from branched and star polymers to highly cross-linked networks. In the chemorheology and processing of reactive systems, addition polymerization and often adventitious free-radical reactions play a major role in defining the rate of a process. The controlled free-radical graft copolymerization and crosslinking reactions are among the most widely employed means in reactive processing to achieve useful mechanical properties. In Section 1.2.2 the architectures achieved through step polymerization were considered and, in this section, the same categories will be considered.

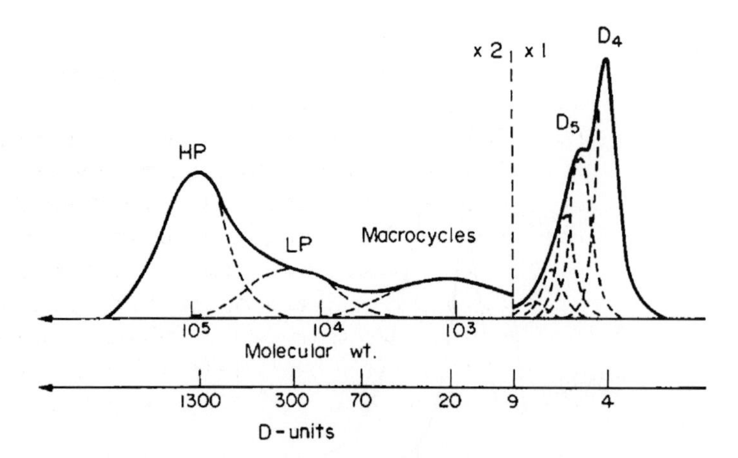

Figure 1.25. A GPC trace for the polymer from the polymerization of D_4 with CF_3SO_3H, showing products ranging from cyclics through to linear (LP) and high polymer (HP). Reproduced with permission from Kendrick, Parbhoo *et al.* (1989). Copyright 1989, Pergamon Press.

Cyclization

The same thermodynamic and ring-stability considerations for the formation of large cyclic polymer molecules as discussed in Section 1.2.2 will apply to polymers formed through addition polymerization. It has been noted that in siloxane polymers cyclic oligomers may constitute a significant component of the final polymer. This depends on the polymerization mechanism and, in the case of the base-and acid-catalysed ring-opening polymerization of the cyclotetrasiloxane D_4 (by anionic and cationic reactions, respectively), there is a continuous population of cyclic oligomers in equilibrium with the linear polymer up to D_{400} (Kendrick *et al.*, 1989). The isolation and identification of oligomers beyond D_{25} is challenging, but GPC and mass-spectral methods such as MALDI-TOF-MS enable these to be precisely identified (Hunt and George, 2000). In the cationic polymerization of D_3 and D_4 with CF_3SO_3H macrocyles with molar mass up to 10 000 g/mol are formed, as seen from the resolution of these macrocycles from the high linear polymer at one end of the distribution and the small cycles at the other end in the GPC trace (Figure 1.25).

It is of interest to determine whether these macrocycles form by addition of the small cyclic species, e.g. D_3, directly to the growing macrocycle or the reaction proceeds by closure of the linear chain after addition (Kendrick *et al.*, 1989). Although the latter seems mechanistically improbable for large cycles, the actual root mean separation of the ends (R_{rms}, Section 1.1.1) of the polymer increases as the square root of the degree of polymerization, so it has been noted that the end groups for a 50-siloxane-unit chain will be only 4 nm apart (Kendrick *et al.*, 1989). This suggests that a significant fraction will be able to react, so cyclization competes directly with linear propagation. Analysis of the conformational entropy of the cyclic oligomers by Stockmayer and Jacobson led to the prediction that the concentration, $[D_n]$, of the n-oligomer varied as $n^{-5/2}$. The kinetic analysis gave a prediction of $n^{-3/2}$. In either case the end-closure mechanism is favoured over the ring-expansion mechanism for the production of the siloxane macrocycles.

The importance of cyclization resides in the difference between the properties of linear and cyclic oligomers of the same molar mass. The cyclics have a lower viscosity and a higher diffusion coefficient and, in applications of siloxane polymers, the role that the highly mobile fraction of oligomers may play in determining the properties, particularly of the surface, of even a crosslinked PDMS elastomer should be considered (Hunt *et al.*, 2002).

Copolymerization

Many of the polymers of industrial importance are copolymers consisting of units of more than one type of monomer **in the same polymer chain**, and addition polymerization is one of the major routes for copolymer formation. The properties that result depend on how the monomers are distributed along the polymer chain as well as on their concentration.

Considering two monomers M_1 and M_2, the possibilities are shown below.

1. **Random or statistical copolymer**: $-M_1M_1M_1M_2M_2M_2M_1M_2M_1M_1M_1-$
   In this case the distribution of M_1 and M_2 along the copolymer chain cannot be predicted or controlled and arises when the propagating radicals (or ions) of the different monomers have a finite probability of reaction with either monomer.
2. **Alternating copolymer**: $-M_1M_2M_1M_2M_1M_2M_1M_2M_1M_2-$
   In this case, the propagating radical of one monomer will react only with the second monomer, not with its own monomer. After initiation, the system will develop an alternating composition regardless of the initial ratio of the two monomer concentrations.
3. **Block copolymer**: $-(M_1M_1M_1M_1M_1M_1)(M_2M_2M_2M_2M_2M_2)-$
   The block copolymer cannot readily be synthesized from an initial mixture of the two monomers, but requires a strategy such as the reaction of preformed oligomers with functional end groups or the polymerization of a sequence of M_1 followed by a sequence of M_2. The living polymerization methods described above in Section 1.2.3 are a useful route to such systems.
4. **Graft copolymer:**

$$
\begin{array}{l}
\qquad\qquad\qquad\qquad\qquad -M_1 \\
\qquad\qquad\qquad\qquad\qquad\;\;\; | \\
\qquad\qquad\qquad\qquad\qquad\;\; M_1 \\
\qquad\qquad\qquad\qquad\qquad\;\;\; | \\
\qquad\qquad\qquad\qquad\qquad\;\; M_1 \\
\qquad\qquad\qquad\qquad\qquad\;\;\; | \\
\qquad\qquad\qquad\qquad\qquad\;\; M_1M_2M_2M_2M_2M_2M_2M_2- \\
\qquad\qquad\qquad\qquad\qquad\;\;\; | \\
\qquad\qquad\qquad\qquad\qquad\;\; M_1 \\
\qquad\qquad\qquad\qquad\qquad\;\;\; | \\
\qquad\qquad\qquad\qquad\qquad\;\; M_1 \\
\qquad\qquad\qquad\qquad\qquad\;\;\; | \\
-M_2M_2M_2M_2M_2M_2M_1 \\
\qquad\qquad\qquad\qquad\qquad\;\;\; | \\
\qquad\qquad\qquad\qquad\qquad\;\; M_1-
\end{array}
$$

In graft copolymerization, there are also sequences of M_1 and sequences of M_2, but they are formed at branch points, so all graft copolymers are, by definition, branched. These are also formed in sequential reactions, and graft copolymerization is frequently employed in reactive processing.

The composition of a copolymer formed in an addition-polymerization reaction will not simply be the composition of the feed M_1/M_2 since the reactivity of the two monomers to the initiating and propagating species (whether free radical, anion or cation) may differ. The kinetics of copolymerization is a suitable route for the introduction of the concepts since this then allows the composition of the copolymers to be described systematically. Free-radical reactions are those mostly encountered in reactive processing and are considered below.

Table 1.12. Reactivity ratios for copolymerization of two monomers M_1 and M_2

Reaction	Rate coefficient	Reactivity ratio
$–M_1\cdot + M_1 \rightarrow –M_1M_1\cdot$	k_{11}	
$–M_1\cdot + M_2 \rightarrow –M_1M_2\cdot$	k_{12}	$r_1 = k_{11}/k_{12}$
$–M_2\cdot + M_2 \rightarrow –M_2M_2\cdot$	k_{22}	
$–M_2\cdot + M_1 \rightarrow –M_2M_1\cdot$	k_{21}	$r_2 = k_{22}/k_{21}$

Free-radical copolymerization kinetics and copolymer composition

The polymerization of a mixture of two monomers of concentrations $[M_1]$ and $[M_2]$ may be described by four possible propagation reactions with their associated rate coefficients as shown in Table 1.12. The ratio of the rate coefficient for the reaction of a monomer radical with its *own* monomer to that with the *other* monomer is the reactivity ratio. There are two such ratios, r_1 and r_2, defined in Table 1.12.

By analogy with Equation (1.46) for a single monomer and radical species, two rate equations may now be defined for the disappearance of the two monomers by the above reactions:

$$-\mathrm{d}[M_1]/\mathrm{d}t = k_{11}[M_1][M_1\cdot] + k_{21}[M_1][M_2\cdot], \tag{1.77}$$

$$-\mathrm{d}[M_2]/\mathrm{d}t = k_{12}[M_2][M_1\cdot] + k_{22}[M_2][M_2\cdot]. \tag{1.78}$$

By simply dividing one of these two equations by the other, it is possible to obtain an expression for the instantaneous composition of the copolymer:

$$\frac{\mathrm{d}[M_1]}{\mathrm{d}[M_2]} = \frac{k_{11}[M_1][M_1\cdot] + k_{21}[M_1][M_2\cdot]}{k_{12}[M_2][M_1\cdot] + k_{22}[M_2][M_2\cdot]}. \tag{1.79}$$

To solve these equations and eliminate the reactive intermediates, we note the stationary-state approximation and the definition of the reactivity ratios r_1 and r_2 above. Then

$$\frac{\mathrm{d}[M_1]}{\mathrm{d}[M_2]} = \frac{[M_1](r_1[M_1\cdot] + [M_2])}{[M_2](r_2[M_2\cdot] + [M_1])}. \tag{1.80}$$

This is the copolymer equation, which may also be derived by statistical means without invoking the stationary-state approximation (Odian, 1991).

Inspection of the copolymer equation shows the importance of the reactivity ratios in determining the type of copolymerization reaction that will occur. Thus, we consider the following possibilities.

(a) $r_1 = r_2 = 0$. This means that $k_{11} \ll k_{12}$ and $k_{22} \ll k_{21}$, i.e. each monomer will react preferentially with the other monomer radical rather than homopolymerize. This will result in an **alternating copolymer**, i.e. $\mathrm{d}[M_1]/\mathrm{d}[M_2] = 1$.

(b) $r_1 = r_2 = 1$. The propagating radicals have an equal reactivity with each of the monomers. This will result in a **random copolymer** with composition identical to that of the feed, i.e. $\mathrm{d}[M_1]/\mathrm{d}[M_2] = [M_1]/[M_2]$.

(c) $r_1 = 1/r_2$. This is termed an **ideal copolymer** since each of the propagating species has an equal probability of reaction with each of the monomers (for the same concentration of each). The composition is then

$$\mathrm{d}[M_1]/\mathrm{d}[M_2] = r_1[M_1]/[M_2].$$

Table 1.13. Copolymer composition for the ideal copolymerization, for which $r_1 = 1/r_2$; f_1 and f_2 are the molar feed ratios of monomers, M_1 and M_2, and F_1 and F_2 are the mole fractions of the repeat units in the copolymer (Equation (1.82))

r_1	r_2	f_1	f_2	F_1	F_2
10	0.1	0.9	0.1	0.99	0.01
		0.5	0.5	0.91	0.09
		0.1	0.9	0.53	0.47
5	0.2	0.9	0.1	0.98	0.02
		0.5	0.5	0.83	0.17
		0.1	0.9	0.36	0.64
1	1	0.9	0.1	0.9	0.1
		0.5	0.5	0.5	0.5
		0.1	0.9	0.1	0.9
0.5	2	0.9	0.1	0.82	0.18
		0.5	0.5	0.33	0.67
		0.1	0.9	0.05	0.95

A consequence of this is that in many systems $r_1 \ll 1$ so $r_2 \gg 1$ and the copolymer consists of a much greater proportion of M_2 than there was in the feed. In the extreme, this monomer dominates the copolymer composition until it is consumed and then the other monomer homopolymerizes. The result is a copolymer with extended blocks of the two components. Thus simply increasing the feed composition does not improve the copolymer composition, as may be seen in Table 1.13, which shows the effect of the different values of r on the composition for the different mole fraction of feed. $f_1 = [M_1]/([M_1] + [M_2])$ and F_1 is the mole fraction of M_1 in the copolymer:

$$F_1 = d[M_1]/(d[M_1] + d[M_2]) \tag{1.81}$$

$$= r_1 f_1/(r_1 f_1 + f_2) \tag{1.82}$$

when $r_1 = 1/r_2$.

Note that Table 1.13 also includes the special case $r_1 = r_2 = 1$, for which $F_1 = f_1$ and $F_2 = f_2$. Another condition for such *azeotropic copolymerization* in which the copolymer composition is equal to the feed composition is when $f_1/f_2 = (r_2 - 1)/(r_1 - 1)$. The desirable range of r_1 to produce copolymers with a reasonable range of both monomers present is from 0.5 to 2.

(d) $0 < r_1, r_2 < 1$. In this case, each radical will react with the other to give a **random copolymer** where the composition in the copolymer is given by the copolymer equation.

(e) r_1, $r_2 > 1$. In this case, for both monomers there is a greater tendency to homopolymerize than to copolymerize and the resulting copolymer will have extended blocks of the individual monomer units. This is not a controlled **block copolymer** of the type that may be prepared using a living polymerization system as discussed in Section 1.2.3.

Copolymer sequence-length distributions

The copolymer equation enables the mole fractions of the respective monomers that are incorporated into a copolymer to be determined for any feed composition, provided that the reactivity ratios are known. This equation does not indicate the way in which they are incorporated, other than for the cases for which there is a special relationship, such as

Table 1.14. Sequence-length distributions $(N_1)_x$ for copolymerization of monomers M_1 and M_2 with reactivity ratios r_1 and r_2 and feed compositions f_1 and f_2

r_1	r_2	f_1	f_2	$(N_1)_1$	$(N_1)_2$	$(N_1)_3$	$(N_1)_4$	$(N_1)_5$	$(N_1)_6$
1	1	0.5	0.5	0.5	0.25	0.13	0.06	0.03	0.02
1	1	0.75	0.25	0.75	0.19	0.05	0.01		
5	0.2	0.5	0.5	0.17	0.14	0.12	0.10	0.08	0.07
5	0.2	0.75	0.25	0.375	0.23	0.15	0.09	0.06	0.04
0.1	0.1	0.5	0.5	0.91	0.08	0.01			
0.1	0.1	0.75	0.25	0.97	0.03				

$r_1 = r_2 = 0$, when the monomer units alternate along the backbone. Of interest is the distribution of the lengths of the extended sequences of M_1 and M_2. The probabilities of a sequence of two monomer units, (M_1M_1) or (M_2M_2), occurring are P_{11} and P_{22}, respectively, and that for the termination of the sequence is for the cross-reaction, i.e. P_{12} or P_{21}. These may be related to the reactivity ratios and concentrations in the feed by (Hamielic *et al.*, 1989, Tirrell, 1989):

$$P_{11} = r_1/(r_1 + [M_2]/[M_1]), \quad (1.83)$$

$$P_{22} = r_2/(r_2 + [M_1]/[M_2]), \quad (1.84)$$

$$P_{12} = [M_2]/(r_1[M_1] + [M_2]) = 1 - P_{11}, \quad (1.85)$$

$$P_{21} = [M_1]/(r_2[M_2] + [M_1]) = 1 - P_{22}. \quad (1.86)$$

The number fraction $(N_1)_x$ of sequences of M_1 of length x is given by the product of the probabilities:

$$(N_1)_x = (P_{11})^{x-1}P_{12}. \quad (1.87)$$

Similarly,

$$(N_2)_x = (P_{22})^{x-1}P_{21}. \quad (1.88)$$

Using these equations, it is thus possible to draw up Table 1.14 that gives sequence length distributions for various compositions and reactivity ratios. It is noted that when r_1 and r_2 are small (0.10) the system has a narrow distribution and approaches the alternating sequence since $(N_1)_1$ is dominant in the distribution. In contrast, the sequence length distribution for $r_1 = 1/r_2 = 5$ is very broad, with sequences up to and beyond the hexad, $(N_1)_6$, for both of the feed compositions shown.

Carbon-13 NMR spectroscopy may be used for detailed studies of the microstructure to test the predictions of the sequence lengths from the simple copolymer equation. Refinements to the equation include the consideration of the effect of the penultimate group in the chain on the reactivity of the terminal radical with the monomers (Tirrell, 1989). These studies do not provide information regarding the molecule-to-molecule variation in composition that may occur, since again these sequence distributions are the average over the whole population.

Some typical values of reactivity ratios for monomers that will copolymerize via a free-radical route are shown in Table 1.15 (Tirrell, 1989, Allcock and Lampe, 1981). These have been selected in order to span the range of values shown in the preceding tables. It may be noted that the copolymerization of vinyl monomers with maleic anhydride will invariably lead to alternating copolymers since $r_1 \approx r_2 \approx 0$. The mechanism of alternating

Table 1.15. Selected values of free-radical copolymerization reactivity ratios, r_1 and r_2

M_1	M_2	r_1	r_2
Vinyl acetate	Vinyl chloride	0.24–0.98	1.03–2.3
Vinyl acetate	Acrylonitrile	0.04–0.07	4.05–6.0
Styrcnc	Acrylonitrile	0.29–0.55	0.0–0.17
Styrene methacrylate	Methyl	0.52–0.59	0.46–0.54
Styrene	Vinyl chloride	12.4–25	0.005–0.16
Methyl methacrylate	Vinyl chloride	8.99	0.07
Methyl methacrylate	Acrylonitrile	1.32	0.14
Maleic anhydride	Acrylonitrile	0.0	6.0
Maleic anhydride	Styrene	0.0–0.02	0.0–0.097

$$M_1 + M_2 \leftrightarrow [M_1^{+}M_2^{-}] \rightarrow \text{----}(M_1M_2)_n\text{-----}$$

Scheme 1.36. Formation of a donor–acceptor complex that then homopolymerizes as a route to formation of an alternating copolymer.

copolymerization has been studied in detail (Cowie, 1989b) and the best-characterized example, and the one that seems most applicable for the examples shown above, is through a 1 : 1 donor–acceptor complex. Thus styrene and acrylonitrile are electron donors, which, when combined with maleic anhydride as an electron acceptor, form a charge-transfer complex with a characteristic UV–visible absorption spectrum. It has been suggested (Cowie, 1989b) that this complex may then homopolymerize as shown in Scheme 1.36.

This mechanism accounts for several features of the polymerization of monomers that form alternating copolymers (Odian, 1991) but in particular the absence of any sequences of $-M_1M_1-$ or $-M_2M_2-$. If the mechanism of polymerization may be changed, then a pair of monomers that might otherwise form random copolymers may form alternating copolymers. Thus, from Table 1.14, methyl methacrylate and acrylonitrile should form a random copolymer with 25% diads (M_1M_1) and 14% triads ($M_1M_1M_1$). However, if the monomers are polymerized in the presence of a Lewis acid such as zinc chloride, then there is a tendency towards alternation, which suggests that complexation between the electron-donor monomer (methyl methacrylate) and the electron-acceptor monomer (acrylonitrile) is facilitated by the Lewis acid. This will often be restricted to a limited temperature and composition range, outside which the two mechanisms may occur simultaneously.

Block copolymers

Block copolymers were defined above, and it was noted that there was a tendency for blocks to form when both of the reactivity ratios, r_1 and r_2, were greater than unity. This approach results in a random copolymer with extended blocks of M_1 and M_2 rather than a pure block copolymer with well-defined block length and composition. More precise control can be achieved through **living polymerization**, since on exhaustion of the first monomer added, M_1, the reactive sites are still able to initiate polymerization. Thus addition of M_2 will allow continued chain growth to a DP given by the ratio $[M_2]$/[initiator]. These sequential methods may be anionic or living free-radical techniques such as ATRP and RAFT (Section 1.2.3).

The limitations to living block copolymerization arise when the anionic addition site is unable to initiate the polymerization of the second monomer. An example is the methacrylate anion formed with lithium, which is unable to initiate the polymerization of styrene. This contrasts with the reverse situation in which the polymerization of methyl methacrylate may be initiated by polystyryl lithium (Cowie, 1989a). Thus the synthesis of a PS–PMMA block copolymer by an anionic route requires *first* the polymerization of styrene and then addition of methyl methacrylate.

If both monomers have high electrophilicity there will be mutual initiation at the anionic sites. An example is butadiene and styrene, which form industrially important diblock and triblock polymers by sequential addition of monomer. It has been noted that, because of the difference between the reactivity ratios of styrene and butadiene in anionic polymerization, if a *mixture* of the two monomers is initiated, there is rapid reaction of butadiene so that it is almost totally polymerized before styrene adds to the chain. The chain sequence then consists of a copolymer of styrene and butadiene until the latter has been exhausted and then the styrene homopolymerizes, to give a styrene block. The resulting 'tapered' copolymer has blocks of polybutadiene and polystyrene separated by a segment with a progressive change in composition. In another example, the styrene–butadiene–styrene triblock copolymer, which has short segments of styrene and a long segment of butadiene, forms a thermoplastic elastomer. In this case the high-T_g styrene blocks act as physical crosslinks at ambient temperature, separating the flexible butadiene segment so elastomeric behaviour is observed. On heating above T_g of polystyrene, the entire system softens, elastomeric behaviour is lost and the polymer may be melt-processed. The importance of phase separation in this and related examples is discussed later.

Alternative approaches involve the reaction together of preformed blocks, such as in telechelic polymers and coupling reactions. The latter allow highly branched architectures such as star polymers to be formed by linking living anionic chains to a multifunctional core such as $SiCl_4$ (Cowie, 1989a).

Telechelic polymers are end-capped oligomers in which the capping group may react with other end groups to form a block copolymer. They have the general formula $Y–(M)_n–Z$ and are formed by the polymerization of the monomer, M, with a *telogen*, Y–Z, so that the telogen both reacts with the initiator and acts as a chain-transfer agent (Boutevin, 2000). The resulting oligomer has a low DP ($1 < n < 10$) and, if the end groups, are vinyl groups, is referred to as a macromonomer. These may function in crosslinking reactions, as well as undergoing polymerization.

Living free-radical polymerization methods provide a versatile approach to block copolymer formation such that polymers unachievable by other routes have been synthesized (Hawker *et al.*, 2001). Using *nitroxide-mediated polymerization*, it is possible to prepare a range of block copolymers, two examples of which are shown in Scheme 1.37. Here (A) shows an example of a sequential polymerization of two different monomers using the nitroxide I–N (as in living anionic polymerization discussed above) and (B) shows formation of a macro-initiator by using an alkoxy amine X–N to terminate the polymerization of a monomer that cannot react by a living free-radical route. This is then reacted by a living free-radical route using monomer M_2 and allows a functional group, X, to be introduced at the junction between the two blocks.

$$(A):\quad I\text{—}N + nM_1 \rightarrow I\text{—}(M_1)_n\text{—}N$$

$$+\, mM_2 \rightarrow I\text{—}(M_1)_n\text{—}(M_2)_m\text{—}N$$

$$(B):\quad R\cdot + nM_1 \rightarrow R\text{—}(M_1)_n\cdot$$

$$+\, X\text{—}N \rightarrow R\text{—}(M_1)_nX\text{—}N$$

$$mM_2 \rightarrow R\text{—}(M_1)_nX(M_2)_m\text{—}N$$

Scheme 1.37. Formation of block copolymers using nitroxide-mediated polymerization.

An alternative approach uses *reversible addition–fragmentation chain transfer (RAFT)*, which has fewer limitations in the selection of monomers (Chong *et al.*, 1999) so that monomers with hydroxyl, t-amino and acid functionality may be polymerized into narrow-polydispersity block copolymers. The RAFT agent is a dithioester of the general formula S=C(Z)S–R, where Z is usually –Ph (phenyl) and R is chosen from a group of alkyl phenyls (e.g. $–C(CH_3)_2Ph$).

The system operates in a sequential mode with the first polymer, $S{=}C(Z)S–(M_1)_n$, having the RAFT agent terminal to most of the polymer chains. This is then available to allow polymerization of the second monomer, M_2, provided that fresh initiator is provided. It is noted (Chong *et al.*, 1999) that in order to form a narrow-polydispersity block copolymer it is necessary to ensure that these chain ends with the transfer agent have a high transfer constant to the new propagating radical $M_2\cdot$. This means that, for example, in the preparation of a poly(styrene-block-methyl methacrylate) the MMA block should be prepared first since the styrene group is not a good leaving group from the RAFT agent. This is the **reverse** of the limitations noted for the anionic polymerization of the same block copolymer, for which it was noted above that the styrene block should be polymerized first because of the different reactivity of the anion.

It is also noted that it is possible to prepare ABA triblock and star copolymers by using RAFT agents with difunctionality or multifunctionality. Tapered composition in the blocks may be achieved by exploiting the differences in monomer reactivity ratios as described earlier for non-living free-radical polymerization. Again, as for nitroxide-mediated synthesis, blocks formed by non-radical routes may be modified by coupling of dithioester RAFT agent to the chain end. This may then be used in the living free-radical polymerization of the other block. Thus a polymer such as poly(ethylene oxide) that was not amenable to efficient block copolymerization with styrene by other living free-radical routes (e.g. nitroxide-mediated or ATRP) could be functionalized by coupling an acid-functionalized dithioester such as $S{=}C(Ph)SC(CH_3)(CN)CH_2CH_2COOH$ to the terminal –OH groups by using dicyclohexyl carbodiimide (DCC). As shown in Scheme 1.38, the resulting RAFT agent was then used in an AIBN-initiated styrene polymerization at 60 °C and produced a block copolymer with a polydispersity <1.10 and a minimal amount of homopolymer polystyrene (Chong *et al.*, 1999). These authors note that block copolymers of this type could not be prepared with good yield and low polydispersity by other living free-radical polymerization methods such as ATRP and nitroxide-mediated polymerization. The use of trithiocarbonates as RAFT agents for the preparation of ABA triblock copolymers in two steps has also been demonstrated (Mayadunne *et al.*, 2000).

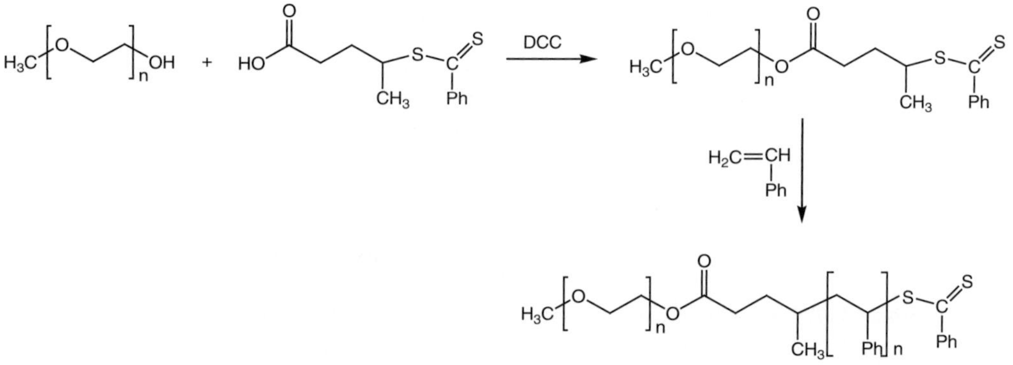

Scheme 1.38. Use of RAFT agent for (styrene–block PEO) copolymer formation starting with poly(ethylene oxide).

An effective, if uncontrolled, method for producing block copolymers during reactive processing involves the formation of mechanoradicals (see Section 1.4.1) from the mastication of a homopolymer and the addition of another monomer which is then initiated to homopolymerize at the radical site:

$$(\text{—M}_1\text{M}_1\text{—}) \xrightarrow{\textit{mastication}} \text{—M}_1\cdot + \cdot\text{M}_1\text{—} + \text{M}_2 \rightarrow \text{—M}_1\text{M}_2\text{M}_2\text{M}_2\text{M}_2\text{M}_2\text{—}$$

Incomplete conversion of all homopolymer means that the resulting polymer will be a blend of homopolymer of M_1 and block copolymer $(M_1)_n(M_2)_m$.

If two homopolymers are masticated together then terminal macro-radicals $(–M_1\cdot)$ and $(–M_2\cdot)$ are formed that can cross-terminate to give a block copolymer. In this case there will be a blend of two homopolymers as well, since formation of the cross-termination product will compete with recombination of the macro-radicals. There will also be incomplete chain scission of the homopolymers. This strategy has also been employed to produce compatibilization and enhanced interfacial adhesion of immiscible homopolymers since the block copolymer will be soluble in each phase and thus able to bridge the phase boundary. This and other topics concerning polymer blends are discussed in Section 1.3.

Graft copolymers

Graft copolymerization has some of the features of block-copolymer formation except that the radical centres are not at the end of the chain but *in-chain*. The polymer to which the graft is attached is the *backbone* polymer. Cowie (1989a) has recognized three different ways in which graft copolymers may be formed:

- grafting *from* the backbone, in which a backbone reactive site is created and this initiates polymerization, as described above;
- grafting *onto* the backbone, in which a separately initiated polymer chain reacts with a site on the backbone polymer and becomes grafted to it; and
- coupling of a macromonomer that has a known chain length to a single reactive group that can then react with the active sites on the backbone to create branches of precise length.

Of these, the first is the most common and the graft copolymer is formed on the backbone polymer by creating the in-chain radical by means of

- chain transfer during polymerization of the comonomer,
- hydrogen abstraction initiated by a free-radical initiator,
- ionizing radiation, or
- a redox reaction on a suitable functional group.

Ionic routes such as formation of the polymer anion by reaction with a strong base or the direct reaction of a polyamide with sodium are less likely to be used in reactive processing than is free-radical initiation. The process of 'self-graft' polymerization by chain transfer to polymer, when it occurs in a single monomer/polymer system during polymerization, is an example of chain branching that is discussed in the next section.

High-temperature grafting of polyolefins

In reactive processing, most graft reactions occur at high temperatures, such as under conditions of reactive extrusion. In this case the system consists of molten polymer, a free-radical initiator, a monomer and other additives to enhance grafting or prevent degradation during processing. In the following, just the chemistry of the radical-formation and polymerization reactions is considered, since the chemorheology of the reacting systems is discussed later. The substrates (backbone polymers) that have been studied the most are the polyolefins, and the temperatures are those which ensure that the polymers are molten. The processes occurring under the conditions of reactive processing may be very different from those occurring at lower temperatures and in solution (Moad, 1999, Russell, 2002). It is also necessary to separate out from the graft copolymerization reactions the degradation and crosslinking reactions that may occur in the polyolefin at the high temperatures and pressures. In many cases the reactions of polymerization, cross-linking and degradation may occur in competition, with the balance of these reactions being very dependent on the polyolefin being grafted (Section 1.4.1). There is the added complication that the stabilizers added to limit degradation will interfere with the grafting process (Section 1.4.3). The challenge in these systems is to achieve reproducibility and control of the grafting kinetics and of the chain-length distribution of the products. Most recently nitroxide-mediated graft polymerization has been applied in order to achieve a lower-polydispersity graft (Russell, 2002).

Formation of backbone radicals. The source of free radicals for the formation of the initiation site on the backbone is normally the decomposition of a dialkyl peroxide, since these systems have the longest half-lives at elevated temperatures. Radical formation by shear (mechanochemistry) results in chain scission and thus block-copolymer formation at the chain end, so it is not a widely used method for initiating graft copolymerization. The temperatures for a half-life of one hour for a range of free-radical initiators were given previously in Table 1.5. The choice of peroxide is governed by the nature of the decomposition products, as well as the half-life, and the widely used dicumyl peroxide produces acetophenone as a result of fragmentation (β-scission) of the alkoxy radical, liberating two methyl radicals (Scheme 1.39).

Either the original alkoxy radical or the methyl radical by-product may initiate the grafting reaction, although it is noted that the cumyloxy radical has a higher tendency than the methyl radical for hydrogen-atom abstraction from the polyolefin backbone, and this is enhanced at higher temperatures. This is, however, counteracted by the greater tendency

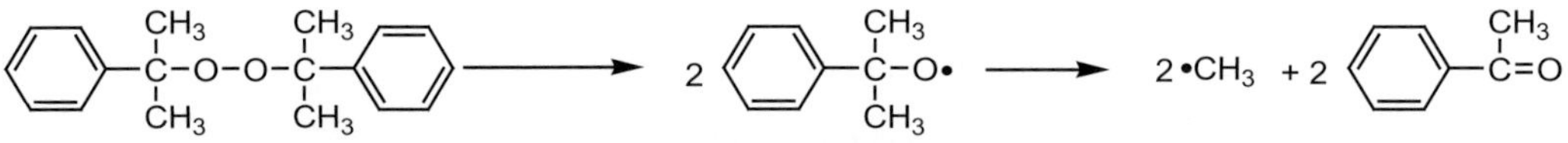

Scheme 1.39. Chemical structure and decomposition of DCP to form either cumyloxy radicals or methyl radicals and acetophenone.

of the alkoxy radical to undergo β-scission to form the methyl radical at higher temperature (Moad, 1999). Other reactions of the alkoxy radical that compete with backbone radical formation are addition to monomer and hydrogen-atom abstraction from monomer, but it has been noted that these are less important at elevated temperatures and at the low concentrations of monomer present than are the reactions with the substrate (Russell, 2002). Other factors noted in the choice of a free-radical initiator (Moad, 1999) are the following:

- the solubility in the polymer melt, and the partition coefficient in the different phases if it is a multicomponent blend;
- the physical form and vapour pressure at the reaction temperature;
- the dependence of the free-radical escape efficiency (f) on the viscosity; and
- the tendency towards (radical) induced decomposition and other side reactions that lower the radical yield.

The formation of the backbone radical site depends on the nature of the polyolefin. The presence of tertiary carbon atoms as in polypropylene and branch points in low-density polyethylene, LDPE, would be expected to provide the preferred site for hydrogen-atom abstraction. However, it has been noted (Russell, 2002) that the difference between the reactivity of tertiary and secondary carbon atoms decreases at elevated temperatures.

Addition of monomer to backbone radical. The addition of the monomer to the backbone radical R· may be considered as equivalent to the copolymerization of M_2 with the radical from monomer M_1. Thus, in the terminology of the copolymerization equation the reaction rate coefficient of importance is k_{12} and the estimates of the magnitude of this place it sufficiently high to ensure that the monomer will add to the radical site in polyethylene (Russell, 2002). The monomer will add preferentially to tertiary sites over secondary sites in ethylene copolymers with a high concentration of propylene, but in LDPE there seems to be little preference for branch points.

Propagation. The reaction of the radical $RM_2\cdot$ with further monomer M_2 will depend on the availability of monomer versus other competing termination or chain-transfer reactions. The kinetic chain length of the graft depends on the ratio of these rates, with the most significant process at the higher graft temperatures being chain transfer to another polymer backbone site. This may restrict the graft chain length to a few monomer repeat units before either intermolecular or intramolecular chain transfer occurs. The availability of intramolecular sites tends to favour these reactions unless bulky substituents inhibit the process. Another feature of a high-temperature reaction is the likely importance of depropagation limiting the graft length. Thus when the graft polymer is near the ceiling temperature, the equilibrium will favour monomer over polymer. This has an important application in the grafting of maleic anhydride (MA) to polyolefins, which is an industrially important reactive processing operation. The ceiling

R· + R· → R-R

RM_2· + R· → RM_2R

RM_2· + RM_2· → RM_2M_2R

Scheme 1.40. Various possible crosslinking reactions during grafting.

temperature of the polymer is 100 °C so the graft reaction will lead to molecular addition and there is no homopolymer formed. This is an advantage in ensuring that there are no homopolymer side reactions, which can be significant in grafting styrene for example (Russell, 2002).

Termination. The presence of small reactive radical species such as the methyl radical produced by fragmentation of the DCP initiator means that there may be a range of termination routes available. The bimolecular termination of graft radicals on different chains may result in a crosslink unless they disproportionate. The termination of the primary backbone radicals, R·, will be competitive with monomer addition at low concentrations of M_2 and again produce a crosslink. The possible competing crosslinking reactions are shown in Scheme 1.40.

Macromolecular chain branching

The chain transfer to polymer during free-radical polymerization was noted (Section 1.2.3) as resulting in the formation of branches and broadening of the molar-mass distribution. It does not result in a change in the degree of polymerization (DP), but the resulting polymer will have very different rheological and solid-state properties since the branches will affect the chain entanglements and also the ability of the polymer to crystallize. In the polymerization of most monomers, it will be more important at high extents of conversion, and the measurement of the chain-transfer constant, C_{pol} for this process is difficult (Odian, 1991). Estimated values for styrene polymerization are $C_{pol} \approx 10^{-4}$ and that about one chain in ten is branched when conversion reaches 80%.

For certain monomers such as vinyl acetate and ethylene, branching is much more significant. The free-radical (high-pressure) polymerization of low-density polyethylene (LDPE) includes a back-biting internal chain-transfer reaction that results in the formation of a short branch. It is this branching that results in an upper limit for the crystallinity of LDPE of about 60%–70% and a melt temperature of 110 °C; the back-biting reaction preferentially occurs with the formation of an intramolecular six-membered ring that results in preferential formation of a C_4 short-chain branch as shown in Scheme 1.41.

In addition, short-chain branches of C_5 and C_7 are seen at levels of ~1 per 1000 carbon atoms, compared with a maximum of ~15 n-butyl branches. This also arises from a back-biting reaction of the propagating radical and the resultant intramolecular chain transfer, and the relative amounts of the branches of various lengths may vary depending on the conditions of synthesis.

The use of Ziegler–Natta catalysts for the low-pressure polymerization of high-density polyethylene (HDPE) eliminated branching and produced a higher crystallinity and melting point (135 °C). However, the same co-ordination polymerization may be used to produce linear low-density polyethylene (LLDPE) by copolymerizing ethylene with an α-olefin such as 1-butene or 1-hexene to produce a polymer with the equivalent branching of LDPE but through a more efficient polymerization process than the high-pressure free-radical route.

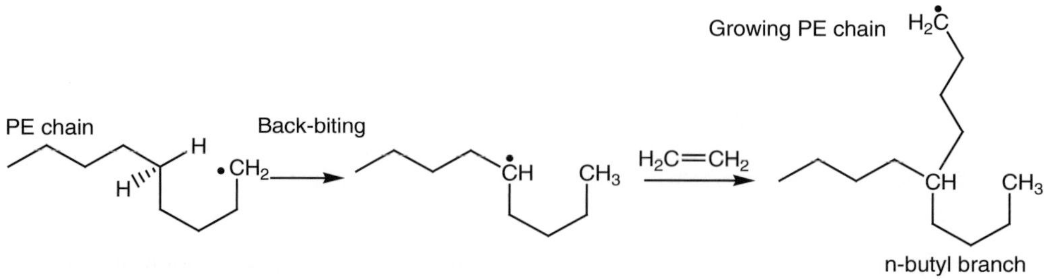

Scheme 1.41. Back-biting during high-pressure polymerization of ethylene to LDPE results in n-butyl branch formation.

Short-chain branching also is important in polyvinyl chloride, in which the dominant branch is the chloromethyl group $-CH_2Cl$ (~4–6 per 1000 repeat units), while other branches seen are 2-chloroethyl ($-CH_2CH_2Cl$) and 2,4-dichloro-n-butyl ($-CH_2CHClCH_2CH_2Cl$) at ~1 per 1000 monomer units.

In addition to short branches from back-biting reactions, long branches are seen to be formed both in LDPE and in PVC from intramolecular chain transfer to polymer. These are much less frequent (~1 per 2000 repeat units) but have a great effect on the viscosity of the melt and thus the processing. This is discussed in more detail in later chapters.

Hyperbranched polymers and dendrimers

It was noted earlier that dendrimer and hyperbranched polymers were formed by very carefully controlled step polymerization by either a divergent or a convergent synthetic strategy. The formation of star and hyperbranched polymers by an addition-polymerization route requires careful control of the chain length of polymerization. It was noted that it was possible to form a star polymer by anionic living polymerization by reacting the chain with a tetrafunctional core. The living free-radical routes provide another approach, and some examples are given below.

Hawker has discussed the strategies involving nitroxide-terminated multifunctional initiators that may be either preformed or produced in the reaction (Hawker *et al.*, 2001). In the latter case, Frechet *et al.* have shown that a monomer that contains a polymerizable styryl group (A) and a nitroxide (B) that controls the polymerization will produce an AB_2 monomer analogous to that used in condensation reactions (Hawker *et al.*, 1995). The subsequent reaction is one of self-condensing vinyl polymerization with the production of a hyperbranched system (Voit, 2000). The advantage of living free-radical polymerization is the low concentration of radicals present at any point in time so that radical-coupling reactions are infrequent and crosslinking thus negligible (at least up to a branch length of about 20 repeat units). It was noted that nitroxide-mediated synthesis is not as versatile as ATRP, in which the radical population may be controlled by the addition of catalyst (Hawker *et al.*, 2001).

Another pathway to highly branched systems is through RAFT by using precursors with multiple-fragmentation chain-transfer agents (e.g. thiocarbonyl thio groups). Thus it has been shown that the agents shown in Scheme 1.42) are able to produce four- and five-armed polymers (Chong *et al.*, 1999), but these are star systems rather than hyperbranches.

When styrene free-radical polymerization was carried out in the presence of a polymerizable dithioester, benzyl 4-vinyldithiobenzoate (Wang *et al.*, 2003), branched polystyrene

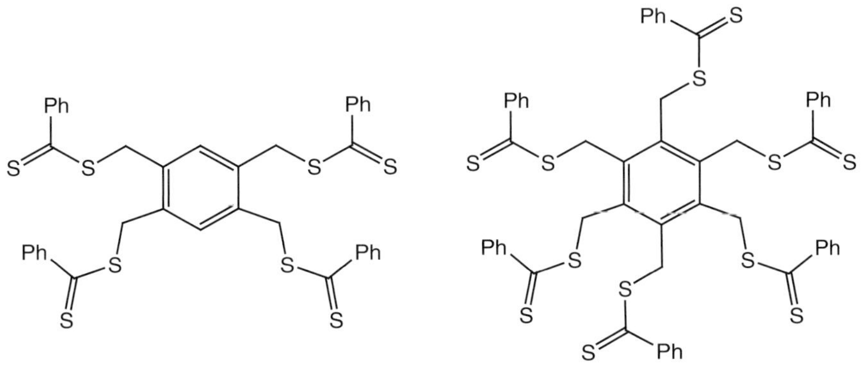

Scheme 1.42. Multi-arm RAFT agents for the formation of star polymers. After Chong (1999).

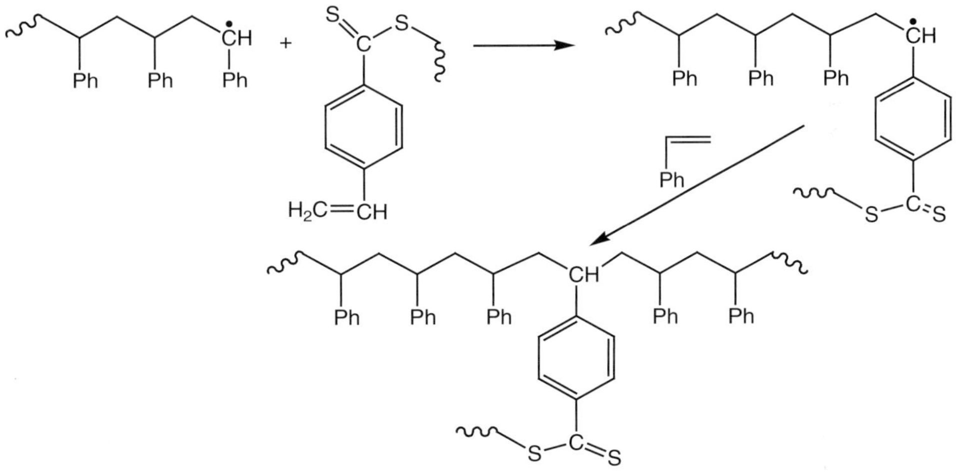

Scheme 1.43. Branching from RAFT polymerization with a polymerizable thio-ester. After Wang (2003).

that contained a dithiobenzoate C(=S)S moiety at each branch point was obtained (Scheme 1.43). After cleavage of the branches with amine, the molar mass of the branches was narrow and increased linearly with monomer conversion, which is consistent with living free-radical polymerization.

It has been noted (Guan *et al.*, 1999) that polyethylene is not just branched (as in linear low-density polyethylene, where branching is controlled by the copolymerization of 1-hexene, or in low-density polyethylene, where it is uncontrolled due to back-biting) but may, under certain circumstances, also be hyperbranched. The mechanism for hyper-branching is to create the branch point by controlled isomerization of the active site by an appropriate choice of both co-ordination catalyst and pressure.

1.2.5 Networks from addition polymerization

The development of polymers with the required modulus and durability for application as composite materials, coatings, elastomers etc. has required the formation of a three-dimensional network by using multifunctional monomers. These systems require an understanding of the chemorheology of the system as reaction takes place in order to achieve optimum properties for the particular application. The resins for many critical applications such as aerospace composites are formed by stepwise reactions of thermosetting polymers such as epoxy resins; however, there are many free-radical and ionic addition reactions that achieve high crosslink densities in short times. These exploit the high reactivity of addition polymerization compared with step polymerization.

The detailed physical process of an increase in viscosity followed by gelation and vitrification of network polymers that results from the following crosslinking chemistry is discussed in Section 2.4.

Difunctional crosslinkers

The most widely studied addition-polymerization reaction for crosslinking is the free-radical polymerization of difunctional monomers (such as divinyl benzene and ethylene glycol dimethacrylate) in the presence of the corresponding monofunctional monomer. This may be thermally or photochemically initiated, and the latter application is widely used for coatings and dental composites. This is shown in Figure 1.26.

The polymerization reaction results in the formation of an infinite network so the system gels and then with further reaction vitrifies as the glass-transition temperature, T_g, of the network reaches the reaction temperature. The polymerizing system, which had become diffusion-controlled between gelation and vitrification, then ceases chemical reaction until it is heated further to $T > T_g$. The rates of gelation and vitrification depend both on the reactivity and on the concentration of the difunctional monomer.

The system may be modelled initially by considering a system in which all the functional groups are the same type chemically and so have the same reactivity. Some typical systems would be monomer styrene ($PhCH{=}CH_2$) and crosslinker divinyl benzene ($CH_2{=}CH{-}Ph{-}CH{=}CH_2$), and monomer methyl methacrylate ($CH_2{=}CMeCOOMe$) and crosslinker ethylene glycol dimethacrylate ($CH_2{=}CMeCOOCH_2CH_2OOCMeC{=}CH_2$).

Thus, labelling the monomer M_1 and the difunctional crosslinker M–M and taking the reactivity ratios $r_1 = r_2$, the number of crosslinks is the number of molecules of M–M in which **both** double bonds have reacted. The difunctional monomer may also homopolymerize as well as participate in crosslinking, i.e. only one of the M groups reacts. The respective concentrations of reactive groups are $[M_1]$ and $[M_2] = 2[M\text{–}M]$.

To determine the extent of reaction of the monomers at gelation, p_c, it is necessary to determine the critical concentrations of M_1 and M_2 when the number of crosslinks per chain is 0.5. If the polymer had not contained the difunctional reagent, it would have attained a weight-average degree of polymerization DP_w under the conditions of polymerization. The gelation condition is then

$$p_c = ([M_1] + [M_2])/([M_2]\,DP_w). \tag{1.89}$$

Even for very low concentrations of crosslinker $[M\text{–}M] \approx 5\times10^{-3}$ mol/dm^3 the gel point occurs at extents of conversion, p_c, of 12%–20%. The process of network formation is shown in Figure 1.26.

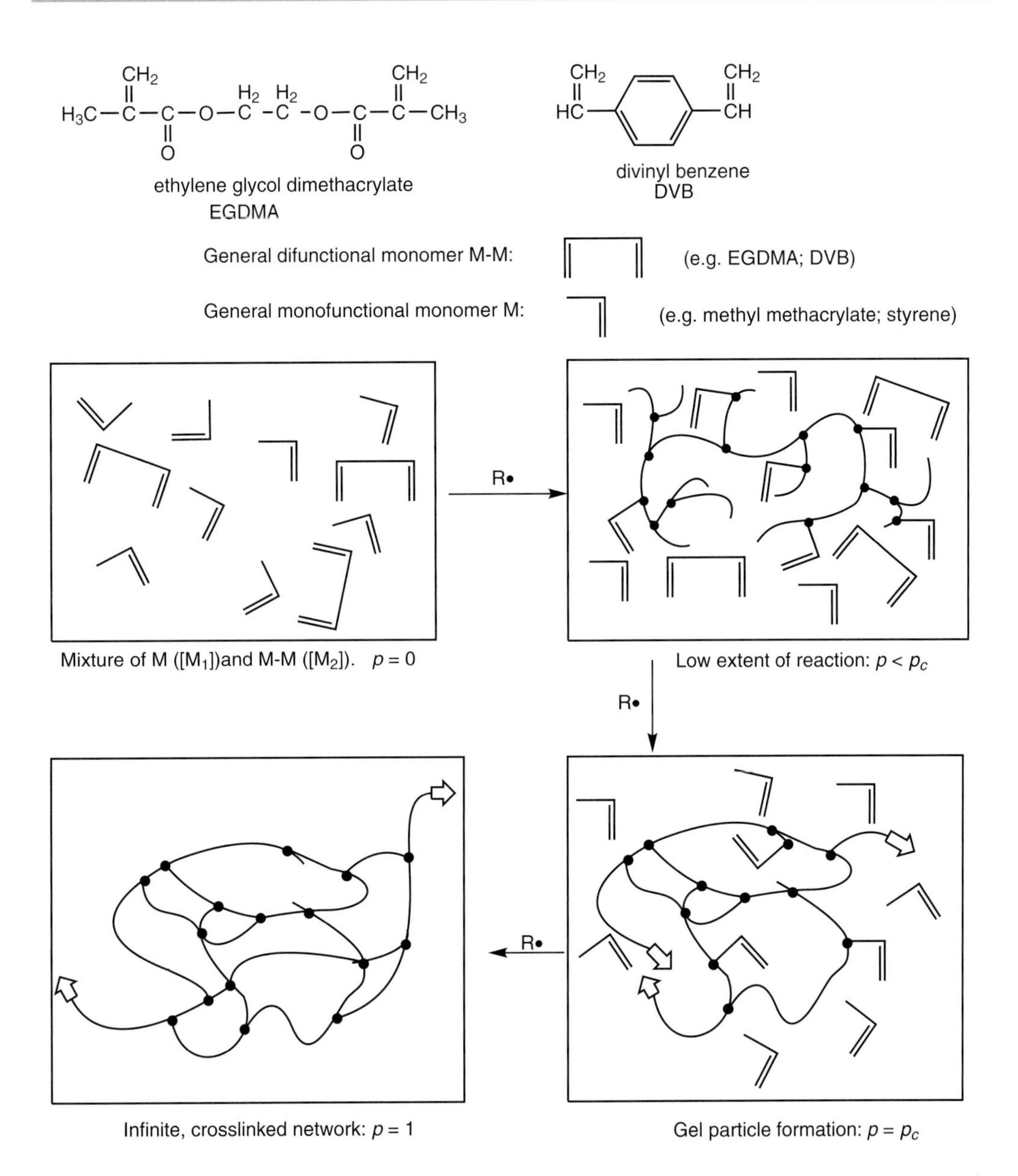

Figure 1.26. A schematic diagram for the formation of a crosslinked network by the polymerization of a monofunctional monomer, e.g. methyl methacrylate (MMA), in the presence of a difunctional crosslinker, such as ethylene glycol dimethacrylate (EGDMA). Gelation occurs at low conversion ($p_c \sim 12\%–20\%$). Adapted from Pascualt *et al.* (2002).

Other cases with different reactivity ratios as well as other functional group reactivities have been considered (Odian, 1991).

Unsaturated polyester resins

The styrene crosslinking of unsaturated polyester resins is perhaps one of the most widely used reactions to form the matrix for fibreglass-reinforced composite materials. (The

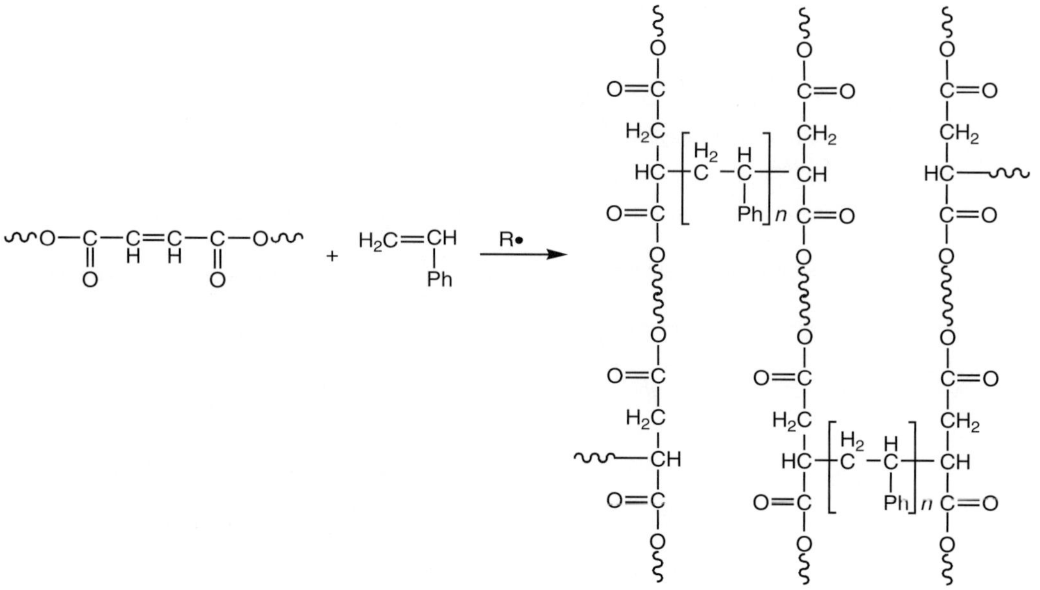

Scheme 1.44. Crosslinking reaction of styrene with fumarate group in an unsaturated polyester resin. After Bracco *et al.* (2005).

synthesis of these resins was shown earlier in Scheme 1.7 of Section 1.2.2.) The crosslinking reaction differs from the system that uses a polyfunctional reagent (e.g. DVB in Figure 1.26) in that the crosslinker is a monofunctional monomer, styrene, that copolymerizes with the in-chain fumarate groups that were previously step-copolymerized in a polyester oligomer, as shown in Scheme 1.44.

The properties depend on the crosslink density, the length of the crosslink as well as the molecular structure of the crosslinker. When styrene is used, the degree of polymerization, n, at a crosslink is low (about 1–2), so the network is stiff. The low DP reflects the low value of the reactivity ratio of styrene with the fumarate group $r_1 \approx r_2 \approx 0$, so the system tends to alternation and thus can form only a short crosslink (Odian, 1991). The final modulus also depends on the number of fumarate groups along the polyester backbone between the saturated aromatic acid groups, since this controls the crosslink density. It is noted (Odian, 1991) that, if methyl methacrylate is used as the crosslinking monomer, then longer chains are formed, so the network has a lower modulus.

Kinetics and heterogeneity of free-radical-crosslinked networks

The formation of networks by addition polymerization of multifunctional monomers as minor components included with the monofunctional vinyl or acrylic monomer is industrially important in applications as diverse as dental composites and UV-cured metal coatings. The chemorheology of these systems is therefore of industrial importance and the differences between these and the step-growth networks such as amine-cured epoxy resins (Section 1.2.2) need to be understood. One of the major differences recognized has been that addition polymerization results in the formation of 'microgel' at very low extents of conversion ($<10\%$) compared with stepwise polymerization of epoxy resins, for which the gel point occurs at a high extent of conversion (e.g. 60%) that is consistent with the

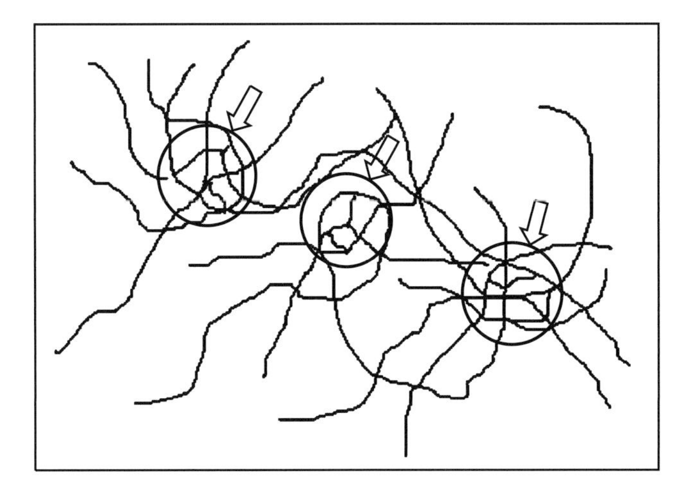

Figure 1.27. A schematic diagram of a network crosslinked by addition polymerization showing regions of high crosslink density (microgel particles, arrowed) in regions of resin of lower crosslink density. Chains from the latter region that connect gel particles are also shown. Adapted from Pascualt *et al.* (2002).

development of a more homogeneous network (Pascault *et al.*, 2002). This is shown schematically in Figure 1.27.

The process of macroscopic gelation that is sensed as a change in the rheological properties (a rapid increase in the steady-state shear viscosity) then involves the aggregation and crosslinking between these preformed gel particles to form an infinite network.

Crosslinked polyethylene

The crosslinking of polyethylene is of great practical importance in order to increase the softening temperature and maintain dimensional stability at elevated temperature. There are various ways of achieving this through using chemical initiators as well as radiation to create macro-radicals that may recombine intermolecularly or undergo reaction with end groups in the polyolefin to create a crosslink. Scheme 1.45 shows a sequence of radical reactions that lead to 'H'-type crosslinks in polyethylene. However, the available free volume for such a reaction to take place is limited and the actual crosslink need not be of 'H'-type but may instead involve reaction of functional groups such as vinyl at chain ends. This produces a 'Y'-type crosslink as proposed for γ-radiation-initiated crosslinking in UHMWPE (Bracco *et al.*, 2005), as shown in the third reaction of Scheme 1.45.

An alternative approach is to graft the polyethylene with a reactive monomer that may then undergo a separate reaction to crosslink the chains. An example is the hydrolytic crosslinking of triethoxy silane groups that had previously been grafted onto PE during reactive processing. Scheme 1.46(a) shows the free-radical-initiated grafting reaction of vinyl trimethoxy silane (VTMS) which is performed in the extruder, for example in the coating of a metal cable. This is followed by the slow reaction with water, Scheme 1.46(b), in a separate step that is diffusion-controlled because reaction is occurring in the rubbery state (unlike the grafting reaction in the molten polymer). The condensation reaction to give the crosslinks may be catalysed by an organotin compound such as dibutyltin dilaurate.

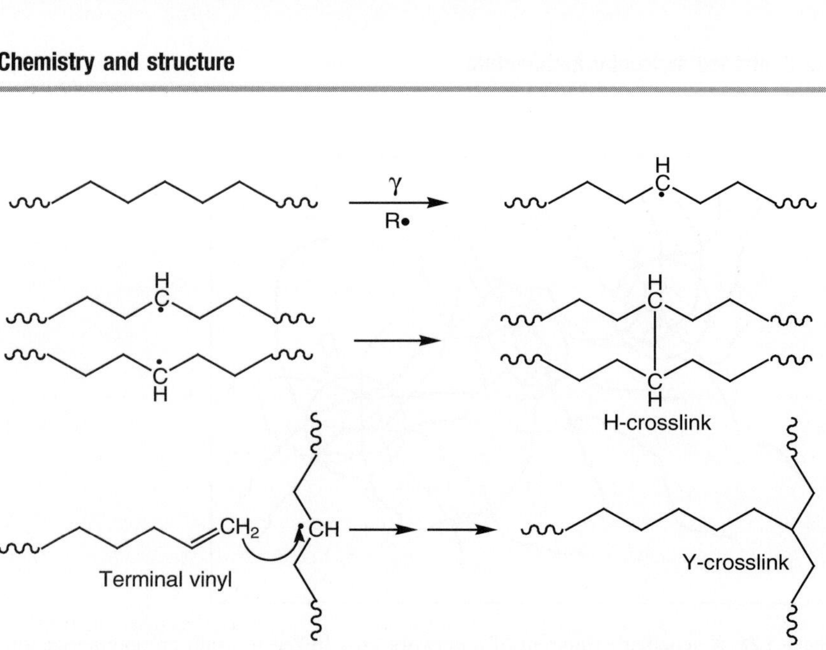

Scheme 1.45. The reaction sequence for radical crosslinking of polyethylene.

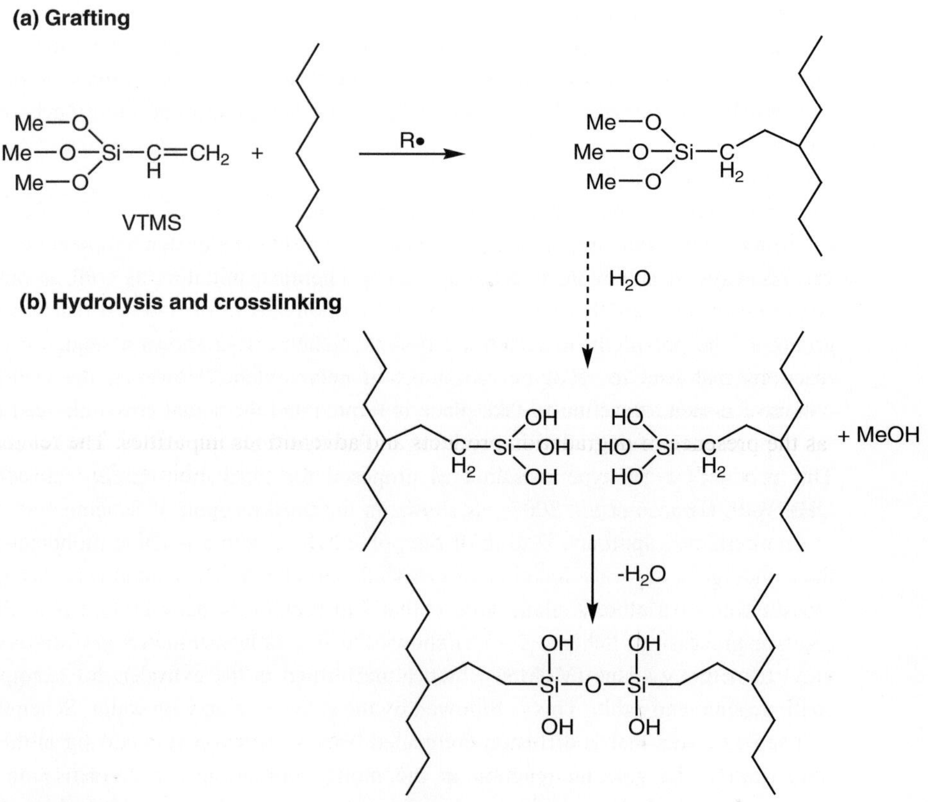

Scheme 1.46. (a) VTMS grafting and (b) hydrolytic crosslinking.

In this case the final crosslinking reaction is not an addition polymerization reaction but the hydrolysis and condensation reaction of alkoxy silanes incorporated through a separate step of addition graft copolymerization.

1.3 Polymer blends and composites

Reactive processing is often employed in order to achieve the formation of a blend between two (or more) different polymers that are fed into a twin-screw extruder or similar reactor. These polymers may be totally immiscible and then the question of how a stable and useful polymer with good properties and performance may be achieved arises. If the polymer systems are totally immiscible, then, depending on the relative proportions of the components there will be a hetero-phasic system with phase-separated material dispersed in a continuous phase of the majority component. The formation of these blends often occurs after processing at a temperature at which the component polymers may in fact be miscible. Phase separation then occurs on cooling after processing. The problem of polymer miscibility and phase separation is a fundamental one in physical polymer science and will be addressed here at a level necessary for employing the underlying principles in the later discussion of reactive processing. The reactions that occur when processing the two polymers may result in grafting of the polymer chains, so the net result may be a material with different local composition and properties from those of material formed by casting the blend from solution, for example. Not only immiscible blends exhibit phase separation. Block copolymers may contain sequences of immiscible polymers that will be miscible during melt processing or in a solvent at elevated temperatures and undergo phase separation on cooling. Examples include thermoplastic elastomers and tapered block copolymers used for viscosity modification in lubricants (Sperling, 2001).

In some cases, the material being processed may contain one or more components that maintain their physical state while the other component polymer melts and reacts. The best-known examples occur when one of the components is reinforcing, such as short ceramic or glass fibres, and the end result is a fibre-reinforced composite. In this case, particular attention is paid to the interface between the polymer and the fibre. The properties of the composite depend on the interfacial adhesion between the fibre and the polymer matrix, and this can be greatly affected by the conditions of processing as well as the presence of degradation products and adventitious impurities. The related issue in polymer–polymer blends occurs with the use of compatibilizers for enhancing interphase adhesion. Again reactive processing is a route for optimizing this. In this case the rheological properties will be affected by the filler, but the changes in the chemistry of the system with time should be very similar to those for the matrix alone. The only difference is in the region of the filler or fibre, where there will be a zone that extends from the surface of the fibre into the matrix for a distance of perhaps a few micrometres where the composition and thus the reactivity may differ from those of the bulk. This will depend on the type of surface treatment given to the fibre. In the case of glass fibres this will involve a coupling agent designed to react both with the fibre and with the matrix. The role that interfacial adhesion, wetting and interfacial energy play in determining the performance is considered in more detail when discussing composite materials.

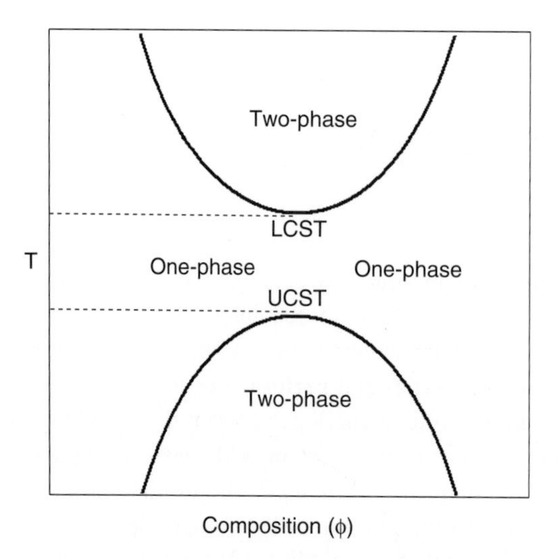

Figure 1.28. A phase diagram for a (hypothetical) polymer–solvent or polymer–polymer system showing both lower (LCST) and upper (UCST) critical solution temperatures. The boundary may be determined by locating the cloud point as a function of temperature for a fixed composition as the system moves from being a single-phase system to being a two-phase system and vice versa.

1.3.1 Miscibility of polymers

The understanding of the formation of miscible and immiscible polymer blends requires the application of the principles of phase chemistry. A miscible blend may be regarded as a solution of one polymer in the other. The thermodynamic criteria for the miscibility of liquids are well known and may be applied to polymers as a first approximation. The added complexity comes from the long-chain nature of polymers. In addition to the entropic factors there are kinetic factors to be considered. Since in reactive processing the reactions are occurring within a short time, they will very often be a long way from equilibrium.

Liquid–liquid, liquid–polymer and polymer–polymer miscibility

In a 'miscible' system of two liquids it is not expected that they will remain as a single phase over all ranges of composition and temperature. Instead there will be cloud points where two phases occur and the composition of the separated material will be close to that of the original pure components. Even in so-called two-phase systems there will be a partitioning of one component in the other corresponding to a small solubility at the particular temperature. An example of a phase diagram that may occur for two liquids, a polymer solution or a polymer blend is shown in Figure 1.28.

This figure clearly shows the temperature and composition windows where it is either a two-phase system or a single-phase system. The characteristic features of an upper critical solution temperature (UCST) and a lower critical solution temperature (LCST) corresponding to the phase transition are identified. For a particular composition of two immiscible polymers, if the temperature is increased, the UCST is the highest temperature at which two phases may co-exist in the blend. There is then a window of miscibility as the temperature is increased further, followed by phase separation again at the LCST. This type of diagram is often seen for polymer solutions, e.g. polystyrene in cyclohexane. Often polymer blends show

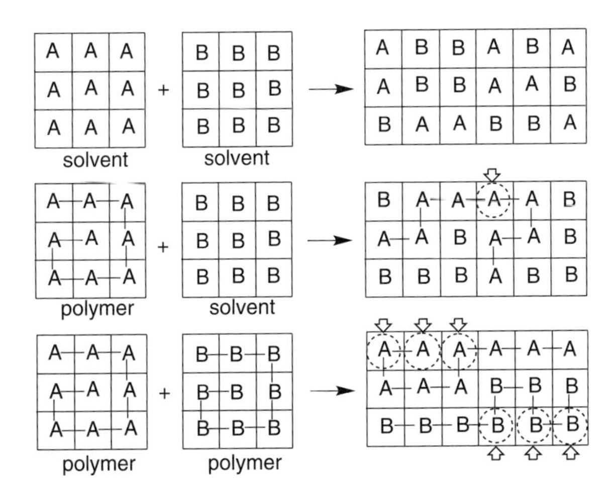

Figure 1.29. Illustration of the difference in mixing of two solvents, A and B. Mixing and solvation of a polymer, $[A\text{–}A]_n$ by a low-molar-mass solvent, B, and mixing of two polymers $[A\text{–}A]_n$ and $[B\text{–}B]_n$. The segments that are not solvated are highlighted. Adapted from Olabesi *et al.* (1979).

either a UCST or an LCST, and it has been noted (Sperling, 2001) that, because of the effect of polymer chain length on the entropy of mixing of the components, a UCST is seen with shorter chains and an LCST with higher-molar-mass polymer blends.

It is noted that ideal solutions occur rarely even for two liquids, since this requires that there be no volume change (and thus no enthalpy change) on mixing. The dissolution process arises solely due to the entropy change arising from the increased degree of freedom on mixing. A schematic way of comparing the entropy change on mixing two liquid solvents, a polymer and a solvent, and two polymers is shown in Figure 1.29 (Olabisi *et al.*, 1979). This firstly shows the situation when the segments A and B are uncorrelated, i.e. low molar mass. The unrestricted mixing process of the liquids ensures that every A segment has a B segment alongside it, but this becomes progressively more difficult as firstly the A and then the B segments are covalently linked in the polymer A + solvent B (second case) and the polymer A + polymer B (third case) systems as depicted. The segments that are isolated from the other component are circled and clearly are greatest in the polymer–polymer example.

This combinatorial entropy of mixing for low-molar-mass species is given by

$$\Delta S^{\mathrm{M}} = -k_{\mathrm{B}}(N_1 + N_2)(x_1 \ln x_1 + x_2 \ln x_2), \tag{1.90}$$

where k_{B} is the Boltzmann constant, N_i is the number of molecules of component i and x_i is the mole fraction.

In a real solution, the interaction between the mixing species is considered (i.e. $\Delta H^{\mathrm{M}} \neq 0$) but the entropy terms are unchanged. Hydrogen bonding and dipole interactions are an important source of the energy of interaction in polymer blends. The free energy of mixing ΔG^{M} is then the ideal value of $-T\Delta S^{\mathrm{M}}$ plus the enthalpy term.

Progressing to the polymer–solvent system (Figure 1.29), the ideal entropy term above must be replaced by a relationship that recognizes the features of interconnection of segments in the polymer chain. This is the Flory–Huggins relation:

$$\Delta S^{\mathrm{M}} = -k_{\mathrm{B}}(N_1 \ln \varphi_1 + N_2 \ln \varphi_2), \tag{1.91}$$

where the mole fraction x_i has been replaced by the volume fraction φ_i. Here $i=1$ for the solvent and 2 for the polymer.

The enthalpy of mixing ΔH^{M} is also calculated for the solvent–polymer interactions taking into account the occupancy of the lattice sites by the polymer. If w is the energy of interaction between solvent and polymer and z is a co-ordination number, then (Olabisi *et al.*, 1979) the free energy per unit volume is

$$\Delta G^{\mathrm{M}}/V = k_{\mathrm{B}}T[(\varphi_1/V_1)\ln\varphi_1 + (\varphi_2/V_2)\ln\varphi_2 + \chi(\varphi_1\varphi_2)/V_1], \tag{1.92}$$

where χ is the Flory–Huggins interaction parameter $zw/(k_{\mathrm{B}}T)$.

The qualitative extension of this concept from polymer–solvent to polymer–polymer miscibility is straightforward. It may be noted that for a high polymer there are few molecules per unit volume, so ΔS^{M} becomes small and makes a negligible contribution to the free energy of mixing. Thus polymer miscibility depends on the forces of interaction between the polymer segments, which in turn depend on the extent of organization that the chains may undergo.

Forces of interaction between polymer chain segments

The forces that contribute to the enthalpy of mixing and thus the value of the Flory–Huggins parameter χ are generally of short range. These have been considered in turn in terms of their relative contribution to the total energy of interaction (Olabisi *et al.*, 1979).

London or dispersion forces. These decrease with distance as r^{-6} and depend on the polarizability of the functional group in the segment (e.g. aromatic rings and unsaturation).

Dipole–dipole forces. These may be from either two permanent dipoles or one induced and one permanent dipole. The halogen-containing polymers such as PVC have large dipoles, which make a significant contribution to the interaction energy.

Ion–dipole interactions. These arise when there are ionic species on the polymer chain and are much longer-range forces than the London and dipole interactions (which are typically close-range, Van der Waals interactions). The force–distance relationship is comparable to that for electrostatic forces. However, the presence of counter-ions in many systems means that the forces may be shielded.

Hydrogen bonding. These interactions, which are typically between a non-bonding electron on an oxygen atom (e.g. in C=O) and a hydrogen atom on a segment with an electron-withdrawing group (e.g. H–N–) result in a high local force that approaches 50 kJ/mol. While this is still less than that of a conventional chemical bond, it is much greater than any of those above. In many miscible polymer blends it is the presence of hydrogen bonds that provides the high enthalpy of mixing that is necessary to counter the positive entropy terms in the free-energy equation. Examples include soft/hard segmented polyurethanes and PVC–polycaprolactone.

Acid–base interactions. These are not as well characterized as hydrogen-bonding interactions, but examples where these forces should be important include the blending of polymers that have pendant groups that are of acid and base character, e.g. salts of sulfonic acid with salts of tertiary amines (Olabisi *et al.*, 1979).

Charge-transfer or donor–acceptor interactions. The development of polymers for non-linear optics has been concerned with substituents such as nitro groups that are electron acceptors. When these are blended with polymers containing

Table 1.16. Solubility parameters for common polymers

Polymer	Solubility parameter, δ (MPa)$^{1/2}$
Poly(ethylene)	16.2
Poly(vinyl chloride)	19.9
Poly(ethylene terephthalate)	21.9
Poly(vinylidene chloride)	25.1
Poly(hexamethylene adipamide) (nylon-6,6)	28.0

electron-donor groups charge transfer can occur. This is often seen as the formation of a coloured complex, indicating the formation of a new absorption band due to the new species.

The extent to which these polymer–solvent interactions may control the miscibility of a system has been described for some time in terms of the **solubility parameter**, δ. In simple liquids, this has a direct link to the cohesive energy density ($\Delta E/V$) of the material:

$$\delta = (\Delta E/V)^{1/2}, \tag{1.93}$$

where ΔE is the molar energy of vaporization of the solvent and is thus a measure of the forces of interaction between the molecules. The temperature dependence of the vapour pressure enables this to be determined for a low-molar-mass solvent. The units of the solubility parameter will be $(\mathrm{J/cm^3})^{1/2}$ or $\mathrm{MPa}^{1/2}$.

In order to extend the concept to the polymer component, an indirect method of measurement is required. This is done by dissolving the polymer in a range of solvents and measuring the viscosity. Since the maximum expansion of the chain on dissolution will occur with a solvent having the same value of δ as the polymer, and thus give the highest viscosity for a particular concentration, this is the value taken for the solubility parameter of the polymer. This approach may also be applied to the segments of a crosslinked polymer that may be swollen by the solvent. Some values for polymers determined in this way are given in Table 1.16. In order to gain an understanding of the different forces that make up the cohesive energy density, δ^2, the separate components may be considered:

$$\delta^2_{\text{total}} = \delta^2_{\text{non-polar}} + \delta^2_{\text{polar}} + \delta^2_{\text{hydrogen bonding}}. \tag{1.94}$$

Considering Table 1.16, only the first polymer, polyethylene, has non-polar contributions alone; the next three have also polar components and the last, nylon-6,6, has contributions from all three forces. The largest solubility parameter for this polymer also corresponds to the highest melting point and stiffness, reflecting the importance of cohesive energy density as a measure of intermolecular forces.

Solubility parameters may also be calculated by a molar-additivity approach if the chemical composition is known. Tables of group molar attraction constants, G, obtained from molar heats of vaporization (Sperling, 2001) may be used. The solubility parameter, δ, is then given by summing these for all species ($\sum G$) in the monomer repeat unit of molar mass M and density ρ:

$$\delta = \rho\left(\sum G\right)/M. \tag{1.95}$$

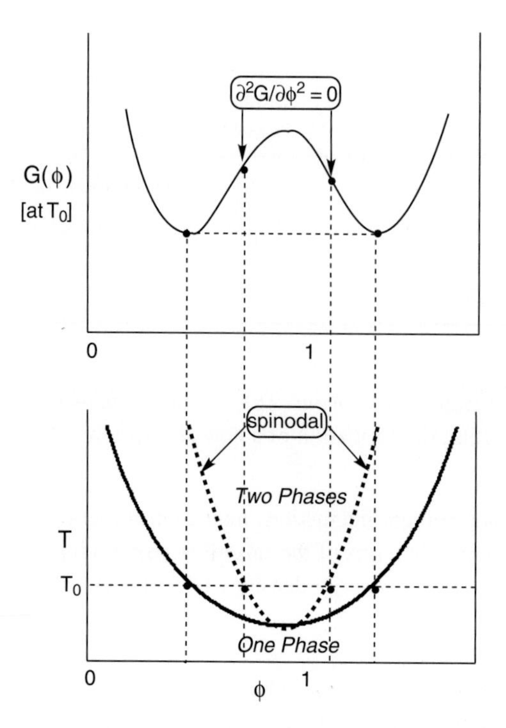

Figure 1.30. A plot of the change in Gibbs free energy, G, as a function of composition, φ, for a system that a LCST that shows spinodal as well as binodal decomposition on phase separation. Adapted from Kwei and Wang (1978).

The Flory–Huggins parameter, χ although it is a dimensionless parameter, is related to the heat of mixing, ΔH^{M}, and may be determined from the solubility parameters δ_1 and δ_2 of the two components (e.g. a solvent and a polymer):

$$\chi = \beta_1 + V_1/[RT(\delta_1 - \delta_2)^2], \tag{1.96}$$

where β_1 is the lattice constant of entropic origin and is ≈ 0, V_1 is the volume fraction of the solvent and R is the gas constant. Typical values of χ for non-polar polymer–solvent combinations at low concentration of polymer are within the range 0.3–0.4 (Sperling, 2001). If $\chi = 0.5$ the solvent is a θ-solvent at that temperature.

While the concept of solubility parameter has a physical basis, it is essentially an empirical parameter that may be related to the Flory–Huggins parameter. As such it has value in understanding polymer blends in that the conditions for phase separation may be rationalized.

More detailed understanding of the process of mixing of polymer chains has been developed from theories of equations of state. While these are founded in the equations of state for gases (starting with the fundamental equation $pV = nRT$ and then applying corrections for molecular volume and forces of interaction as discussed earlier in this section), the major difference is in the inclusion of free volume in the system. Detailed discussions of this approach may be found in Olabisi *et al.* (1979) and Sperling (2001).

1.3.2 Phase-separation phenomena

The separation of two, initially miscible, polymers will occur under conditions of temperature and composition given in a phase diagram such as Figure 1.30. It may be recognized that the process of passing from one phase to another when the temperature change moves the system across a phase boundary cannot occur instantaneously. The most common example is the growth of the solid phase from the liquid. In this case **nucleation** of the crystalline state is followed by growth of the phase in three dimensions. The formation of spherical domains provides the minimization of the surface energy in the new system. In liquid–liquid transitions the process of **spinodal decomposition** leads to phase separation. This is also seen in polymer blends.

Figure 1.30 shows a phase diagram for such a system that has a lower critical solution temperature; the change in free energy as a function of composition is also shown (Kwei and Wang, 1978). This allows the condition for spinodal decomposition to be identified as the points of inflection on the free energy (G)–composition (φ) diagram. The condition for this inflection is $\partial^2 G/\partial\varphi^2 = 0$. Between these spinodal points the value of the second derivative is negative and the system is unstable. In this region, small fluctuations in concentration lead to phase separation by a spontaneous and continuous process. The line corresponding to $\partial^2 G/\partial\varphi^2 = 0$ may be constructed on the temperature–composition phase diagram for all temperatures and is shown as the dotted line. Between this spinodal line and the binodal line of the phase diagram, the system is metastable. It requires larger fluctuations in composition to effect phase separation by nucleation and growth.

In spinodal decomposition, phase separation does not arise due to nucleation and growth of spherical domains, but rather arises from fluctuations in concentration such that phase separation arises in interconnected regions. Sperling (2001) has highlighted the differences between the two methods of phase separation in terms of

- *activation energy*: nucleation is activated; spinodal decomposition is not;
- *diffusion coefficient*: nucleation is positive (migration towards the nucleation site); spinodal decomposition is negative (migration away from the domain); and
- *connectivity*: nucleation produces discrete, spherical domains; spinodal decomposition produces (initially) interconnected domains with no sharp interface between domains until later stages of phase separation.

Figure 1.31 contrasts the two processes and the local changes in concentration as phase separation takes place. Also the different phase structures for nucleation and growth (approximately spherical domains) and spinodal decomposition (interconnected phases) are shown. The difference in the sign of the diffusion coefficient, D, is fundamental to the two processes.

In the case of nucleation, there is a small nucleus that then grows, fed by a positive concentration gradient, so D is positive. In spinodal decomposition the domains form by small-amplitude modulation of the concentration, which, because the energy barrier is low, leads to formation of the new phase. The sign of the diffusion coefficient is negative since the growth occurs against the concentration gradient created by the small fluctuations. The formal theory of spinodal kinetics is well developed (Kwei and Wang, 1978) and, since the actual morphology may be affected by external factors such as applied shear, this is the only unambiguous test of spinodal decomposition.

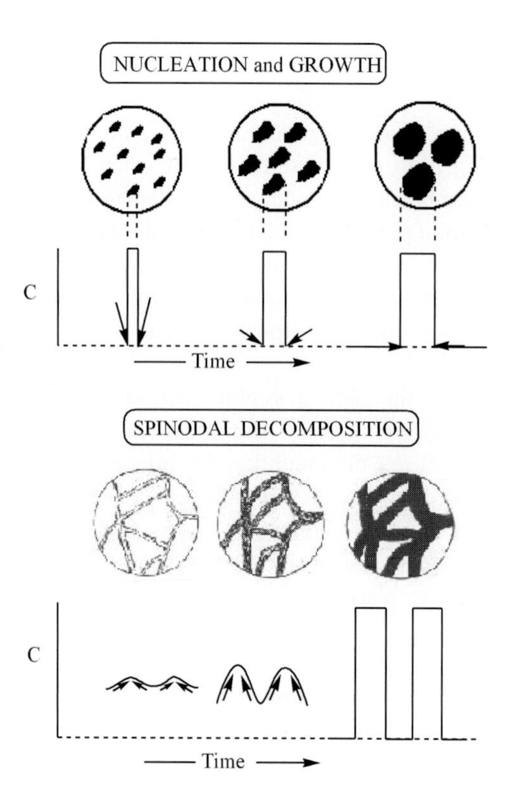

Figure 1.31. The development of the characteristic morphology of polymer blends that have phase-separated by nucleation and growth compared with the case of spinodal decomposition. The local fluctuations in concentration leading to phase separation are also shown. Adapted from Olabisi (1979).

The further development of the separate phases from spinodal decomposition occurs by a viscous-flow process that may lead to layered structures or coarsening, giving other partially dispersed arrangements of the phases. The simple phase diagram such as Figure 1.28 with an LCST has been modified by inclusion of the separate regions for spinodal decomposition and nucleation (Figure 1.30) since these lead to different morphologies. These morphologies may readily be studied by electron microscopy.

Experimental evidence for phase separation and miscibility

The appearance of two phases is signalled by changes in physical properties. In the most common method of identifying phase separation, cloud points are measured using a hot-stage microscope. By measuring this point for samples of different compositions held at different temperatures for a length of time sufficient for the equilibrium structure to be observed (e.g. 12 hours or more), it is possible to construct the phase diagram and determine the critical temperatures.

Criteria for miscibility have been developed using various experimental techniques that have specific sensitivities to the scale of miscibility (or phase separation). An example is the glass-transition temperature (Section 1.1.4). Only a single glass-transition temperature should be detected for a single-phase, fully-miscible blend. As phase separation occurs there will be two glass-transition temperatures, but, if each component is partially miscible with the other, as often occurs, the resultant T_g values are shifted away from those of the pure components and towards the single value for the miscible system. Sperling (2001) has shown that this may be used to calculate the composition of the phase-separated material.

In order to observe the resulting morphology, transmission and scanning electron microscopy (TEM and SEM) have been employed. Care is required in sample preparation so as not to distort the morphology, in addition to which changes occurring during measurement due to the electron-beam irradiation and localized heating must be understood (Olabisi *et al.*, 1979). Extension of this to atomic-force microscopy (AFM) enables the smaller length scales for phase separation to be probed and, in some systems, has shown that separation occurs at a lower temperature than that measured in terms of cloud points (Viville *et al.*, 2001). Other considerations in AFM include the possibility that the phase detected is unique to the surface and is not representative of the bulk as well as effects of the probe itself. Photophysical methods, such as energy transfer and excimer fluorescence, similarly probe miscibility over much smaller scales and lead to detailed information on inter-chain interactions. It should be noted that the phases accessed from the melt may be very different from those obtained when a solvent is used to prepare a 'miscible' solution that is then evaporated to give a thin-film sample for photophysical study. The resulting structures will often differ from the true equilibrium structure unless the film is annealed.

Morphologies of phase-separated systems

Phase-separated polymers are of practical importance. An example is the way that a brittle polymer such as polystyrene may be toughened by the addition of an incompatible rubber that undergoes phase separation. Provided that this is done with attention to the interface between the two phases, a rubber-toughened high-impact polystyrene (HIPS) will result.

For many blends, depending on the conditions for phase separation, very different morphologies and therefore different polymer properties may result. This reflects the time that is taken to reach an equilibrium structure, so that, if a system is interrupted by quenching it to below T_g, the morphology that is developed depends on whether it was in the spinodal region or the binodal region (Figure 1.30). There are ways in which the phase-separated system may be stabilized and the process facilitated by the use of compatibilizers or related interfacial agents that affect the surface energetics as the two phases separate (Paul, 1978). Block copolymers that have compatibility with each of the phases provide a particular example.

An example of a block copolymer is polyisoprene-*block*-polystyrene. This system has the ability to undergo phase separation into a range of different morphologies depending on the relative amount of the two blocks present. If a single polymer chain can separate into two phases then there must be an interfacial region where the composition changes from one block to the other. This is shown schematically in Figure 1.32. The domains that form must be extremely small to allow separation of the two phases, although it has been noted that the segments of a block copolymer of a particular composition are more miscible than for the blend of the same composition. The effect of the chemical bond between the two phases is to affect the kinetics of the phase separation of the block copolymer itself so that a number of unstable morphologies may be seen.

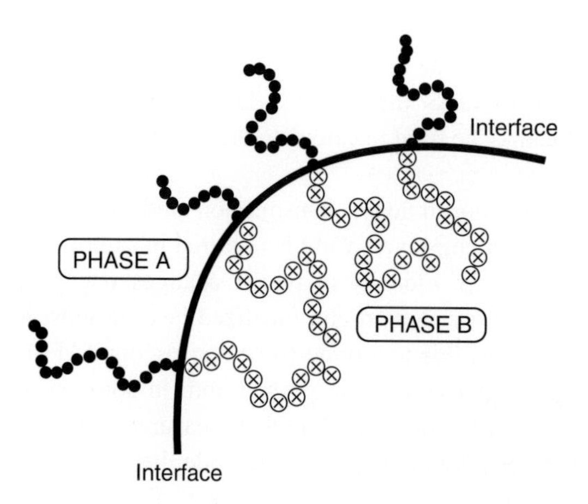

Figure 1.32. A schematic diagram showing the phase-separated structure of a block copolymer comprising chemically distinct blocks and a (theoretical) sharp interface between the phases.

On changing the length of the two blocks, the morphology changes from spheres to cylinders to lamellae as follows:

- *spheres*: short blocks up to $\varphi = 0.25$ separate as spheres in the continuous phase of the longer block
- *cylinders*: unequal-length blocks with $\varphi = 0.25$–0.4 separate as cylinders of the minor component in the longer block as continuous phase
- *lamellae*: equal-length blocks with $\varphi = 0.5$ form alternating layers with the interface separating the layers

In addition to these morphologies, perforated layers, bicontinuous layers and other unstable morphologies have been described (Sperling, 2001). It has also been noted that, unlike polymer blends that become less miscible at higher temperatures, block copolymers become more miscible due to the temperature dependence of the Flory–Huggins parameter, χ.

The importance of these block copolymers is realized when they are used as minor components with normally immiscible homopolymers (each having the composition of one of the blocks). In the absence of the block copolymer (compatibilizer), the resulting two-phase system might not be of practical usefulness. The often high interfacial tension between the two phases results in poor dispersion of the minor phase in the other, continuous phase. On processing, this will lead to macroscopic separation, so the result is a crumbly, useless material. The presence of a minor third component, the block copolymer, can enhance adhesion between the two components and so stabilize the morphology of the system. This process is discussed in more detail later.

This compatibilization relies on the chemical bonding between the two blocks, so alternative routes such as grafting may be used to achieve the same effect. Thus the usual route to HIPS is to graft the rubber toughener with styrene and the resulting graft copolymer is then blended with polystyrene. The resulting morphology as revealed by TEM is shown in Figure 1.33 (Sperling, 2006). This shows that the morphology is quite complex, with the phase-separated domains containing polystyrene (PS, light zones)

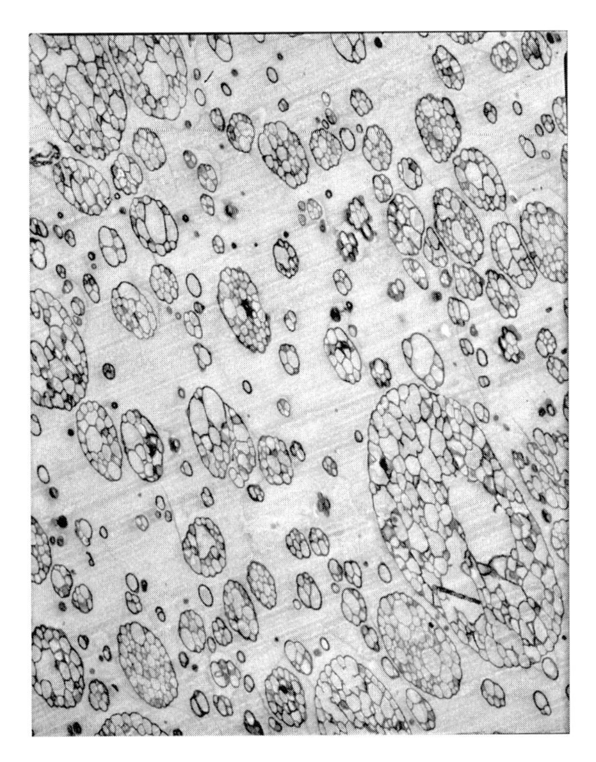

Figure 1.33. Transmission electron micrograph of HIPS after staining with OsO_4 to enable differentiation of the PB domains from the PS. Micrograph courtesy of Bob (J. S.) Vastenhout of Dow Benelux B.V., The Netherlands.

grafted to the elastomer (dark zones of polybutadiene, PB, from OsO_4 staining of unsaturation) and the particles adhering strongly to the continuous polystyrene phase. This has been described as a phase-within-a-phase-within-a-phase organization (Sperling, 2006), and in HIPS the continuous phase is poly(styrene). Development of this structure is controlled by mixing during polymerization, and it has been noted (Paul, 1978) that the role of the compatibilizer is multifunctional in controlling the dispersion of the minor phase and the adhesion between the phases. The comparison of the impact properties shows that there is little gain in toughness of the ungrafted blend, compared with toughening by a factor of 4–5 for the HIPS containing well-dispersed rubber particles with a diameter of $\approx 0.5\,\mu m$.

In many systems, such as epoxy resins, the rubber toughener may be soluble in the other phase, such as epoxy resin, so the phase separation must be achieved during cure. This is an example of phase separation during reactive processing.

Phase separation during reactive processing

The achievement of enhanced toughness in a brittle polymer such as an epoxy resin is vital in applications such as adhesives, for which peel strength and resistance to interfacial debonding through crack propagation are important criteria for performance. This may be achieved by incorporation of elastomeric or other phase-separated particles into the network. These may be present before reaction or they may form during the process of

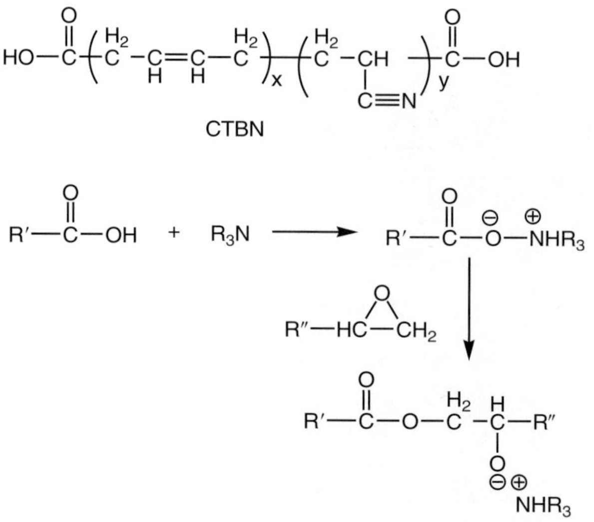

CTBN

Scheme 1.47. The structure of CTBN elastomer and reaction of the terminal carboxyl groups with epoxy in the presence of amine curing agent.

crosslinking and three-dimensional network formation. In the latter process it has been recognized that there may be three possibilities (Pascault *et al.*, 2002):

- a dispersion of elastomer particles in the matrix material
- a phase-inverted dispersion of the matrix as particles in an elastomer-rich continuous phase
- two co-continuous phases

The separate phases will be rich in one component but may have the other present as a minor component. In order to control compatibility the elastomer may have reactive end groups to enhance interfacial adhesion. A common example in epoxy-resin technology is the carboxy-terminated butadiene–acrylonitrile copolymer (CTBN). The structure is shown in Scheme 1.47. In this resin the solubility in the epoxy resin is conferred by the acrylonitrile group, and an increase in the fraction present decreases the upper critical solution temperature, with 26% acrylonitrile conferring total miscibility of CTBN with a DGEBA-based epoxy resin (Pascault *et al.*, 2002).

The process of phase separation during cure arises from the change in the phase diagram as the cure reaction of the epoxy resin progresses. This is shown schematically in Figure 1.34 (Pascault *et al.*, 2002) for a system with an upper critical solution temperature (UCST) in which the lower curve represents the system miscible at room temperature, with the fraction φ_{R0} of elastomer corresponding to the initial composition of the rubber–epoxy-resin system before any cure reaction has taken place.

The material is raised to the cure temperature T_{cure} and reaction commences. There is an increase in the molar mass of the epoxy-resin network as the oligomers react, so that in the equation for the entropy of mixing there will be a decrease in the number of molecules, N_2, and thus the entropic component (ΔS^M) to the free energy of mixing (ΔG^M). The resultant decrease in the solubility of the rubber in the network produces phase separation **before** the resin has undergone gelation (characterized by the formation of an infinite network, i.e.

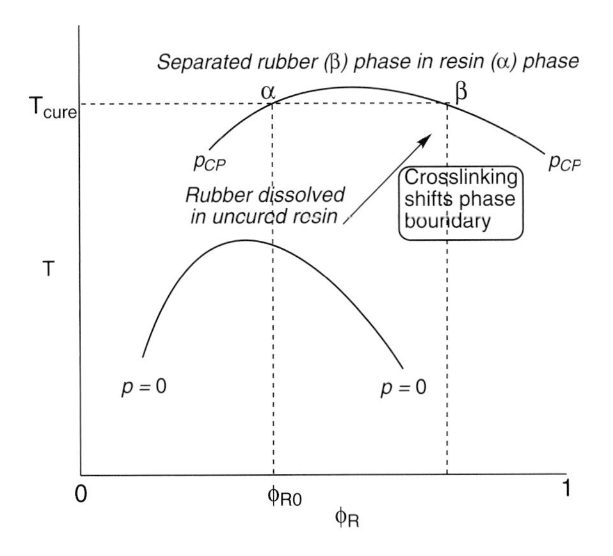

Figure 1.34. Change in the phase boundary (UCST) during the cure reaction for a system of a rubber (e.g. CTBN) dissolved in an epoxy resin. Toughening requires that phase separation be achieved during the cure cycle. After Pascualt *et al.* (2002).

$M_w \rightarrow \infty$). This new phase boundary for the cloud point of the rubber is shown as the upper curve in Figure 1.34 and will correspond to an extent of conversion $p_{\rm CP} < p_{\rm gel}$. There will now be the formation of the rubber (β-)phase dispersed in the thermoset (α-)phase and, as shown in Figure 1.34, the theoretical composition of the β-phase will be given by the position on the cloud-point boundary as shown. (It has been noted that this is not strictly correct due to polydispersity effects and the actual composition lies outside this line (Pascault *et al.*, 2002)).

This is a simplification of the process occurring in a curing resin-hardener system and a detailed discussion may be found in Pascault *et al.* (2002), Williams *et al.* (1997) and Inoue (1995). The main parameter that it is important to control in the reactive phase separation is the diameter of the elastomer particle. This is because the toughness of the resulting network is controlled by the energy-absorbing mechanisms such as particle cavitation and rubber bridging of cracks. Also of importance is the limitation of the effect of the rubber dispersed phase on the critical properties of the cured epoxy resin such as the stiffness and T_g. This will be affected by the extent to which the rubber dissolves in the matrix-rich phase.

The composition of the α- and β-phases will continue to change with conversion beyond $p_{\rm CP}$, with them becoming richer in their main components. However, there will be low-molar-mass species of the other phase present in each, which may lead to secondary phase separation, e.g. the presence of a thermoset phase with a lower extent of conversion inside the rubber particles. Such sub-structures may be seen by TEM of sections of the cured thermoset. The final morphology that develops will then depend on the temperature profile used in cure and post-cure, but is largely controlled by that achieved by the onset of gelation of the α-phase.

If the cloud-point curve is plotted as a function of conversion at constant cure temperature, the possible morphologies may be recognized and the composition of the original

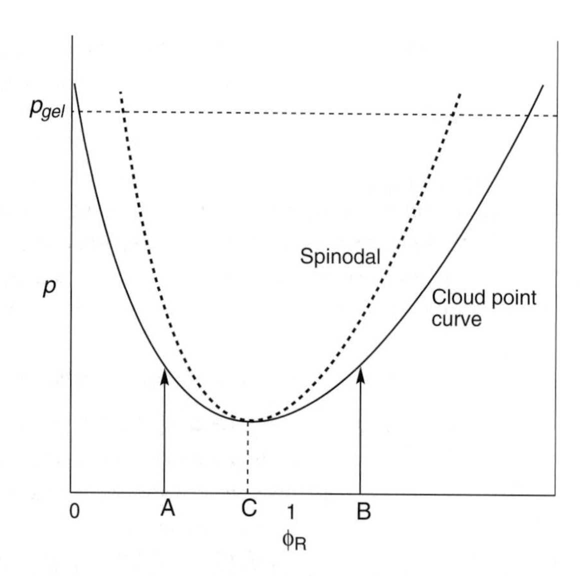

Figure 1.35. Phase separation of the rubber phase from the epoxy phase plotted as conversion, p, against composition of rubber in the two phases at the cure temperature.

mixture necessary to achieve a β-phase rich in rubber readily determined. This is seen in Figure 1.35 for an (epoxy resin + rubber) system such as that which was shown in Figure 1.34 in temperature–composition co-ordinates for just two conversions ($p = 0$ and $p = p_{CP}$) now plotted in conversion–composition co-ordinates for a single cure temperature.

The phase boundaries may be recognized that

- define the stable region for the dissolution of the two components in each other (below the binodal line);
- define a metastable region between the binodal and the spinodal line corresponding to nucleation and growth of spherical domains (β-phase) dispersed in the predominant phase (α-phase) (where the composition of the elastomer is below the critical composition, point C, e.g. point A, the morphology will be approximately spherical rubber particles in the thermoset; in contrast, when the elastomer composition is above point C, e.g. point B, the composition will be phase-inverted, i.e. spherical thermoset particles dispersed in an elastomer continuous phase); and
- define an unstable region within the spinodal curve (this is directly accessible at point C, the critical point, and will lead to a co-continuous morphology due to spinodal demixing (Section 1.3.2)).

All of the above morphologies may be recognized by SEM of the resulting networks after reactive processing. For a toughened network the morphology accessed from point A is the one desired, and this is achieved by ensuring that a low composition of the elastomer (below point C in Figure 1.35) is added to the epoxy resin before cure.

An alternative approach to using a rubber such as CTBN as the disperse phase is to use an engineering thermoplastic such as polyether sulfone or a polyimide in which the mechanical properties and T_g are more closely matched to those of the resin. These systems exhibit LCST behaviour, so the cure temperature T_{cure} is located below the initial miscibility gap and the phase boundary shifts to lower temperature with conversion until the cloud-point conversion is reached. It is noted that a phase-inverted structure may occur in the course of phase separation of thermoplastics such as polyether sulfone (PES) from an epoxy resin during the early stages of cure because of the high molar mass of PES compared with that of the elastomers such as CTBN (Williams *et al.*, 1997). This can contribute to improved fracture toughness, provided that the final phase-separated structure with the dispersed thermoplastic is achieved.

The thermodynamics of the phase-separation process depends both on changes in the Flory–Huggins parameter, χ, as the chemistry changes and on the entropic effects as the molar mass increases. Consequently, any functionalization of the original thermoplastic (or elastomer) in order to enhance interfacial adhesion between the phases will affect the phase-separation trajectory as given in Figure 1.35 and thus the final morphology. At the extreme of not achieving particles of the correct size distribution, the efficiency of the system in toughening the epoxy-resin phase may be compromised. It has been noted (Williams *et al.*, 1997) that the phase-separation process described is thermodynamic in origin and does not depend on any chemical reactivity between the functional groups of the dispersed and continuous phases. The kinetics of cure, however, have a role to play insofar as it is important that the rate of phase separation exceeds the rate of the crosslinking reaction that leads to gelation. If the reaction rate is too high, the trajectory will pass to the spinodal region and a co-continuous structure will be obtained.

In a fundamental study of the factors affecting the growth of the rubber domains in a CTBN-toughened DGEBA epoxy resin cured by piperidine (Manzione and Gillham, 1981) the kinetics of phase separation were linked to the diffusivity, D_{AB}, of the rubber (A) dissolved in the epoxy resin (B). The relevant dependence on the molar volume of the rubber (proportional to the radius R_A of the rubber molecules when dissolved in the epoxy resin) at the viscosity η_B for the temperature T of reaction is given by the Stokes–Einstein equation:

$$D_{AB} = k_B T/(6\pi R_A \eta_B). \tag{1.97}$$

From this, the time scale for diffusion of the rubber particles to coalesce to form a particle may be estimated from the SEM data showing the distance, L, between phase-separated particles. This was found to be of the order of 10 μm in the DGEBA–CTBN system (Manzione and Gillham, 1981). Since

$$t_{diff} = L^2/(2D_{AB}), \tag{1.98}$$

t_{diff} is of the order of minutes, which indicates a lower bound for the processing window to allow the kinetics of phase separation to take place. The processing window may be regarded as $t_{gel} - t_{cl}$: the difference between the time to cloud point (t_{cl}) and the gel time (t_{gel}), which in the system studied was $>$100 min at the cure temperature. This is a very large processing window and in industrial processing this may need to be reduced by at least an order of magnitude to achieve efficient production (bearing in mind, however, that slow reaction times generally produce better control of the cure exotherm and thus minimize shrinkage stresses).

From the above discussion, the choice of both composition and cure schedule is essential in the successful reactive processing of a thermoset toughened by the separation of dispersed elastomeric or thermoplastic (spherical) domains during the curing reaction. The cure kinetics and thus t_{gel} may be controlled through the choice of curing agent as well as the stoichiometry in addition to the temperature–time profile. The effect of the phase separation of a thermoplastic polyimide toughener on the cure rate of a DGEBA resin cured with an aromatic diamine has been studied (Bonnet *et al.*, 1999), and the main effect is the change in the local concentration of the reacting components as phase separation occurs. At concentrations of polyimide above 30 wt.% (which is used in some compositions) there is a sudden acceleration in the global reaction rate. This is attributed to the formation of an epoxy–amine-rich phase that has a higher reaction rate. Parallel studies of the rheology of the phase-separating system (Bonnet, 1999) showed that the viscosity profile reflected the nature of the phase-separation process for compositions corresponding to A, B and C on Figure 1.35. For low concentrations of thermoplastic (point A) there was a decrease in viscosity on phase separation, followed by a rise to gelation. For the critical composition, C, there was a sharp increase followed by a decline and then a steady increase to gelation attributed to the percolation of the thermoplastic-rich phase in the co-continuous structure. At composition B, where phase inversion is possible, there was a steady increase in viscosity to gelation and then vitrification. For DGEBA cured with an aromatic diamine the compositions for A, C and B correspond to 10%, 20% and 33% thermoplastic polyetherimide, respectively.

Hyperbranched polymers (HBPs) provide an approach intermediate between thermoplastics and CTBN elastomers as toughening agents (Boogh *et al.*, 1999) insofar as they have a large increase in fracture toughness for a low additive level (about 5 wt.%). The branched structure (see Figure 1.18 and also Section 1.2.4) results in a high surface functionality and a low viscosity (Aerts, 1998) because of the compact molecular architecture (Lederer *et al.*, 2002). In terms of the above equations, the low values of both R_A and η_B result in efficient phase separation. It has also been noted that it is this unique chemical functionality of the shell that enables a wide processing window to be achieved without compromising the ultimate properties. The near-spherical architecture of the hyperbranched molecule is a factor in achieving optimum nucleation and growth rates of the particles and avoiding spinodal decomposition. The good toughening properties at low levels of addition were attributed to the high connectivity in the phase-separated particles (Boogh *et al.*, 1999). It has been shown that there is a low concentration of the continuous (epoxy) phase dissolved in the HBP particle, but this appears to be graded in composition, producing an effective interface with a gradation in modulus, rather than a step function. In a related study the improvement afforded by HBPs over the comparable linear polyester was questioned (Wu *et al.*, 1999) and the only advantage was that noted above, viz. the ability to tailor the interfacial behaviour.

Even when phase separation of a rubber such as CTBN or a HBP is optimized to produce an effective material, there will be the small fraction of the disperse phase that is soluble in the continuous phase, and trapped there at gelation. As noted earlier, this can affect the T_g if an elastomer is used, but is not a problem with the high-performance thermoplastics such as PES.

One way of overcoming this is to have the second phase already separated before cure, i.e. the toughener is insoluble in the epoxy resin before cure. Examples are found in toughened thermoplastics, where core–shell particles that optimize adhesion and compatibility and have a particle-size distribution to maximize toughness have been synthesized. These particles can be added at the desired volume fraction to achieve toughness without compromising performance rather than relying on the phase trajectory to achieve the desired

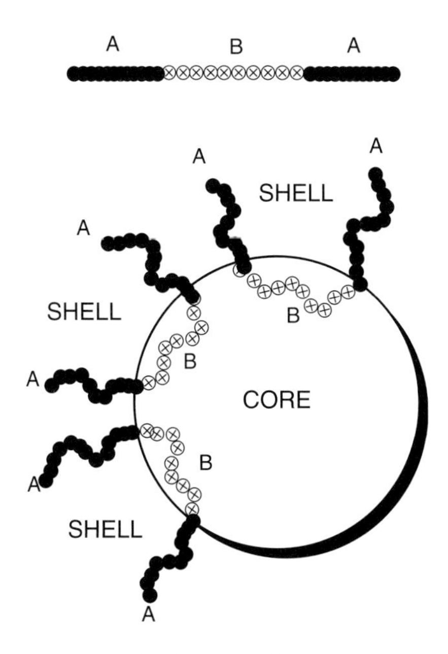

Figure 1.36. A schematic diagram of an ABA block copolymer and the core–shell morphology of toughening particles preformed for dispersion in an epoxy resin.

volume fraction and size distribution. Figure 1.36 illustrates the morphology and the chemical composition of some core–shell particles. Typical systems are ABA block copolymers that have A as poly(styrene–butadiene) and B as poly(methyl methacrylate). The A block forms a particle that is insoluble in the resin and the B block is soluble at all extents of conversion, so giving compatibility.

One disadvantage is the need to ensure that there is a uniform dispersion of the particles in the epoxy resin up to gelation and the corresponding much higher viscosity that the two-phase system will have even before cure. The difficulty in processing is heightened if agglomeration occurs. The compatibility and thus high interfacial adhesion between the phase-separated materials is vital in the application of the final material since the efficient transfer of stress between the toughener and the more brittle thermoset matrix is vital in the achievement of a high value for fracture toughness, for example. The process of compatibilization has become an important area in the reactive processing of polymers. This is considered in the next section before the effect of shear on the process of phase separation is discussed.

Interfacial adhesion and compatibilization

It was noted in the previous section that the carboxyl end groups on the CTBN elastomer affected the final performance of the material as a toughener since these groups would co-react with the epoxy resin and facilitate stress transfer from the brittle matrix to the phase-separated elastomer. Without this adhesion the particles could debond prematurely, which would lead to poor dissipation of the energy of the growing crack. It has also been noted that excessive adhesion between an epoxy resin and a thermoplastic could be deleterious to the performance (Williams *et al.*, 1997). The process of toughening of a thermoset is complex, involving a range of energy-dissipation mechanisms that can occur at the crack tip, such as cavitation in the rubber or in the rubber–matrix interface, as well as localized shear yielding in the matrix due to stress concentration in the rubbery particles (Kinloch, 1985).

The actual reaction between the carboxyl group on the CTBN and the epoxy resin in an amine-cured epoxy system is complex due to the low concentration of carboxyl groups compared with that of amine groups (a ratio of about 1 : 10), so the first step in the reaction is the formation of a salt with the amine which then proceeds to react with the epoxy groups (Scheme 1.47). As noted earlier, there will be solubility of the rubber in the resin, so there will be a copolymerization reaction as the matrix cures and the conditions for phase separation are achieved (Bucknall, 1977). It was recognized that this interfacial-reaction chemistry did not control the process of phase separation since this was governed by thermodynamic considerations (Pascault *et al.*, 2002).

An extension of the use of CTBN has been noted in the use of it as a 'compatibilizer'. In this case it is used as a minor component that is soluble in the thermoplastic toughener (e.g. poly(ether sulfone)) and is able to react at the interface with the epoxy resin, thereby promoting interfacial adhesion (Pascault *et al.*, 2002). It has been recognized that when a small amount of such a compatibilizer is used, there may be the possibility of creating a separate 'interphase' at the interface that is not a separate thermodynamic phase but is a region with properties intermediate between the two phases, thereby promoting adhesion and other properties that are desirable in the end application of the material. This has been studied in detail in composite materials science, where the use of coupling agents is mandatory in order to modify the surface of glass fibres and control both the mechanical properties (particularly toughness) and the environmental durability of the composite by resisting debonding due to the ingress of moisture (Ghotra and Pritchard, 1984). It has been suggested that this interphase extends for several hundred nanometres from the fibre into the resin and will have a gradation in composition and properties that is important in order to avoid having a sharp stress boundary due to the abrupt change in properties.

The definition of compatibility has been differentiated from miscibility since it is concerned with phase-separated polymers and is approached through the attainment of optimum properties for the blend (Bonner and Hope, 1993). Two of the main technologies used to achieve it are the addition of a third component (as discussed above) and reactive blending. The target in using a compatibilizer is the control of the interfacial tension between the components in the melt, translating to interfacial adhesion in the blend after processing.

The reduction of the interfacial tension in the melt results in the formation of lower-diameter particles as phase separation occurs. These are also stabilized by the compatibilizer and resist growth during subsequent thermal treatment such as annealing. Thirdly, the enhanced interfacial adhesion results in mechanical continuity such as stress transfer between the phases. This is a fundamental requirement for a useful polymer blend.

Block copolymers as compatibilizers

The concept of the solubility parameter (Section 1.3.1) leads to the conclusion that the ideal block-copolymer compatibilizer would have components that were identical to the two phases that were to be stabilized. Ideally, the chain length of each block would also match that of the corresponding phase, so ensuring total interpenetration of the copolymer block into each homopolymer. However, it has been demonstrated (Bonner and Hope, 1993) that this is not required and practical considerations dominate, such as

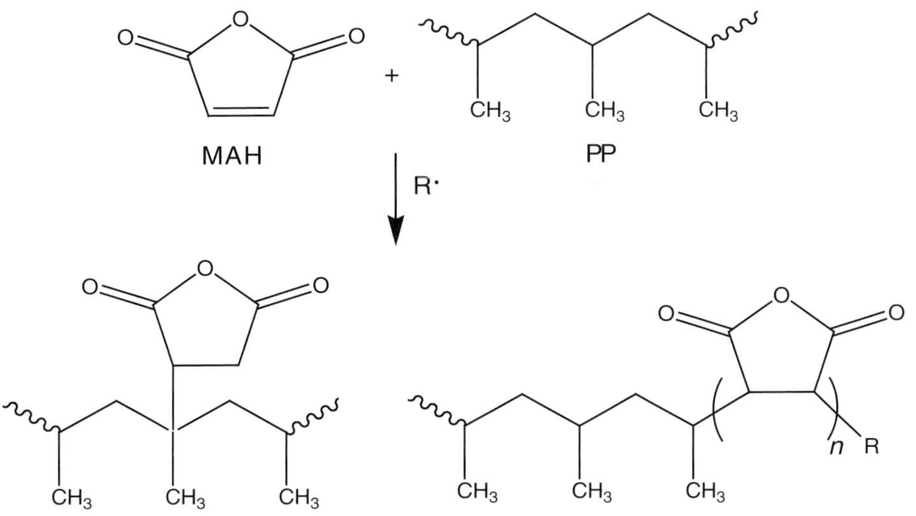

Scheme 1.48. Grafting of maleic anhydride to polypropylene (PP) to give either in-chain isolated succinic anhydride groups or a terminal grafted chain of poly(succinic anhydride).

- the total amount of compatibilizer that may be economically added and
- matching the time required for the phase-separation process to occur with the time taken for diffusion of the compatibilizer block to the interface.

In an assessment of the relative importance of the block-copolymer architecture, the compatibilizer effectiveness was ranked in the following order:

$$\text{tapered diblock} > \text{pure diblock} > \text{triblock} > \text{graft copolymers}$$

It was noted also that satisfactory compatibilization could be achieved by local miscibility (given by solubility parameters) rather than direct matching of the chemical and macromolecular composition. This may require the copolymer to have several components, such as poly(styrene-co-ethylene)-b-poly(butene-co-styrene) (SEBS), which has been found useful in compatibilization of HDPE and PET (Bonner and Hope, 1993).

If chemical reaction can take place between the functional groups on the compatibilizer and the two phases, then this will result in high interfacial adhesion and the miscibility is important only insofar as the reacting groups need to approach one another in order to enable reaction. Often this functionalization is achieved in a separate reactive-processing step, such as the grafting of maleic anhydride to polyolefins, Scheme 1.48 (Moad, 1999).

In spite of this being one of the most studied functionalization reactions, the detailed structure of the graft copolymer has been shown to be complicated (Moad, 1999) by

- the distribution of the anhydride functionality along and between polyolefin chains
- the possibility of oligo-(maleic anhydride) grafts and sequences of grafts
- the effect of polyolefin structure on the location of the graft site

The industrial importance of this compatibilizer has meant that the technology of preparation has been studied widely (and patented), but the detailed information addressing the structure of the resultant graft copolymer has not been obtained.

Self-compatibilization by reactive blending

As the term implies, no separate compatibilizer is required, and the process of melt blending of the two normally immiscible polymers results in reaction so that the phase-separated system is stabilized. The polymers are chosen so that one or more of the following types of reaction may take place (Bonner and Hope, 1993).

- Chain scission under mechanical shear and heat, resulting in macro-alkyl radicals (R·) that may graft to the other chain or terminate with macro-alkyl radicals from the other chain (R′·) to give the block copolymer RR′.
- Ester-exchange reactions in polyesters (and analogous amide reactions in nylons) that result in a new block copolymer. An example is the formation of a blend between polycarbonate and poly(butylene terephthalate) in which the compatibilizer is formed by a transesterification reaction at elevated temperatures (350 °C), but, when processed at lower temperatures, the systems are not compatible.
- Direct chemical reaction between functional groups, so that graft or block copolymers are formed. For example, a polyolefin may be maleated so that it may then react with a polyamide during melt blending to give an imide that functions as the compatibilizer.

In systems such as these in which the phase-separated system is being formed and stabilized in the process of the reactive blending (often reactive extrusion) the effect of mechanical shear itself on the phase-separation process must be considered.

The effect of mechanical shear on phase separation

The practical preparation of polymer blends involves the application of shear during processing. This applies both to reactive and to non-reactive processing. A full study of the multiphase flow that can occur in the extrusion of blends has been carried out by Han (1981). The analysis begins with the determination of the deformation of single droplets in shearing and extensional fields, and then is extended to emulsions and then blends in flow fields encountered in fibre spinning and injection moulding.

If the composition and temperature of the blend corresponds to a position on the phase diagram close to a phase boundary then it is likely that flow-induced mixing or de-mixing will occur through a shift of the phase boundary (Higgins *et al.*, 2000).

In a study of initially incompatible blends of SAN and PC (30 : 70) extrusion at a high shear rate of up to $10^7\,s^{-1}$ produced a blend in the first extrusion run in which the minor constituent (SAN) was present as small spherical particles in PC (Takahashi *et al.*, 1988). The apparent volume fraction of the spherical SAN decreased with the shear rate and repeated extrusion. The DMA measurements showed that the maximum of $\tan\delta$ for PC shifted to lower temperatures with repeated extrusion, which is consistent with some SAN being mixed with PC in a compatible form. This result suggests that compatibility is enhanced in extremely-high-shear-rate processing. Because of the rapid quench from an elevated temperature that occurs after processing, this new structure may be retained.

The effect of shear is to lower the cloud-point curve, and it was suggested that the large number of heterogeneities in melt-mixed samples may provide multiple nucleating sites once the sample is in the metastable region. The kinetics of phase separation have been studied for a range of systems to generalize this process further (Rojanapitayakorn *et al.*, 2001), and it has been demonstrated that a theoretical model in which the Gibbs free energy of mixing also includes the energy that is stored in the sheared fluids may explain the effect

(Higgins *et al.*, 2000). It is recognized that a flowing mixture is a steady-state system, not an equilibrium thermodynamic system, so it must be treated as a perturbation and only a qualitative explanation is possible.

A real-time study of the process of shear-induced mixing and de-mixing was performed using on-line Raman spectroscopy followed by microscopic examination of the extrudate (Adebayo *et al.*, 2003). Shear flows were performed through a slit conduit starting in the one-phase and two-phase regions of blends of PMMA and poly(α-methyl styrene-co-acrylonitrile) that exhibit an LCST. For a high melt viscosity sample near the (equilibrium) cloud point, the shear-heating effects over-ride all other effects. In contrast, for a lower-viscosity system, shear-induced de-mixing is observed at low shear rates, while with increased shear rate the system underwent mixing prior to the onset of shear heating that resulted in the system entering the two-phase region and optically observable phase-separated domains corresponding to the maxima in shear heating across the extrudate. Processing a system without detailed understanding of the phase properties and the shear heating will result in complex shear-dependent morphologies.

The above examples do not involve reactive processing (or degradation, even when carried out in a twin-screw extruder (Papathanasiou *et al.*, 1998)), and the introduction of the possibility of interfacial reaction for *in situ* compatibilization will have an effect on the morphology. One approach to the study of the dynamics of these processes involves the freezing of the reaction in the twin-screw extruder and the analysis of the resultant morphology (Van Duin *et al.*, 2001). This has provided a method for studying several of the most common reactive-processing operations:

- processing of polyolefins with peroxides either to crosslink or to create branch and graft sites in PE, or to degrade PP to alter the melt-flow properties
- grafting of maleic anhydride (MA) onto polyolefins to create polar sites for compatibility
- reactive blending of nylon-6 with these MA-grafted polyolefins, such as an ethylene–propylene rubber (EPR)

The technique involves the uses of rotatable sampling ports that enable the polymer to be removed and quenched as it passes particular zones (corresponding to different values of *L*/*D*) as shown in Figure 1.37 (Van Duin *et al.*, 2001). In this way 2 g of polymer can be collected for off-line analysis in about 3–5 s. For the above polymers, the most common profile for product formation along the axis of the extruder was one that followed the profile for the decomposition of the peroxide with time (and thus distance along the extruder). Reaction of MA was very rapid, and there was an exponential growth in the amount of maleated polyolefin as the MA was consumed.

Of particular relevance here was the study of the reactive blending of the MA-grafted rubber with the polyamide and the morphology as a function of distance along the barrel. In the melt zone a graft copolymer formed between the nylon-6 and the EPR-g-MA that consumed the MA sites and formed a layer around the nylon-6 pellets at the first sampling point. As the nylon-6 then melted, by the second sampling port the quenched polymer blend had EPR domains a few micrometres in diameter. These decreased in diameter on kneading and exhibited a narrower particle-size distribution. No phase inversion was observed, although the EPR particles did contain inclusions of nylon-6, presumably from the graft copolymer that had acted as an *in situ* compatibilizer to give a stable morphology to the blend. One consequence of the grafting reaction between

(a)

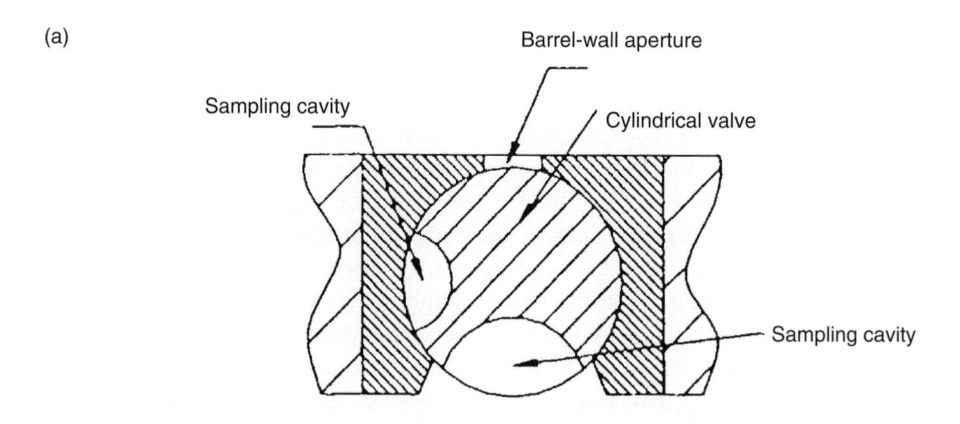

(b)

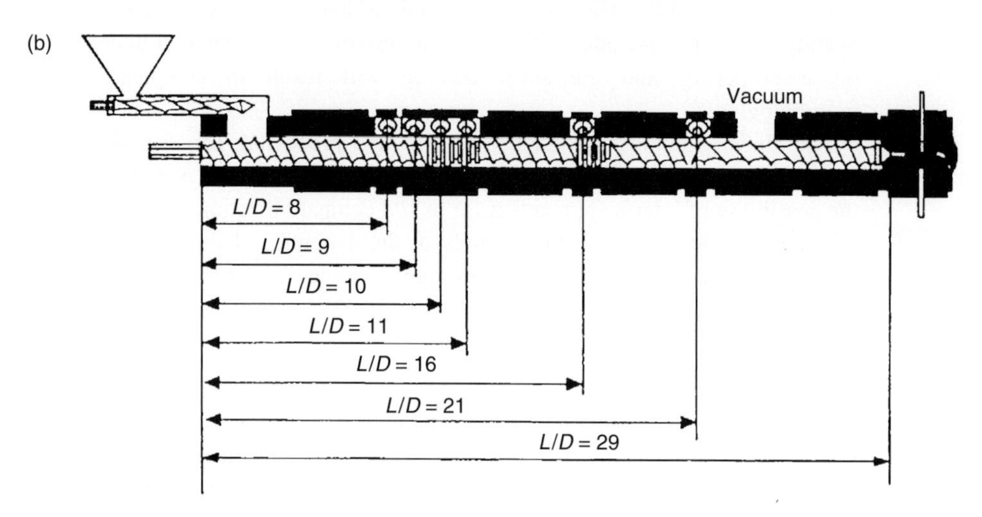

Figure 1.37. (a) Detail of the sampling valve and (b) location at various *L*/*D* positions in a twin-screw reactive extruder to enable monitoring of the chemical and morphological changes during reaction in the extruder barrel. Reproduced with permission from Van Duin *et al.* (2001). Copyright Wiley-VCH Verlag GmbH & Co. KGaA.

amide and anhydride that formed the compatibilizing copolymer was the liberation of water that at the extrusion temperature led to hydrolysis and a decrease in molar mass of the nylon-6. This example illustrates well the complexity of reactions and side reactions that can occur during reactive modification and blending in a reactive extruder (Van Duin *et al.*, 2001).

1.3.3 Interpenetrating networks

It has been noted above that phase separation in thermoplastics is a common occurrence when two or more polymers are mixed and that miscibility is the uncommon event. This is exploited in toughening of thermosets by elastomers when phase separation occurs during the reaction that leads to three-dimensional network formation. If macroscopic phase separation is *not* desired then it is possible to achieve a different microscopic morphology and in some cases maintain some features of miscibility

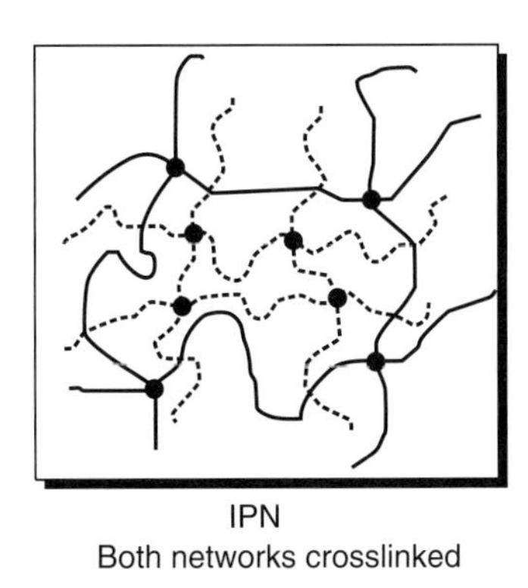

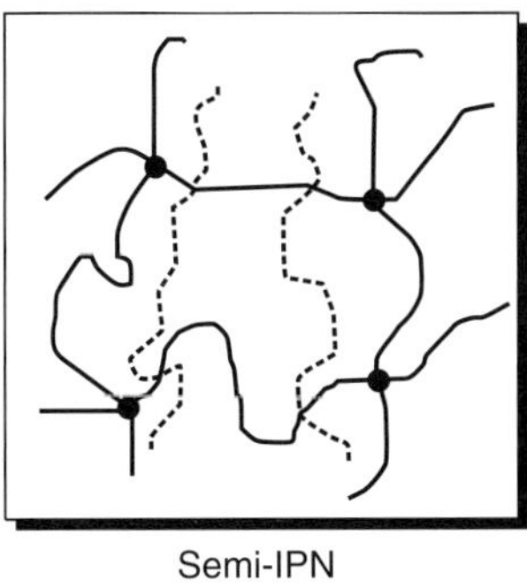

Figure 1.38. Schematic architectures of interpenetrating networks. After Sperling (2006).

against the thermodynamic driving force by forming an interpenetrating network (IPN) (Sperling, 1989).

The characteristic feature of an IPN is the presence of two crosslinked networks that are intimately mixed and have reacted to freeze the structure in place. While the crosslinks are most usually covalent and the crosslink density is sufficiently high for one to class the product as a thermoset, it is possible to have an IPN based on physical crosslinks that is thermoplastic. An example is an ionomer in which the charges on the polymer chain produce high inter-chain forces that enable the formation of the crosslinks. On heating, such as during processing, the physical crosslinks are broken and the material flows; on cooling the crosslinks are formed again. Thermoplastic elastomers from block copolymers form physical crosslinks by virtue of the crystalline blocks of one component that bridge the amorphous second component (as shown earlier in Figure 1.17).

The method of formation of thermosetting IPNs governs the nomenclature for the system and some examples are the following.

- Synthesis, in the one pot, of two separately reacting systems that form networks independently of one another but, because of their proximity, interpenetrate. This is a **simultaneous** *interpenetrating network.*
- Synthesis of the first network, into which are then swollen the reagents to form the second network, which are then reacted. This is a **sequential** *interpenetrating network.*
- Synthesis of a linear polymer that is then trapped in a crosslinked network formed from a second component. Since only one polymer is crosslinked, this is a ***semi**-interpenetrating network.* This is also known as a **pseudo**-*interpenetrating network.*

Cartoons that illustrate the two principal types of IPN are given in Figure 1.38. The properties of the IPN and the morphology that results are governed by the kinetics and thermodynamics of phase separation.

1.4 Degradation and stabilization

The melting of a polymer for chemorheological measurements and processing often involves subjecting the material to shear in the presence of oxidizing agents, such as air.

These conditions, if uncontrolled, will lead to changes in the chemical structure of the polymer, so the understanding of this process is essential both to accurate chemorheology and to reactive processing. In many cases the degradation or crosslinking of the polymer is desirable in order to modify the final molar-mass distribution or intermolecular interactions in the material, and the goal in this section is to indicate the strategies that may be employed to **control** degradation.

In this section the fundamental process underlying the changes to molar mass and its distribution through either chain scission or crosslinking, namely the formation of volatile products, either by exceeding the ceiling temperature or by chemical reaction, will be considered by using the polyolefins as the starting point and then extending the study to more complex polymers. Following the attainment of an understanding of reaction mechanisms, the tool which has been employed to quantify the oxidation of polymers has been chemical kinetics. The theory of free-radical chain-reaction kinetics in the gas phase or well-mixed liquid state has long been established. While molten polymers differ from low-molar-mass compounds in solution because of persistent entanglements and significant viscosity effects, the theory for oxidation and stabilization of the polymer melt during processing is consistent with homogeneous chemical kinetics. In this system, the concentration of the reactants may be averaged over the volume to produce a representative value for kinetic modelling.

1.4.1 Free-radical formation during melt processing

The processing of polymers and many high-shear rheological measurements occur at high temperature in the presence of large mechanical deformations. The first question to be answered is that of whether this treatment results in the formation of free radicals that can lead to changes in the polymer's molar-mass distribution and overall structure. In the history of polymer science and technology, the first polymer to be studied in detail was *cis*-poly (isoprene), natural rubber. It had been discovered that the viscosity of raw rubber decreased dramatically after heating under the shearing conditions of a roll mill. This process, termed mastication, resulted in a plasticized, viscous material that was amenable to further reaction and crosslinking to form an elastomer. The process of mastication was found to require oxygen in order for it to occur efficiently, and the material was able to initiate the graft polymerization of vinyl monomers to give copolymers. It was proposed that the route was through the formation of macro-alkyl radicals, which then reacted with oxygen to form hydroperoxides, the initiating species (Scott, 1995). In support of this, it had been found that little mastication occurred in the absence of air. The use of free-radical formation in (uncontrolled) grafting during reactive processing has been noted earlier in Section 1.2.3 (Scheme 1.39).

Radicals from mechanical chain scission

Of interest in the understanding of the mastication of rubber was whether free radicals were formed mechanochemically, i.e. by direct scission of the backbone of the polymer during processing, or arose purely from thermal events. This required sophisticated experiments using a spectroscopic method specifically able to identify the type and concentration of free radicals: electron spin resonance (ESR) spectroscopy. In fundamental studies using ESR, free-radical formation had clearly been demonstrated during the grinding, stressing and fracture of solid polymers at temperatures sufficiently low that the species were stabilized

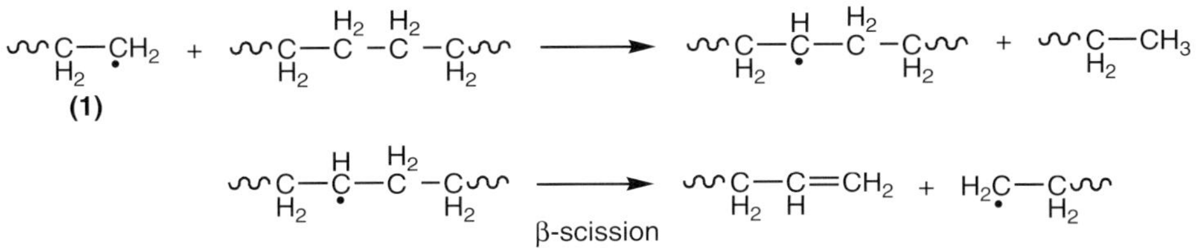

~CH_2–$CH_2^{\bullet}$ (**1**) + ~CH_2–CH_2–CH_2–CH_2~ ⟶ ~CH_2–$\dot{C}H$–CH_2–CH_2~ + ~CH_2–CH_3

~CH_2–$\dot{C}H$–CH_2–CH_2~ ⟶ (β-scission) ~CH_2–CH=CH_2 + $H_2\dot{C}$–CH_2~

Scheme 1.49. Possible reactions of the main-chain mechanoradical (1) of polyethylene.

(e.g. 77–200 K) (Sohma, 1989a, 1989b). The temperature dependence of the species formed also assisted in determining the mechanism of radical formation. Scheme 1.49 illustrates the parent free radical (**1**) formed on cryogenic milling of polyethylene and the subsequent abstraction to give an in-chain radical that will compete with radical recombination. The further reaction of this radical may also result in a decrease in molar mass through β-scission and formation of terminal unsaturation, as well as regenerating a terminal radical equivalent to (**1**) as shown in Scheme 1.49.

Extension of these studies to the polymer in solution and the measurement of the attendant decrease in molar mass showed that it was the chain length which was important, not the solid-state properties. In particular the solid-state studies showed that there was a critical molar mass before mechanochemical effects could be observed (for polyethylene (PE), a degree of polymerization of about 100). In solution a similar effect was observed since the degradation of poly(methyl methacrylate) (PMMA) could not continue when the chain length reached about 4000 repeat units (Sohma, 1989a, 1989b). This was shown by a slow decrease in molar mass to the asymptotic value as shown in Figure 1.39.

This difference between the behaviour of the solid PE and the solution of PMMA was attributed to the fundamental mechanism of scission, such that in the solid state there was amplification of the effect of crack growth from the initial site of chain scission whereas in PMMA it was chain entanglement that produced the mechano-scission.

In order to rationalize this result for PE, a simple calculation of the energy required for scission of main-chain bonds ($E_{C\text{–}C}$) with the frictional dissipation between chains due to mechanical action was performed. The latter was calculated from the activation energy for viscosity of the monomer repeat unit, E_η, determined from a low-molar-mass analogue of polyethylene. This was then used to determine the number of repeat units, n, necessary to exceed the C–C bond energy of 349 kJ/mol. From the value of 4.22 kJ/mol for E_η, an estimate for n of 83 was made, which was considered a good approximation to the experimental result of 100 (Sohma, 1989b). However, such a simple approach is not amenable to extension to systems of structural complexity such as chain scission during the mastication of rubber.

Casale and Porter (1978) brought together much of the early work on the theory of mechanical generation of free radicals. Much of the theory for radical formation in the rubbery state has developed from the ideas of Bueche regarding the scission of entangled chains. The entangled chain does not permit the rotation of the entire chain to dissipate energy when shear is applied and, furthermore, tensions reach a maximum at the centre of the chain. The number of scissions will decline exponentially with distance from the centre of the chain such that at link q from the centre the ratio of the number of scissions, n_q, to that at the centre, n_0, is

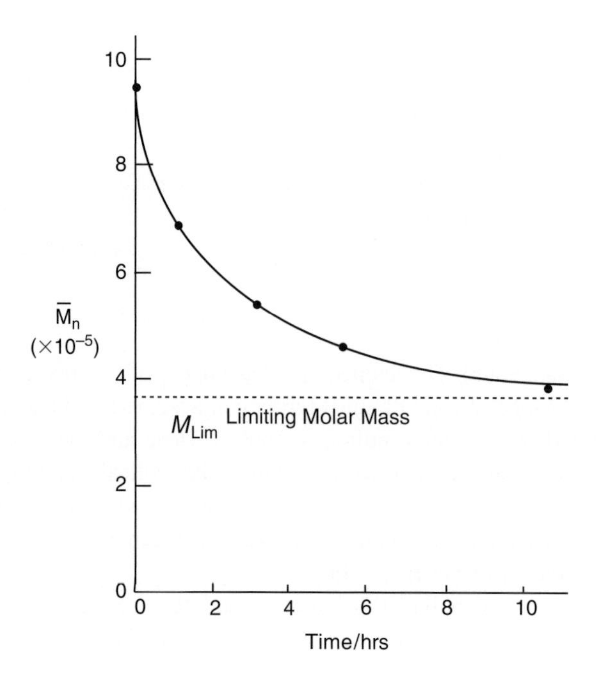

Figure 1.39. Change in molar mass of PMMA due to shear in solution for times up to 11 hours. Adapted from Sohma (1989).

$$n_q/n_0 = \exp[-F_0\delta/(k_B T)(4q^2/n^2)], \tag{1.99}$$

where F_0 is the force acting on the bond, δ is the bond elongation before breaking, and n is the degree of polymerization. A typical value for $F_0\delta/(k_B\mathrm{T})$ is 20 so that, on substituting that into this equation, the number of chains that break in the centre is 150 times the number which break a quarter of the way from the chain end. This relationship also shows that the longer chains have a greater probability of undergoing chain scission. This conclusion has been confirmed experimentally from studies on a number of polymers, e.g. polybutadiene and *cis*-polyisoprene. These studies do not provide any details of the kinetics of the process, and it is necessary to measure the rate of change of molar mass under mechanical shear.

As noted earlier, on mechanical degradation of chains with an initial number-average molar mass, M_{In}, there is a limiting molar mass M_{Lim}, below which no further mechanical chain scission can occur (as shown by the asymptotic value in Figure 1.39). The molar mass at time, t, M_t, is then

$$M_t = M_{Lim} + (M_{In} - M_{Lim})\exp(-kt), \tag{1.100}$$

where k is a rate coefficient that is unique for the polymer type. This may be extended to determine the number of radicals, Z, produced per gram of polymer as

$$Z = 2N(1/M_{Lim} - 1/M_{In}), \tag{1.101}$$

where N is Avogadro's number. A typical value of Z is 10^{-6} mol/l for the mastication of rubber (Scott, 1993b).

The approach of Zhurkov and Bueche is an activation-volume argument in which the presence of the applied stress lowers the activation energy for chain scission, so increasing the value of the rate coefficient, k. The effect is to increase the probability, P, of scission of the central bond from P_0 to (Casale and Porter, 1978)

$$P = P_0 \exp[-(E - F_0\delta)/(k_B T)], \tag{1.102}$$

where E is the bond energy in the absence of the applied force F_0 to the bond which has an elongation to breakage of δ.

Other equations have been proposed to relate the experimental observations of the decrease in molar mass and the observations of free radicals to the input of mechanical energy into the system. The complexity in energy dissipation in many systems ensures that the relations are often specific for a particular polymer and set of conditions, since these define M_{Lim}.

The important results are that mechanoradicals are formed, but their reactivity ensures that they are affected by the temperature of the system as well as by the presence of trace amounts of radical-scavenging chemicals. One of the most efficient radical scavengers is molecular oxygen, which will react with alkyl radicals at the diffusion-controlled rate for the partial pressure of oxygen in the system. The scavenging of radicals produced either mechanically or thermally is considered in the following sections.

Radical formation and depolymerization on thermal degradation

Processes that produce mechanoradicals are often accompanied by a rise in temperature of the system, so it is important to separate the contributions of thermal degradation and mechanical scission to radical formation and reactivity in the absence of oxygen. The pure thermal degradation of polymers is of importance in the determination of the limiting temperature (ceiling temperature, T_c) in the polymerization of vinyl monomers and all polymerizations that involve equilibria. While polymers should never be processed or their rheological properties studied at temperatures approaching the ceiling temperature, it is useful to examine the thermodynamics of depolymerization because it is one route for the formation of low-molar-mass products through free-radical reactions in the polymer. This may occur at $T < T_c$, due to defects in the chain or the formation of free radicals through reactions of trace impurities, so in a practical application the evolution of volatiles through chain-unzipping reactions can have severe consequences in processing.

The process of unzipping of the chain to produce monomer is dependent on the polymer type and also the temperature of reaction. For example, polymers that may potentially produce 100% monomer by depolymerization include poly(methyl methacrylate) (PMMA) and poly(α-methyl styrene). However, if PMMA is heated at 220 °C only 50% of the chains unzip, and these contain unsaturated end groups (Allcock and Lampe, 1981). A higher temperature of 350 °C, beyond those encountered in processing, is required for complete decomposition to monomer, and in this case the chain ends are fully saturated. In contrast to PMMA, poly(styrene) produces oligomers in addition to monomer on heating at 350 °C. This is shown in Scheme 1.50, and the oligomers range up to tetramers through a back-biting process of the terminal polystyryl radical, as shown, with the oligomer having a terminal double bond.

Thus polystyrene is not a depolymerizing monomer due to the more favourable reaction of abstraction of the hydrogen at the tertiary carbon site. This is in contrast to poly(α-methyl styrene), which has this position blocked by the methyl group, so unzipping occurs.

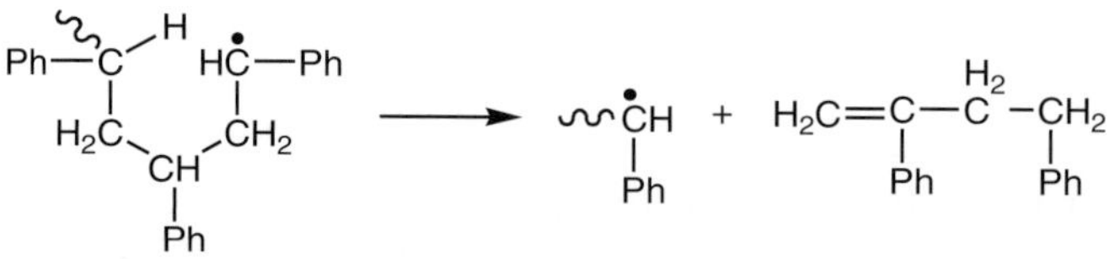

Scheme 1.50. Oligomer formation from poly(styrene) scission by back-biting.

Addition polymerization may be viewed as an equilibrium between monomer and polymer, and the extent to which long chains can form for a particular monomer at the polymerization temperature can be simply understood as a consequence of the changes in entropy during polymerization. From Equation (1.64) in Section 1.2.3, for polymer to form the overall change in free energy, ΔG, must be negative. The loss of entropy, ΔS, in forming a long polymer chain of one molecule from several thousand monomer molecules must be matched by a large and negative change in enthalpy, ΔH, if the free-energy change, ΔG, is to be negative at temperature T and formation of the polymer thermodynamically favoured. Should ΔG be positive under the conditions of the polymerization, then the equilibrium will shift to favour monomer, i.e. depolymerization will occur. As noted in Section 1.2.3, Table 1.9 summarized the values of the ceiling temperature, T_c, above which the polymerization cannot occur because the depolymerization to monomer is then favoured. Examination of values of ΔH^0 and ΔS^0 for a number of monomers undergoing addition polymerization (Allcock and Lampe, 1981) shows that there is a wide variation in ΔH^0, reflecting steric and substituent effects on reactivity, but little variation in ΔS^0, since the loss of translational entropy on formation of a chain is little affected by the nature of the monomer. Thus the polymerization of α-methyl styrene is much less exothermic than that of styrene ($\Delta H^0 = -35$ kJ/mol), but the change in ΔS^0 is almost identical at $-104\,\mathrm{J\,K^{-1}\,mol^{-1}}$. This results in a ceiling temperature for poly(α-methyl styrene) of only 66 °C. It is noted that the ceiling temperature is the upper limit for polymerization since there will be weak links and other sites for initiation of depolymerization. Formally, the ceiling temperature is subject to the fundamental thermodynamic requirement that $\Delta G = 0$ in a closed system, i.e. when there is no loss of volatiles. This also is not achieved in practical applications.

Random thermal chain scission

Depolymerization in many systems is only a minor contribution to degradation and is replaced by random chain scission and the evolution of volatile hydrocarbons in addition to the original monomer. Thus polyethylene above 300 °C eliminates a series of unsaturated hydrocarbons up to 70 carbon atoms in length. In addition to the free-radical initiation process of random scission, there will be subsequent free-radical reactions such as abstraction, disproportionation and chain transfer. The study of the thermal degradation of linear polymers such as vinyl and acrylate polymers has been facilitated by molar-mass measurements to determine the number of scission events and the mechanism has been elucidated from the determination of product distribution, including by thermo-gravimetric analysis (TGA) and pyrolysis gas-chromatography mass spectrometry (GC-MS) in an inert atmosphere. Coupling a Fourier-transform infrared (FT-IR) spectrophotometer or a mass spectrometer to a TGA also allows direct identification of volatiles corresponding to the weight-loss regime. McNeill (1989) has highlighted the pitfalls of deducing kinetic

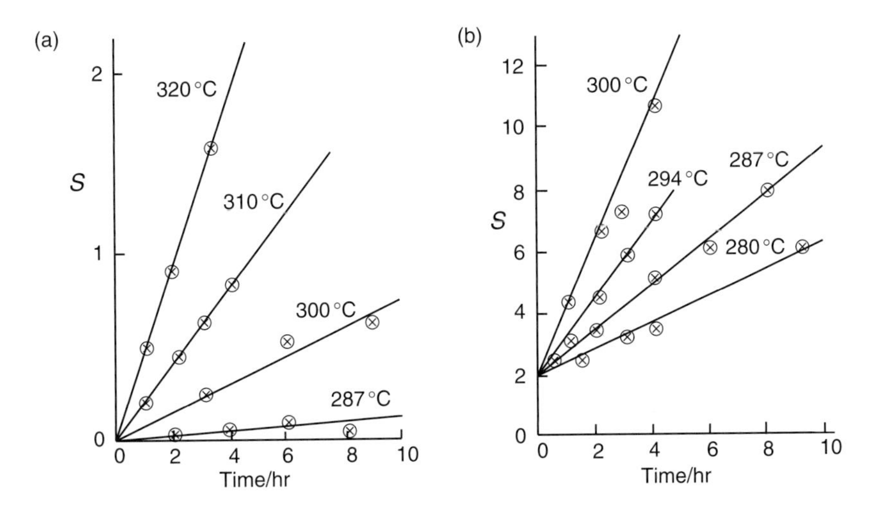

Figure 1.40. Chain scission (given as number of scissions, S, per number-average molecule) during thermal degradation of poly(styrene) polymerized anionically (a), $M_n = 2.3 \times 10^5$, or by free-radical initiation (b), $M_n = 1.5 \times 10^6$. Adapted from McNeill (1989).

information from these analyses because of the sensitivity of the reactions to sample size. This may arise from the diffusion of volatiles and the self-catalysis of degradation, e.g. dehydrochlorination and polyene formation in PVC is catalysed by evolved hydrogen chloride as discussed later. Thus the flow rate of inert gas can have an effect on the rate of removal of synergists. Another mass effect will occur when the degradation reaction may be highly exothermic, so the heat rise in the sample will depend directly on the mass and inversely on its surface area. These considerations are of greatest importance in kinetic studies.

The number of chain-scission events during thermal degradation is most readily determined by the change in degree of polymerization as a function of time. The number of chain scissions $S(t)$ at any time t per number-average molecule (M_n^0) is given by the relation

$$S(t) = M_n^0/M_n(t) - 1, \tag{1.103}$$

where $M_n(t)$ is the relative molar mass at time t. This is plotted in Figure 1.40 for polystyrene polymerized by two different methods, (a) anionically, with M_n^0 of 229 000, and (b) free-radically with M_n^0 of 1 490 000 (McNeill, 1989).

There are several features immediately apparent.

1. The absolute number of scissions, S, is much lower for the anionically polymerized polymer and increases linearly with time.
2. The higher-molar-mass polymer polymerized by a free-radical route is much less stable, e.g. the number of scissions at 300 °C is greater by a factor of seven.
3. The plot for the free-radical-produced polymer shows an intercept at zero time of ~2 scissions per molecule. This indicates that there are weak links in this polymer that degrade very rapidly at all temperatures studied. These are believed to be peroxide groups in the polymer chain, formed by the rapid scavenging of the styryl radical by oxygen during polymerization, and are primary initiation sites for thermal degradation.

Table 1.17. Elementary reactions and the related rate coefficients and rates for the thermal degradation of a polyolefin, P_n

Process	Elementary reaction	Rate coefficient	Rate of reaction
Initiation	$P_n \rightarrow R_r\cdot + R_{n-r}\cdot$	k_i	$r_i = 2k_i[P_n]$
Propagation			
intermolecular	$P_n + R_r\cdot \rightarrow P_r + R_j\cdot + P_{n-j}$	k_1	$r_p = k_1[P_n][R_r\cdot]$
intramolecular	$R_n\cdot \rightarrow R_{n-r}\cdot + P_r$	k_2	$r_2 = k_2[R_n\cdot]$
Termination by disproportionation	$R_n\cdot + R_r\cdot \rightarrow P_n + P_r$	k_t	$r_t = 2k_t[R_n\cdot][R_r\cdot] \sim 2k_t[R_r\cdot]^2$

4. The plot for the anionically polymerized polystyrene has a zero intercept indicating that there are no potential weak links. The initiation in this case must involve homolysis of the backbone followed by disproportionation.

The role of weak links has been considered in all thermal-initiation possibilities (Scott, 1995), and the tendency of many monomers to copolymerise with trace amounts of oxygen in free-radical polymerization (in the case of styrene, to form a 1 : 1 copolymer) until it has all been scavenged is well known.

The polyolefins, particularly polyethylene (PE) and polypropylene (PP), have been studied in detail because of their importance both in processing and in recycling and energy recovery. Low-density PE is highly branched, which this has a bearing on the thermal degradation since these sites constitute weak links, resulting in scission at lower temperatures. Polypropylene is extremely sensitive to the presence of oxygen but even in the absence of air, volatiles are evolved at temperatures 100 °C lower than for PE. The products are a series of alkenes due to intramolecular reactions of the secondary alkyl radical following random chain scission. In a study of the effect of multiple extrusion cycles on the evolution of volatile organic compounds, it was found for PP that the levels and rate of generation were closer to mild thermo-oxidation rather than pure thermal degradation, indicating the difficulty of achieving an oxygen-free environment under processing conditions. It was found that the changes in composition and evolution of volatiles were consistent with changes in rheological data generated at the same time (Xiang *et al.*, 2002).

The kinetics of the degradation of PP have been studied by following the molar-mass distribution rather than just the chain scissions as a function of degradation time for temperatures from 275 to 315 °C (Chan and Balke, 1997b). A simplified model was established to account for these changes and the minimum number of parameters was chosen to fit the data. Initiation is assumed to be by random scission into two free radicals, and the only radical reactions are scission and chain transfer. Termination is assumed to be by disproportionation, not combination (i.e. there is no increase in molar mass through the formation of crosslinks or other combination reactions). The mechanistic equations are summarized in Table 1.17. It is assumed that a stationary state is reached and also that the reaction sequence may be further simplified by assuming that the probability of intramolecular transfer is very low compared with that of intermolecular transfer, i.e. $k_2 \sim 0$. The system becomes a simple free-radical kinetic solution for a single initiation, propagation and termination step, in which the rate of polymer degradation in the initiation step is equal to that for recovery in the termination step (which is a consequence of the stationary-state assumption that $d[R_r\cdot]/dt = 0$) and thus the stationary-state concentration of radicals $R_r\cdot$ is

$$[R_r\cdot] = [k_i/(2k_t)]^{1/2}[P_n]^{1/2}, \quad (1.104)$$

and on substituting into the intermolecular propagation reaction, the rate of degradation of the polymer is given by

$$-d[P_n]/dt = k_1[k_i/(2k_t)]^{1/2}[P_n]^{3/2}. \quad (1.105)$$

It is possible to determine the change to the molar-mass distribution by integrating this relation over time. Also the overall apparent activation energy for the degradation will be given by the following relation for the activation energies for the propagation, initiation and termination reactions, respectively: $E_{app} = E_1 + \frac{1}{2}E_i - \frac{1}{2}E_t$, which is a form common to many free-radical reactions (Dainton, 1956). The model provided a good fit to the molar-mass-distribution data over the temperature range from 250 to 375 °C (Chan and Balke, 1997a). The simplifications of this model should be noted, however. For example, the assumption that homolysis of the macro-alkyl radical (rate coefficient k_2, above) is unimportant compared with the intermolecular propagation reaction neglects the importance of β-scission as a route in PP. This reaction is one of the reasons why PP undergoes negligible crosslinking in comparison with scission since the radical is too short-lived to react with another radical to form a crosslink before it undergoes scission to give the products $-CH_3C{=}CH_2$ and $\cdot CCH_3CH_2-$.

Other thermal reactions of linear polymers

In addition to the depolymerization reaction discussed earlier, other reactions may be favoured. These are elimination from a side chain and cyclization. For example, propylene is eliminated from the side chain of poly(isopropyl acrylate) as shown in Scheme 1.51(a), leaving poly(acrylic acid). This occurs in all polymers having ester side groups with β-hydrogens available to form a six-membered transition state, as shown. Thus both acrylate and methacrylate polymers will undergo this reaction and, since depolymerization is the dominant thermal-degradation reaction in methacrylates, elimination of alkenes is more important in the poly(acrylates).

There have been concerns with the facile degradation during melt processing of the biodegradable polymer poly(hydroxy butyrate) (PHB). Thermal degradation of PHB occurs at the processing temperature of 180 °C by the β-elimination of crotonic acid, $CH_3CH{=}CHCOOH$, as shown in Scheme 1.51(b), as well as the formation of oligomers of PHB (Billingham *et al.* 1987). Unlike the situation for the ester side groups illustrated in Scheme 1.51(a), the six-membered transition state now occurs in the backbone, so elimination results in chain scission.

The material therefore follows the kinetics of random chain scission:

$$1/DP - 1/DP_0 = kt, \quad (1.106)$$

with an activation energy of 247 ± 19 kJ/mol, which is similar to that for the degradation of simple esters. Copolymers with β-hydroxy pentanoic acid (valeric acid) at levels of 5%–25% to give PHB-co-VA have a lower crystallinity and can be melt-processed at lower temperatures (<170 °C) without degradation. It should be noted that these elimination reactions are not free-radical in origin, so the stabilization strategies discussed later will not be applicable.

The cyclization reaction of poly(acrylonitrile) results in the formation of an extended conjugated backbone through the intramolecular imidization reaction which occurs without the elimination of any product. Reactions of this type are favoured by the chain conformation

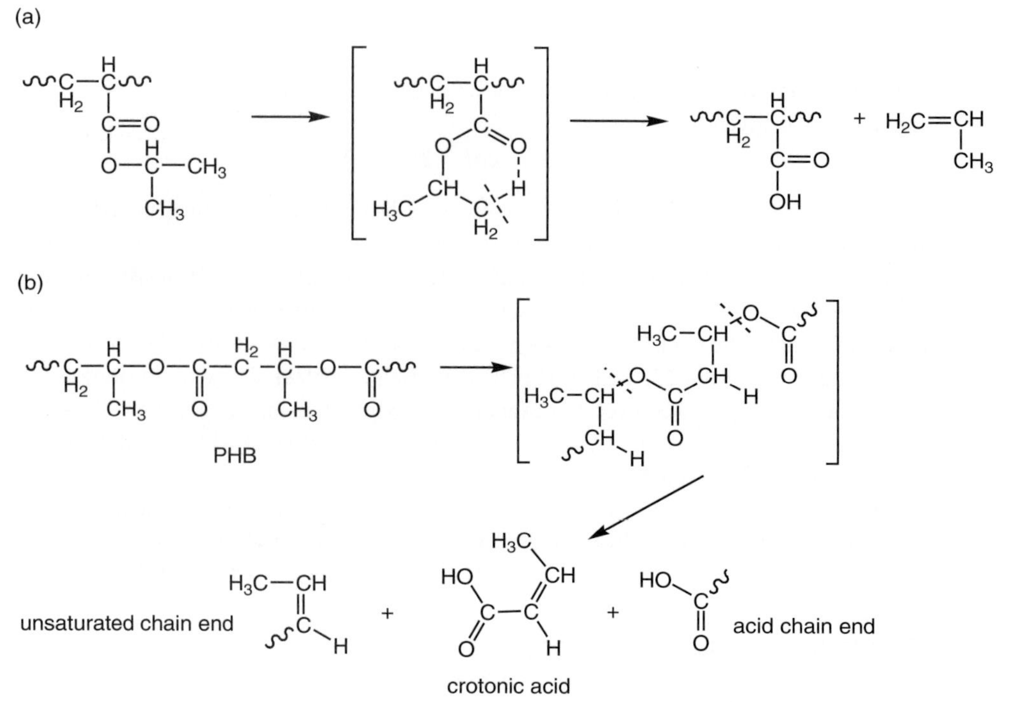

Scheme 1.51. (a) Elimination of ester side groups through a six-membered transition state: propylene from thermolysis of poly(isopropyl acetate). (b) PHB elimination of crotonic acid and accompanying chain scission.

producing adjacent nitrile groups. This reaction is exploited in the production of carbon fibres by the cyclized product being pyrolysed at high temperature in an oxidizing environment to eliminate hydrogen from the backbone, so producing aromatization, and, if the PAN precursor is a fibre, the result is a carbon (graphite) fibre (Bansal and Donnet 1989) (Scheme 1.52).

These reactions when studied by FT-IR emission spectroscopy gave little production of oxidation products such as carbonyls under the graphitization conditions (Celina *et al.* 1997).

Intermolecular reactions between carbon-centred radicals on adjacent chains will produce crosslinking, which will compete with the chain-scission and intramolecular reactions. The effect on the molar-mass distribution will depend on the competition between these reactions. For example the poly(acrylates) will preferentially crosslink rather than degrade and the recovery of monomer from these polymers is less than 1%, compared with close to 100% for poly(methacrylates). Poly(vinyl chloride) (PVC) is also a polymer that undergoes crosslinking in preference to scission, but the major observation of importance is the rapid elimination of hydrogen chloride by an auto-catalysed reaction as shown in Scheme 1.53. This is discussed in more detail later.

Random thermal crosslinking

In the free-radical kinetic Scheme for the thermal degradation of polymers, it was considered that the termination reaction resulted in disproportionation so that there was no increase in molar mass of the polymer in the termination step. At moderate temperatures, particularly in polymers such as PE and PVC for which unsaturation plays an important role

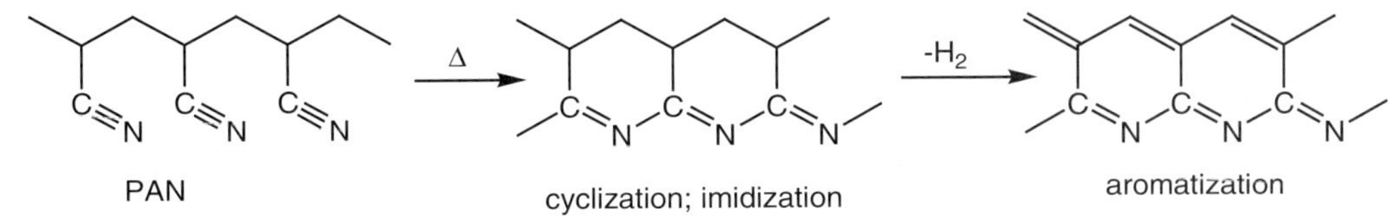

Scheme 1.52 Intramolecular cyclization of PAN to give a carbon-fibre precursor.

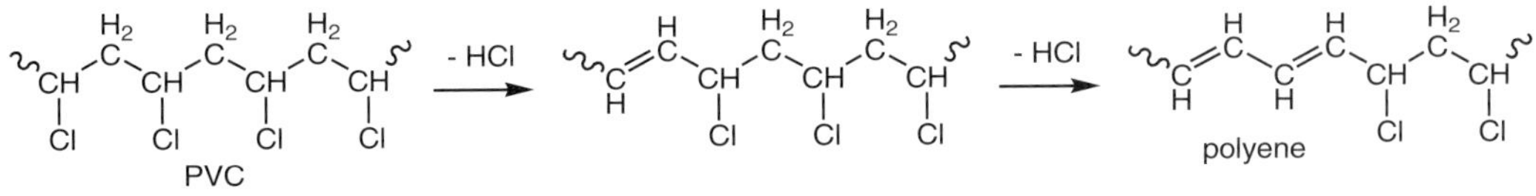

Scheme 1.53. Auto-catalysed elimination of HCl from PVC to give a polyene sequence.

in the degradative pathway, this will not be the case and combination of the macro-alkyl radicals is an important process leading to the formation of a crosslink.

The control of this process is technologically important in the formation of elastomeric networks, the first example of which was the vulcanization of rubber (Brydson 1988). Thus, in natural rubber, the two processes of mastication to lower the molar mass of the *cis*-poly (isoprene) and then the use of sulfur or peroxides to form a network of effectively infinite molar mass are exploited. Crosslinking is one of the commonly performed reactive processing operations and the control of the chemorheology underpins successful elastomer production. The process for crosslinking of polyethylene has been described earlier (Scheme 1.45) and the use of specific additives (e.g. silane grafting agents prior to hydrolytic crosslinking) was discussed (Scheme 1.46) and is also considered later. Here we are concerned with the process only in the presence of a free-radical generator. Thus the starting point is the polymer free radical formed by the abstraction of a hydrogen atom from the backbone by an initiator, which may be deliberately added, such as the free radical, I·, from the decomposition of a peroxide, or a macro-alkyl radical from a chain-scission event, Scheme 1.54.

The radical R_n· may be a backbone radical, not a terminal radical as results from scission, so if there is reaction with another chain having a backbone radical, R_r·, the crosslink will form at random between two chains. If there is a vinyl group on the other chain then an addition reaction is possible and the nature of the crosslink will depend on the location of the unsaturation (Scheme 1.45). Abstraction of hydrogen allyl to the double bond will also compete with this, but the resultant main-chain macro-alkyl radical may then also undergo crosslinking.

When crosslinking predominates over disproportionation or scission, the molar-mass distribution will broaden and M_w will shift to higher values. The process of crosslinking may be followed by extracting the polymer to determine the residual soluble (sol) fraction, S, i.e., the amount of material not yet incorporated into the infinite network. The equation describing this is the modified Charlesby–Pinner equation (Boyd and Phillips 1993):

$$S + S^{1/2} = p_0/q_0 + 1/(q_0 \mathrm{DP}_n r), \tag{1.107}$$

where p_0 and q_0 are the fractions of monomer units involved in a crosslink or chain-scission event, respectively, per free radical and r is the radical yield (proportional to the initiator concentration); DP_n is the number average-degree of polymerization of the original polymer. Thus, in an application of this equation, a plot would be made of $(S + S^{1/2})$

$R_nH + I\cdot \rightarrow R_n\cdot + HI$

$R_n\cdot + R_r\cdot \rightarrow P_n\text{--}P_r$

Scheme 1.54. Free-radical formation and crosslinking initiated by a radical I· from an external source (radiation or a labile molecule, e.g. peroxide).

against the inverse of the initiator concentration, 1/[I], and the ratio of the intercept to the slope of this linear plot will be (p_0DP$_n$) so that both p_0 and q_0 may be determined. There are several assumptions in this equation, and an alternative approach is via measurement of equilibrium swelling, but then the polymer–solvent-interaction parameter must be known (Boyd and Phillips 1993).

The technology of controlled crosslinking during reactive extrusion has been surveyed by Brown (1992) and there have been many reviews of the technology of elastomer vulcanization (Brydson, 1988, Morrison and Porter, 1989).

Thermal degradation of PVC

Many of the above features of free-radical thermal reactions (and the need to inhibit them) are seen in the degradation of PVC. This commences at temperatures as low as 150 °C with the evolution of hydrogen chloride and darkening of the polymer as noted above in Scheme 1.53. Both of these processes occur at much lower temperatures than would be predicted from studies of model compounds. Initiation reactions have been the subject of intense research, but it is agreed that there are weak links arising from the polymerization reaction (Braun, 1981, Starnes, 1981). These may be chloroallyl (–CH=CHCHCl–) groups, branch points, chain ends with initiator fragments, head-to-head linkages and oxidized polymer segments such as the carbonylallyl (–COCH=CHCHClCH$_2$–) group. The role of the β-chloroallyl group as an initiator of dehydrochlorination was noted earlier (Tudos *et al.*, 1979), but the mechanism by which this occurs is most probably not free-radical but ionic. The concentration of these groups in the polymer is small, but, since the reaction is auto-catalytic in evolved HCl, facile dehydrochlorination to form polyenes with a length of 4–10 repeat units soon follows. Free-radical formation does occur, but leads to crosslinking rather than chain scission. Because of the rapid formation of unsaturation, PVC is a crosslinking polymer on degradation. Further free-radical reactions occur during the thermo-oxidative reactions of PVC (Tudos *et al.*, 1979), which are discussed in Section 1.4.2.

Most attention has been paid to the dehydrochlorination reaction during processing because of the production of a dark product as well as the damaging effect of the evolved HCl. Certain copolymers and blends are even less stable than PVC, which fact is generally linked to the presence of carbonyl groups (such as in *N*-vinyl-2-pyrrolidone) in the blend that associate with the chlorine atom and lower the onset temperature for dehydrochlorination. This also has the effect of increasing the conjugation length and resulting in extended *trans*-polyenes with $n > 20$ (Dong *et al.*, 1997, Kaynak *et al.*, 2001). Similar results have been reported with blends of PVC with poly(vinyl acetate) (PVAc), in which the PVC sensitizes the elimination of acetic acid from PVAc. This acid then catalyses the dehydrochlorination of the PVC, so that, together, both components of the blend degrade more rapidly than if they were degraded separately (McNeill, 1989). Processing of PVC and its copolymers and blends is possible only if the HCl or other acidic products evolved are scavenged to prevent the auto-catalytic elimination reactions that result in polyene formation and darkening. The stabilization strategies are discussed later.

1.4.2 Free-radical formation in the presence of oxygen

Since it is often impossible to avoid the presence of atmospheric oxygen in the study of the rheological properties and/or processing of a molten polymer, the effect of oxygen on the chemical composition of the polymer is considered first. It is then possible to consider the ways in which this may be exploited by accelerating the oxidation with pro-oxidants or hindered by adding stabilizers to inhibit or at least retard the reactions. The starting point is the free-radical chemistry of an oxidizing, molten hydrocarbon polymer. Initially the polymer is treated as a simple saturated hydrocarbon polymer, PH, which focusses attention on the fact that it is the presence of an abstractable hydrogen atom, somewhere in the repeat unit of the polymer, which is significant. The reaction sequence differs from that for thermal and mechanical degradation in an inert atmosphere due to the rapid scavenging of the macro-alkyl radicals by oxygen to give alkoxy and peroxy radicals and the formation of the hydroperoxide as a reactive intermediate by further abstraction reactions.

A suitable system to consider initially is polypropylene, which has a hydrogen atom on the tertiary carbon atom that is available for abstraction to form a polymer-backbone radical. This is also of practical significance since controlled scission of polypropylene has been used to produce polymers with controlled molar-mass distribution and rheological properties (Brown, 1992).

Free-radical oxidation of polypropylene

The fundamental radical processes during the thermal oxidation of polyolefins during processing are, in principle, identical, with only minor differences due to variations in the initiation mechanism or secondary reactions that change the final product distribution (Zweifel, 1998, Al-Malaika, 1989). Evidence for the reaction scheme and the products which may be formed has been obtained from studies of model compounds as well as analysis of polypropylene at often high extents of oxidation. The determination of specific species, e.g. p-, s- or t-hydroperoxides and higher oxidation products such as per-acids, has required specific derivatization reactions coupled to spectroscopic analytical tools, principally involving infrared spectroscopy (Scheirs *et al.*, 1995b). As the sensitivity and specificity of these techniques have been improved, the identification of the products from different possible reaction steps that occur during the thermal oxidation of unstabilized polypropylene has meant that the mechanism has increased in complexity. In spite of this, the free-radical oxidation mechanism has generally been regarded us consisting of the steps shown in Scheme 1.55 (Al-Malaika, 1989, Scott, 1993b).

In thermal oxidation, initiation (1) results from the thermal dissociation of chemical bonds that may arise from intrinsically weak links formed as by-products of the polymerization reaction (e.g. head-to-head links) or impurities formed in the polymerization reactor such as hydroperoxides, POOH, or in-chain peroxides as occur in polystyrene from oxygen scavenging. Reaction (1′) shows that POOH may produce peroxy and alkoxy radicals that may subsequently form alkyl radicals via reaction (3).

The key reaction in the propagation sequences is the reaction of polymer alkyl radicals (P·) with oxygen to form polymer peroxy radicals (2). This reaction is very fast. The next propagation step, reaction (3), is the abstraction of a hydrogen atom by the polymer peroxy radical (POO·) to yield a polymer hydroperoxide (POOH) and a new polymer alkyl radical (P·). In polypropylene (Scheme 1.56), hydrogen abstraction occurs preferentially from the tertiary carbon atoms since they are the most reactive, as shown in reaction (11).

Initiation:

Polymer (defect)	$\rightarrow$	$\alpha P\bullet$	(1)
2 POOH	$\rightarrow$	$PO_2\bullet + PO\bullet + H_2O$	(1′)

Propagation:

$P\bullet + O_2$	$\rightarrow$	$POO\bullet$	(2)
$POO\bullet + PH$	$\rightarrow$	$POOH + P\bullet$	(3)

Chain branching:

POOH	$\rightarrow$	$PO\bullet + \bullet OH$	(4)
$PH + \bullet OH$	$\rightarrow$	$P\bullet + H_2O$	(5)
$PH + PO\bullet$	$\rightarrow$	$P\bullet + POH$	(6)
$PO\bullet$	$\rightarrow$	chain scission	(7)

Termination (leading to non-radical products):

$POO\bullet + POO\bullet$	$\rightarrow$	POOOOP ($\rightarrow$ POOP + O_2)	(8)
$POO\bullet + P\bullet$	$\rightarrow$	POOP	(9)
$P\bullet + P\bullet$	$\rightarrow$	PP	(10)

Scheme 1.55. A simplified free-radical oxidation scheme for a polyolefin.

$$POO\bullet + \sim\!\!CH_2\!-\!CH(CH_3)\!-\!CH_2\!\!\sim \ (\text{PP}) \longrightarrow \sim\!\!CH_2\!-\!\dot{C}(CH_3)\!-\!CH_2\!\!\sim + POOH \quad (11)$$

$$POO\bullet + \sim\!\!CH_2\!-\!CH_2\!\!\sim \ (\text{PE}) \longrightarrow \sim\!\!CH_2\!-\!\dot{C}H\!\!\sim + POOH \quad (12)$$

Scheme 1.56. Reactions of the chain-carrying peroxy radical with polypropylene (11) or polyethylene (12) to give tertiary- or secondary-alkyl radicals, respectively.

Hydrogen abstraction is known to occur from secondary carbon atoms in polyethylene (12) and may also occur in polypropylene, but with lower reaction rates. For polypropylene it was shown that intramolecular hydrogen abstraction in a six-ring favourable stereochemical arrangement will preferentially lead to the formation of sequences of hydroperoxides in close proximity (Scheirs *et al.*, 1995b, Chien *et al.*, 1968, Mayo, 1978). Infrared studies of polypropylene hydroperoxides showed that more than 90% of these groups were intramolecularly

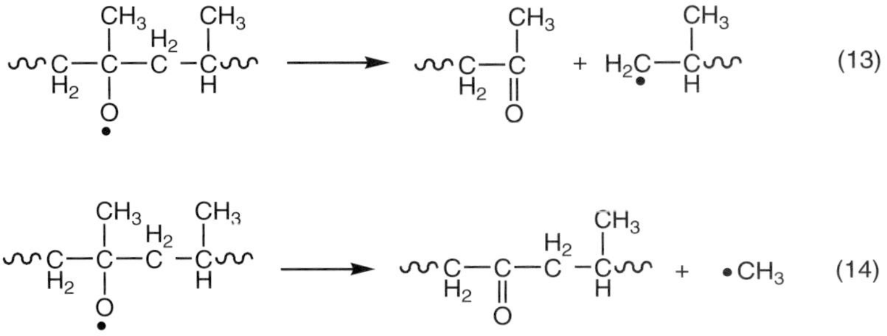

$$\sim\!\!CH_2{-}C(CH_3)(O^{\bullet}){-}CH_2{-}CH(CH_3)\!\!\sim \;\longrightarrow\; \sim\!\!CH_2{-}C(CH_3){=}O \;+\; {}^{\bullet}CH_2{-}CH(CH_3)\!\!\sim \qquad (13)$$

$$\sim\!\!CH_2{-}C(CH_3)(O^{\bullet}){-}CH_2{-}CH(CH_3)\!\!\sim \;\longrightarrow\; \sim\!\!CH_2{-}C({=}O){-}CH_2{-}CH(CH_3)\!\!\sim \;+\; {}^{\bullet}CH_3 \qquad (14)$$

Scheme 1.57. Chain scission (13) or elimination (14) reactions of alkoxy radicals in polypropylene resulting in terminal (13) or in-chain (14) ketone groups as well as alkyl radicals that may react further with oxygen.

hydrogen-bonded. The results indicated that the peroxide groups were preferentially present in sequences of two or more, which supported intramolecular hydrogen abstraction during the oxidation (Chien *et al.*, 1968). It was also concluded that the cleavage reaction accompanied the oxidation propagation rather than the termination reaction.

As shown in Scheme 1.55, chain branching by thermolysis or photolysis, reaction (4), of polymer hydroperoxides (POOH) results in the formation of very reactive polymer alkoxy radicals (PO·) and hydroxyl radicals (·OH). The highly mobile hydroxyl and polymer alkoxy radical can abstract hydrogen atoms from the same or a nearby polymer chain by reactions (5) and (6), respectively (Zweifel, 1998). These polymer oxy radicals can react further to result in β-scission, by reaction (13), or the formation of in-chain ketones by reaction (14), as shown in Scheme 1.57, or can be involved in termination reactions.

The termination of polymer radicals occurs by various bimolecular recombination reactions. When the oxygen supply is sufficient the termination is almost exclusively via reaction (8) of Scheme 1.55. At low oxygen pressure other termination reactions may take place (Zweifel, 1998). The recombination is influenced by cage effects, steric control, mutual diffusion and the molecular dynamics of the polymer matrix. In melts the recombination of polymer peroxy radicals (POO·) is efficient due to the high rate of their encounter.

Secondary reactions and product formation

As shown in Scheme 1.58, polymer acyl radicals can be produced by reaction (15) via hydrogen abstraction from chain-end aldehyde groups. Polymer acyl radicals are easily oxidized further to polymer peracyl radicals in reaction (16), and these can abstract hydrogen to form peracids in reaction (17).

The cleavage of peracids can lead to polymer carboxy and hydroxyl radicals, reaction (18), with the carboxy radicals again able to abstract hydrogen to form carboxylic groups by reaction (19).

Alternative sources of acidic species during the oxidation of isotactic polypropylene have been suggested from mass-spectrometric analysis of thermal-decomposition products from polymer hydroperoxides (Commerce *et al.*, 1997). Acetone, acetic acid and methanol comprised 70% of the decomposition products, suggesting either a high extent of oxidation involving secondary hydroperoxides or direct reactions of hydroxyl radicals with ketones (derived through reactions discussed in the next section).

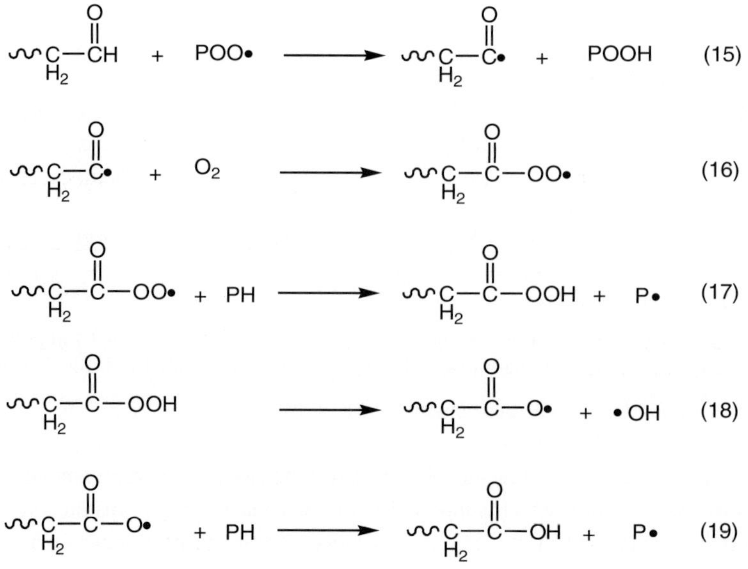

$$\sim C(H_2)-C(=O)H + POO\bullet \longrightarrow \sim C(H_2)-C(=O)\bullet + POOH \quad (15)$$

$$\sim C(H_2)-C(=O)\bullet + O_2 \longrightarrow \sim C(H_2)-C(=O)-OO\bullet \quad (16)$$

$$\sim C(H_2)-C(=O)-OO\bullet + PH \longrightarrow \sim C(H_2)-C(=O)-OOH + P\bullet \quad (17)$$

$$\sim C(H_2)-C(=O)-OOH \longrightarrow \sim C(H_2)-C(=O)-O\bullet + \bullet OH \quad (18)$$

$$\sim C(H_2)-C(=O)-O\bullet + PH \longrightarrow \sim C(H_2)-C(=O)-OH + P\bullet \quad (19)$$

Scheme 1.58. Formation and reactions of peracids ultimately producing carboxylic acids during the oxidation of polypropylene.

$$\sim C(H_2)-C(H)(CH_3)-CH_2\bullet + O_2 + PH \longrightarrow \sim C(H_2)-C(H)(CH_3)-CH_2OOH + P\bullet \quad (20)$$

$$\sim C(H_2)-C(H)(CH_3)-CH_2OOH + O_2 \longrightarrow \sim C(H_2)-C(H)(CH_3)-C(=O)OOH + H_2O \quad (21)$$

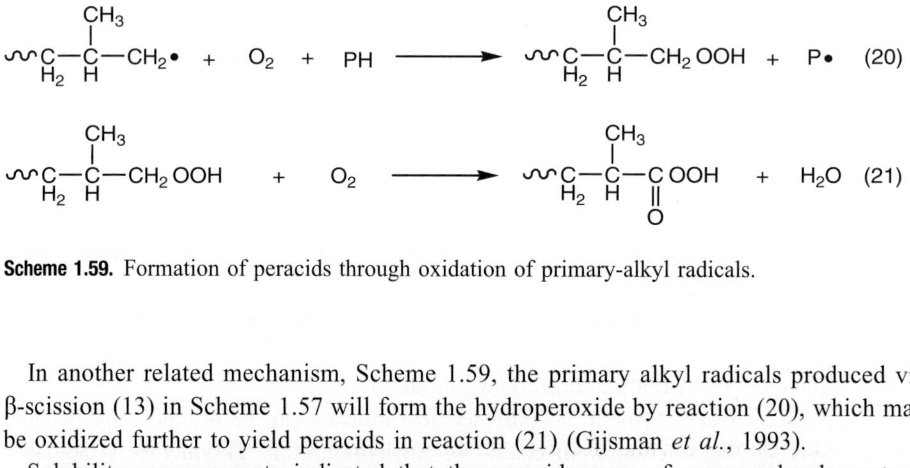

Scheme 1.59. Formation of peracids through oxidation of primary-alkyl radicals.

In another related mechanism, Scheme 1.59, the primary alkyl radicals produced via β-scission (13) in Scheme 1.57 will form the hydroperoxide by reaction (20), which may be oxidized further to yield peracids in reaction (21) (Gijsman *et al.*, 1993).

Solubility measurements indicated that the peracids were of macromolecular nature, being formed by the oxidation of macroaldehydes, presumably via radical attack on methyl hydrogens, subsequent peroxidation and disproportionation (Zahradnikova *et al.*, 1991). The difficulty of analysis of these reactive species has meant that unequivocal quantitation of the reactions has not been achieved.

Classical free-radical oxidation kinetics

The starting point in the kinetic analysis of the oxidation of polypropylene is the measurement of the extent of oxidation of the polymer as a function of time. In rheological and processing studies it is difficult to obtain real-time chemical information, and many of the chemical changes are elucidated after processing by analysis of the polymer product withdrawn after various durations of reaction. Classically, the fundamental measurement is that of the uptake of oxygen by the polymer when in a controlled environment (Scheirs *et al.*, 1995a), but it may also involve oxidation-product analysis by FT-IR and related spectroscopic techniques as discussed in Section 4.3 as well as GPC analysis of the molar-mass distribution. In-line rheometry may also give information on molar mass for these analyses, and remote spectroscopy with fibre optics provides a method for obtaining chemical data in real time.

The fundamental reaction mechanism for the free-radical oxidation of hydrocarbons has been used to relate the consumption of oxygen to the formation of oxidation products in polypropylene. A kinetic interpretation is based on the steady-state approximation equating the rates of the initiation and termination reactions. With this approach it is possible to derive mathematical equations describing the consumption of oxygen or the formation of specific oxidation products. To solve the equations it is necessary to determine the most likely route for initiation of oxidation. The initiation mechanism chosen is the bimolecular reaction of hydroperoxides, reaction (1′) of Scheme 1.55, with a rate coefficient k_1'.

On simplifying the oxidation of polypropylene to the propagation and termination reaction sequences (2), (3), (8), (9) and (10) of Scheme 1.55, the consumption of oxygen may be related to the formation of hydroperoxides at high and low oxygen pressures: Equations (1.108) and (1.109) below, respectively. Both equations were derived using the steady-state approximation between initiation and termination, with reaction (9) and (10) being neglected at high oxygen pressure, and reactions (8) and (9) being redundant at low oxygen pressure.

At high oxygen pressure

$$-\mathrm{d}[\mathrm{O_2}]/\mathrm{d}t = k_3(k_1'/k_8)^{1/2}[\mathrm{ROOH}][\mathrm{RH}]. \tag{1.108}$$

At low oxygen pressure

$$-\mathrm{d}[\mathrm{O_2}]/\mathrm{d}t = k_2(k_1'/k_{10})^{1/2}[\mathrm{ROOH}][\mathrm{O_2}]. \tag{1.109}$$

A more complex relationship was presented with Equation (1.110) below, obtained with the assumption that the kinetic chain is not too short and that $\mathrm{k}_9{}^2 = \mathrm{k}_{10}k_8$ (and with the initiation rate $r_\mathrm{i} = k_1'[\mathrm{POOH}]^2$ from (1′)):

$$-\mathrm{d}[\mathrm{O_2}]/\mathrm{d}t = \ k_2k_3[\mathrm{PP}][\mathrm{O_2}]r_i^{1/2}/(k_2k_8^{1/2}[\mathrm{O_2}] \ + \ k_3k_{10}^{1/2}[\mathrm{PP}]). \tag{1.110}$$

One of the obvious features of the oxidation of polypropylene is the formation of hydroperoxides (reaction (3) in Scheme 1.55) as a product. The initiation of the oxidation sequence is usually considered to be thermolysis of hydroperoxides formed during synthesis and processing (shown as the bimolecular reaction (1′) in Scheme 1.55). The kinetics of oxidation in the melt then become those of a branched chain reaction as the number of free radicals in the system continually increases with time (ie the product of the oxidation is also an initiator). Because of the different stabilities of the hydroperoxides (e.g. p-, s- and t-; isolated or associated) under the conditions of the oxidation, only a fraction of those formed will be measured in any hydroperoxide analysis of the oxidizing polymer. The kinetic character of the oxidation will change from a linear chain reaction, in which the steady-state approximation applies, to a branched-chain reaction, for which the approximation might not be valid since the rate of formation of free radicals is not constant and the oxidation demonstrates *auto-acceleration*. Such reactions have long been studied in the high-temperature gas-phase oxidation of hydrocarbons, and the formation of hydroperoxides as a degenerate branching agent results in the observation of cool flames in which the chain reaction is linear until a critical concentration of hydroperoxide is reached (Dainton, 1956). It then becomes branched, an exponentially increasing chain reaction (flame or explosion) ensues, which consumes the branching agent, and the system then becomes a linear chain again. This alternating process continues until the fuel is exhausted. By altering the conditions of pressure and temperature, the termination rate may be altered and the system may be held in the explosion (branched) regime or in the linear regime in which hydroperoxide accumulates.

Obviously, under the conditions of processing and chemorheology, the oxidation of the polymeric hydrocarbon cannot reach the above critical condition, and homogeneous kinetic treatments of the oxidation of polypropylene involve perturbations of the steady-state approximation. The application of models involving degenerate branching has been a particular feature of kinetic treatments of the oxidation of polypropylene by many Russian authors (Emanuel and Buchachenko, 1987). In one such study, it was suggested that, under conditions for degenerate branching, the maximum rate of oxygen consumption may be expressed by the following formula, where δ is the probability of degenerate chain branching, which was derived again by neglecting reactions (9) and (10) at high oxygen pressure and assuming the oxidation chain to be sufficiently long:

$$[-\mathrm{d}[\mathrm{O}_2]/\mathrm{d}t]_{\max} = \delta k_3[\mathrm{RH}]^2/(2k_8)^{1/2}. \tag{1.111}$$

It is noted that, in many of these studies, the basic kinetic information is obtained from an oxygen-uptake curve. If the consumption of oxygen has been measured by a pressure transducer (Scheirs *et al.*, 1995b) then one limitation is the lack of specificity of the measurement. The total pressure in the system will depend on the partial pressure of the products (for example carbon dioxide and water) as well as on that of oxygen. This may be overcome by undertaking a total analysis of the gases present in the apparatus.

From the slope of the oxygen-uptake curve in the linear region, a 'steady-state' rate of O_2 consumption has often been used to compare the oxidation of various polymer samples (Dainton, 1956). The determination of rate coefficients and kinetic information has been repeatedly obtained from studies of the oxidation of model hydrocarbons in solution at relatively low conversions. Studies of model substances, such as 2,4-dimethylpentane and 2,4,6-trimethylheptane, as well as of polypropylene enabled the evaluation of rate constants for propagation and termination (Chien and Wang, 1975). Any attempt to establish precise rate constants for the initiation, propagation and termination in the auto-oxidation of PP is, however, complicated by the fact, as pointed out by Mayo (1972), that intermediate oxidation products of a saturated hydrocarbon are 10–100 times more susceptible to oxidation than is the starting material. Rate constants as measured may be those for a complex co-oxidation of a highly degraded polymer and its oxidation products (Mayo, 1972). These include the range of by-products outlined in Schemes 1.57, 1.58 and 1.59, viz. carboxylic acids, peracids and peresters, from the oxidation of the aldehyde produced as a primary product of the free-radical oxidation of the polymer. The diversity of products can be significant in determining the subsequent performance of the polymer since the chemical products will be photoinitiators of oxidation, and this is one of the reasons for the incorporation of stabilizers in the melt during processing.

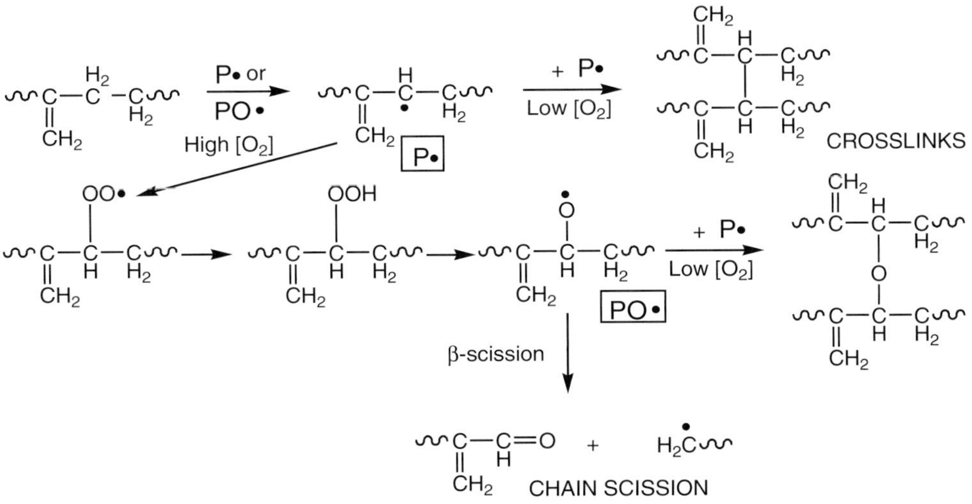

Scheme 1.60 Competitive chain scission and crosslinking with various $[O_2]$ in thermo-oxidation during processing of LDPE. After Al-Malaika (1989).

Free-radical oxidation of other linear polymers

It was noted in the section on thermal degradation of polymers that there were significant differences in behaviour depending on the nature of radical and non-radical intermediates responsible for chain-scission events. For example, polyethylene is principally a crosslinking polymer, whereas polypropylene tends to undergo chain scission. Careful studies of the oxidative degradation of all commercially important polymers have been undertaken with the aim of understanding the mechanism of oxidation and thus ways to control the process. Prior to a consideration of combined thermal, mechanical and oxidative degradation in the melt phase, it is appropriate to look at the reasons for these differences.

Polyethylene

In polyethylene, the tertiary carbon atom, which dominated the chemistry of the oxidative degradation of PP, is present only at branch points. This suggests that there may be a difference among LDPE, LLDPE and HDPE in terms of the expected rates of oxidation. This is complicated further by the presence of catalyst residues from the Ziegler–Natta polymerization of HDPE that may be potential free-radical initiators. The polymers also have differences in degree of crystallinity, but these should not impinge on the melt properties at other than low temperatures at which residual structure may prevail in the melt. Also of significance is residual unsaturation such as in-chain *trans*-vinylene and vinylidene as well as terminal vinyl, which are 'defects' in the idealized PE structure.

The reaction sequence (Scheme 1.60) shows the competition between chain scission and crosslinking reactions as well as the participation of unsaturation in the thermo-oxidation of LDPE in the melt (Al-Malaika, 1989). The process of formation of H and Y crosslinks involving terminal unsaturation (Scheme 1.45) and the formation of terminal unsaturation during β-scission (Scheme 1.49) in the absence of oxygen have been discussed previously.

It has been shown that the type of catalyst used in HDPE polymerization affects the flow behaviour and that crosslinking is favoured at higher temperatures (Zweifel, 1998). As in

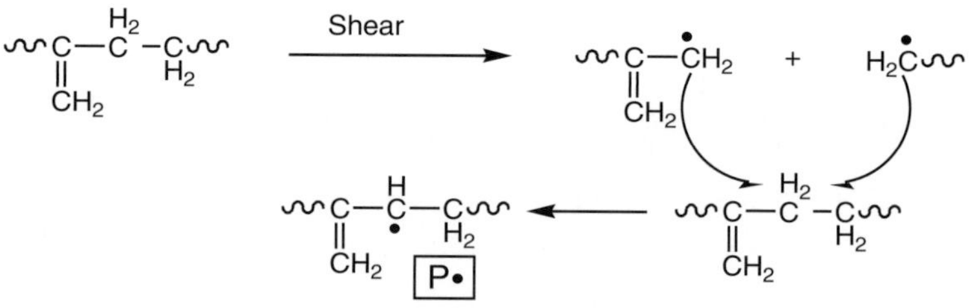

Scheme 1.61 Initiation of the mechano-oxidation of PE and subsequent in-chain radical (P·) formation by abstraction by a terminal alkyl radical after the initial chain-scission event. After Scott (1993b).

PP, the polymer hydroperoxide is the key intermediate which controls the oxidative performance in the melt. In an analysis of the published data on the processing performance of PE, Gugumus (1999) noted the following.

1. The induction period for the oxidation was inversely proportional to the vinyl content of the polymer, suggesting the importance of initiation by primary attack by oxygen of the allylic C–H bond.
2. The differentiation between PP and HDPE in the processing behaviour was linked to steric effects that limited direct reaction of a macro-alkyl (or alkoxy or peroxy) radical with the alkene $CH_2{=}C(CH_3)$– formed in PP, in contrast to the unhindered reaction of the radical with the –C=C– group in HDPE. *Thus crosslinking was favoured in HDPE over PP.*
3. The terminal vinyl groups appeared to be those responsible for crosslinking, rather than the *trans*-vinylene or vinylidene groups known to be present in LLDPE. This conclusion has been supported by results from later studies on γ-initiated crosslinking (Bracco *et al.*, 2005).

In his further studies Gugumus (2000) showed that, for PE, the contribution of direct thermal production of free radicals during mechanochemistry in the melt was negligible in comparison with mechanical effects such as shear. This was based on the dependence of the initial rate of hydroperoxide formation being sensitive to the melt viscosity as well as the high activation energy for pure thermal initiation. Scott (1993b) also reached this conclusion, and proposed a scheme similar to that in Scheme 1.60 in terms of competition between chain scission and crosslinking for mechano-oxidation in the limits of high and low oxygen concentration. The major difference is that the initiating step in mechano-oxidation is the scission of the main chain to produce two terminal primary alkyl radicals (Scheme 1.61) rather than an in-chain secondary-alkyl radical (P· in Scheme 1.60) as in thermo-oxidation. The subsequent free-radical chemistry and the scission and crosslinking reactions are identical for thermo- and mechano-degradation since the terminal primary radicals then abstract protons from the PE chain to produce secondary-alkyl radicals.

An interesting result was the way in which the viscosity, η, had another effect, namely to control the diffusion of the free radicals out of the initial cage in which they were formed. Geminate recombination of radicals is therefore important and the formation of chain-carrying radicals was dependent on $\eta^{-1/3}$. Also of interest is the observation that the decomposition reactions at higher temperatures (170–200 °C) are different from those at lower processing temperatures (150–160 °C). This may reflect the importance of clusters of associated hydroperoxides, which decompose at the higher temperatures by a bimolecular reaction (Gugumus, 2002a, 2002b).

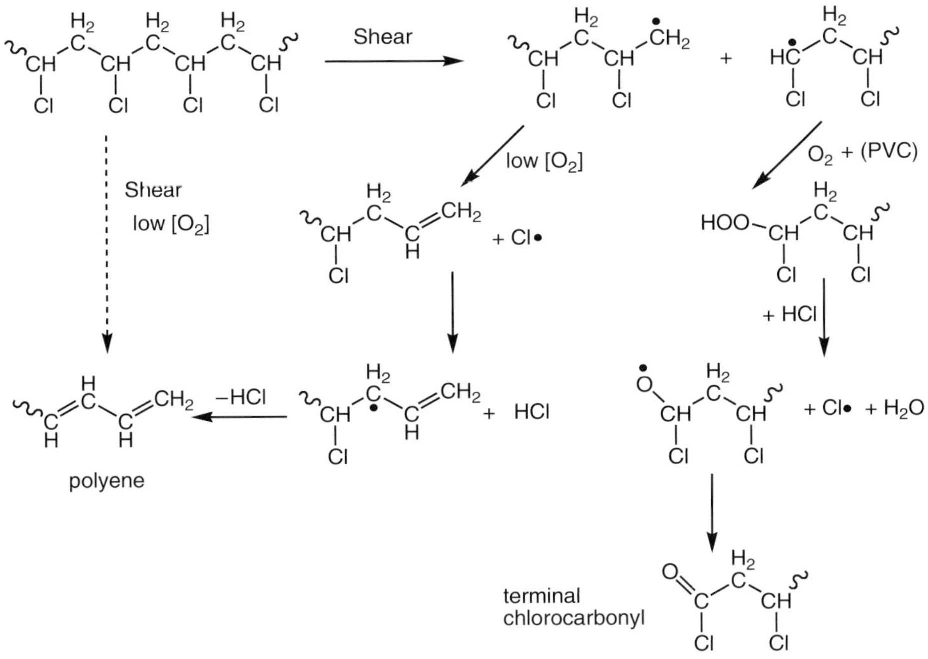

Scheme 1.62. The reaction sequence for PVC oxidation during mechanical working.

Poly(vinyl chloride)

While the polyene formation characteristic of PVC degradation will occur both in an oxidizing and in an inert atmosphere as described above, the presence of oxygen leads to oxidation of the polyenes to both cyclic peroxides and hydroperoxides (Tudos *et al.*, 1979). This is seen as the 'bleaching' of the polyenes in thermally degraded PVC after the admission of oxygen because of the loss in conjugation. The delocalized nature of the polyenyl radical leads to a large number of possible products, as shown in Scheme 1.62.

The products formed may be rationalized as a competition between polyene formation by dehydrochlorination and hydroperoxide formation through direct radical reaction as well as oxidation of the formed polyenes. Unlike the case for PE or PP, the decomposition of the hydroperoxides is catalysed by the hydrogen chloride present, so alkoxy-radical formation is enhanced. This is seen in the reaction sequence on the right of Scheme 1.62. The alkoxy radical, RO·, rearranges to form a terminal chlorocarbonyl and the chlorine atom abstracts hydrogen to form hydrogen chloride and continue the cycle. The simultaneous crosslinking and chain scission has been analysed by use of a Charlesby–Pinner plot (Equation (1.107)) and the ratio of scission to crosslinking was found to range from 0.4 to 0.7 at 180 °C in oxygen, depending on the type of PVC (Tudos *et al.*, 1979).

Aliphatic polyamides

Most of the polymers discussed so far have been addition polymers but there are important condensation polymers that have relevant mechanochemistry. This may involve reactions in addition to free-radical oxidation (Levchik *et al.*, 1999), and it has been shown that the carboxy-terminated Nylon 6 ε-caprolactam

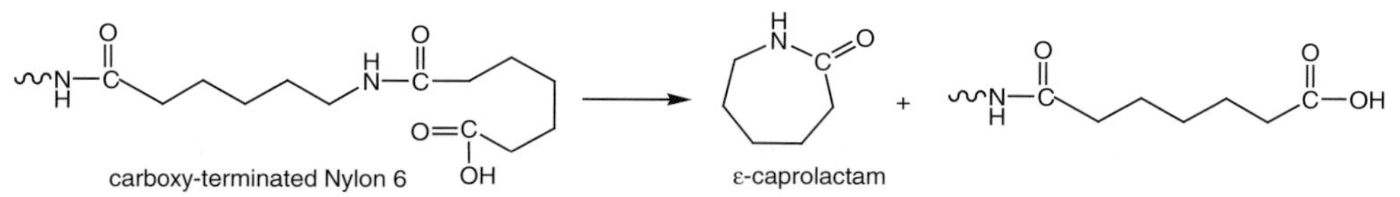

Scheme 1.63. Formation of ε-caprolactam by depolymerization of nylon 6. After Levchik (1999).

thermal and thermo-oxidative degradation of aliphatic nylons include several parallel and consecutive reactions that are sensitive to the temperature and conditions, particularly the presence of impurities. The aliphatic nylons of commercial significance are nylon-6, synthesized by the ring-opening of caprolactam, and nylon-6,6, which is synthesized by the condensation of hexamethylene diamine and adipic acid. This difference in synthesis carries over to the difference in degradation pathways.

Nylon-6 will undergo re-equilibration with the cyclic monomer as well as with larger cyclics at elevated temperature. This is the reverse of the polymerization process, which occurs at 200 °C and takes place through an intermediate carboxy-terminated hydrolysis fragment that undergoes intramolecular (or intermolecular) reaction to generate the cyclic monomer ε-caprolactam as shown below in Scheme 1.63.

This may also occur through the amino end group if this can be formed via an aminolysis reaction. Caprolactam formation is favoured by the presence of amino end groups. This is an example of a more general reaction for all polylactams that are in equilibrium with the cyclic monomer and the larger oligomers. There are many secondary reactions that may occur and one of key importance is the formation of precursors to crosslinks. This would include secondary amines from the condensation of amine end groups to eliminate ammonia and form an in-chain amine, which can then condense with an acid end group to form a crosslink. While the detailed chemistry of these reactions has been questioned, gelation from reactions such as this can occur on prolonged heating at 300 °C. It is known that residual catalyst from anionic polymerization of caprolactam will result in depolymerization. Base-catalysed decomposition is the more efficient route for recovery of caprolactam from nylon-6.

Nylon-6,6 undergoes gelation through crosslinking more readily than do other aliphatic nylons, through a mechanism similar to that for nylon-6. Other reactions are also observed, and different crosslinks have been proposed (Levchik *et al.*, 1999). In the presence of oxygen the polymer undergoes yellowing and discolouration through a free-radical chain reaction involving the macroradical below formed by abstraction from the *N*-vicinal methylene.This may arise from the rearrangement and reaction of the radicals from primary main-chain scission of the amide bond due to shear, as shown in Scheme 1.64. The acyl radical (R·) and amino radical (R′ ·) will abstract the hydrogen of the methylene group α to the amide group in another chain of nylon-6. The resulting alkyl radical is rapidly scavenged by oxygen to form a peroxy radical and then a hydroperoxide. Decomposition of this to the alkoxy radical results in an α-hydroxy amide, which is thermally unstable and results in chain scission.

It has been found that the end groups play a major role in the rate of oxidation since carboxylic acid groups will catalyse decomposition of hydroperoxide and lower the temperature for the maximum rate of degradation by 60 °C. In contrast polyamides terminated with amine end groups are more stable.

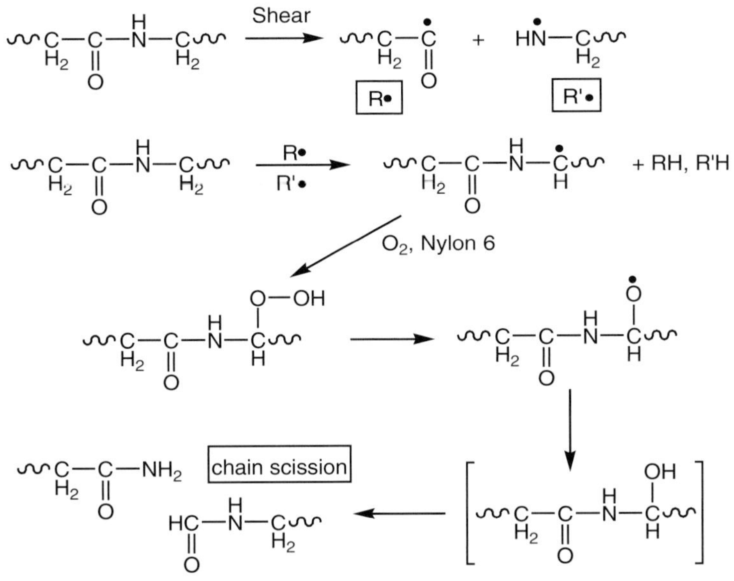

Scheme 1.64. The reaction sequence for shear-initiated oxidation and chain scission of nylon-6.

1.4.3 Control of free-radical reactions during processing

It is apparent from the above reactions that may occur during processing that these must be controlled if the product after processing is to have reasonable properties. This is particularly so during recycling where successive extrusion steps result in the accumulation of oxidation products. While a significant effort has been directed to the prevention of these free-radical reactions, there are many applications (e.g. lowering viscosity by changing the molar-mass distribution of polypropylene) in which the degradation reactions are exploited. This section will therefore discuss the **control** of free-radical reactions rather then just their inhibition. The control of free-radical reactions in polymers is important at several stages of the life of the polymer and the strategies employed are linked to the physical state of the polymer, the temperature and other potential degradative agents as well as the local concentration of oxygen.

In a kinetic model of the degradation which can occur during extrusion of polyethylene (El'darov *et al.*, 1996) it was shown that the radical reactions depend on the zones of the extruder and there are variations in the concentration profiles of dissolved oxygen, peroxidic products and inhibitors along its length. There is a zone where there is no oxygen remaining in the polymer and no formation of peroxide. Where mechanical action is responsible for free-radical formation in the absence of oxygen a different stabilization strategy is required. The different stabilizers for each of these conditions are considered separately below. There has been development of a strong technology for the stabilization of important polymers such as polypropylene in the melt, and it has been noted that very often the optimum melt stabilization system is not appropriate for the long-term oxidative stability of the polymer when it is in service (Henman, 1979). This arises because of the importance of alkyl-radical reactions under oxygen-starved conditions.

Free-radical scavenging

The strategy to be employed in free-radical scavenging is to intercept the radical species responsible for the propagation step in the degradation, since this reaction is the main route to produce products (in this case a degraded polymer through chain scission or crosslinking). For all free-radical reactions, the kinetic chain length, ν, of the reaction is defined as the number of products produced per initiating event or the rate of the propagation reaction r_p divided by the rate of initiation r_i. For a simple linear chain reaction of a hydrocarbon polymer of concentration [PH] as shown earlier in Table 1.17, Equation (1.105) may be written in a form that does not explicitly identify the initiation reaction so that

$$\nu = r_p/r_i = k_p[\mathrm{PH}]/(k_t r_i)^{1/2}, \tag{1.112}$$

where k_p and k_t are the rate coefficients for the propagation and termination reactions, respectively. The kinetics of free-radical oxidation reaction as described in Equations (1.108) and (1.109), Section 1.4.2, were sensitive to the partial pressure of oxygen since this defined the nature of the propagation reaction as well as the most likely termination steps

Depending on the conditions of the reaction, ν may range from slightly greater than unity to several thousand. For example, in the initiated oxidation of the model hydrocarbon, tetralin (tetrahydronaphthalene) at 30 °C the kinetic chain length, ν, is calculated from published values (Kamiya and Niki, 1978) to be about 2500. If this reaction is to be inhibited, then the value of ν must be reduced to <1. To achieve this level of inhibition, an additive that can compete with the propagation reaction by providing alternative termination routes for the radicals other than the usual recombination reaction is required. This problem is routinely met in the inhibition of polymerization of a liquid monomer such as styrene, and in that case a free-radical scavenger such as hydroquinone or the hindered phenol BHT is used. The purpose of a free-radical-scavenging stabilizer is to reduce the kinetic chain length by providing a more readily abstracted hydrogen atom and producing a free-radical product that is stable at the temperature of the reaction.

In the inhibition of the oxidation of polymer melts, the types of antioxidant may be characterized by their mode of action.

Chain-breaking acceptor antioxidants (CB-A)

The simplest examples of CB-A antioxidant are quinones, and benzoquinone has widely been used to inhibit alkyl-radical reactions (such as the thermal polymerization of styrene on storage). They react with polymer alkyl radicals (R·) formed on chain scission or through attack on the backbone to give the reduced form of the radical as shown in Scheme 1.65.

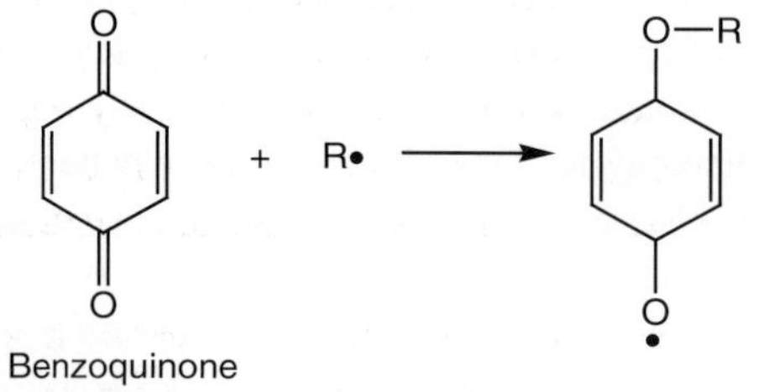

Benzoquinone + R• → (O–R, O•)

Scheme 1.65. Alkyl-radical scavenging by benzoquinone. After Al-Malaika (1989).

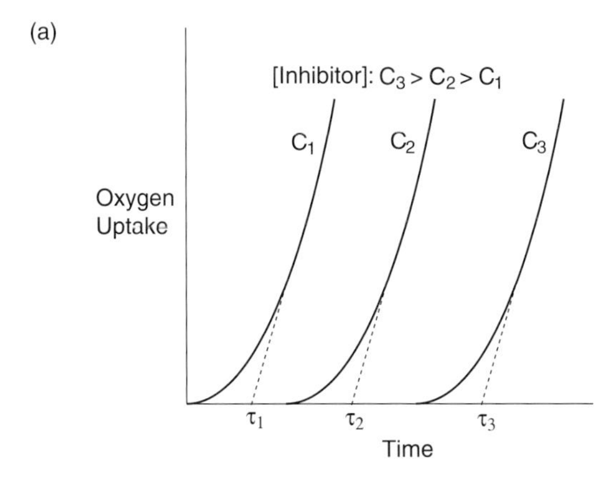

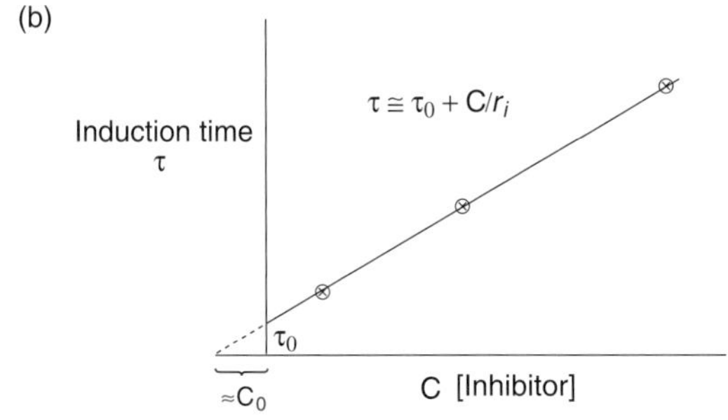

Figure 1.41. (a) Oxidation induction time, τ, for various concentrations of oxidation inhibitor (antioxidant) as determined from oxygen uptake (or carbonyl index) plots. (b) Estimation of intrinsic inhibitor concentration, C_0, from a plot of τ versus [Inhibitor]. The inverse of the slope of the plot reflects the rate of initiation.

These stabilizers react stoichiometrically with the alkyl radicals and are thus consumed in the process, unless there is a subsequent reaction to liberate the quinone again. In the absence of this, the kinetic relation for the CB-A stabilizer may be determined by simply replacing the usual termination reaction of the free-radical reaction scheme with that for the reaction of the inhibitor with the alkyl radical, R·, which would otherwise be propagating the chain. Thus, if reaction with the inhibitor can occur at the diffusion-controlled rate, then the rate of termination, r_t is dominated by the reaction with inhibitor of concentration [Inh] and becomes, in the steady state,

$$r_t = k_{inh}[\mathrm{Inh}][\mathrm{R}\cdot] = r_i. \tag{1.113}$$

The kinetic chain length, ν, is then reduced to

$$\begin{aligned}\nu = r_p/r_i = r_p/r_t &= k_p[\mathrm{RH}][\mathrm{R}\cdot]/k_{inh}[\mathrm{Inh}][\mathrm{R}\cdot] \\ &= k_p[\mathrm{RH}]/(k_{inh}[\mathrm{Inh}]).\end{aligned} \tag{1.114}$$

Since the reaction of Inh with the radicals has a low activation energy, k_{inh} is a diffusion-controlled rate constant for the reaction (about $10^9\,\mathrm{l\,mol^{-1}s^{-1}}$, depending on the viscosity of the melt) and thus, even for low concentrations of inhibitor, $\nu \ll 1$. The oxidation rate, given by r_p, will be very low until $[\mathrm{Inh}] < k_p[\mathrm{RH}]/\ k_{inh}$ and thereafter the oxidation will proceed as if there were no additive. This is shown schematically in Figure 1.41(a), and it may be seen that the induction period increases with the concentration, C, of inhibitor. If the reaction of the inhibitor is efficient and destroys one radical at each encounter, then the rate of consumption of inhibitor must be a measure of the rate of formation of free radicals in the system. In this case the induction period, τ, of the chain reaction is given by

$$\tau \cong [\mathrm{Inh}]/r_\mathrm{i}. \tag{1.115}$$

If the inhibitor is able to destroy more than one radical (as described below for chain-breaking redox antioxidants) then the efficiency of inhibition will be increased by this ratio.

A plot of the induction period against the concentration of inhibitor should be a straight line passing through the origin and with a slope depending on the efficiency of the antioxidant at that temperature. If the polymer already contains a free-radical scavenger or has intrinsic radical-scavenging capability, then the intercept on the induction period axis is non-zero and the system contains an equivalent inhibitor concentration given by the negative intercept on the concentration axis, as shown in Figure 1.41(a).

Chain-breaking donor antioxidants (CB-D)

These antioxidants include the hindered phenols and are considered to be most effective when the chain-carrying (propagation) radical is an -oxy radical such as alkyl peroxy, $RO_2\cdot$. Thus, in reactive processing, they would be expected to be of value in suppressing the oxidation reactions which can occur in the earlier zones of a reactive extruder. The chemistry of these systems has been studied in detail (Al-Malaika, 1989, Scott, 1993b), and it has been found in the case of hindered phenols that the effectiveness of these stabilizers is dependent on the chemistry of the oxidation product rather than the simple donor reaction of the phenol hydrogen atom to the propagating radical.

Thus while it may be considered, at the simplest level, that the product of the reaction of the peroxy radical with BHT is a stable hindered phenoxy radical [Ph(t-Bu)O·], it is the subsequent reaction of the phenoxy radical which allows a broader range of stabilizer reactions. The chemistry is shown in Scheme 1.66, and it can be seen that there is a key

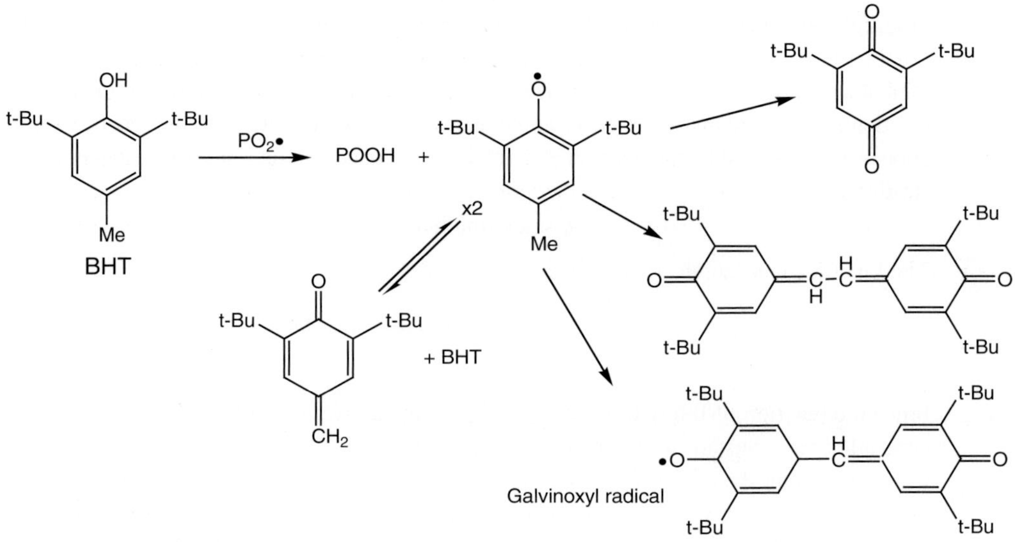

OH
t-Bu
t-Bu
Me
BHT
$PO_2\bullet$
POOH +
t-Bu
t-Bu
Me
x2
O
t-Bu
t-Bu
CH_2
+ BHT
O
t-Bu
t-Bu
O
t-Bu
t-Bu
O
H
C-C
H
O
t-Bu
t-Bu
t-Bu
t-Bu
•O
H
C
O
t-Bu
t-Bu
Galvinoxyl radical

Scheme 1.66. Some BHT stabilization reactions with peroxy radicals. After Scott (1993b).

reactive intermediate formed from the coupling together of quinone methide fragments from the rearrangement of the phenoxy radical after the initial donor reaction.

The resulting galvinoxyl radical was found to be fluctuating in concentration throughout the processing of polypropylene (PP) in a closed mixer, and also the addition of quinone methide oxidation products of BHT was more effective than the antioxidant itself in stabilizing PP. The quinone products operate as CB-A antioxidants as discussed in the previous section. The mechanism involves a cycle between the phenoxyl radical and its reduced form, which is formed by accepting a hydrogen atom in the reaction with an alkyl radical. This feature common to many additives, namely that they can function as two different types of stabilizer by cycling through the oxidized and reduced forms, is discussed below. This is the reason why a stabilizer that would appear to be effective only as a CB-D antioxidant against oxy radicals (and therefore not apparently able to scavenge alkyl radicals) is an effective processing stabilizer.

The molecules able to be efficient H-atom donors to alkyl radicals are not restricted to hindered phenols. It has long been noted in the rubber industry that secondary aromatic amines are effective radical scavengers, and study of these materials has shown that they produce a range of coupling products that function in a way similar to the phenols. The problem of the intense colour of these by-products is limited when they are used in carbon-black-filled rubber but a major limitation in other polymers (Zweifel, 1998). The exceptions are the polyamides, in which aromatic amines have a greater stabilization efficiency than do hindered phenols. This may be related to the general observation in polyamides (e.g. nylon-6) that the amine-terminated polymer is more stable than the carboxylic acid-or methyl- terminated polymer.

Chain-breaking redox antioxidants

The behaviour of the galvinoxyl radical as described in the previous section is the result of a redox cycle incorporating features of both CB-A and CB-D antioxidant steps. This is shown in Scheme 1.67.

The redox reaction also extends to the participation of hydroperoxides, but their efficient decomposition depends on the formation of a non-radical product such as an alcohol. Another example of a redox couple is found in the behaviour of the nitroxyl radical (R′NO·). Depending on the structure of R′, these are efficient radical scavengers and a redox couple between the radical and the hydroxyl amine (R′NO·/R′NOH) is formed (which is analogous to the galvinoxyl radical G·/GH). It is noted that the hindered amine stabilizers (e.g. Tinuvin® 770 and the monomeric and polymeric analogues) are ineffective as melt antioxidants, possibly because of reaction with hydroperoxides or their sensitivity to acid.

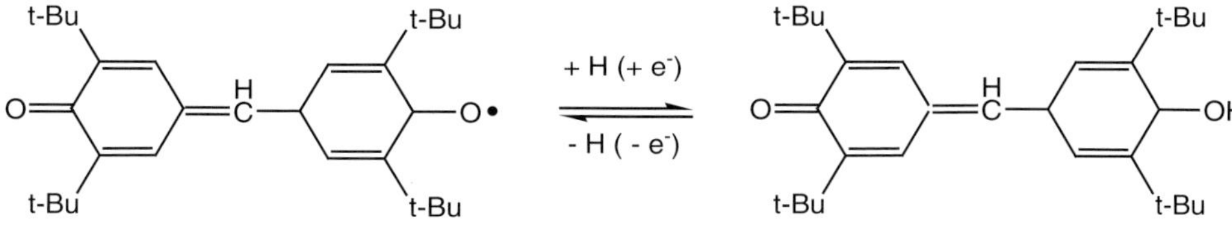

Scheme 1.67. The redox reaction of galvinoxyl radical, *G*, and the parent phenol, GH. After Al-Malaika (1989).

$$—CH_2CONHCH\cdot + Cu^{2+} \rightarrow —CH_2CON{=}CH— + Cu^+ + H^+$$

$$ROO\cdot + Cu^+ + H^+ \rightarrow ROOH + Cu^{2+}$$

Scheme 1.68. The redox reaction of copper iodide as exploited for stabilizing polyamides.

$$ROOH + P\text{-}[O\text{-}Ar]_3 \rightarrow ROH + P{=}[O\text{-}Ar]_3$$

$$ROOH + [CH_3(CH_2)_{11}OCO(CH_2)_2]_2\text{-}S \rightarrow ROH + [CH_3(CH_2)_{11}OCO(CH_2)_2]_2\text{-}S{=}O$$

Scheme 1.69. Reduction of hydroperoxide to alcohol products by phosphite or thiodipropionate synergisitic stabilizers.

An interesting stabilizer system that may involve a redox couple is the use of copper iodide in the stabilization of polyamides. This is often the melt stabilizer of choice, and the reaction (Scheme 1.68) to remove the main chain-carrying radical is proposed, followed by a redox reaction with peroxy radicals to regenerate the cuprous ion.

It was noted that the efficiency is increased in the presence of amines (Scott, 1993b), so the effectiveness in conjunction with aromatic amine free-radical (CB-D) stabilizers or the amine end groups of nylon would be expected. It is also noted that cuprous salts are often used as the iodide, and the effective alkyl-scavenging ability of iodo compounds has been demonstrated (Henman, 1979). Copper salts are also effective in the melt stabilization of polyesters, but in this case the effectiveness is improved by using a hindered phenol antioxidant. This performance is in contrast to the strong prodegradant effect of copper and other transition metals on the polyolefins as discussed later.

Melt stabilization by non-radical mechanisms

The prevention of secondary reactions of free radicals during the processing step, and the removal of secondary products of the stabilization reactions as well as by-products formed by non-radical routes require the addition of compounds that enhance the total stabilization performance of the system. The compounds that enhance the performance of a chain-breaking stabilizer are termed synergists.

Hydroperoxide decomposers

The efficient decomposition of hydroperoxides by a non-radical pathway can greatly increase the stabilizing efficiency of a chain-breaking antioxidant. This generally occurs by an ionic reaction mechanism. Typical additives are sulfur compounds and phosphite esters. These are able to compete with the decomposition reactions (either unimolecular or bimolecular) that produce the reactive alkoxy, hydroxy and peroxy radicals and reduce the peroxide to the alcohol. This is shown in the first reaction in Scheme 1.69 for the behaviour of a triaryl phosphite, $P(OAr)_3$ in reducing ROOH to ROH while itself being oxidized to the phosphate.

The structure of the aryl groups is designed to increase the hydrolytic stability of the compound. The sulfur compounds which are the most efficient hydroperoxide decomposers are thiopropionate esters, such as DLTDP (dilauryl thiodipropionate), as shown in the second reaction in Scheme 1.69, which forms a sulfoxide in the first step and then is progressively oxidized to liberate sulfur dioxide, with each step providing a peroxide decomposition reaction. Thus these are very efficient, but it has been noted that the effectiveness drops off after 150 °C, so they are of limited value in melt stabilization, in contrast to the phosphite esters, which are effective at high temperatures (Zweifel, 1998). However, DLTDP does provide effective thermal stabilization in service at lower temperatures when used with a phenolic antioxidant.

Optimum stabilizer levels

Zweifel (1998) has summarized the practical levels of additive which are appropriate for the melt stabilization of polyolefins. For example, in the melt stabilization of polypropylene at 280 °C, the combination of 0.04% of a tetrafunctional hindered phenol (Ciba Geigy Irganox 1010) and 0.06% of a tri-aryl phosphite (Ciba Geigy Irgafos 168) provided stabilization against chain scission far superior to that achieved with 0.10% of the hindered phenol alone. It is noted that only low concentrations of additives (~0.1 wt.% in total) are required because of the efficient mixing of the stabilizers in the melt. A similar combination of these additives at a slightly higher concentration (~0.2 wt.% in total) was able to bring about total suppression of the crosslinking of high-density polyethylene. The tendency of unsaturated elastomers such as styrene-butadiene (SBR) and ethylene-propylene-diene (EPDM) to crosslink and gel during processing may be similarly eliminated with similar stabilizer combinations of hindered phenols and phosphites. In certain cases the residual catalyst and its carrier can have an effect on the degradation rate. For example, the $MgCl_2$ carrier of the transition-metal catalyst can result in acid by-products and thereby affect the rate of decomposition of hydroperoxides. This can be overcome by the addition of an acid scavenger as described in the next section for the stabilization of PVC.

Acid scavengers

In the degradation of PVC during processing, the most important factor is the liberation of HCl. This results in the formation of polyenes and thus secondary oxidation involving hydroperoxides, and then catalyses the decomposition of these peroxides to form further free radicals. The removal of HCl is the target of the main group of stabilizers in PVC technology, namely the metal bases and inorganic salts. Some of the more effective of these, such as lead and cadmium stearates, are toxic and their use is no longer recommended. Zinc stearate is less effective, but non-toxic, so it is used in combination with other stabilizers. While it has been suggested that there may be reactions of zinc stearate with allylic chlorine to inhibit the initiation of dehydrochlorination, there is no evidence for this being significant compared with the direct scavenging reaction to produce zinc chloride and stearic acid.

There have been many different compounds used as acid scavengers, but the products must be stable and bind the chlorine at the temperature of processing. Such compounds include

- epoxides (as in epoxy resins), which will undergo ring-opening to give the chlorohydrin; and
- dibutyltin maleate, which, as well as scavenging HCl, is considered also to undergo a Diels–Alder reaction with the unsaturation on the PVC following dehydrochlorination (Scott, 1993b).

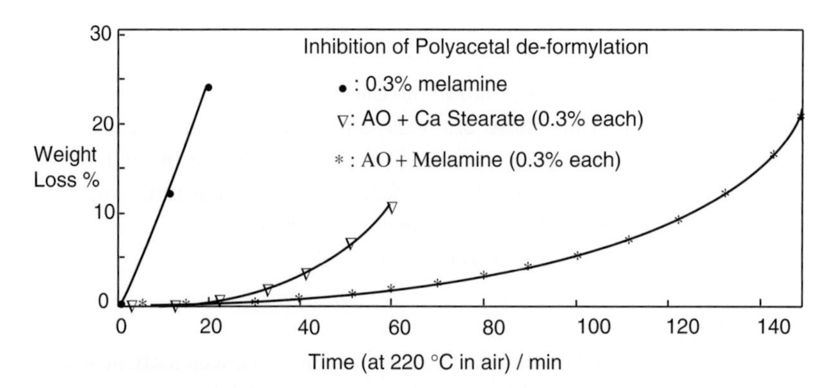

Figure 1.42. Weight loss from polyacetal during processing due to liberation of formaldehyde and inhibition by an acid scavenger combined with a hindered phenol antioxidant. Adapted from Zweifel (1998).

Other polymers undergo acid-catalysed depolymerization during melt processing and can be processed only in the presence of an acid scavenger. Thus calcium stearate and aromatic bases such as melamine are essential during the melt processing of polyacetal, which has a relatively low ceiling temperature. The effectiveness of stabilization of polyacetal against weight loss due to liberation of formaldehyde in the melt by these additives (when combined with a hindered phenol antioxidant, Irganox 245) is shown in Figure 1.42 (Zweifel, 1998).

Accelerated degradation during processing

The mastication of rubber was the first and perhaps the most important example of controlled polymer degradation through mechanochemistry. The reactions are important since they illustrate the importance of the use of prodegradants (peptizers) to enhance the chain-scission reactions. It was noted earlier that the controlled degradation of polypropylene was desirable in order to change the molar-mass distribution and thereby alter the processibility of the polymer. In this case the conditions for free-radical formation through shear are maximized and a further increase in the rate of degradation is achieved through addition of free-radical generating prodegradants such as peroxides to the system. Similarly, in the functionalization of polymers in reactive processing, the reaction may be enhanced through a change in the end-group chemistry of the polymer or oligomer.

All of these systems operate to increase the rate of the free-radical generation reaction or introduce new mechanisms for radical formation not seen in the original degradation-reaction scheme for that polymer (e.g. they enhance scission in an otherwise crosslinking polymer).

Free-radical generation

The addition of peroxides, R′OOR″, during reactive processing has become a routine aspect of reactive extrusion technology for polypropylene and polyethylene (Xanthos, 1992). In particular, the dialkyl peroxides at a level ranging from 0.001 to 1 wt.% have been used and have half-lives ranging from 1 to 3 minutes at 180 °C, so 99% consumption will occur in 7–20 minutes at that temperature. The reaction steps depend on whether oxygen is present

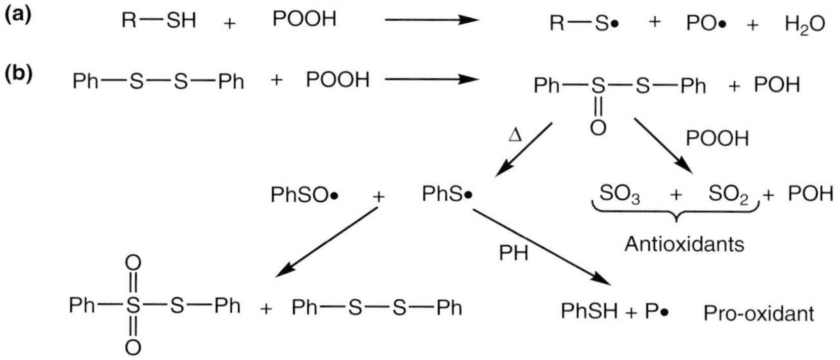

(a) R—SH + POOH ⟶ R—S• + PO• + H_2O

(b) Ph—S—S—Ph + POOH ⟶ Ph—S(=O)—S—Ph + POH

Ph—S(=O)—S—Ph —Δ⟶ PhSO• + PhS•

Ph—S(=O)—S—Ph —POOH⟶ SO_3 + SO_2 + POH (Antioxidants)

PhSO• + PhS• ⟶ Ph—S(=O)(=O)—S—Ph + Ph—S—S—Ph

PhS• —PH⟶ PhSH + P• Pro-oxidant

Scheme 1.70. Pro-oxidant activity of alkyl mercaptans (a) and both pro-oxidant and delayed antioxidant activity of diphenyl disulfides (b) during melt processing of polymer PH containing hydroperoxide groups POOH. After Scott (1993b).

and will follow the radical reaction schemes in Section 1.4.1 for an inert environment and Section 1.4.2 for low and high oxygen partial pressures. The rate-determining step in the reaction is the decomposition of the peroxide to the active alkoxy radical, which is a first-order reaction with a characteristic half-life. The kinetic schemes will apply with a higher rate of initiation and a *lower chain length of oxidation* (since $\nu = r_p/r_i$). Loss of volatile peroxides during processing can affect the degradation or crosslinking rate, as can the presence of free-radical-scavenging stabilizers.

In the subsequent use of the polymer it is important to recognize that many of the products of processing degradation will be photoinitiators (as will any residual peroxide), and the incorporation of suitable stabilizers that do not affect the controlled degradation may be necessary.

Peptizers and peroxide decomposers

The role of a peptizer is to enhance the mechanochemical degradation of a polymer, such as the mastication of natural rubber, and known peptizers are thiols and diarylsulfides. The mechanism of action is believed to be the redox decomposition of hydroperoxides (Scott, 1993b), as shown in Scheme 1.70(a):

The alkoxy radical RO· will produce further scission, and the alkyl sulfide radical can abstract another hydrogen atom to produce a macro-alkyl radical and so further propagate the chain scission. Depending on the conditions of the processing, the diarylsulfides may act initially as stabilizers and then as prodegradants. This is a common occurrence with many thio compounds and has been researched by Scott and co-workers (Scott, 1993b, 1995). The following reaction scheme illustrates such a system in which a diarylsulfide is both a pro-oxidant in the early stage of processing and then an antioxidant since the sulfide is oxidifed to SO_2 and SO_3, which will decompose peroxides ionically. This enables total control of degradation and stabilization as processing is completed.

Transition-Metal ions

It was noted (Scheme 1.68) that the redox cycle of Cu I/Cu II was an effective way to ensure the continued decomposition of alkyl radicals in the inhibition of the oxidation of aliphatic polyamides. The *converse* is observed with certain systems such as polypropylene, where the metal ion can decompose hydroperoxides catalytically and is a pro-oxidant. This is a problem in service when polyolefins are used in contact with copper, and it has been demonstrated that the contact with metal causes degradation remote from the metal, suggesting that there is migration of the ions in the solid or the production of a volatile pro-oxidant that can spread the oxidation. In the molten state it has been recognized that the extrusion of polyolefins through dies containing various levels of copper will affect the extensional rheology and the morphology of the polymer when it cools. This has been linked to the change in the molar-mass distribution of the polymer. The role of transition-metal catalyst, notably titanium residues from the Ziegler–Natta polymerization, in the initiation of free-radical reactions and the spreading of infectious oxidation in the solid polymer has also been recognized (George and Celina, 2000). In this case the redox chemistry involves Ti III/Ti IV.

The dual role of transition metals as either prodegradant or stabilizer is rationalized by the observation that the free-radical chemistry which dominates depends on the concentration of the metal ion. At low concentrations Cu I/Cu II may be a pro-oxidant, whereas at high concentrations it may be a stabilizer. The explanation lies in the complexation of hydroperoxides with transition metals (Black, 1978). Thus, taking the example of Co II/Co III, the reactions in Scheme 1.71 are recognized (Black, 1978), which together give the usual reaction for the metal-catalysed bimolecular decomposition of hydroperoxides to alkyl peroxy and alkoxy radicals.

When the reaction is written in this shortened form it neglects the role of the complexes $(ROOHCo)^{2+}$ and $(ROOHCo)^{3+}$ in the reactions since, depending on the concentration of hydroperoxide, the free ions available for the radical-scavenging redox reactions in Scheme 1.72 will be limited.

$$Co^{2+} + ROOH \rightarrow (ROOHCo)^{2+}$$

$$(ROOHCo)^{2+} \rightarrow Co^{2+} + ROOH$$

$$(ROOHCo)^{2+} \rightarrow RO\cdot + Co^{3+} + OH^-$$

$$Co^{3+} + ROOH \rightarrow (ROOHCo)^{3+}$$

$$(ROOHCo)^{3+} \rightarrow Co^{3+} + ROOH$$

$$(ROOHCo)^{3+} \rightarrow RO_2\cdot + Co^{2+} + H^+$$

This gives the overall pro-oxidant reaction:

$$2\ ROOH \xrightarrow{Co^{2+}/Co^{3+}} RO\cdot + RO_2\cdot + H_2O$$

Scheme 1.71. Redox reactions of Co II/Co III with hydroperoxides.

$$Co^{2+} + RO_2\cdot \rightarrow Co^{3+} + \text{inert products}$$

$$Co^{3+} + R\cdot \rightarrow Co^{2+} + \text{inert products}$$

Scheme 1.72. Radical scavenging reactions of Co II/Co III at high concentrations.

Thus, only when the transition-metal-ion concentration is very much greater than that of the hydroperoxide, ROOH, will there be a stabilizing effect. Therefore, at the low levels obtained from a metal due to trace acid attack for example, there will be a pronounced pro-oxidant effect with the bimolecular reaction above. If there is a need to control this reaction, then it is possible to add complexing agents such as oxanilide, which form metal complexes of greater stability than the hydroperoxides (Zweifel, 1998). The efficiency of the above reactions differs from metal to metal, depending on the stability constants for the complexes and the detailed kinetics of the reactions. For example, in the case of Cu I/Cu II it has been shown that the inhibitor is Cu II, which must then compete with oxygen for alkyl-radical scavenging and stabilization. This has a lower rate coefficient than has cobalt, so the pro-degradant behaviour is greater in non-polar media (Black, 1978). The solubility of oxygen in the polyamides is much less than that in polyolefins, which fact and this may have a bearing on the differentce in performance of copper salts in the two polymers.

The metal-catalysed decomposition of hydroperoxides is exploited in the formation of crosslinked polymer networks in composite fabrication, where cobalt octoate and cobalt naphthenate are used to catalyse the decomposition of cumene hydroperoxide and methyl ethyl ketone hydroperoxide at ambient temperature to provide the initiating radicals for the styrene crosslinking reaction of unsaturated polyester networks. The reaction sheme for this process was given earlier in Scheme 1.44. The use of drying oils as surface coatings involves the addition of metal ions to catalyse the decomposition of hydroperoxides formed on atmospheric oxidation of the unsaturated fatty acid esters (triglycerides) and so facilitate crosslinking and network formation (Scott, 1993a). The same reactions as shown above are applicable and show the generality of the use of metal ions as either prodegradants or stabilizers.

Non-oxidative chain-scission processes

The processing of polymers should occur with dry materials and with control of the atmosphere so that oxidative reactions may be either avoided, to maintain the polymer's molar mass, or exploited to maximize scission events (in order to raise the melt-flow index). The previous sections have considered the oxidative degradation of polymers and its control in some detail. What has not been considered are reactions during processing that do not involve oxidation but may lead to scission of the polymer chain. Examples include the thermal scission of aliphatic esters by an intramolecular abstraction (Scheme 1.51) (Billingham *et al.*, 1987) and acid- or base- catalysed hydrolysis of polymers such as polyesters and polyamides (Scheirs, 2000). If a polymer is not dry, the evolution of steam at the processing temperature can lead to physical defects such as voids. However, there can also be chemical changes such as hydrolysis that can occur under these conditions.

Hydrolytic chain scission of polyesters, polyamides and polycarbonates

Polymers that have been synthesized by a stepwise condensation route (Section 1.2.1) that involves the evolution of water are susceptible to degradation by hydrolysis since this involves the reversal of the polymerization reaction. The conditions of high temperature and the presence of traces of an acidic or basic catalyst are easily met in an extruder. The reaction is often auto-catalytic since the product of hydrolysis will be an acid end group. This shown in Scheme 1.73 for the acid-catalysed hydrolysis of poly(ethylene terephthalate) (PET).

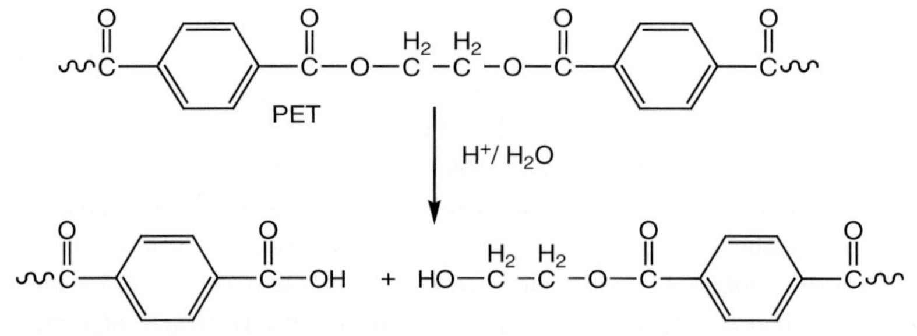

Scheme 1.73. Acid hydrolysis of PET producing chain scission and generation of acid and alcohol end groups. The carboxylic acid catalyses further hydrolysis.

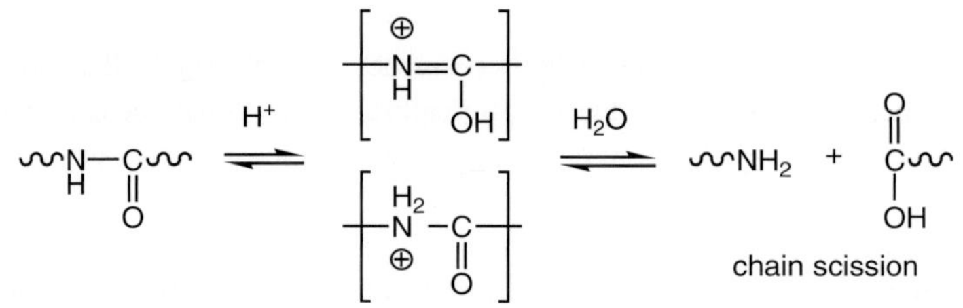

Scheme 1.74. The mechanism for acid hydrolysis of polyamide leading to chain scission and amine and carboxylic acid end groups. After Zaikov (1985).

It has been noted (Scheirs, 2000) that this leads to a dramatic decrease in viscosity that renders the polymer unprocessable or, at the best, results in defects such as haze due to crystallites that nucleate more readily from the lower-molar-mass, degraded polymer. The rate of loss of properties due to hydrolysis is orders of magnitude faster than oxidative degradation at the same temperature. To avoid these effects, the moisture level in an aromatic polyester such as PET must be kept below 0.02%.

The sensitivity of the polymer to processing with an ambient level of moisture may be determined by monitoring the change in the melt viscosity at the processing temperature (Seo and Cloyd, 1991). This is of importance in assessing the effect of repeated recycling on the performance of a polymer. It has been noted that the combined hydrolytic and thermomechanical chain scission result in a dramatic decline in the properties of the final PET product after three repeated extrusions, but this extent of reduction in properties is not apparent after further reprocessing (La Mantia and Vinci, 1994). An important observation in PET blends with ABS has been that the blend undergoes degradation during processing more rapidly than does the PET alone, and this has been attributed to the presence of catalytic impurities in the ABS that accelerate chain scission of the PET (Cook *et al.*, 1996).

Polyamides, particularly nylon-6, are very susceptible to hydrolysis during both processing and service, so the resulting mechanical properties of tensile strength and modulus are compromised. The mechanism of this process at low concentrations of acid is believed to involve protonation of the nitrogen of the amide followed by a bimolecular reaction with water leading to chain scission and generating amine and acid end groups, Scheme 1.74 (Zaikov, 1985)

The properties of polyamides are particularly sensitive to agents that may disrupt the hydrogen bonding between chains and this includes trace-metal (such as zinc and cobalt) salts as well as water.

Polycarbonates are processed at very elevated temperatures (>300 °C) so that hydrolysis of the carbonate linkage to yield carbon dioxide and a phenol end group can occur if the polymer is not rigorously dried (residual water <0.02 wt.%). Complex chemistry can follow (Scheirs, 2000), resulting in the formation of low-molar-mass species and darkening of the polymer in addition to chain scission. The degradation is very sensitive to trace acidic or basic impurities, and it has been noted that in blends (e.g. PC/ABS) there may be initiation of the degradation by impurities in the ABS (Cook *et al.*, 1996).

Chain scission by elimination reactions

In polyalkylene terephthalates and also in fully aliphatic polyesters, such as polyhydroxybutyrate (PHB), random intramolecular abstraction of a β-hydrogen atom will result in chain scission with the formation of a carboxylic acid end group and an alkene end group (Scheme 1.51) (Pilati, 1989). This reaction results in the formation of macrocyclic degradation products from both PET and aliphatic polyesters (Tighe, 1984). These reactions are more likely under processing conditions in which oxygen is excluded, otherwise free-radical chain reactions will dominate (Section 1.4.2), producing a diverse range of oxidation products.

It is the β-elimination reaction that has rendered the processing of the biologically synthesized and biodegradable polyester poly(hydroxy butyrate) (PHB) extremely difficult from the melt. The high melting temperature of the homopolymer results in extended β-elimination with the production of iso-crotonic acid upon extended exposure to processing temperatures above 200 °C (Scheme 1.51(b)). The inclusion of 5%–15% hydroxyvalerate as a comonomer, by changing the feed to the bioreactor, results in a copolymer that is lower-melting and able to be processed without extensive degradation by β-elimination (Billingham *et al.*, 1987).

Degradation of heterogeneous polymer systems

In the processing of multicomponent systems, such as incompatible blends and composites, the question of whether the multicomponent system is more or less stable than the individual component(s) arises. In systems that maintain heterogeneity during all stages of elevated-temperature processing, the only place at which interaction between the components can take place is at the phase boundaries. In a study of the kinetics and mechanism of pyrolysis of several different blends it was found that heterogeneous blends (e.g. PMMA/HDPE and PMMA/PVC) degrade predominantly within their phase-separated regions but, if small radical species (e.g. Cl·) or molecules (e.g. HCl) can migrate across the phase boundaries, then cross-products can form (Bate and Lehrle, 1998). If abstraction reactions can occur at the phase boundary this may stabilize one phase at the expense of the other. Since an important application of reactive extrusion is the compatibilization of otherwise-immiscible blends, the understanding of the free-radical chemistry at the interphase between the two (or more) components is vital.

The formation of compatibilizer frequently occurs by interfacial reaction of the modified polymers. An example is the use of maleic anhydride grafted to a polyolefin elastomer (EPM/EPM-g-MA) as the interfacial additive for reactive blending with a polyamide, nylon-6 (Van Duin *et al.*, 2001), in a twin-screw extruder. It was found that the consumption of MA occurred rapidly and different blend morphologies were produced. However, it was found that the nylon-6 degraded during processing, due to the reaction of anhydride with the amide, liberating water. The sensitivity of the polymer to hydrolysis in the presence of acid catalyst was noted earlier. This interfacial effect was superimposed on the degradation of the nylon-6 itself that occurred under the reactive-extrusion conditions (which may have been linked to residual moisture).

Many commercial polymer formulations contain pigments, stabilizers, processing aids etc. and the chemistry that may occur during high-temperature processing is complex. There has been recent interest in the use of clay nanofillers to produce **thermoplastic nanocomposites** in which there is firstly intercalation and then exfoliation of the clay by the polymer. This has resulted in polyamides with greatly elevated heat-distortion temperatures. These may be formed either by *in situ* polymerization of caprolactam (Kojima *et al.*, 1993) or by melt mixing of the polyamide with a clay that has been modified with an organic reagent such as an alkyl quaternary ammonium salt or long-chain fatty acid to assist in intercalation. It has been observed that when injection moulding of these nanocomposites occurred at 295 °C there was a decrease in the molar mass of the nylon-6 (Davis, 2003). From a consideration of the degradation route for polyamides under various conditions (Levchik *et al.*, 1999) it was concluded that the main reason for degradation was scission of the amide bond due to the release of water from the surface of the clay and/or the polymer when held for long times (12.5 min) at about 300 °C. This water could also come from dehydroxylation of the aluminosilicate clay. The catalytic role that the clay may play has been hypothesized as involving sites such as Fe^{2+} and Fe^{3+} or, alternatively, the cation sites that had not exchanged with the organic modifier that facilitates intercalation (Davis, 2003). No conclusive evidence for a mechanism was provided.

Nylon-6 without added montmorillonite did not degrade under the same conditions, and it was unambiguously the case that the clay had a particular role to play in promoting the hydrolytic degradation through the different product distribution. In the absence of the clay (and nucleophile) only thermal degradation took place, with caprolactam being the principal product above 200 °C and larger cyclics forming only above 390 °C. The latter temperatures are unrealistic for reactive processing. These studies focussed on conditions of injection moulding.

In a related study of nylon-6 nanocomposite processing with a twin-screw extruder, it was found that the level of degradation was related both to the type of nylon and to the surfactant used in the treatment of the clay (Fornes *et al.*, 2003). The higher rate of degradation and discolouration in higher-molar-mass polymer was linked to the increase in clay-platelet exfoliation. The temperature employed (240 °C) was somewhat lower than in the above example of injection moulding and this may be responsible for the elimination of hydrolysis as a major contributor to degradation and attention being focussed on the reactions of the surfactant used for exfoliation. Interestingly, it was found that these degradation reactions did not appreciably affect the morphology and physical properties of the nanocomposite. In an extension of this study to polycarbonate nanocomposites processed at 260 °C, it was also found that the presence of unsaturation or hydroxy-ethyl groups in the surfactant used in the clay increased the darkening of the polymer. The presence of iron sites in several of the clays was also linked to this darkening.

References

Adebayo, G., Koombhongse, R. & Cakmak, M. (2003) *Int. Polym. Processing*, **18**, 260–272.

Aerts, J. (1998) *Comput. Theor. Polym. Sci.*, **8**, 49–54.

Al-Malaika, S. (1989) Effects of anti-oxidants and stabilizers, in Allen, G. (Ed.) *Comprehensive Polymer Science*, Oxford: Pergamon.

Allcock, H. R. & Lampe, F. W. (1981) *Contemporary Polymer Chemistry*, Englewood Cliffs, NJ: Prentice-Hall.

Bansal, R. C. & Donnet, J. B. (1989) Pyrolytic formation of high-performance carbon fibres, in Allen, G. (Ed.) *Comprehensive Polymer Science*, Oxford: Pergamon.

Barson, C. A. (1989) Chain transfer, in Allen, G. (Ed.) *Comprehensive Polymer Science*, Oxford: Pergamon.

Bate, D. M. & Lehrle, R. S. (1998) *Polym. Deg. Stab.*, **62**, 57–66.

Bauer, R. S. (1990) Epoxy resins, in Lee, S. M. (Ed.) *International Encyclopedia of Composites*, New York: VCH.

Berry, G. C. & Cotts, P. M. (1999) Static and dynamic light scattering, in Pethrick, R. A. & Dawkins, J. V. (Eds.) *Modern Techniques for Polymer Characterisation*, Chichester: John Wiley and Sons.

Billingham, N. C., Henman, T. J. & Holmes, P. A. (1987) Degradation and stabilisation of polyesters of biological and synthetic origin, in Grassie, N. (Ed.) *Developments in Polymer Degradation – 7*, London: Elsevier Applied Science.

Black, J. F. (1978) *J. Am. Chem. Soc.*, **100**, 527–535.

Bonner, J. G. & Hope, P. S. (1993) Compatibilisation and reactive blending, in Folkes, M. J. & Hope, P. S. (Eds.) *Polymer Blends and Alloys*, London: Blackie Academic and Professional.

Bonnet, A., Pascault, J.-P., Sautereau, H. & Taha, M. (1999) *Macromolecules*, **32**, 8517–8523.

Bonnet, A., Pascault, J-P, Sautereau, H. & Camberlin, Y. (1999) *Macromolecules*, **32**, 8524–8530.

Boogh, L., Petterson, B. & Manson, J.-A. E. (1999) *Polymer*, **40**, 2249–2261.

Boor, J. (1979) *Ziegler–Natta Catalysts and Polymerizations*, New York: Academic Press.

Bosman, A. W., Janssen, H. M. & Meijer, E. W. (1999) *Chem. Rev.*, **99**, 1665–1688.

Boutevin, B. (2000) *J. Polym. Sci. A: Polym. Chem.*, **38**, 3235–3243.

Boyd, R. H. & Phillips, P. J. (1993) *The Science of Polymer Molecules*, Cambridge: Cambridge University Press.

Bracco, P., Brunella, V., Luda, M. P., Zanetti, M. & Costa, L. (2005) *Polymer*, **46**, 10 648–10 657.

Braun, D. (1981) Thermal degradation of poly(vinyl chloride), in Grassie, N. (Ed.) *Developments in Polymer Degradation – 3*, London: Elsevier Applied Science.

Braun, D. & Ritzert, H.-J. (1989) Urea–formaldehyde and melamine–formaldehyde polymers, in Allen, G. (Ed.) *Comprehensive Polymer Science*, Oxford: Pergamon.

Brown, S. B. (1992) Reactive extrusion: a survey of chemical reactions of monomers and polymers during extrusion processing, in Xanthos, M. (Ed.) *Reactive Extrusion: Principles and Practice*, Munich: Hanser-Verlag.

Brydson, L. A. (1988) *Rubbery Materials and their Compounds*, Essex: Elsevier Applied Science.

Bucknall, C. B. (1977) *Toughened Plastics*, London: Elsevier Applied Science.

Budd, P. M. (1989) Sedimentation and diffusion, in Allen, G. (Ed.) *Comprehensive Polymer Science*, Oxford: Elesvier Pergamon.

Casale, A. & Porter, R. S. (1978) *Polymer Stress Reactions. Volume 1. Introduction*, New York: Academic Press.

Celina, M., Ottesen, D. K., Gillen, K. T. & Clough, R. L. (1997) *Polym. Deg. Stab.*, **58**, 15–31.

Chan, J. H. & Balke, S. (1997a) *Polym. Deg. Stab.*, **57**, 113–125.

Chan, J. H. & Balke, S. T. (1997b) *Polym. Deg. Stab.*, **57**, 127–134.

Chiefari, E., Mayadunne, R. T. A., Moad, C. L., *et al.* (2003) *Macromolecules*, **36**, 2273–2283.

Chien, J. C. W., Vandenberg, E. J. & Jabloner, H. (1968) *J. Polym. Sci. A-1*, **6**, 381.

Chien, J. C. W. & Wang, D. S. T. (1975) *Macromolecules*, **8**, 920.

Chong, Y. K., Krstina, J., Le, T. P. T. *et al.* (2003) *Macromolecules*, **36**, 2256–2272.

Chong, Y. K., Le, T. P. T., Moad, G., Rizzardo, E. & Thang, S. H. (1999) *Macromolecules*, **32**, 2071–2074.

Chynoweth, K. (1989) Glass transition and crystallization, in Cook, W. D. & Guise, G. B. (Eds.) *Polymer Update: Science and Engineering*, Melbourne: Royal Australian Chemical Institute.

Ciriscioli, P. R. & Springer, G. S. (1990) *Smart Autoclave Cure of Composites*, Lancaster, PA: Technomic.

Clagett, D. C. & Shafer, S. J. (1989) Polycarbonates, in Allen, G. (Ed.) *Comprehensive Polymer Science*, Oxford: Pergamon.

Cole, K. C., Hechler, A. A. & Noel, D. (1991) *Macromolecules*, **24**, 3098–3110.

Commerce, S., Vaillant, D., Phillippart, J. L., Lacoste, J. & Carlsson, D. J. (1997) *Polym. Deg. Stab.*, **57**, 175–182.

Cook, W. D., Moad, G., Fox, B., *et al.* (1996) *J. Appl. Polym. Sci.*, **62**, 1709–1714.

Corradini, P. & Busico, V. (1989) Monoalkene polymerization: stereospecifity, in Allen, G. (Ed.) *Comprehensive Polymer Science*, Oxford: Pergamon.

Cowie, I. A. (1989a) Block and graft copolymers, in Allen, G. (Ed.) *Comprehensive Polymer Science*, Oxford: Pergamon.

Cowie, J. M. G. (1989b) Alternating copolymerization, in Allen, G. (Ed.) *Comprehensive Polymer Science*, Oxford: Pergamon.

Crivello, J. V. (1999) *J. Polym. Sci. A: Polym. Chem.*, **37**, 4241–4254.

Dainton, F. R. (1956) *Chain Reactions*, London: Methuen.

Darling, T. R., Davis, T. P., Fryd, M. *et al.* (2000) *J. Polym. Sci. A: Polym. Chem.*, **38**, 1706–1708.

Davis, R., Gilman, G.J., Van Der Hart, D.I. (2003) *Polym. Deg. Stab.*, **79**, 111–121.

Dawkins, J. V. (1989) Size exclusion chromatography, in Allen, G. (Ed.) *Comprehensive Polymer Science*, Oxford: Pergamon.

De Bakker, C. J., St John, N. A. & George, G. A. (1993) *Polymer*, **34**, 716–726.

Doi, M. & Edwards, S. F. (1986) *The Theory of Polymer Dynamics*, Oxford: Clarendon Press.

Dong, J., Fredericks, P. M. & George, G. A. (1997) *Polym. Deg. Stab.*, **58**, 159–169.

Dusek, K. (1986) Network formation in curing of epoxy resins, in Dusek, K. (Ed.) *Advances in Polymer Science 78: Epoxy Resins and Composites III*, Berlin: Springer-Verlag.

El'darov, E. G., Mamedov, F. V., Gol'dberg, V. M. & Zaikov, G. E. (1996) *Polym. Deg. Stab.*, **51**, 271–279.

Emanuel, N. M. & Buchachenko, A. L. (1987) *Chemical Physics of Polymer Degradation and Stabilization*, Utrecht: VNU Science.

Fischer, H. (2001) *Chem. Rev.*, **101**, 3581–3610.

Flory, P. J. (1953) *Principles of Polymer Chemistry*, New York: Cornell University Press.

Flory, P. J., Allcock, H. R., Eisenberg, H., Feldman, R. J. & Yoon, D. Y. (1982) The science of macromolecules, in Seitz, F. (Ed.) *Outlook for Science and Technology: The Next Five Years*, San Francisco, CA: W. H. Freeman.

Fontanille, M. (1989) Carbanionic polymerization: general aspects and initiation, in Allen, G. (Ed.) *Comprehensive Polymer Science*, Oxford: Pergamon.

Fornes, T. D., Yoon, P. J. & Paul, D. R. (2003) *Polymer*, **44**, 7545–7556.

Fradet, A. & Arlaud, P. (1989) Unsaturated polyesters, in Allen, G. (Ed.) *Comprehensive Polymer Science*, Oxford: Pergamon.

Frechet, J. M. & Tomalia, D. A. (2001) Introduction to the dendritic state, in Frechet, J. M. & Tomalia, D. A. (Eds.) *Dendrimers and other dendritic polymers*, Chichester: John Wiley and Sons.

Frechet, M. J. & Hawker, C. J. (1996) Synthesis and properties of dendrimers and hyperbranched molecules, in Allen, G. (Ed.) *Comprehensive Polymer Science*, Oxford: Pergamon.

Frisch, K. C. & Klempner, D. (1989) Polyurethanes, in Allen, G. (Ed.) *Comprehensive Polymer Science*, Oxford: Pergamon.

Gaymans, R. J. & Sikkema, D. J. (1989) Aliphatic polyamides, in Allen, G. (Ed.) *Comprehensive Polymer Science*, Oxford: Pergamon.
George, G. A. & Celina, M. (2000) Homogeneous and heterogeneous oxidation of polypropylene, in Halim Hamid, S. (Ed.) *Handbook of Polymer Degradation*, 2nd edn, New York: Marcel Dekker.
Ghotra, J. S. & Pritchard, G. (1984) Osmosis in resins and laminates, in Pritchard, G. (Ed.) *Developments in Reinforced Plastics – 3*, London: Elsevier Applied Sceince.
Gijsman, P., Hennekens, J. & Vincent, J. (1993) *Polym. Deg. Stab.*, **42**, 95–105.
Goto, A., Sato, K., Tsujii, Y. *et al.* (2001) *Macromolecules*, **34**, 402–408.
Guan, Z., Cotts, P. M., McCord, E. F. & McLain, S. J. (1999) *Science*, **283**, 2059–2062.
Gugumus, F. (1999) *Polym. Deg. Stab.*, **66**, 161–172.
Gugumus, F. (2000) *Polym. Deg. Stab.*, **68**, 337–352.
Gugumus, F. (2002a) *Polym. Deg. Stab.*, **75**, 55–71.
Gugumus, F. (2002b) *Polym. Deg. Stab.*, **75**, 131–142.
Hall, A. J. & Hodge, P. (1999) *Reactive Functional Polym.*, **41**, 133–139.
Hamielic, A. E., Macgregor, J. F. & Penlidis, A. (1989) Copolymerization, in Allen, G. (Ed.) *Comprehensive Polymer Science*, Oxford: Pergamon.
Han, C. D. (1981) *Multiphase Flow in Polymer Processing*, New York: Academic Press.
Hawker, C., Frechet, J., Grubbs, R. B. & Dao, J. (1995) *J. Am. Chem. Soc.*, **117**, 10 763–10 764.
Hawker, C. J., Bosman, A. W. & Harth, E. (2001) *Chem. Rev.*, **101**, 3661–3688.
Heath, R. E., Wood, B. R. & Semlyen, J. A. (2000) *Polymer*, **41**, 1487–1495.
Heck, B., Hugel, T., Ijima, M. & Strobl, G. R. (2000) *Polymer*, **41**, 8839–8848.
Henman, T. J. (1979) Melt stabilisation of polypropylene, in Scott, G. (Ed.) *Developments in Polymer Stabilisation – 1*, London: Elsevier Applied Science.
Hepburn, C. (1982) *Polyurethane Elastomers*, London: Elsevier Applied Science.
Higgins, J., Gerard, H., Vlassopoulos, D., Horst, R. & Wolf, B. A. (2000) *Macromol. Symp.*, **149**, 165–170.
Hobbs, J. K., Humphries, A. D. L. & Miles, M. J. (2001) *Macromolecules*, **34**, 5508–5519.
Hodd, K. (1989) Epoxy resins, in Allen, G. (Ed.) *Comprehensive Polymer Science*, Oxford: Pergamon.
Hodge, P. & Peng, P. (1998) *Polymer*, **39**, 981–990.
Hunt, B. J. & James, M. I. (1999) Vapour pressure osmometry/ membrane osmometry/ viscometry, in Pethrick, R. A. & Dawkins, J. V. (Eds.) *Modern Techniques for Polymer Characterisation*, Chichester: John Wiley and Sons.
Hunt, S., Cash, G., Liu, H., George, G. & Birtwhistle, D. (2002) *J. Macromol. Sci – Pure Appl. Chem.* A, **A39**, 1007–1024.
Hunt, S. M. & George, G. A. (2000) *Polym. Int.*, **49**, 1505–1512.
Inoue, T. (1995) *Prog. Polym. Sci.*, **20**, 119–153.
Jenkins, A. D. (1999) Terminology for polymer chemistry, in Pethrick, R. A. & Dawkins, J. V. (Eds.) *Modern Techniques of Polymer Characterisation*, Chichester: John Wiley and Sons.
Jikei, M. & Kakimoto, M. (2001) *Prog. Polym. Sci.*, **26**, 1233–1285.
Kamiya, Y. & Niki, E. (1978) Oxidative Degradation, in Jellinek, H. H. G. (Ed.) *Aspects of degradation and Stabilization of Polymers*, Amsterdam: Elsevier.
Kaynak, A., Bartley, J. P. & George, G. A. (2001) *J. Macromol. Sci. – Pure Appl. Chem.* A, **38**, 1033–1048.
Keller, A. (2000) *Polymer*, **41**, 8751–8754.
Kendrick, T. C., Parbhoo, B. M. & White, J. W. (1989) Polymerization of cyclosiloxanes, in Allen, G. (Ed.) *Comprehensive Polymer Science*, Oxford: Pergamon.
Kim, Y. H. (1998) *J. Polym. Sci. A: Polym. Chem.*, **36**, 1685–1698.
Kinloch, A. J. (1985) *Adv. Polym. Sci.*, **72**, 45–67.

Knop, A., Bohmer, V. & Pilato, L. A. (1989) Phenol–formaldehyde polymers, in Allen, G. (Ed.) *Comprehensive Polymer Science*, Oxford: Pergamon.
Koenig, J. L. (1999) *Spectroscopy of Polymers*.
Kojima, Y., Yusuki, A., Kawasumi, M. *et al.* (1993) *J. Polym. Sci.: Polym. Chem.*, **31**, 1755–1758.
Kozielski, K. A., George, G. A., St John, N. A. & Billingham, N. C. (1994) *High Perform. Polym.*, **6**, 263–286.
Kwei, T. K. & Wang, T. T. (1978) Phase separation behavior of polymer–polymer mixtures, in Paul, D. A. N. (Ed.) *Polymer Blends*, New York: Academic Press.
La Mantia, F. P. & Vinci, M. (1994) *Polym. Deg. Stab.*, **45**, 121–125.
Lambla, M. (1992) Reactive processing of thermoplastic polymers, in Allen, G. (Ed.) *Comprehensive Polymer Science*, Oxford: Pergamon.
Lederer, A., Voigt, A., Clausnitzer, C. & Voit, B. (2002) *J. Chromatogr.* A, **976**, 171–179.
Levchik, S. V., Weil, E. D. & Lewin, M. (1999) *Polym. Int.*, **48**, 532–557.
Manaresi, P. & Munari, A. (1989) Step polymerization – general aspects, in Allen, G. (Ed.) *Comprehensive Polymer Science*, Oxford: Pergamon.
Mandelkern, L. (1989) Crystallization and melting, in Allen, G. (Ed.) *Comprehensive Polymer Science*, Oxford: Pergamon.
Manzione, L. T. & Gillham, J. K. (1981) *J. Appl. Polym. Sci.*, **26**, 889–905.
Matejka, L. & Dusek, K. (1989) *Macromolecules*, **22**, 2911–2917.
Matejka, L., Spacek, P. & Dusek, K. (1991) *Polymer*, **32**, 3190–3194.
Matyjaszewski, K. (1999) Similarities and discrepancies between controlled cationic and radical polymerizations, in Puskas, J. E. (Ed.) *Ionic Polymerizations and Related Processes*, Dordrecht: Kluwer Academic.
Mayadunne, R. T. A., Rizzardo, E. *et al.* (1999) *Macromolecules*, **32**, 6977–6980.
Mayadunne, R. T. A., Rizzardo, E. *et al.* (2000) *Macromolecules*, **33**, 243–245.
Mayo, F. R. (1972) *J. Polym. Sci. Polym. Lett. Edn.*, **10**, 921–923.
Mayo, F. R. (1978) *Macromolecules*, **11**, 942–946.
McKenna, G. B. (1989) Glass formation and glassy behaviour, in Allen, G. (Ed.) *Comprehensive Polymer Science*, Oxford: Pergamon.
McNeill, I. C. (1989) Thermal degradation, in Allen, G. (Ed.) *Comprehensive Polymer Science*, Oxford: Pergamon.
Mijovic, J. & Andjelic, S. (1995) *Macromolecules*, **28**, 2787–2796.
Moad, G. (1999) *Progress in Polymer Science*, **24**, 81–142.
Morrison, N. J. & Porter, M. (1989) Crosslinking of rubbers, in Allen, G. (Ed.) *Comprehensive Polymer Science*, Oxford: Pergamon.
Muller, A. H. E. (1989) Carbanionic polymerization: kinetics and thermodynamics, in Allen, G. (Ed.) *Comprehensive Polymer Science*, Oxford: Pergamon.
Nagahata, R., Sugiyama, J. J., Goyal, M. *et al.* (2001) *Polymer*, **42**, 1275–1279.
Nuyken, O. & Pask, S. D. (1989) Carbocationic polymerization: alkenes and dienes, in Allen, G. (Ed.) *Comprehensive Polymer Science*, Oxford: Pergamon.
Odian, G. (1991) *Principles of Polymerization*, New York: John Wiley and Sons.
Oh, S. J., Lee, S. K. & Park, S. Y. (2006) *Vib. Spectrosc.*, **42**, 273–277.
Olabisi, O., Robeson, L. M. & Shaw, M. T. (1979) *Polymer–Polymer Miscibility*, New York: Academic Press.
Painter, P. C. & Coleman, M. M. (1994) *Fundamentals of Polymer Science*, Lancaster, PA: Technomic.
Papathanasiou, T. D., Higgins, J. S. & Soontaranum, W. (1998) *Polym. Polym. Composites*, **6**, 223–227.
Pascault, J.-P., Sauterau, H., Verdu, J. & Williams, R. (2002) *Thermosetting Polymers*, New York: Marcel Dekker.

Paul, D. R. (1978) Interfacial agents ("compatibilizers") for polymer blends, in Paul, D. R. A. N. (Ed.) *Polymer Blends*, New York: Academic Press.

Penczek, S. & Kubisa, P. (1989) Cationic ring-opening polymerization: acetals, in Allen, G. (Ed.) *Comprehensive Polymer Science*, Oxford: Pergamon.

Pilati, F. (1989) Polyesters, in Allen, G. (Ed.) *Comprehensive Polymer Science*, Oxford: Pergamon.

Podzimek, S., Dobas, I. & Kubin, M. (1992) *J. Appl. Polym. Sci.*, **44**, 1601–1605.

Queslel, J.-P. & Mark, J. E. (1989) Rubber elasticity and characterization of networks, in Allen, G. (Ed.) *Comprehensive Polymer Science*, Oxford: Pergamon.

Richards, R. W. (1989) Scattering properties: neutrons, in Allen, G. (Ed.) *Comprehensive Polymer Science*, Oxford: Pergamon.

Richardson, M. J. (1989) Thermal analysis, in Allen, G. (Ed.) *Comprehensive Polymer Science*, Oxford: Pergamon.

Rojanapitayakorn, P., Thongyai, S., Higgins, J. S. & Clarke, N. (2001) *Polymer*, **42**, 3475–3487.

Russell, K. E. (2002) *Progress in Polymer Science*, **27**, 1007–1038.

Ryan, A. J. & Stanford, J. L. (1989) Polyureas, in Allen, G. (Ed.) *Comprehensive Polymer Science*, Oxford: Pergamon.

Sauvet, G. & Sigwalt, P. (1989) Carbocationic polymerization: general aspects and initiation, in Allen, G. (Ed.) *Comprehensive Polymer Science*, Oxford: Pergamon.

Scamporrino, E. & Vitalini, D. (1999) Recent advances in mass spectrometry of polymers, in Pethrick, R. A. & Dawkins, J. V. (Eds.) *Modern Techniques of Polymer Characterisation*, Chichester: John Wiley and Sons.

Scheirs, J. (2000) *Compositional and Failure analysis of Polymers*, Chichester: John Wiley and Sons.

Scheirs, J., Bigger, S. W. & Billingham, N. C. (1995a) *Polym. Test.*, **14**, 211–241.

Scheirs, J., Carlsson, D. J. & Bigger, S. W. (1995b) *Polym. Plast. Technol. Eng.*, **34**, 97–116.

Scott, G. (1993a) Autoxidation and antioxidants: historical perspective, in Scott, G. (Ed.) *Atmospheric Oxidation and Antioxidants*, Amsterdam: Elsevier.

Scott, G. (1993b) Oxidation and stabilisation of polymers during processing, in Scott, G. (Ed.) *Atmospheric Oxidation and Antioxidants*, Amsterdam: Elsevier.

Scott, G. (1995) *Polym. Deg. Stab.*, **48**, 315–324.

Seo, K. S. & Cloyd, J. D. (1991) *J. Appl. Polym. Sci.*, **42**, 845–850.

Sohma, J. (1989a) Mechanochemical degradation, in Allen, G. (Ed.) *Comprehensive Polymer Science*, Oxford: Pergamon.

Sohma, J. (1989b) *Prog. Polym. Sci.*, **14**, 451–596.

Somani, R. H., Yang, L. & Hsiao, B. S. (2002) *Physica* A, **304**, 145–157.

Sperling, L. (1989) Interpenetrating polymer networks, in Allen, G. (Ed.) *Comprehensive Polymer Science*, Oxford: Pergamon.

Sperling, L. (2001) *Introduction to Physical Polymer Science*, New York: Wiley Interscience.

Sperling, L. (2006) *Introduction to Polymer Science*, New York: Wiley Interscience.

St John, N. A. (1993) Spectroscopic Studies of the Cure and Structure of an Epoxy Amine Network, Thesis, Department of Chemistry. St Lucia, University of Queensland.

St John, N. A. & George, G. A. (1994) *Prog. Polym. Sci.*, **19**, 755–795.

St John, N. A., George, G. A., Cole-Clarke, P. A., Mackay, M. E. & Halley, P. J. (1993) *High Perform. Polym.*, **5**, 212–236.

Stachurski, Z. (1987) *Engineering Science of Polymeric Materials*, Melbourne: Royal Australian Chemical Institute.

Starnes, W. H. (1981) Mechanistic aspects of the degradation and stabilisation of poly(vinyl chloride), in Grassie, N. (Ed.) *Developments in Polymer Degradation – 3*, London: Elsevier Applied Science.

Strobl, G. R. (1996) *The Physics of Polymers*, Berlin: Springer-Verlag.
Suter, U. W. (1989) Ring–Chain equilibria, in Allen, G. (Ed.) *Comprehensive Polymer Science*, Oxford: Pergamon.
Tait, P. J. T. (1989) Monoalkene polymerization: Ziegler–Natta and transition metal catalysts, in Allen, G. (Ed.) *Comprehensive Polymer Science*, Oxford: Pergamon.
Tait, P. J. T. & Watkins, N. D. (1989) Monoalkene polymerization: mechanisms, in Allen, G. (Ed.) *Comprehensive Polymer Science*, Oxford: Pergamon.
Takahashi, H., Matsuoka, T., Ohta, T., *et al.* (1988) *J. Appl. Polym. Sci.*, **36**, 1821–1831.
Tighe, B. (1984) The thermal degradation of poly-alpha-esters, in Grassie, N. (Ed.) *Developments in Polymer Degradation – 5*, London: Elsevier Applied Science.
Tirrell, D. A. (1989) Copolymer composition, in Allen, G. (Ed.) *Comprehensive Polymer Science*, Oxford: Pergamon.
Tudos, F., Kelen, T. & Nagy, T. T. (1979) Thermo-oxidative degradation of poly(vinyl-chloride), in Grassie, N. (Ed.) *Developments in Polymer Degradation – 2*, London: Elsevier Applied Science.
Van Duin, M., Machado, A. V. & Covas, J. (2001) *Macromol. Symp.*, **170**, 29–39.
Vaughan, A. S. & Bassett, D. C. (1989) Crystallization and morphology, in Allen, G. (Ed.) *Comprehensive Polymer Science*, Oxford: Pergamon.
Viville, P., Biscarini, F., Bredas, J. L. & Lazzaroni, R. (2001) *J. Phys. Chem.* B, **105**, 7499–7507.
Voit, B. (2000) *J. Polym. Sci. A: Polym. Chem*, **38**, 2505–2525.
Wang, Z., He, J., Tao, Y.*et al.* (2003) *Macromolecules*, **36**, 7446–7452.
Williams, R. J. J., Rozenberg, B. A. & Pascault, J.-P. (1997) *Adv. Polym. Sci.*, **128**, 95–156.
Wu, H., Xu, J., Liu, Y. & Heiden, P. (1999) *J. Appl. Polym. Sci.*, **72**, 151–163.
Xanthos, M. (1992) Process analysis from reaction fundamentals.Examples of polymerization and controlled degradation in extruders, in Xanthos, M. (Ed.) *Reactive Extrusion: Principles and Practice*, Munich: Hanser-Verlag.
Xiang, Q., Xanthos, M., Mitra, S., Patel, S. & Guo, J. (2002) *Polym. Deg. Stab.*, **77**, 93–102.
Zahradnikova, A., Sedlar, J. & Dastych, D. (1991) *Polym. Deg. Stab.*, **32**, 155–176.
Zaikov, G. E. (1985) Polymer stability in aggressive media, in Grassie, N. (Ed.) *Developments in Polymer Degradation – 6*, London: Elsevier Applied Science.
Zweifel, H. (1998) *Stabilization of Polymeric Materials*, Berlin: Springer-Verlag.

2 Physics and dynamics of reactive polymers

2.1 Chapter rationale

This chapter focusses on the physical properties and models of network and reactively modified polymers. Understanding changes in physical properties during curing, in tandem with changes in chemical properties (Chapter 1), and chemorheological properties (Chapter 4), is essential to fully characterizing network and reactively modified polymer systems. This chapter will first give a brief introduction to polymer physics and dynamics before focussing on redefining network and reactively modified polymer systems. Then it will focus on defining the key changes in physical properties during cure. Finally this chapter will focus on key experimental techniques for describing changes in physical properties during cure.

2.2 Polymer physics and dynamics

Chapter 1 has already introduced basic concepts of polymers relating to their physical nature, such as crystalline and amorphous regions, molar mass, glass transition and rubbery regions. This section will focus on developing further basic polymer-physics and polymer-dynamics concepts that will be pertinent to reactive polymer systems. Specifically we will be interested in examining the physics behind polymer dynamics – to understand how to characterize the dynamics and stress behaviour of polymers under deformation and flow. This will be essential background for the chemorheology of polymer systems.

2.2.1 Polymer physics and motion – early models

Polymer chains consist of large molecules (macromolecules), which are composed of multiple repetition of one or more species of atoms or groups of atoms that are interlinked. In general polymers exhibit a random chain (or Gaussian) conformation; that is a random distribution of *trans* and *gauche* states (Gedde, 1995). This can occur in polymer solutions (in good solvents), polymer melts and amorphous polymers. However, semi-crystalline polymers can have more ordered energetically favourable chain configurations due to intermolecular interactions. Configurations of a set of polymer chains are typically described by the spatial distribution function, which is characterized typically by the radius of gyration (R_g) (or root mean square distance of molecules from their centre of gravity) or the end-to-end distance (R_0) (or distance between polymer-chain ends) as described in Section 1.1.2. These characteristics can then be used to describe polymer configurations under a range of environmental conditions (for example under changing temperature, stress or flow). Gedde (1995) provides an excellent description from first principles of the effects of intra-chain and inter-chain thermodynamic interactions on polymer configurations, and their relationship to macroscopic stresses. Here the change in conformation (and subsequent stresses) is related to entropic and enthalpic changes via statistical mechanics.

An early model based on crosslinked rubbers put forward by Flory and Rehner (1943) assumed that chain segments deform independently and in the same manner as the whole sample (affine deformation) where crosslinks were fixed in space. James and Guth (1943) then described a phantom-network model that allowed free motion of crosslinks about the average affine deformation. The stress (σ) described from these theories can be described in the following equations:

$$\sigma = \psi v RT/[2(\lambda^2 - 1/\lambda)] \tag{2.1}$$

for the Flory model, and

$$\sigma = (1 - 2/\psi)\psi v RT/[2(\lambda^2 - 1/\lambda)] \tag{2.2}$$

for the James and Guth model, where σ is the stress, ψ is the functionality of the crosslink, v is the number of crosslinks, R is the gas constant, T is the temperature and λ is the deformation.

Modifications to these classical statistical models can also be made, such as by the incorporation of loose chains (non-load-bearing chains), physical crosslinks (temporary or permanent) and intramolecular crosslinks (loops). At higher deformations (strains) or increasing crosslink densities, it may be necessary to use a non-Gaussian statistical treatment that considers the finite extensibility of the chain. Non-Gaussian models are reviewed extensively by Treloar (1975).

In reality, of course, polymer systems are varied and consist of polymer solutions, polymer melts, and polymer networks and gels, and the polymer dynamics of these systems should be examined in detail.

2.2.2 Theories of polymer dynamics

Theories of dynamics for polymer systems, which can range from neat polymers to reactive filled polymer composites, require a combined discussion of the dynamics of suspensions, polymer solutions and polymer melts. However, prior to this discussion, one should give a brief introduction to rheological terminology. (For a detailed introduction to rheological techniques, please refer to Chapter 3.)

Introduction to rheology

Rheology is the study of fluid flow and deformation, and provides the specific science to characterize polymer dynamics. There are many types of flow, but generally they are classified into shear or elongational flow. In a localized sense shear flow involves shearing of molecules or polymer chains over one another, whereas elongation flow involves localized stretching of molecules or polymer chains. In shear or elongational rheology one can further subclassify systems into steady (time-independent and steady-state), dynamic (time-independent and oscilliatory) and transient (time-dependent) flows.

Steady-state shear rheology typically involves characterizing the polymer's response to steady shearing flows in terms of the steady shear viscosity (η), which is defined by the ratio of shear stress (σ) to shearing rate (γ'). The steady shear viscosity is thus a measure of resistance to steady shearing deformation. Other characteristics such as normal stresses (N_1 and N_2) and yield stresses (σ_Y) are discussed in further detail in Chapter 3.

Dynamic shear rheology involves measuring the resistance to dynamic oscillatory flows. Dynamic moduli such as the storage (or solid-like) modulus (G'), the loss (or fluid-like) modulus (G''), the loss tangent ($\tan\delta = G''/G'$) and the complex viscosity (η^*) can all be used to characterize deformation resistance to dynamic oscillation of a sinusoidally imposed deformation with a characteristic frequency of oscillation (ω).

Transient shear flows involve examining the shear stress and viscosity response to a time-dependent shear. The stress build up at the start of steady flow (σ_+) and at the cessation of steady flow (σ_-) and the stress decay ($\sigma(t)$) after a dynamic instantaneous impulse of deformation strain (γ) can be used to characterize transient rheological behaviour.

Elongational behaviour under steady and transient conditions can also be characterized, and more information is provided in Chapter 3. This short introduction to rheology should provide enough background for further interpretation of polymer-dynamic theories of suspensions, polymer solutions and polymer melts.

Suspensions

A suspension is a dispersion of particles within a solvent (usually a low-molar-mass liquid). Thermodynamics (Brownian motion and collisions) favours the clumping of small particles, and this can be increased by flow. However, particles over 1 μm tend to settle under gravity, unless stability measures have been considered (matching the density of the particle to that of the medium, increasing the Brownian/gravitational force ratio, electrostatic stabilization, steric stabilization). Other complications can occur in the dynamics of suspensions, such as particle migration across streamlines, particle inertial effects and wall slip (Larson, 1999).

Early work (Einstein, 1906) showed that the shear viscosity, η, of a suspension at low volume fractions ($\varphi < 0.03$) can be described by

$$\eta = \eta_s(1 + 2.5\varphi), \tag{2.3}$$

where η_s is the solvent's viscosity and φ is the volume fraction of particles.

If one considers hydrodynamic interactions from neighbouring spheres (at higher concentrations of particles)

$$\eta = \eta_s \exp(5\varphi/2). \tag{2.4}$$

Or, for arbitrarily shaped particles, one gets

$$\eta = \eta_s \exp([\eta]\varphi), \tag{2.5}$$

where $[\eta]$ is the intrinsic viscosity defined by

$$[\eta] = \lim_{\varphi\to 0} (\eta - \eta_s)/(\varphi\eta_s). \tag{2.6}$$

At very high concentrations, however, these equations fail to account for the overgrowing, and one should use the empirical Kreiger–Dougherty equation (Kreiger and Dougherty, 1959)

$$\eta = \eta_s(1 - \varphi/\varphi_m)^{-[\eta]\varphi_m}, \tag{2.7}$$

where φ_m is the maximum packing fraction

Of course, many suspensions are also shear thinning. For example, when $\varphi > 0.3$, viscosity becomes a function of shear rate, since the shear rate is high enough to disturb the inter-particle spaces from their equilibrium. Kreiger (1972) developed the following expression to account for shear thinning:

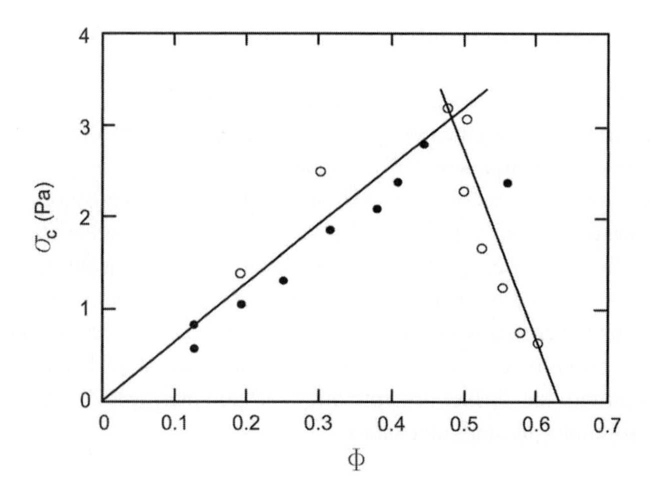

Figure 2.1. The dependence of the critical stress for shear thinning on the volume fraction of sterically stabilized hard spheres. From Larson (1999, Figure 6.6) by permission of Oxford University Press, Inc.

$$\eta_r = \eta_{r\infty}(\eta_{r0} - \eta_{r\infty})/(1 + b\sigma_r), \tag{2.8}$$

where $\eta_r = \eta/\eta_s$ is the reduced viscosity, η_{r0} is the low-shear-rate viscosity, $\eta_{r\infty}$ is the high-shear-rate viscosity, $\sigma_r = \sigma a^3/(k_B bT)$ is the reduced shear stress and b is a fitting parameter. Deviations from this equation at high shear rates can occur due to shear thickening (Larson, 1999). Also a critical stress for shear thinning to occur, σ_c, has been found (Larson, 1999) and is shown in Figure 2.1. Note that the effect of high concentrations is to reduce the critical stress to shear thinning.

An excellent discussion in Larson (1999) examines additional effects such as steric stabilization, particle-size distributions, shear thickening, and deformable particles and droplets. It is interesting to examine the equation for deformable droplets,

$$\eta = \eta_s\left(1 + \frac{5M + 2}{2M + 1}\varphi\right), \tag{2.9}$$

where $M = \eta_d/\eta_s$ is the ratio of the viscosity of the droplet to the viscosity of the solution. One can see that, in the limit as M approaches ∞ (hard sphere), the equation reverts back to Equation (2.3) (Einstein's equation). An excellent section follows in Larson (1999) about droplet break up and coalescence and the application of these equations to foams, emulsions and immiscible liquids.

Polymer solutions

A polymer solution is a mixture of a solvent (usually a low-molar-mass liquid) and a solute (usually a high-molar-mass polymer). For dilute polymer solutions (from suspension theory and $[\eta] = 2.5\varphi$) it is simple to show from geometric considerations that the viscosity varies as

$$[\eta] = k M_w^{0.5}. \tag{2.10}$$

In fact experimentally the viscosity of dilute polymer solutions follows

$$[\eta] = K M_w^a, \tag{2.11}$$

where $a = 0.5$–0.8, and this equation is known as the Mark–Houwink equation.

Theories for polymer dynamics of dilute polymer solutions include the elastic (Hookean) spring model (Kuhn, 1934) which considers that the system is mechanically equivalent to a set of beads attached with a spring. The properties are then based on a spring constant between beads and the friction of beads through solvent. The viscosity of a Hookean system is then described by

$$\eta = \eta_s + n_0 k_B T \lambda, \tag{2.12}$$

where, n_0 is the number of dumbells per unit volume, k_B is the Boltzmann constant, T is the temperature and λ is the relaxation time. Note that no shear thinning is predicted by this theory.

The Zimm model (Zimm, 1956) extends the spring model by considering intermolecular forces such as hydrodynamic forces (perturbations of the velocity field near beads by other beads), reduced excluded-volume effects (coil expansion and reduced contacts), non-linear spring forces (finitely extendable springs) and internal viscosities (coil sluggishness). One can obtain the following expression for the viscosity from the Zimm model:

$$\eta = (\eta_s + n_0 k_B T \lambda)/(1 + \lambda^2 \omega^2), \tag{2.13}$$

where ω is the dynamic shear rate (or frequency). Note that this expression does account for shear thinning.

The Rouse model (Rouse, 1953) extends these theories to multiple beads and springs (or multiple-relaxation modes). Here the expression for the viscosity becomes

$$\eta = \sum_{i=1}^{N} (\eta_s + n_0 k_B T \lambda_i)/(1 + \lambda_i^2 \omega^2). \tag{2.14}$$

For more concentrated solutions one must begin to consider interactions with neighbouring polymers. The expression for viscosity for concentrated solutions is

$$\eta = \eta_s(1 + [\eta]c + k'[\eta]^2 c^2 + \cdots). \tag{2.15}$$

For the coil-overlap region (where polymer coils overlap) the viscosity is

$$(\eta - \eta_s)/(\eta_s c) = [\eta] \exp(kc[\eta]), \tag{2.16}$$

where coil overlap occurs when $c[\eta]$ ($= c{M_w}^a$) $\sim$ 1–10. Figure 2.2 further characterizes concentrated systems. The dilute region is defined as that with concentration less than the critical overlap concentration (c^*), the semi-dilute region is that with concentration above the overlap concentration but below the critical packing concentration (c_{++}) and the concentrated region is that with concentration above the critical packing concentration. Physically what is happening in these regions is shown in Figure 2.3.

Here at low concentration ($c < c^*$) the chain dimension, r is unaffected by its neighbours, but at higher concentrations ($c > c^*$) the chain dimension is reduced. Eventually, at very high concentrations ($c > c_{++}$), the polymers are well entangled and no further reduction in spacing can occur. Also, as shown in Figure 2.4, the viscosity behaviour becomes different above a critical concentration times molecular weight (cM_w).

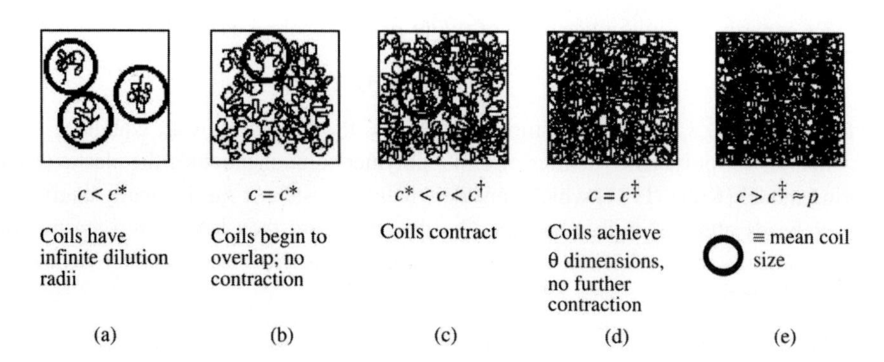

Figure 2.2. Concentration regimes in good solvents. From Macosko (1994, Figure 11.3.2). Copyright (1994). Reprinted with permission of John Wiley and Sons, Inc.

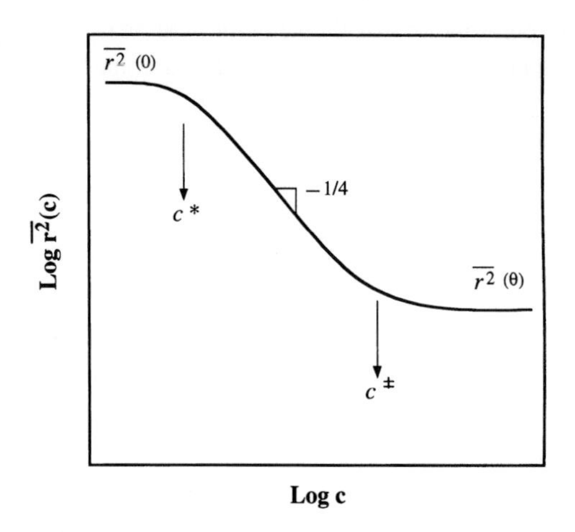

Figure 2.3. Chain dimension versus concentration. Adapted from Macosko (1994, Figure 11.3.5). Copyright (1994). Reprinted with permission of John Wiley and Sons, Inc.

This difference in behaviour necessitated the development of polymer-dynamics theories for concentrated polymer systems or melts.

Polymer melts

In some respects the dynamics of melts is simpler than the dynamics of polymer solutions. In melts, for example, the excluded-volume effect of solutions and hydrodynamic interactions of molecules within the same chain are screened by surrounding chains (hydrodynamic screening). However, the complication of melts is that the motion of each chain is affected by the entanglements in the surrounding chain. Grassley (1982) noted that there was a co-operative relaxation process in polymer melts that was characterized by a relaxation time (λ). De Gennes (1971) noted that the entanglements of surrounding chains restricted the motion of a polymer chain, and described a polymer chain as surrounded by an effective tube, where the polymer chain would have to undergo reptation or snake-like motion to relax (or get out of tube). The time taken to reptate out of the tube was

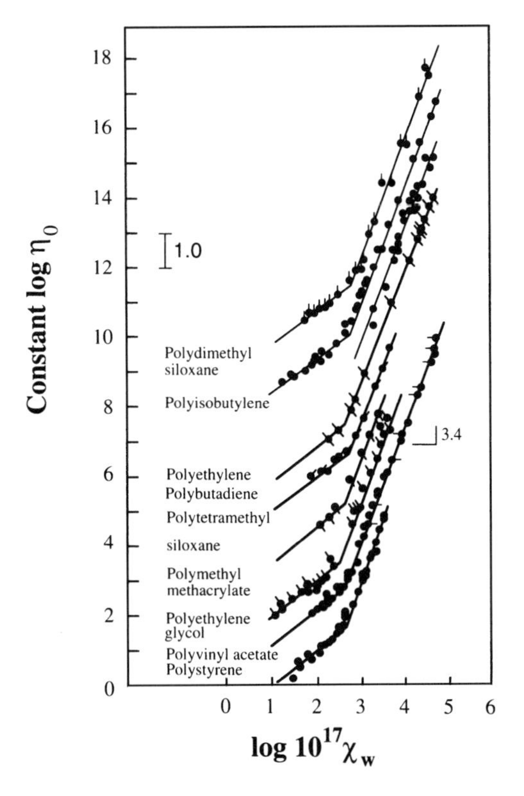

Figure 2.4. Typical viscosity–molar-mass dependence for molten polymers. From Macosko (1994, Figure 11.5.4). Copyright (1994). Reprinted with permission of John Wiley and Sons, Inc.

characterized by a reptation relaxation time (λ_{rep}). It can be shown from this theory that the viscosity should scale with the weight-average molar mass as $\eta \sim M_w^{\ 3}$; however, experimentally this relationship has been shown to be $\eta \sim M_w^{3.4}$ (Berry and Fox, 1968).

Doi and Edwards (1978a, 1978b, 1978c) extended the work of de Gennes by incorporating non-linear deformation at higher rates. Shear thinning and elasticity are predicted with the Doi–Edwards model, but overpredictions of shear thinning can occur at high shear rates. Extensions to this work include incorporation of primitive path fluctuations, constraint release, tube stretching, polydispersity, star molecules and long-chain branching, and are well summarized in Larson (1999).

2.3 Introduction to the physics of reactive polymers

As discussed in the previous section, this book focusses on reactive polymer systems, from network-forming polymers to reactively modified polymer systems. Network polymers may be defined as polymer systems that contain chemical or physical networks between the constituent molecules of the system. Within the term reactively modified polymers we include polymer systems that have been modified by grafting, chemical reaction or high-energy radiation. Typically network or reactively modified polymers are manufactured by various

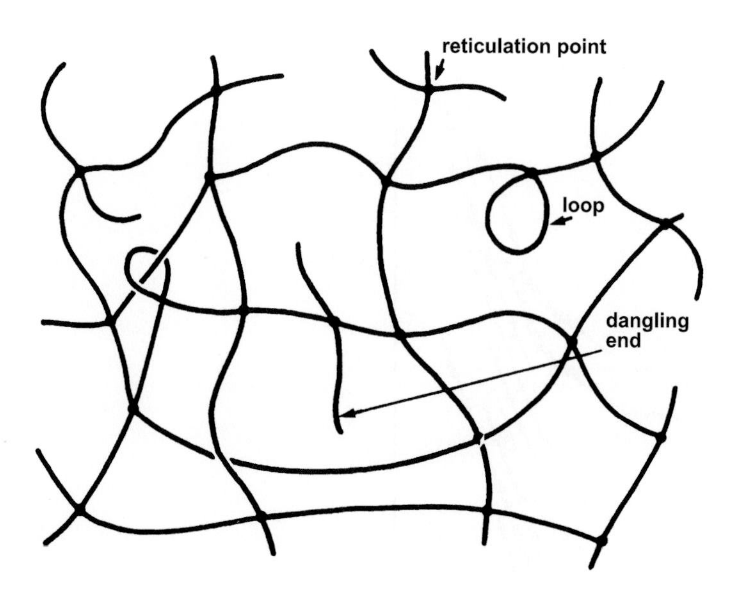

Figure 2.5. A typical polymer gel network. From Larson (1999, Figure 5.1) by permission of Oxford University Press, Inc.

reactive-processing methods such as reactive extrusion, reactive injection moulding and transfer moulding, to name only a few of the available techniques (refer to Chapter 6 for processing methods). In addition, these reactive systems may have added components (such as fillers to reduce cost and improve strength, particles to improve toughness or other polymers acting as solvents or interpenetrating networks) to facilitate further property enhancements.

We will now briefly introduce and differentiate between network and reactively modified polymers.

2.3.1 Network polymers

The structure of a network polymer is sketched in Figure 2.5 (de Gennes, 1979).

Network polymers may be formed via chemical or physical interactions. There are three types of chemical reactions that produce chemical gels (de Gennes, 1979).

- stepwise (including condensation) polymerization (Section 1.2.1)
- addition polymerization (Section 1.2.4)
- vulcanization (Section 1.4.1)

These reactions have been explained in detail in the sections of Chapter 1 indicated and they are all identified by permanent, covalent bonds between polymeric chains. These chemical-network polymer systems are an important sub-class of network polymers and are sometimes referred to as thermoset polymers. Thermoset polymers cannot be re-melted or undergo flow under the influence of heat once they have been polymerized (i.e. they have a thermo*set* morphology, unlike thermo*plastic* polymers, which can be re-melted and flow under application of heat). Owing to their excellent properties at high temperatures (and under extreme environmental conditions) thermoset polymers are suitable for a wide variety of high-performance applications such as electronic packaging, automotive panelling and

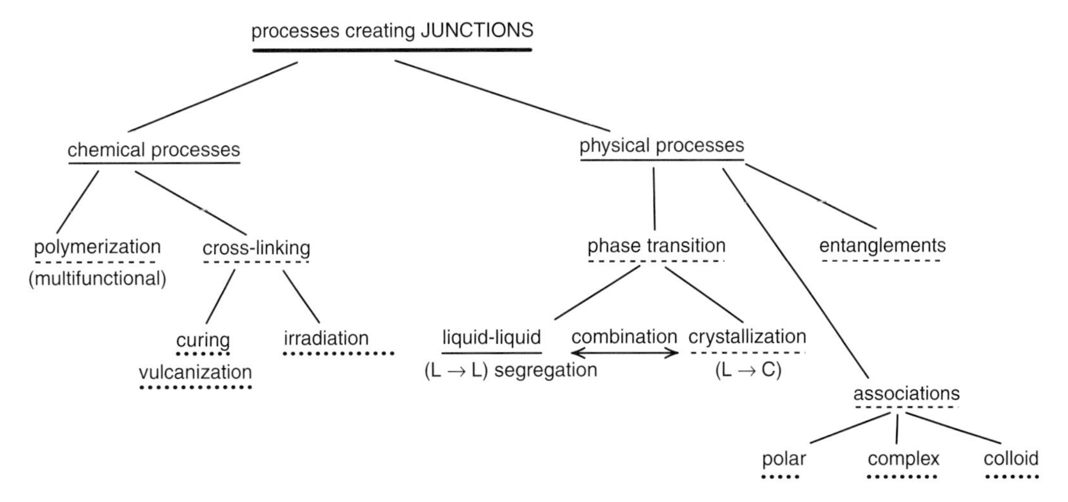

Figure 2.6. Processes causing network junctions. Reproduced by permission of the Royal Society of Chemistry from Figure 1 (Keller, 1995).

aerospace applications. Thermoset polymers have a wide range of applications because their final physical properties may be tailored by control of the initial monomer chemistry, control of cure conditions, addition of fillers and control of the reactive processing.

For physical gelation de Gennes (1979) described the following pathways:

- local helical structures
- microcrystallite junctions
- nodular domains of end groups or parts of chains

Typical intermolecular forces or associations in physical gels are van der Waals forces, electrostatic forces and hydrogen bonding, all of which are characterized by weak, reversible links.

An instructive sketch by Keller (1995) is shown in Figure 2.6. This incorporates a broader definition of network polymers (termed processes creating junctions) that incorporates other forms of reactions, but essentially follows the same lines as the physical and chemical classification of de Gennes.

An important parameter in determining the network structure in a network polymer system is the functionality of the monomers. The functionality (f) of a molecule is the number of bonds it can form with other molecules (Larson, 1999), thus the higher the functionality the more likely it is that the molecules may react to develop a network system. Of course, during cure as molecules interact the functionality of the growing intermediate chains may increase, further accelerating cure.

2.3.2 Reactively modified polymers

Reactively modified polymers are polymer systems that have been chemically or physically changed during processing. Reactively modified polymers may be classified in terms of their production by the following processing techniques:

- reactive extrusion
- reactive batch compounding

Table 2.1. Reactively extruded polymer systems

Type of reaction	Explanation	Example systems
Bulk polymerization	Conversion of low-M_w monomers or prepolymers into high-M_w polymers	Polyetherimides, melamine-formaldehydes, polyurethanes, polyamides, PBT, PET, PS, PMMA
Grafting reaction	Reaction of a monomer and a polymer to form a grafted polymer or copolymer	Grafting vinyl silanes onto polyolefins for crosslinked polyolefins; acrylic acids onto polyolefins for hydrophilicity; maleic anhydride onto polyolefins for compatabilization
Inter-chain copolymerization	Reaction of two or more polymers to form copolymers	Graft copolymers for compatibilization of immiscible polymers; PS + epoxy; PP + epoxy
Coupling or crosslinking reactions	Reaction of a polymer with a coupling, branching or crosslinking agent to increase M_w	Viscosity building for nylon-6,6; thermoplastic elastomers via crosslinking elastomeric phase in thermoplastics; PBT + polyepoxides
Controlled degradation	Controlled breakdown of M_w to low-M_w polymer or monomer	Controlled-rheology polypropylene via peroxide reaction; controlled M_w of PET polymers
Functionalization	Introduction or modification of functional groups	Chlorination/bromination of polyolefins; capping carboxylic groups on polyesters;

Source: Modified from Xanthos (1992, Table 4.1), with permission.

- reactive moulding
- irradiation processing

Reactive extrusion is well surveyed by a recent text (Xanthos, 1992), and has been referred to as 'the deliberate performance of chemical reactions during continuous extrusion of polymers and/or polymerizable monomers' (Xanthos, 1992). This usually represents chemical reactions conducted during a continuous extrusion process at relatively short residence times. Examples of types of reactions and example polymer systems used in reactive-extrusion processes are given in Table 2.1.

Clearly reactive extrusion is attractive for production in that it provides chemical modification of a polymer system in a continuous process. The disadvantages of this process can be lack of process control and the inability to understand fully the degree of modification during the process.

Reactive batch compounding involves the use of batch mixers or kneaders, which are usually employed as long-residence-time chemical reactors in the compounding industry. Reactive compounding can be further categorized, like conventional polymer compounding, in terms of intensive (break-up of agglomerates or droplets) or extensive (reducing the non-uniformity of distribution without disturbing the initial scale of dispersant) mixing. Note that reactive compounding, as opposed to conventional polymer compounding, in addition involves a chemical reaction in parallel with the mixing process. Example systems

Table 2.2. Factors affecting irradiation processing

Parameter	Influences reaction by these means
Type of polymer	Chemistry, crystallinity, density, branching structure
Temperature	Mobility of chains and reaction rates increase with temperature
Type of irradiation source	Type of grafting/crosslinking reaction possible, depth of cure penetration
Dose	Number of reaction sites increases with dose

developed in reactive compounding include internal Banbury-type mixers and subsequent processing for crosslinking natural rubber and elastomers (generally these systems include base elastomers, curatives, fillers, softeners, tackifiers, antioxidants, colourants, flame retardants and blowing agents).

Reactive moulding includes a wide range of processes that involve moulding of reactive thermoset network polymers, such as transfer moulding, reactive injection moulding, reactive compression moulding and fibreglass moulding.

Radiation processing generally refers to continuous polymer processes that are modified by radiation-induced grafting or crosslinking the polymer prior to processing, during processing or after-processing. In general there are three types of radiation sources in use, namely UV radiation, electron-beam radiation and gamma-ray irradiation. The parameters of irradiation that influence reactions are summarized in Table 2.2.

Examples of systems include

- crosslinked wires (PE, PVC, PE-EVA, PU) for increased heat and abrasion resistance
- radiation-crosslinked rubber tyres
- heat-shrinkable (polyolefin and fluoropolymer) tubing and sheets
- crosslinked hydrogels for contact lenses and wound dressings
- medical-supply sterilization (PP syringes)

The aim of this book is not to discuss each of these reactive modification techniques in detail, but rather to show how chemorheology can be a useful tool from fundamental understandings to understanding processing, process control and optimization.

In this chapter we are interested in understanding the physical properties of a curing system. We shall consider this by first describing the important transitions (gelation, vitrification and phase transitions) that occur during curing and then examining techniques used to characterize changes in physical properties in curing systems. We shall address the detailed chemorheological behaviour in Chapter 4.

2.4 Physical transitions in curing systems

To control the final properties of reactive polymers it is essential that the curing of the polymer is well characterized. This means that it is necessary to have a good knowledge of how the materials transform from the original monomeric or polymer matrix to the final cured polymer. Thus it is self-evident that the curing transitions of a system are very

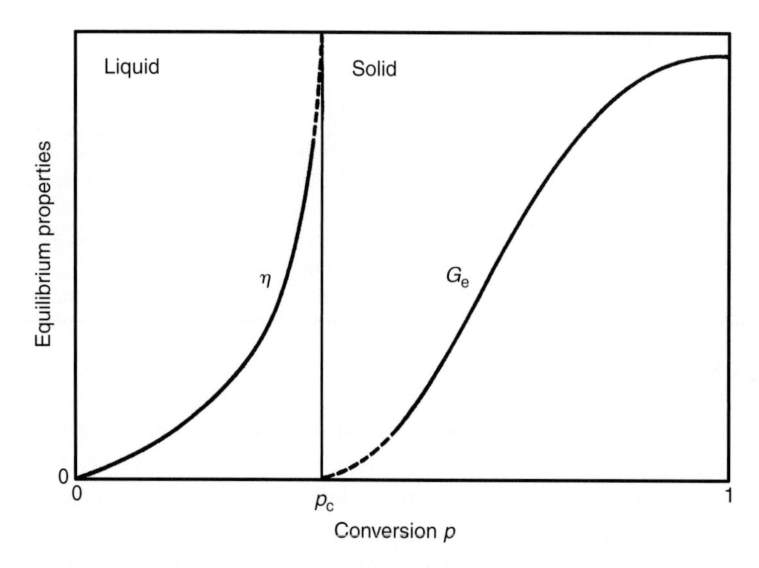

Figure 2.7. The dependence of zero-shear viscosity η_0 and equilibrium modulus G_0 on conversion p for a crosslinking system. Reprinted from Figure 1 p. 344 (Winter 1990). Copyright (1990). Reprinted with permission of John Wiley and Sons, Inc.

important milestones to measure. The chemical reactions underlying these processes are described in Section 1.2.2.

2.4.1 Gelation and vitrification

The curing transitions of most importance in network polymers are gelation and vitrification. Gelation is the transition from soluble material to an insoluble matrix. It is defined as the point at which the insoluble, infinite-molar-mass network is formed (note that the matrix may still contain unreacted soluble intermediates). Larson (1999) defines gelation as the conversion of liquid to solid by formation of a network of chemical or physical bonds between molecules or particles in the fluid. Winter (1990) notes also that at gelation the material has an infinitely broad distribution of molar masses (i.e. M_w/M_n goes to infinity) and that this gives rise to self-similar fractal-like structure. Gelation is also commonly defined as the point at which the material is able to sustain mechanical energy (or the onset of a solid modulus) and a point at which it has an infinite steady shear viscosity and cannot flow. This is shown diagrammatically in Figure 2.7.

It should be noted that further reactions and crosslinking may occur after gelation (via molecules and clusters of molecules reacting with the infinite-M_w gel cluster), but these reactions become governed by diffusion (more so than kinetic) considerations.

Vitrification is defined as the point at which the material is transformed from a gel structure into a glassy structure. When molecules are reacting they are forming larger-M_w chains, which increases the glass-transition temperature (T_g) of the system. At vitrification the T_g of the material has reached the isothermal curing temperature (T_c), and thus the mobility of the system is restricted. Conversely, of course, if the curing temperature (T_c) is kept above the ultimate glass-transition temperature of the fully cured material then the material should not vitrify. In the past vitrification was considered as the point at which the material is fully crosslinked and there is no further reaction (Enns and Gillham, 1983). However, other papers (Huang and Williams, 1994, Barral *et al.*, 1995) indicate that further curing is possible after the vitrification point. Of course, vitrification tends to be a transition region rather than a single transition point due to the fact that there is a distribution of polymer chain lengths in the curing polymer system. That is, vitrification begins when higher-molar-mass species enters the glassy regime (or cannot diffuse) and ends when the lower-molar-mass species reaches the glassy state (or cannot diffuse).

2.4.2 Phase separation

During cure phase separation whereby the solid phase separates from the unreacted liquid phase may also occur (de Gennes, 1979). Thermoplastic-modified thermoset systems are useful as toughened thermosets that have improved ductility over that of conventional thermosets. Reactive-phase separation is thus another important transition in curing systems, and this process is controlled by the interplay between the thermodynamics of phase separation and the cure. For example, other work (Girard-Reydet *et al.*, 1998) has shown that important factors in controlling phase separation in polyetherimide-modified epoxy-resin systems are the proximity of the initial modifier concentration to the critical concentration and the ratio of the phase-separation rate to the polymerization rate. It is known, for phase-separation processes, that for materials near the critical concentration (as determined by mixing theory for reactive blends (Clarke *et al.*, 1995, Inoue, 1995) and based on critical miscibility) phase separation occurs unstably via spinodal de-mixing (as mentioned in Section 1.3.2), whereas for off-critical concentrations, phase separation proceeded via a stable nucleation/growth mechanism. Further excellent studies (Bonnet *et al.*, 1999a, 1999b) have shown the effects of the phase separation on the cure kinetics and rheology of polyetherimide-modified DGEBA/MCDEA epoxies, highlighting again the interdependence of phase separation and curing in determining epoxy-resin properties. Additionally, both cure and post-cure temperatures may control the extent of phase separation due to vitrification halting the evolution of morphologies. Clearly, to optimize the final morphology of phase-separated thermoset systems one needs to monitor carefully the interdependence of phase-separation and cure properties.

Gelation, vitrification and phase-separation transitions in curing systems are very well described in the work of Gillham (1986) via the use of time–temperature-transformation diagrams.

2.4.3 Time–temperature-transformation (TTT) diagrams

Pioneering work by Gillham (Enns and Gillham, 1976) linked important transition phenomena during cure of reactive polymers to diagrams that were easily adapted to processing. These diagrams are known as time–temperature-transformation (TTT) diagrams, and are illustrated in Figure 2.8.

Figure 2.8 (Enns and Gillham, 1983) shows the time–temperature-transformation diagram for a thermoset system undergoing isothermal cure. There are seven distinct regions of matter in this TTT diagram; liquid, sol–gel rubber, gel rubber, sol–gel glass, gel glass, sol glass and char. Isothermal heating of a sample can be represented by a horizontal line on Figure 2.8 (at a certain reaction temperature, T_{rxn}). The sample initially will pass through a particular region depending on the T_{rxn} chosen. T_{g0} is the glass-transition

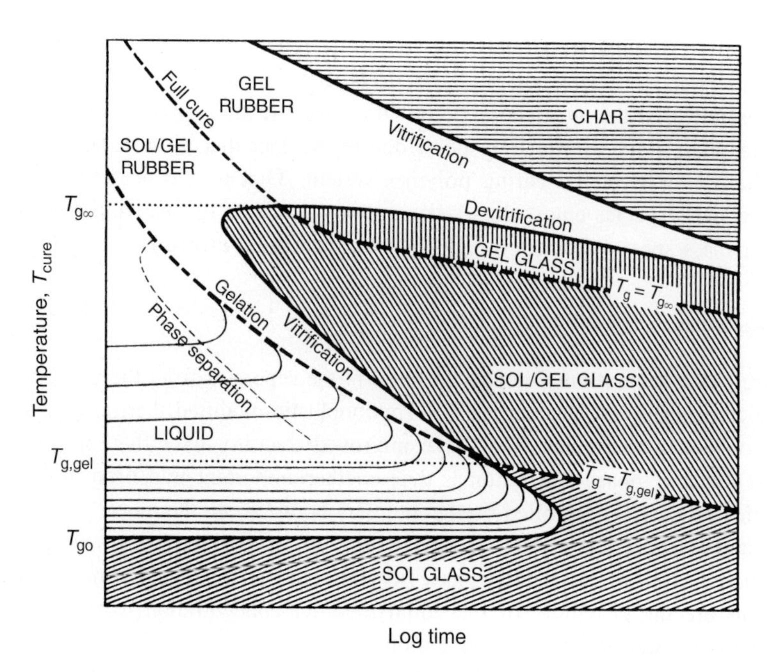

Figure 2.8. A time–temperature-transformation (TTT) isothermal cure diagram for a thermosetting system. Adapted from Figure 1 (Gillham, 1986).

temperature of the initial uncured system, so for any T_{rxn} below this temperature the system will remain unreacted. $T_{g,gel}$ is the temperature at which gelation and vitrification coincide. At T_{rxn} between T_{g0} and $T_{g,gel}$ the system is initially a liquid, but will vitrify before gelling. $T_{g\infty}$ is the maximum T_g of the cured system. At any T_{rxn} between $T_{g,gel}$ and $T_{g\infty}$ the system is initially a liquid and will reach gelation (or the formation of two phases: a finite-molar-mass solvent and an infinite-molar-mass gel). Eventually the T_g of the system will approach the T_{rxn} and vitrification will occur. The full cure line represents the time required at any T_{rxn} for T_g to equal $T_{g\infty}$, and is important for allowing comparative analyses of systems. Above $T_{g\infty}$ the material will gel but will not vitrify until degradation (charring).

It should be noted that in Gillham's work he utilized a torsional braid analyser (TBA), which determines the gelation and vitrification points via the change in resistance to torsional motion of a fixed braid support in an epoxy-resin matrix (more details are given later in Chapter 3). Thus gelation and vitrification were denoted by two peaks in the damping function – the first peak was gelation and the later peak was vitrification. Note that Stutz and Mertes (1989) found for their epoxy-resin systems that the first peak was not caused by gelation of the resin, but occurred at approximately the same viscosity and was observed only when a fibre braid was used as support in a TBA. Subsequently Cadenato *et al.* (1997) developed TTT diagrams based on TMA and DMTA tests that reveal discrepancies between the methods for determining gel and vitrification times. Mijovic *et al.* (1996) utilized FT-IR, dielectric and rheological measurements to determine the kinetics and transition data for a TTT diagram of a trifunctional epoxy–aniline system, and found good agreement among results obtained with FT-IR, dielectric and rheometric methods to determine transitions. Overall it is recommended that any work on developing TTT diagrams should utilize several different techniques to confirm kinetics, gelation and vitrification, and these methods will be discussed in more detail later in this chapter and in Chapters 3 and 4.

The TTT diagram has been utilized for many reactive systems. Wissanrakkit and Gillham (1990) show the kinetics and TTT diagram for a stoichiometrically balanced DGEBA/TMAB epoxy–amine system, whilst Simon and Gillham (1992) show the TTT diagram for an off-stochiometric DGEBA/TMAB epoxy–amine system. Pang and Gillham (1989) and Wang and Gillham (1993a) show the usefulness of using a TTT diagram to describe the effects of physical ageing of high-T_g epoxy–amine systems. Pang and Gillham (1989) also show that the density of an epoxy–amine system has a maximum at gelation with respect to conversion. This is stated to be due to the effects of physical ageing prior to gelation, which increases density, competing against the chemical ageing after gelation (which decreases density). Vendetti *et al.* (1995a) confirmed that there is a specific volume increase (density decrease) with increasing conversion after gelation, and this was subsequently confirmed by PALS testing in a later paper (Vendetti *et al.*, 1995b).

Such TTT diagrams have also been useful in describing the cure of polyimide systems, as shown in Palmese (1987), which shows the TTT diagram of a polyamicacid/polyimide system. The TTT diagram of a polycyanurate system is developed by Simon (1993) on the basis of FT-IR, DSC and torsional braid measurements. Kim *et al.* (1993) developed a TTT diagram for a thermoset–thermoplastic blend, specifically a tetrafunctional epoxy-resin/poly(ether sulfone)/dicyandiamide thermoset–thermoplastic blend (Figure 2.9).

Note that Figure 2.9 shows the transitions in terms of (1) onset of phase separation, (2) gelation, (3) fixation of the dimension of phase-separated structure, (4) the end of phase separation and (5) vitrification. Lu and Pizzi (1998) investigated the curing of phenol and urea formaldehyde wood adhesives in a TTT diagram. Special mention is made of the effect of the wood substrate on the cure of this composite system.

Other useful diagrams have been developed from the TTT diagram. Wang and Gillham (1993b) show the use of a T_g–temperature–property (T_gTP) diagram for an epoxy–amine system as illustrated in Figure 2.10.

In this diagram the x-axis is the extent of cure (T_g) and the y-axis is the temperature. The properties of the curing system may be separated into distinct regions of the glass transition (T_g), the beta-transition (β-T_g) , the end of the glass transition (ε; T_g) and gelation ($T_{g,gel}$). Properties may also be implied on the third axis out of the plane of the paper. Thus the physical properties (behaviour) of a system with respect to conversion are determined by its temperature and T_g.

Wang and Gillham (1993b) and Simon and Gillham (1994) show the use of a continuous-heating (CHT) diagram as illustrated in Figure 2.11. Essentially the diagram includes the basic events (vitrification, gelation, devitrification) of a TTT diagram; however, the key additions to the diagram are the constant-heating-rate curves (where rate 1 > rate 2 etc.). The heating-rate curves are used to show transitions passed through under non-isothermal heating rates. Simon and Gillham (1994) also incorporate lines of constant T_g (iso-T_g curves) into a TTT diagram to produce the iso-T_g–TTT diagram (refer to Figure 2.12). Essentially these diagrams allow one to determine the time taken to reach a specific T_g at a given isothermal curing temperature.

The conversion-temperature diagram (CTT) is described in Figure 2.13 (Prime, 1997). Here conversion is plotted against temperature and various regions are identified. The gel conversion (a_{gel}) is noted on the y-axis as a constant with respect to temperature (but may

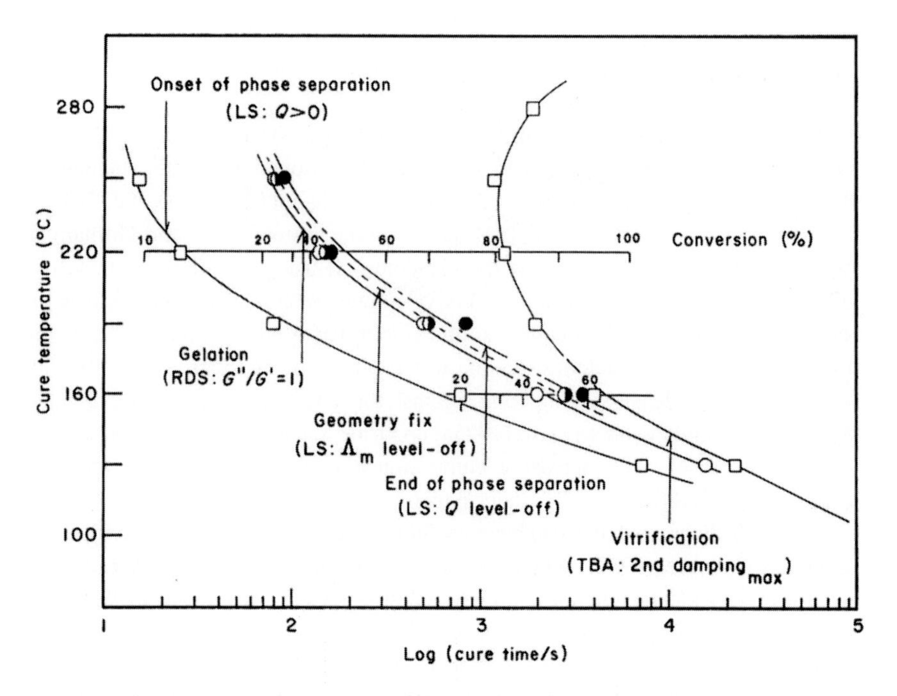

Figure 2.9. A time–temperature-transformation (TTT) diagram for a phase-separating thermoset/thermoplastic blend system. Reprinted from Figure 12 from (Kim *et al.*, 1993). Copyright (1993), with permission from Elsevier.

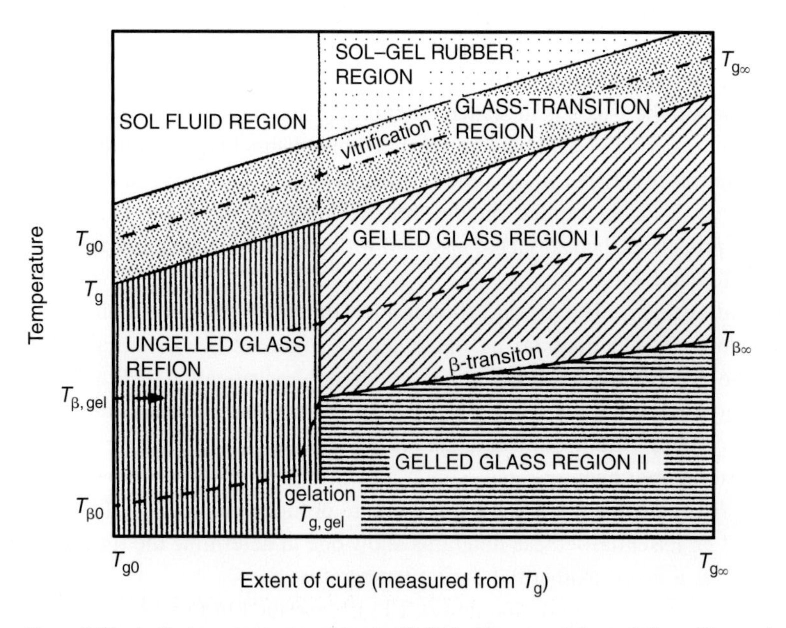

Figure 2.10. A T_g–temperature–property (TgTP) diagram. Adapted from Figure 1 from (Wang and Gillham, 1993b). Copyright (1993), with permission from Elsevier.

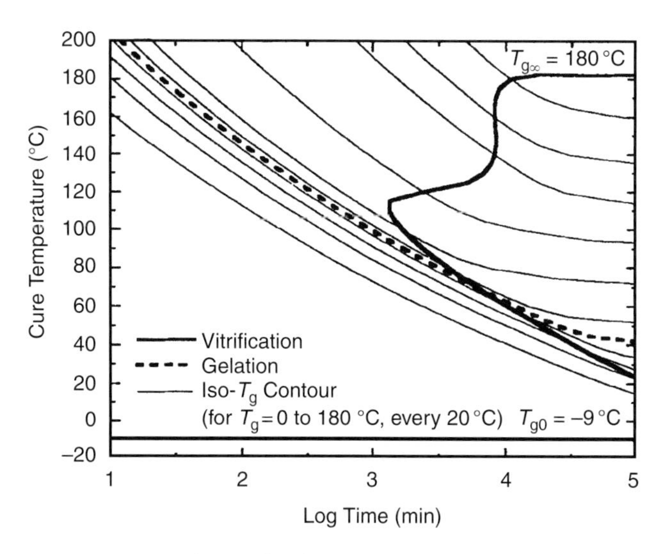

Figure 2.11. A continuous heating temperature (CHT) diagram. Adapted from Figure 23 from (Wang and Gillham, 1993b). Copyright (1993), with permission from Elsevier.

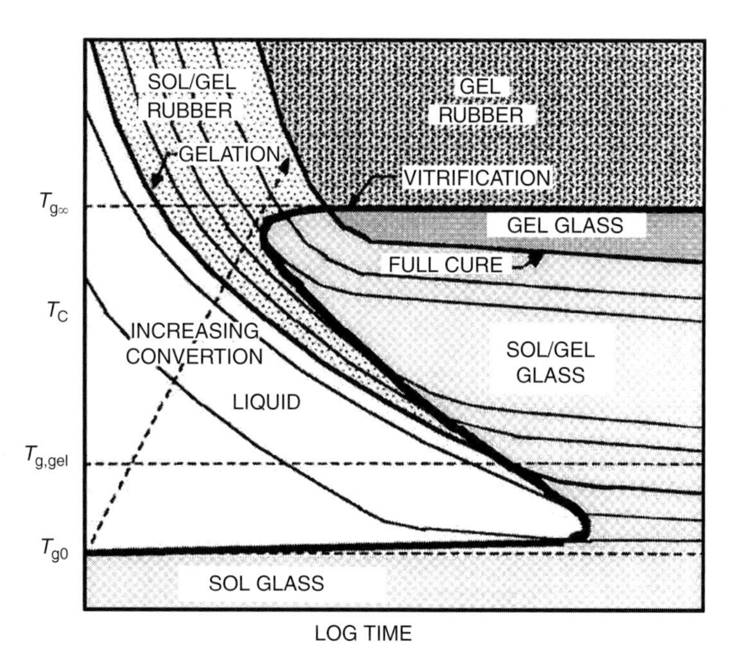

Figure 2.12. An iso-T_g-TTT diagram. Adapted from Figure 7 from (Simon and Gillham, 1994). Copyright (1994), with permission from Elsevier.

change depending on variations of chemical reactions with temperature). The x-axis has $T_{g0,gel}$ $T_{g,gel}$, $T_{g\infty}$ and T_d (the decomposition temperature) noted. The T_g–conversion curve is also plotted in the diagram.

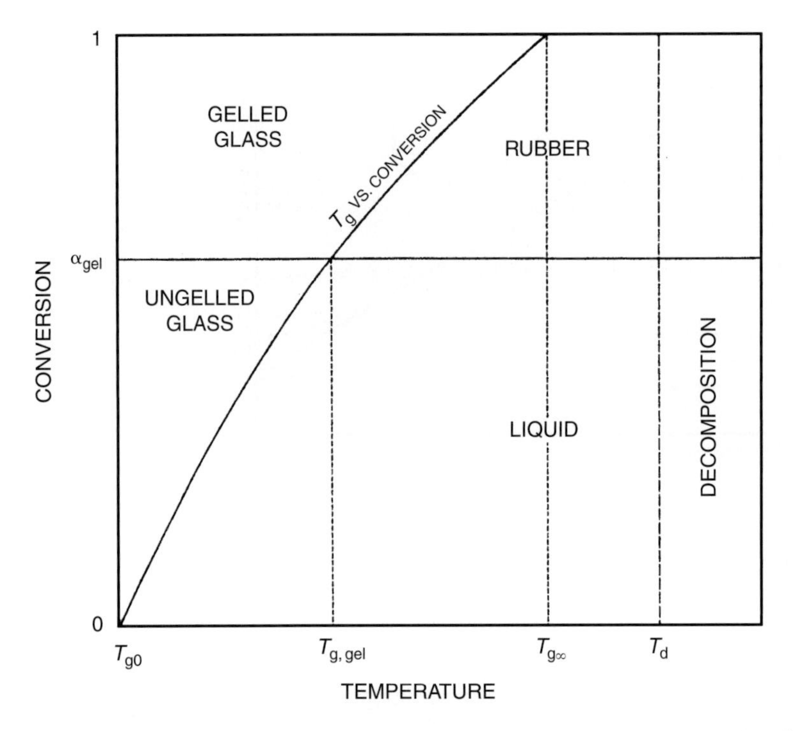

Figure 2.13. A conversion–temperature-transformation (CTT) diagram. Reprinted from (Prime, 1997). Copyright (1997), with permission from Elsevier.

Also Osinski (1993) develops the a–TT or TT–a_T diagram, on which the transitions may be expressed as a function of conversion (a) instead of time (Figure 2.14).

Obviously these curves are flexible in that they may be used for isothermal and non-isothermal cure.

2.4.4 Reactive systems without major transitions

In terms of reactive modification of polymer systems, the goal of the modification need not be to reach gelation or vitrification. The goal may simply be to graft a certain percentage of reactive groups onto the base polymer chain or to achieve light crosslinking of polymer chains. In these systems chemical-composition testing (as detailed in Chapter 1) tends to be more critical than physical-property testing, due to the limited effects of this modification on bulk physical properties. However, some physical tests useful for reactive processing will be detailed later in Chapter 3.

2.5 Physicochemical models of reactive polymers

There has been much work on the development of physicochemical models for network polymers and reactive polymers, and a brief summary is provided here.

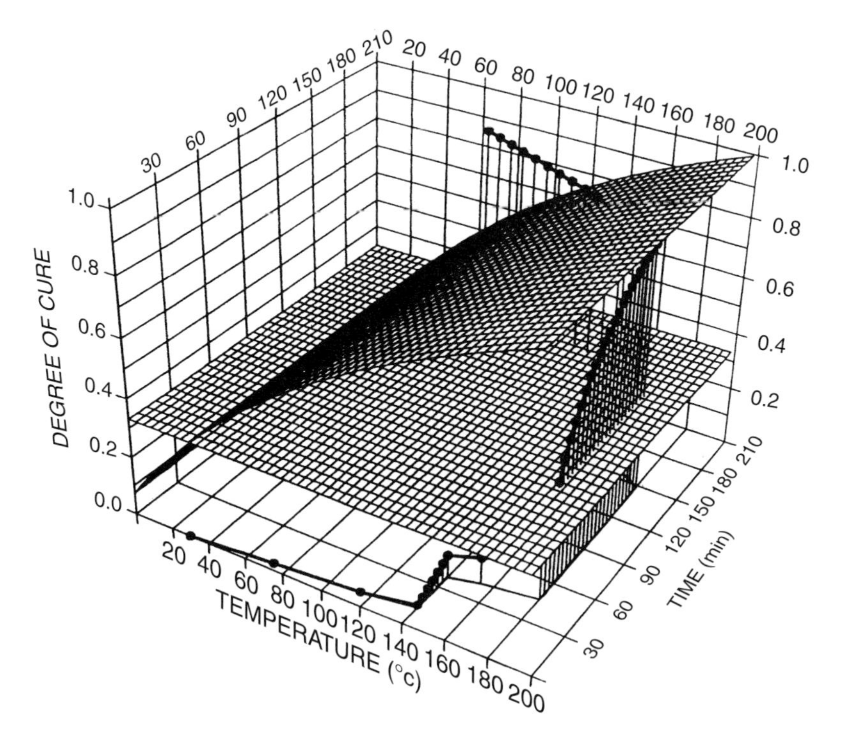

Figure 2.14. A three-dimensional conversion–time–temperature (αTT) diagram. Reprinted from Figure 7 (Osinski, 1993). Copyright (1993), with permission from Elsevier.

2.5.1 Network models

Dusek (1986a) characterizes network-formation models into the following categories: spatially independent and spatially dependent models. Of the spatially independent models, there are statistical models (in which network structure is developed from various interacting monomer units) and kinetic models (in which each concentration of species is modelled by a kinetic differential equation).

Statistical network models were first developed by Flory (Flory and Rehner, 1943, Flory, 1953) and Stockmayer (1943, 1944), who developed a gelation theory (sometimes referred to as mean-field theory of network formation) that is used to determine the gel-point conversions in systems with relatively low crosslink densities, by the use of probability to determine network parameters. They developed their classical theory of network development by considering the build-up of thermoset networks following this random, percolation theory.

Percolation is the process of network formation by random filling of bonds between sites on a lattice. If one increases the fraction of bonds (p) formed, then larger clusters of bonds form until an infinite lattice-spanning cluster (at the percolation threshold, $p = p_c$) is formed. Figure 2.15 (Stanley, 1985) shows the growth of the network corresponding to values of the fraction p of (a) 0.2, (b) 0.4, (c) 0.6 ($p = p_c$) and (d) 0.8.

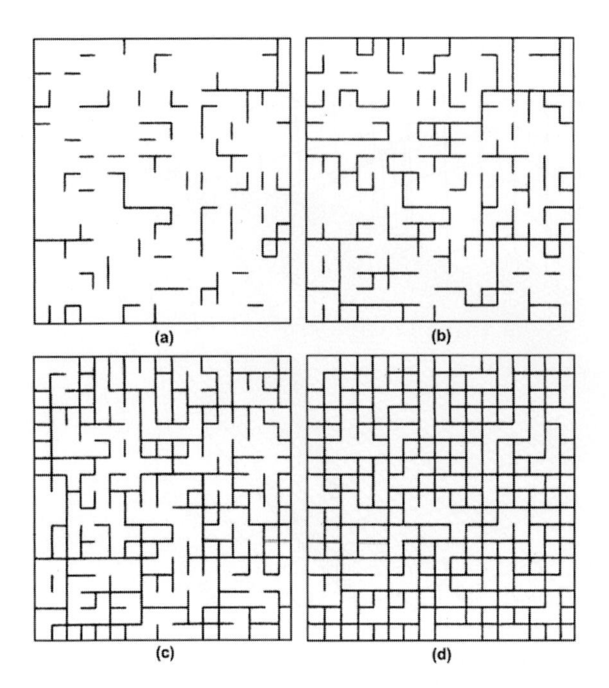

Figure 2.15. The phenomenon of bond percolation from a fraction of links, p, from 0.2 to 0.8. Reprinted from Figure 16 (Stanley, 1985). Reprinted with permission of John Wiley and Sons, Inc. Copyright (1985).

The gel point in this classical theory is the percolation threshold (p_c) given by

$$p_c = 1/(f - 1), \tag{2.17}$$

where f is the co-ordination number of the tree – or the number of bonds that can form at each site (Flory, 1953).

Figure 2.16 (Stanley, 1985) shows the analogy between (a) the gelation threshold and (b) the percolation thershold, and $R(a)$ is the probability that there exists an infinite cluster.

It should be noted that this theory neglects loops or cyclic formations, and this affects the size distribution and other cluster properties. Some of these properties (and their relationships at gelation) are highlighted in Table 2.3 (Larson, 1999), as are their experimental values compared with the classical and three-dimensional-percolation theoretical values at gelation.

Note that power-law behaviour is prevalent at gelation. This has been proposed to be due to a fractal or self-similar character of the gel. Note that the exponent D_f is termed the fractal dimension. For any three-dimensional structure ($D = 3$) the exponent $D_f \leq 3$ (where $D_f < 3$ indicates an open structure and $D_f = 3$ indicates a dense structure). Also Muthukumar (Muthukumar and Winter, 1986, Muthukumar, 1989) and Takahashi *et al.* (1994) show explicitly the relationship between fractal dimension (D_f) and power-law index of viscoelastic behaviour (n). Interestingly, more recent work (Altmann, 2002) has also shown a direct relationship between the power-law behaviour and the mobility of chain relaxations, which will be discussed further in Chapter 6.

Table 2.3. Scaling exponents for classical and percolation theories of gelation

Exponent	Relation	Classical	Three-dimensional percolation		Experimental
λ	$N(m) \sim m^{-\lambda}$	5/2	2.20		2.18–2.3
σ	$M_z \sim \varepsilon^{-1/\sigma}$	1/2	0.45		–
γ	$M_w \sim \varepsilon^{-\gamma}$	1	1.76		1.0–2.7
υ	$R_z \sim \varepsilon^{-\upsilon}$	1	0.89		–
D_f	$R^{D_f} \sim M$	4	2.5		1.98
β	$P \sim \varepsilon^{\beta}$	1	0.39		–
			R–Z	EN	
t	$G \sim \varepsilon^{t}$	3	2.7	1.94	1.9–3.5
ζ	$\zeta \sim \varepsilon^{-\zeta}$	3	4.0–2.7	2.6	3.9
$k = \zeta - t$	$\eta \sim \varepsilon^{-k}$	0	0–1.35	0.75	0.75–1.5

Note: R–Z, Rouse–Zimm theory. EN, electrical network analogy.
Source: From Larson (1999, Table 5.1), used by permission of Oxford University Press, Inc.

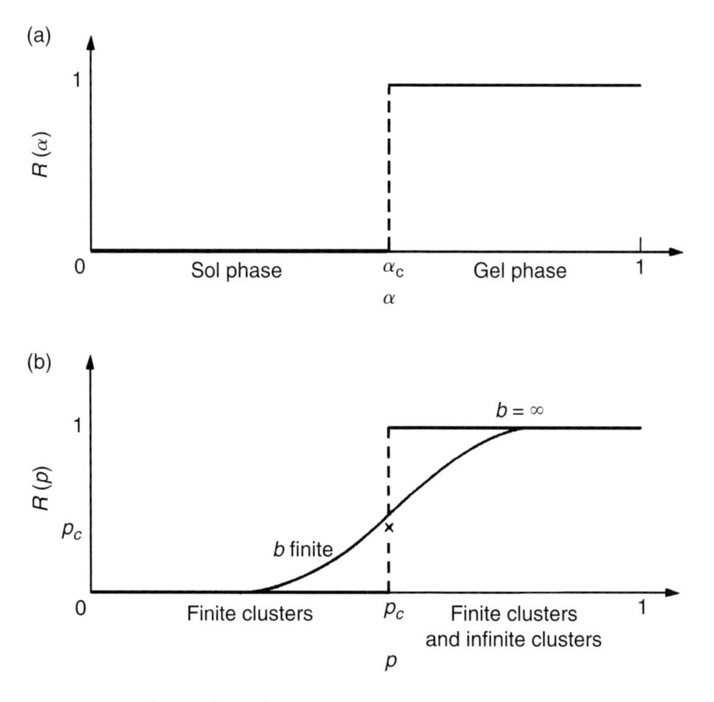

Figure 2.16. The analogy between (a) the gelation threshold and (b) the percolation-theory threshold. Reprinted from Figure 17 (Stanley, 1985). Reprinted with permission of John Wiley and Sons, Inc. Copyright (1985).

Macosko and Miller (1976) and Scranton and Peppas (1990) also developed a recursive statistical theory of network formation whereby polymer structures evolve through the probability of bond formation between monomer units; this theory includes substitution effects of adjacent monomer groups. These statistical models have been used successfully in step-growth polymerizations of amine-cured epoxies (Dusek, 1986a) and urethanes (Dusek *et al.*, 1990). This method enables calculation of the molar mass and mechanical properties, but appears to predict heterogeneous and chain-growth polymerization poorly.

Kinetic models determine the state of cure by predicting the concentration of reacting species from the solution of differential equations for each reacting species. Mikos *et al.* (1986) and Tobita and Hamielec (1989) have developed kinetic models for vinyl and free-radical network systems. Chain-growth polymerization has also been modelled through kinetic simulations by Okay (1994).

Lattice (Stauffer *et al.*, 1981) and off-lattice (Leung and Eichinger, 1984) percolation models are typical examples of spatially dependent network models and are usually used for fast, diffusion-controlled reactions. Anseth and Bowman (1994), Manneville and de Seze (1981), Boots and Pandey (1985) and Simon *et al.* (1989) use another spatially dependent network theory called kinetic gelation modelling, which involves simulation of the network structure in space using percolation simulations. Application of a kinetic gelation model to free-radical polymerization of divinyl monomers (Kloosterboer, 1988) indicates that the structure is well modelled when it begins with regions of inhomogeneity, but becomes more homogeneous at higher conversions.

Overall it is considered that the spatially independent network models, whilst simpler, should strictly be used for lightly crosslinked, homogeneous networks, whereas the spatially dependent models, although computationally intensive and limited by pre-defined lattice structure, provide a better understanding of network heterogeneities in highly crosslinked systems.

Other network models

Labana (1985) classifies gel networks as emanating from step or chain reactions. Table 2.4 summarizes the model estimates of the percolation gel point.

Schiessel and Blumen (1995) describe a mechanical-ladder model based on the algebraic decay (decay $\sim t^{-a}$) profile of viscoelastic properties at gelation. The mechanical model does not relate to the underlying physics, but is based on a mechanical-ladder model that mimics fractional relaxation equations and is useful in determining viscoelastic decay properties of gelled systems.

Barton *et al.* (1997) highlight the use of both group contribution and direct atomistic simulation to predict physico-mechanical properties and highlight the importance of hydrogen bonding in an epoxy-resin system (by virtue of its correlation to T_g).

Dusek and Demjanenko (1986b) describe the cascade theory of branching as a statistical method for describing branching processes that is based on a Flory–Stockmayer tree-like model with uncorrelated circuit closing in the gel. The method uses cascade substitution and probability-generating functions as tools to generate branched structures. Calculations on network development up to and beyond gelation are possible and have been found to provide good agreement with experimental data from epoxy-resin systems. Bauer and Bauer (1990) combine a detailed mechanistic kinetic model with the cascade formalism (which is a statistical development of network formation focussed on describing polymers as rooted trees generated by probability-generating functions) to model the network development of a cyanic

Table 2.4. Gel-point predictions from percolation theory

Step		
Carothers	$P_{gel} = 2/f_{avg}$	P_{gel} is the extent of reaction at gelation: $f_{avg} = \sum_i (n_i f_i)/n_i$ is the average functionality, where n_i is the number of molecules of monomer I with functionality f_i. Useful for systems with equivalent numbers of reactive groups (Carothers, 1936).
Flory	$P_{gel} = 1/(f - 1)$	Neglects unequal reactivities of functional groups, polydisperse oligomers (Flory, 1941).
Flory	$P_{gel}^2 = 1/[r(f_a - 1)(f_b - 1)]$	f_a and f_b are the weight-average functionalities of molecules with reactive groups of types a and b (Macosko and Miller, 1976).
Flory	$P_{gel} = 1/(D_{Pw0} - 1) = 1/D_{Pn0}$	D_{Pw0} is the initial weight-average functionality of starting polymer chains; D_{Pn0} is the initial number-average functionality of starting polymer chains (Miller *et al.*, 1979).
Chain		
Flory	$P_{gel} = 1/(D_{pwav} - 1)f$	D_{pwav} is the weight-average degree of polymerization, f is the fraction of double bonds in the monomer; use only when double bonds have equal reactivity and the copolymer is random (Flory, 1953).

acid ester–glycidyl ether system. Gupta and Macosko (1990) describes a combined model incorporating non-random processes (in this case, initiated living chain-type polymerization) and a random branching model (that recognizes that some processes are statistically random (a naive statistical model)). Results are compared with those derived from existing naive statistical models (Riccardi and Williams, 1986, Bidstrup *et al.*, 1986, Tsou and Peppas, 1986) and an approximate combined model (Dusek, 1986a), and show the importance of including the non-random processes when considering the polymerization of an epoxy-resin system.

Recent work (Altmann, 2002) has focussed on combining a dynamic Monte Carlo percolation-grid simulation for reaction kinetics and an enthalpy-based group-interaction viscoelastic model to develop a model for the chemorheological and network properties of reactive systems. More emphasis will be placed on this model in Chapter 6.

2.5.2 Reactive polymer models

Owing to their relatively recent development, physical models for reactive polymer-processing systems have been studied less rigorously. Of course, modelling of polymerization reactions in an extrusion process will employ modelling techniques similar to those for modelling bulk polymerization reactions (monitoring conversion, selectivity and molar mass), however, the input variables will be different from those for bulk polymerization parameters. For example, Janssen (1998) notes that for reactive extrusion one must consider the inter-relationships of input variables such as reaction speed, homogenization, thermal effects, micro-mixing, residence time, bulk mixing and heat transfer with the output variables of conversion, selectivity and molar mass. In fact there is a dearth of experimental work on examining the effects of processing parameters on the final structure and properties of polymers produced by a wide range of reactive processing techniques such as resin-transfer moulding, reaction injection moulding, reinforced reaction injection moulding and structural reaction injection moulding, reactive extrusion and reactive–rotational moulding as summarized in Brown *et al.* (1994). Here it is clear that there have been only experimental studies of the effects of process parameters on physical properties of the reactive products, and that a fundamental modelling study that incorporates understanding of the changes in chemical and physical properties in these systems is required in order to optimize these systems.

In addition to these complications, Moad (1999) notes that, for typical reactive modifications, the amount of modification can be quite small (0.5–2 mol%) and therefore very difficult to characterize. However, Moad (1999) does suggest some techniques such as chemical methods, FT-IR, NMR and DSC that may be useful to aid characterization. Janssen (1998) also notes complications of thermal, hydrodynamic and chemical instabilities that can occur in reactive extrusion that must be addressed by combining knowledge of the chemistry and of the physics (flow behaviour, mixing) of the reactive extrusion process. Xanthos (1992) presents the importance of understanding both the chemistry and the reaction engineering fundamentals of reactive extrusion, in order better to understand and model the process in practice.

Thus it is clear that the understanding of physical models for reactive processed polymers is less mature than that for network polymer models. However, it is also very clear that useful models that characterize both network and reactively processed polymers require concurrent chemical and physical models. The experimental techniques and models for network and reactive polymers will now be examined in detail in the remaining chapters.

References

Adabbo, H. & Williams, R. (1982) *J. Appl. Polym. Sci.*, **29**, 1327–1334.

Altmann, N. (2002) *A Model for the Chemorheological Behaviour of Thermoset Polymers*, Brisbane: University of Queensland.

Anseth, K. & Bowman, C. N. (1994) *Chem. Eng. Sci.*, **49**, 2207.

Barral, L., Can, J., Lopez, A. J., Nogueira, P., & Ramirez, C. (1995) *Polym. Int.*, **38**, 353–356.

Barton, J., Deazle, A., Hamerton, I., Howlin, B. & Jones, J. (1997) *Polymer,* **38**, 4305–4310.

Bauer, J. & Bauer, M. (1990) *J. Macromol. Sci. – Chem.* A, **27**, 97–116.

Berry, G. & Fox, T. (1968) *Adv. Polym. Sci.*, **5**, 261.

Bidstrup, W., Sheppard, N. & Senturia, S. (1986) *Polym. Eng. Sci.* **26**, 359–361.

Bonnet, A., Pascault, J., Sautereau, H. & Camberlin, Y. (1999a) *Macromolecules*, **32**, 8524–8530.

Bonnet, A., Pascault, J., Sautereau, H., Taha, M. & Camberlin, Y. (1999b) *Macromolecules*, **32**, 8517–8523.

Boots, H. & Pandey, R. (1985) *Polym. Bull.*, **11**, 219–223.

Brown, M., Coates, P. & Johnson, A. (1994) Reactive processing of polymers, in RAPRA (Ed.) *RAPRA Review Reports*, Report 73. vol. 7, no. 1.

Cadenato, A., Salla, J., Ramis, X., *et al.* (1997) *J. Thermal Anal.* **49**, 269–279.

Carothers, W. (1936) *Trans. Faraday Soc.*, **32**, 39.

Clarke, N., McGleish, T. & Jenkins, S. (1995) *Macromolecules*, **28**, 4650.

Couchman, P. (1978) *Macromolecules*, **11**, 117–119.

De Gennes, P. (1971) *J. Chem. Phys.*, **55**, 572.

Degennes, P. (1979) *Scaling Concepts in Polymer Physics*, New York: Cornell University Press.

Di Benedetto, A. (1987) *J. Polym. Sci. B: Polym. Phys.*, **25**, 1949–1969.

Di Marzio, E. (1964) *J. Res. Nat. Bureal Standards* A, **68**, 611–617.

Doi, M. & Edwards, S. (1978a) *J. Chem. Soc. Faraday Trans. II*, **74**, 1789.
Doi, M. & Edwards, S. (1978b) *J. Chem. Soc. Faraday Trans. II*, **74**, 1802.
Doi, M. & Edwards, S. (1978c) *J. Chem. Soc. Faraday Trans. II*, **74**, 1818.
Dusek, K. (1986a) *Adv. Polym. Sci.*, **78**, 1.
Dusek, K. & Demjanenko, M. (1986b) *Radiat. Phys. Chem.*, **28**, 479–486.
Dusek, K., Spirikova, M. & Havlicek, I. (1990) *Macromolecules*, **23**, 1774.
Einstein, A. (1906) *Ann. Phys.*, **19**, 289.
Enns, E. & Gillham, J. (1976) *J. Appl. Polym. Sci.*, **28**, 2562.
Enns, J. & Gillham, J. (1983) *J. Appl. Polym. Sci.*, **28**, 2567–2591.
Flory, P. (1941) *J. Amer. Chem. Soc.*, **63**, 3083.
Flory, P. (1953) *Principles of Polymer Chemistry*, Ithaca, NY: Cornell University Press.
Flory, P. & Rehner, J. (1943) *J. Chem. Phys.*, **11**, 512.
Fox, T. & Loshaek, S. (1955) *J. Polym. Sci.*, **15**, 371–390.
Gan, S., Seferis, J. & Prime, R. (1991) *J. Thermal Anal.*, **37**, 463–478.
Gedde, E. (1995) *Polymer Physics*, London: Chapman and Hall.
Gillham, J. (1986) *Polym. Eng. Sci.*, **26**, 1429–1434.
Girard-Reydet, E., Sautereau, H., Pascualt, J. P. *et al.* (1998) *Polymer*, **39**, 2269–2280.
Grassley, W. (1982) *Adv. Polym. Sci.*, **47**, 68.
Gupta, A., Macosko, C. W. (1990) *J. Polym. Sci. – Part B, Polym. Phys.*, **28**, 2585–2606.
Hale, A., Macosko, C. & Bair, H. (1991) *Macromolecules*, **24**, 2610–2621.
Huang, M. & Williams, J. (1994) *Macromolecules*, **27**, 7423–7428.
Inoue, T. (1995) *Prog. Polym. Sci.*, **20**, 119.
James, H. & Guth, E. (1943) *J. Chem. Phys.* **11**, 455.
Janssen, L. (1998) *Polym. Eng. Sci.*, **38**, 2010.
Keller, A. (1995) *Faraday Disc.*, **101**, 1–49.
Kim, B., Chiba, T. & Inoue, T. (1993) *Polymer*, **34**, 2809–2815.
Kloosterboer, J. (1988) *Adv. Polym. Sci.*, **84**, 1.
Kreiger, I. (1972) *Adv. Colloid Interfacial Sci.*, **3**, 111.
Kreiger, I. & Dougherty, T. (1959) *Trans. Soc. Rheology*, **3**, 137.
Kuhn, P. (1934) *Kolloid Z.*, **68**, 2.
Labana, S. (1985) in Mark, H. (Ed.) *Encyclopedia of Polymer Science and Engineering*, 2nd edn, New York: Wiley.
Larson, R. (1999) *The Structure and Properties of Complex Fluids*, New York: Oxford University Press.
Leung, Y. & Eichinger, B. (1984) *J. Chem. Phys.*, **80**, 3877.
Lu, X. & Pizzi, A. (1998) *Holz Roh- Werkst.*, **56**, 339–346.
Macosko, C. & Miller, D. (1976) *Macromolecules*, **9**, 199–206.
Manneville, P. & De Seze, L. (1981) Numerical methods in the study of critical phenomena, in Della Dora, J., Demongeot, J. & Lacolle, B. (Eds.) *Numerical Methods in the Study of Critical Phenomena, Proceedings of a Colloquium, Carry-le-Rouet, France, June 2–4, 1980*, Berlin: Springer-Verlag.
Mijovic, J., Andjelic, S., Fitz, B. (1996) *J. Polym. Sci., Part B: Polym. Phys.*, **34**, 379–388.
Mikos, A., Takoudis, C. & Peppas, N. (1986) *Macromolecules*, **19**, 2174.
Miller, D., Valles, E. & Macosko, C. (1979) *Polym. Eng. Sci.*, **19**, 272.
Moad, G. (1999) *Prog. Polym. Sci.*, **24**, 81–142.
Muthukumar, M. (1989) *Macromolecules*, **22**, 4656–4658.
Muthukumar, M. & Winter, H. (1986) *Macromolecules*, **19**, 1284–1285.
Okay, O. (1994) *Polymer*, **35**, 1994.
Osinski, B. (1993) *Polymer*, **34**, 752–758.
Palmese, G. & Gillham, J. (1987) *J. Appl. Polym. Sci.*, **34**, 1925–1939.

Pang, K. & Gillham, J. (1989) *J. Appl. Polym. Sci.*, **37**, 1969–1991.
Prime, R. (1997) Thermosets, in Turi, E. (Ed.) *Thermal Characterisation of Polymeric Materials*, 2nd ed, New York: Academic Press.
Riccardi, C. & Williams, R. (1986) *Polymer*, **27**, 913.
Rouse, P. (1953) *J. Chem. Phys.*, **21**, 1272.
Schiessel, H. & Blumen, A. (1995) *Macromolecules*, **28**, 4013–4019.
Scranton, A. & Peppas, N. (1990) *J. Polym. Sci.*, **28** 39–57.
Simon, G., Allen, D., Bennett, D., Williams, E. & Williams, D. (1989) *Macromolecules*, **22**, 3555–3561.
Simon, S. & Gillham, J. (1993) *J. Appl. Polym. Sci.*, **47**, 461–485.
Simon, S. & Gillham, J. (1994) *J. Appl. Polym. Sci.*, **53**, 709–727.
Simon, S. & Gillham, J.K. (1992) *J. Appl. Polym. Sci.*, **46**, 1245–1270.
Stanley, H. (1985) Critical phenomena, in Mark, H. (Ed.) *Encyclopedia of Polymer Science and Engineering*, 2nd edn, New York: Wiley.
Stauffer, D., Coniglio, A. & Adam, M. (1981) *Adv. Polym. Sci.*, **44**, 103.
Stockmayer, W. (1943) *J. Chem. Phys.*, **11**, 45–55.
Stockmayer, W. (1944) *J. Chem. Phys.*, **12**, 125–131.
Stutz, H. & Mertes, J. (1989) *J. Appl. Polym. Sci.*, **38**, 781–787.
Takahashi, M., Yokoyama, K., Masuda, T. & Takigawa, T. (1994) *J. Chem. Phys.*, **101**, 798–804.
Tobita, H. & Hamielec, A. (1989) *Macromolecules*, **22**, 3098–3105.
Treloar, L. (1975) *The Physics of Rubber Elasticity*, Oxford: Clarendon Press.
Tsou, N. & Peppas, A. (1986) *J. Polym. Sci. – Polym. Phys. Edn*, **26**, 2043.
Vendetti, R. & Gillham, J. (1993) *Polym. Mater. Sci. Eng.*, **69**, 434–435.
Vendetti, R., Gillham, J., Jean, Y. & Lou, Y. (1995a) *J. Coatings Technol.*, **67**, 47–56.
Vendetti, R., Gillham, J., Jean, Y. & Lou, Y. (1995b) *J. Appl. Polym. Sci.*, **56**, 1207–1220.
Wang, X. & Gillham, J. (1993a) *J. Appl. Polym. Sci.*, **47**, 447–460.
Wang, X. & Gillham, J. (1993b) *J. Appl. Polym. Sci.*, **47** 425–446.
Winter, H. (1990) Gel point phenomena, in *Encyclopedia of Polymer Science and Engineering*, New York: Wiley.
Wissanrakkit, G. & Gillham, J. (1990) *J. Coatings Technol.*, **62**, 35–50.
Xanthos, M. (1992) *Reactive Extrusion – Principles and Practice*, Munich: Hanser-Verlag.
Zimm, B. (1956) *J. Chem. Phys.*, **24**, 269.

3 Chemical and physical analyses for reactive polymers

3.1 Monitoring physical and chemical changes during reactive processing

Reactive processing involves the production of a novel polymer as a result of chemical reactions that occur during the processing operation. These changes may be deliberate, as in functionalization, or inadvertent, as in chain scission due to undesired thermo- or mechano-chemistry. Section 1.4 gave the chemical changes to be expected when a thermoplastic polymer is subjected to elevated temperature for a period of minutes in the presence or absence of an oxidative atmosphere (Scott, 1993, Zweifel, 1998). For example, there will be the appearance of higher oxidation states of carbon such as in ether, alcohol, ketone and acid groups. These often accompany chain scission, which occurs during the free-radical-initiated oxidation chain reaction. In addition, there may be other chemical changes, such as grafting and functionalization, in which new chemical species are added to the polymer backbone in the course of the processing operation. The initiation of the oxidation, grafting or functionalization reactions will be through added initiator such as an organic peroxide and the novel chemistry will involve a monomer or other grafting agent, neither of which will necessarily be totally consumed during the reaction sequence. The measurement of the concentration profiles of these reagents as a function of time and, if possible, position in the reaction zone is crucial to the development of an understanding of the link between the chemical and rheological changes of the novel polymer system during processing. In this way a high level of control of the process and reproducibility of the product are possible.

The simplest form of reactive processing involves polymerization in the reactor. This may involve the large-scale autoclave environment such as for polyesterification, or the polymerization of caprolactam to polyamide in a reactive extruder. The chemical changes involve the formation of simple functional groups and the increase in molar mass (and hence viscosity of the polymer) is simply linked to the extent of reaction, p. In all cases the number-average degree of polymerization is proportional to $1/(1-p)$, so there is a direct link between the chemistry and viscosity. However, the challenge in analysis may be seen from Table 1.3 by noting that, for a degree of polymerization of 100, p is 0.99, so considerable precision is required in any analysis to be performed on-line. Also the link between viscosity and molar mass will be a power law, so the use of chemical analysis to predict viscosity changes is a challenge.

In the case of a network polymer, the process of crosslinking involves ring-opening, addition or condensation polymerization reactions. In each case there will be a change in the average chemical composition of the system and, from an understanding of the kinetics and mechanisms of these reactions, it is possible to determine the extent of reaction as a function of time as the network passes from a liquid, through the gel, to a glass (Section 2.4). In this case the chemical complexity of polyfunctional reagents means that there may be several reactions competing, controlled initially by the concentration and kinetic parameters of the species and then by the physical state of the system as the viscosity increases rapidly on passage to the gel and glass. In the practical application of these reactions, the polymerization may be occurring in a mould (compression moulding; reaction injection moulding (RIM)), or in an autoclave (fibre-reinforced-composite curing from prepreg), so the possibility of measuring the chemorheological changes by conventional techniques is extremely limited. To this end, novel methods employing combined techniques with fibre-optic probes have been developed and are discussed in a later chapter concerned with processing and quality control (Chapter 6). In this chapter the fundamental analytical techniques and how they may be used under laboratory conditions to generate the instantaneous chemical structure of the polymerizing system are described. This is then extended to the dual physicochemical analysis methods developed for probing the link between the physics and the chemistry of the reacting systems.

3.2 Differential scanning calorimetry (DSC)

Calorimetry is one of the oldest-established techniques for studying the phase changes as well as reactions of polymers. Of particular relevance to chemorheology is that it provides the location of the temperature of phase transition as well as a measure of reaction rates where those reactions or phase changes result in a net heat flow from the sample to the surroundings (exotherm) or from the surroundings to the sample (endotherm). In relation to phase changes, calorimetry has provided transition temperatures, namely the melt temperature T_m, crystallization temperature, T_c, and glass-transition temperature, T_g, as well as a measure of the changes in heat capacity accompanying these transitions.

While classical calorimetry has excellent sensitivity and may be used to measure extremely low reaction rates (Gallagher, 1997, Gmelin, 1997), differential scanning calorimetry (DSC) has become the technique of choice because of the ready availability of commercial apparatus with attached software to enable quantitative data analysis (Richardson, 1989). As will be discussed later, many of these software routines are based on published methods, which may have limited applicability when the diversity of reactions that may occur in chemorheology of, say, thermosets is considered. In reaction processing, such as the reactive extrusion (REX) of crosslinked polyolefins or the autoclave curing of epoxy resins, DSC must be regarded as an off-line technique, but it has general utility in establishing parameters for reaction processing as well as analysing the resulting product.

3.2.1 An outline of DSC theory

The principle of measurement in DSC is shown in Figure 3.1.

The sample, S, is located in a small calorimeter and the reference, R, (usually an inert substance of similar heat capacity) is located in another. In a conventional calorimeter, the heat capacity of the system C_p is measured from the temperature rise accompanying a known amount of (electrical) energy input into a system at constant pressure,

$$C_p = [\partial H/\partial T]_p, \tag{3.1}$$

where H is the enthalpy of the system.

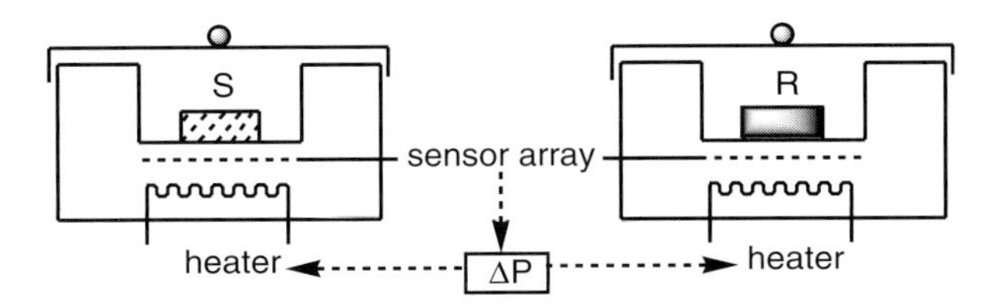

Figure 3.1. A schematic diagram of a differential scanning calorimeter (DSC) showing the sample (S) and reference (R) in separate heating chambers.

In a power-compensated DSC (Figure 3.1) the sample and reference calorimeters are provided with power necessary to maintain them at the same temperature while the polymer passes through a transition or undergoes a chemical reaction. The difference in power, ΔP, supplied to the sample and reference calorimeters is directly related to the heat-capacity difference at that temperature. As the temperature is scanned, ΔP will be either positive or negative depending on whether there is an endotherm, or exotherm respectively. Then the direct conversion to heat flow from power means that

$$\Delta P = \mathrm{d}Q_{\mathrm{s}}/\mathrm{d}t - \mathrm{d}Q_{\mathrm{r}}/\mathrm{d}t = \mathrm{d}\Delta Q/\mathrm{d}t = \mathrm{d}H/\mathrm{d}t = C_p. \tag{3.2}$$

There are several different types of instrument covered under the term DSC, which have evolved from differential thermal analysis (DTA) and measure the temperature difference between sample and reference pans located in the same furnace. This is then converted to heat flow using a calibration factor. A detailed analysis of DSC requires consideration of the various sources of heat loss, and these are generally captured in the calibration routine for the instrument. Absolute temperature calibration is achieved through the use of pure indium (156.6 °C) and tin (231.9 °C) melting-point standards. A comprehensive analysis of the theory of DSC contrasted with DTA may be found in several reference works (Richardson, 1989, Gallagher, 1997).

A more recent development is temperature-modulated DSC (Jones *et al.*, 1997), which is a particular example of the broader area of modulated calorimetry (Gmelin, 1997) and enables deconvolution of kinetic and thermodynamic processes during the reactive curing of polymer networks (Van Assche *et al.*, 1997). This is discussed in more detail later (Section 3.2.3).

3.2.2 Isothermal DSC experiments for polymer chemorheology

Figure 3.2 shows a typical DSC curve for a semi-crystalline polymer such as a polyolefin as it is heated from sub-ambient temperatures until it decomposes.

The heat flows corresponding to the transitions mentioned above are shown, in order, T_{g}, T_{c}, T_{m}, followed by an exotherm corresponding to the thermal oxidation of the polymer. The analysis of this curve is often restricted to noting the temperature of the transition or the onset of oxidation. By ensuring that the system is calibrated as a calorimeter it is possible to obtain quantitative information such as the absolute change in enthalpy and heat capacity. This enables the degree of crystallinity of a semi-crystalline polymer, the relaxation and physical ageing of a network and the effects of residual stress from processing to be determined (Richardson, 1989).

In reactive processing, such as reactive extrusion of thermoplastics, the use of DSC may be seen in determining the stability of the polymer formulation by measuring the oxidation induction time (OIT) (Bair, 1997). This is an isothermal experiment in the presence of air

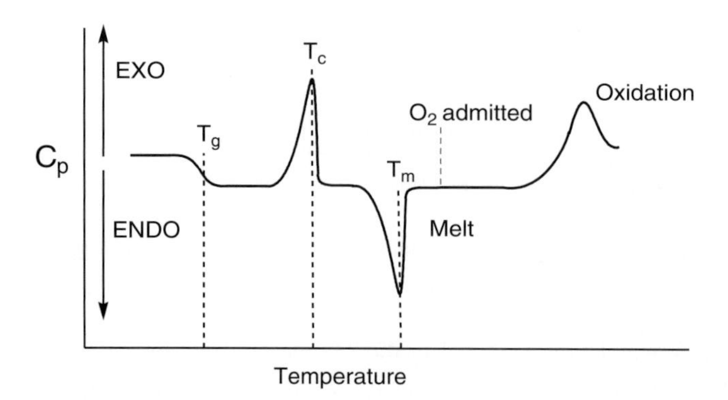

Figure 3.2. Typical output from a DSC when a semi-crystalline polymer is heated through the glass (T_g), crystallization (T_c) and melt (T_m) transitions. The exotherm on oxidation is also shown.

or oxygen, in which the time to the onset of the exotherm is measured. Thus the same information as from oxygen uptake to measure OIT (Figure 1.41) may be obtained by measuring the time after oxygen is admitted to the onset of the exotherm. The limitation of OIT for applications at lower than processing temperatures has been emphasized (Billingham *et al.*, 1981), although for many practical purposes it must also be recognized that a DSC oven with 10 mg of material cannot totally simulate the environment of a reactive extruder in which the shear is responsible for mechanoradical formation as described below (Section 3.3.2). In one example, the contribution to free-radical formation by thermal oxidation was less than that by shear-induced chain scission (Sohma, 1989a), but this depends on the availability of oxygen in the reacting system. The technique is well suited to the comparative ranking of processing stabilizers.

An example of monitoring of reaction rate by DSC is the cure reaction of an epoxy resin. Figure 3.3 shows the DSC curve for the heating of an epoxy resin with its curing agent through the temperature range corresponding to the chemical reactions leading to three-dimensional network formation.

A broad exotherm is obtained, corresponding to a typical heat of reaction (shown here as ΔH_0) of an epoxy resin of 107 ± 4 kJ/mol epoxide (Prime, 1997b). Also shown is the curve obtained by heating the system as mixed in an oven at 60 °C for increasing times and then running the DSC from room temperature to 150 °C. Several features may be noted.

(i) There is a decrease in the area under the curve with time of heating at 60 °C.
(ii) After some reaction has occurred, there is a shift in the baseline indicating an endotherm, which corresponds to the appearance of the T_g of the developing network (shown as the expansion of the early section of the cure curve in Figure 3.3).
(iii) The value of T_g, as given by the midpoint of the change in baseline ($\frac{1}{2}\Delta C_p$), increases with time of cure.
(iv) Isothermal cure of the resin for even long times does not result in complete removal of the residual exotherm on scanning.

These DSC results may be further analysed to give the extent of reaction as a function of cure time by using the assumption that the total area under the exotherm represents the integrated heat of reaction, ΔH_r, for the reaction of epoxide with amine (typically 107 ± 4 kJ/mol epoxide as indicated above). Thus the reduction in this area on reacting the system for a

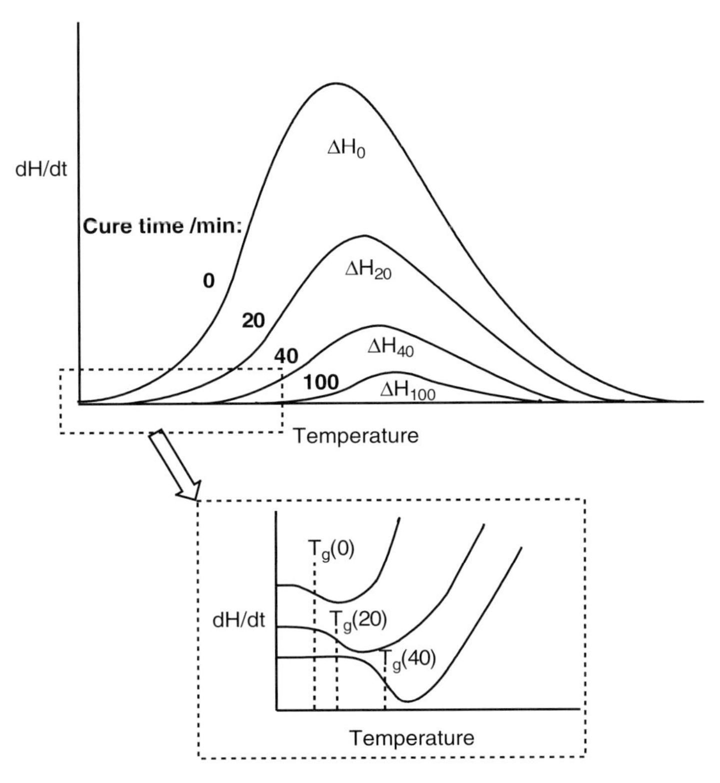

Figure 3.3. Some DSC traces from the exothermic reaction of an epoxy resin after various times of cure, showing the progressive reduction in reaction enthalpy, ΔH. The expanded region shows the change to the baseline due to the enthalpy change at the glass transition. Note the progressive increase in T_g after partial cure.

time t is a measure of the concentration of the epoxide consumed in the curing reaction of the resin up to this time,

$$p = (\Delta H_r - H_t)/\Delta H_r, \tag{3.3}$$

where p is the extent of reaction, given by $1 - c/c_0$; c is the concentration of epoxide remaining at time t and c_0 is the initial concentration of epoxide (or other reacting species). This relationship is the basis of the use of DSC for the analysis of the kinetics of cure of epoxy resins and similar network-forming resins.

Figure 3.4 shows the analysis of the data in Figure 3.3 according to this equation and represents a plot of the extent of reaction with time of heating at 60 °C.

The actual rate of reaction is given by the instantaneous heat flow, so that at any point on a cure exotherm in Figure 3.3 the reaction rate $\mathrm{d}p/\mathrm{d}t$ is given by

$$\mathrm{d}p/\mathrm{d}t = (1/\Delta H_r)\mathrm{d}H/\mathrm{d}t. \tag{3.4}$$

Thus, at the cure temperature of 60 °C, the reaction rate may be obtained either from the heat-flow rate in the DSC trace for the previously unheated resin in Figure 3.3 or from the slope of the extent-of-reaction curve (Figure 3.4) at zero time.

As noted above, several significant features of the network-forming system, which are important for reactive processing, may be learned from this simple analysis. Firstly it may

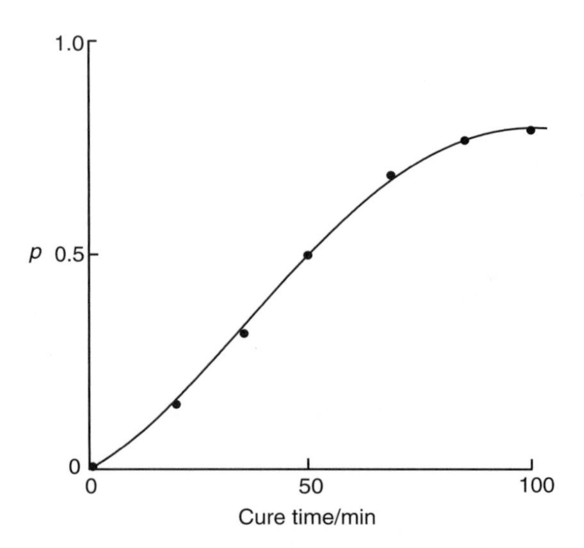

Figure 3.4. The time development of the extent of reaction, p, from the reaction-enthalpy results of Figure 3.3, as analysed by use of Equation (3.3). Note that the system does not reach full cure ($p = 1$) because T_g of the resin reaches the cure temperature and reaction ceases.

be seen that, as the T_g of the resin approaches the cure temperature, the reaction rate becomes vanishingly small, so the final processed resin will have residual reactivity if it is ever heated to above the processing temperature. This is the process of post-curing which results in an increase of T_g and consumption of all reactive species. If this temperature is chosen too high, the network will start to degrade, and this is accompanied by a further exotherm if air is present since oxidation reactions may then dominate.

As an alternative to curing the resin for pre-set times and then measuring the heat flow on ramping the temperature, the system could be cured (in this case at 60 °C) in the DSC in the isothermal mode. The disadvantage of this approach for moderate reaction temperatures that are well below the maximum in the heat-flow curve (which for this resin occurs at 120 °C) is the low heat flow and thus difficulty in achieving a good signal-to-noise ratio (SNR) in the analysis. If the isothermal experiment is run at much higher temperatures, there is the problem of achieving equilibration in the calorimeter before a significant part of the reaction has taken place. This then introduces uncertainty into the data for kinetic analysis.

Figure 3.5 shows an isothermal DSC curve for the curing of TGDDM with the aromatic amine DDS (de Bakker *et al.*, 1993b). Also shown is the NIR analysis of the instantaneous rate of reaction of epoxide. The difference arises due to the change in heat capacity on gelation and vitrification that affects the DSC baseline during cure, as discussed below.

The isothermal DSC curve is a plot of $\mathrm{d}H/\mathrm{d}t$ against time so that, from Equation (3.4), it may be integrated to determine the extent of reaction and so analyse the kinetics of the system. There is still the necessity to perform a scanning DSC experiment in order to determine the residual exotherm from the sample due to the cessation of reaction at the cure temperature as the T_g of the resin reaches the isothermal cure temperature. This just becomes a correction in the form of a scaling factor for the entire curve. A further requirement in this analysis of the isothermal cure exotherm is the determination of the form of the baseline for the integration. In many cases the simple assumption is to choose a flat baseline. However, this ignores the fact that the heat capacity, C_p, of the system will

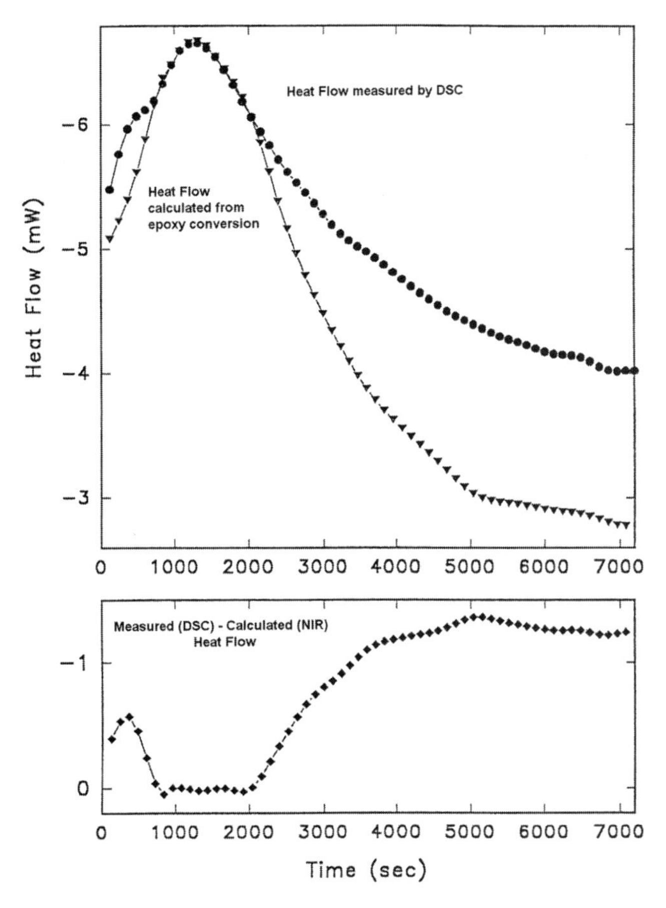

Figure 3.5. Comparison of heat flow as measured by the DSC and calculated heat from the extent of epoxide reaction (measured by NIR Spectroscopy) for TGDDM with 27% DDS at 177 °C for the times indicated. The difference arises from the change in heat capacity change during cure which requires a baseline correction to the DSC trace. Adapted from de Bakker *et al.* (1993b).

decrease as the curing epoxy-resin passes from a liquid through a rubber to a glass. Prior to the advent of modulated DSC, the baseline could only be estimated, and profiles were suggested (Barton, 1985, Gallagher, 1997). An example of the effect of the change in heat capacity on the cure exotherm may be seen by considering simultaneous FT-IR analysis in the NIR of the concentration of epoxide together with the heat-flow data obtained by DSC (de Bakker *et al.*, 1993b). This is discussed in more detail in Section 3.3.6. As shown in Figure 3.5, there is a significant difference between the reaction rate obtained by direct analysis of epoxide and that derived from the DSC heat-flow curve which is attributable to the change in heat capacity on cure. While this simultaneous method gave a clear indication of how the baseline changed during cure, it could not unequivocally show that this change arose solely from the change in C_p. This may be demonstrated more readily by modulated-temperature DSC as discussed in Section 3.2.3.

There is another particular application of isothermal DSC that requires modification of a traditional DSC, and this is in rapid photopolymerization (Scott *et al.*, 2002) such as of multifunctional acrylates (Peinado *et al.*, 2002). In a typical photo-DSC experiment, the sample and reference pans are operated in an open mode and the system is fitted with an ultraviolet illumination system that irradiates both, to ensure that heat flow results solely from the polymerization exotherm. The response time of the system may be improved and quantitative characterization achieved by replacing the DSC with a calorimeter that provides a dynamic range better by an order of magnitude than that of photo-DSC (Roper, 2005). The response time is reduced to 1 s for 90% of maximum heat flow, compared with 4.8 s for photo-DSC. This has enabled the curing of thin films of thickness 163 nm to be measured (Roper *et al.*, 2005), and the results obtained are comparable to those from real-time FT-IR monitoring as discussed in Section 3.3.5.

In the study of the chemorheology of photopolymerizable systems, the use of a traditional rheometer is particularly difficult and the development of improved time resolution from several seconds to better than 1 ms has been noted, together with an increase in the intensity of initiating radiation that may be employed (Schmidt *et al.*, 2005). The challenge is to gather data at the best possible SNR and then use the appropriate algorithm for recovery and analysis of the phase information and intensity to enable calculation of G^* and $\tan\delta$ (Schmidt *et al.*, 2005, Chiou and Khan, 1997). These studies are often augmented by simultaneous real-time spectroscopic studies on the sample in the rheometer in order to provide conversion data as discussed in Section 3.3.

3.2.3 Modulated DSC experiments for chemorheology

The above example of the determination of the change in C_p as the epoxy resin underwent crosslinking is an example of a reversible thermal event (C_p decreases on vitrification) convoluted by a non-reversible chemical change due to the formation of the network by the ring-opening reaction of the epoxy resin (which produced a large exotherm). Experimentally this separation of the events is achieved by applying a small temperature oscillation on top of the isothermal temperature. The conditions are thus quasi-isothermal. The condition may be described (Jones *et al.*, 1997) by

$$dQ_s/dt = C_p\, dT/dt + (1/\Delta H_r)dH/dt. \tag{3.5}$$

The frequency of modulation is adjusted so that there are many cycles over the duration of the transition, which in this case is the glass transition and is therefore broad. A typical modulation condition is 0.5 °C in a 60-s period for studying cure of an epoxy resin (Van Assche *et al.*, 1997). The MDSC instrument provides a modulated heating programme of the form

$$T = T_o + bt + B\sin(\omega t), \tag{3.6}$$

where B is the amplitude and ω the angular frequency of the modulation.

The combination of Equations (3.5) and (3.6) enables the DSC signal to be separated into an underlying component (the first two terms) and a cyclic component (the second two terms):

$$dQ/dt = BC_p + \left(1/\Delta H_r\right)dH/dt + \omega BC_p\cos(\omega t) + C\sin(\omega t), \tag{3.7}$$

where C is the amplitude of the kinetically hindered response to the modulation.

The first two terms are the equivalent of what is measured in a conventional DSC during the cure of an epoxy resin. The software of the instrument enables the non-reversing

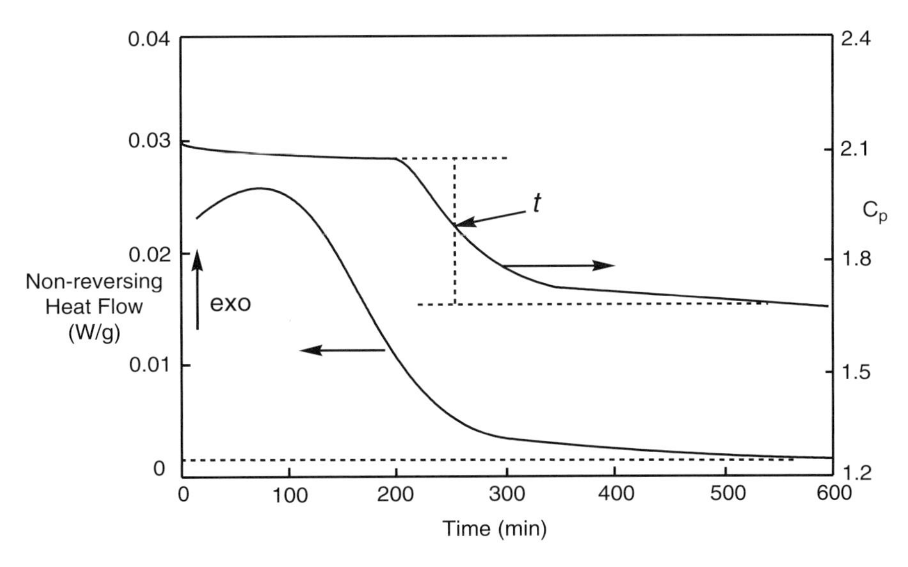

Figure 3.6. Non-reversing heat flow and change in heat capacity (C_p) during cure of an epoxy resin cured with an anhydride at 80 °C. The cure time needed to reach a halving of C_p is marked as *t*. Adapted from Van Assche *et al.* (1997).

component (the reaction exotherm) to be separated from the reversing component (the change in C_p with time). Therefore, provided that the cure reaction is not too rapid and there are many modulations over the total cure time, C_p is readily obtained as a function of temperature. Figure 3.6 shows a separation of the non-reversing heat flow from the heat capacity for the cure of an epoxy resin with an anhydride at 80 °C (Van Assche *et al.*, 1997).

The time for half-decrease in C_p is shown and, in a separate experiment, the resin at this point was found to have a T_g of 80 °C, confirming that the change in C_p arises predominantly due to vitrification of the resin during cure. If cure occurs at a temperature above $T_{g\infty}$ then the change in C_p is small. Figure 3.7 shows the change in C_p as a function of the time of cure of an epoxy resin with an aromatic amine at three different temperatures in a MDSC (Lange *et al.*, 2000).

The three curves correspond to cure below, near and above $T_{g\infty}$, and it may be seen that only in the first case is there a clear step-change in heat capacity. The processing of the resin above $T_{g\infty}$ (150 °C) results in complete chemical reaction since there is no vitrification during cure.

3.2.4 Scanning DSC experiments for chemorheology

The experiments described above for the cure studies of epoxy resins (a typical three-dimensional network) were restricted to isothermal experiments (or quasi-isothermal experiments in the case of MDSC). As discussed later, these are the most reliable methods for generating kinetic information, but suffer from the length of time taken to generate data and, furthermore, in some systems the heat flow may be vanishingly small.

As shown earlier in Section 3.2.2, scanning temperature DSC provides a rapid method for measuring the total heat of reaction for network polymerization, and there are many such applications to provide baseline data for polymer chemorheology. For example, in the

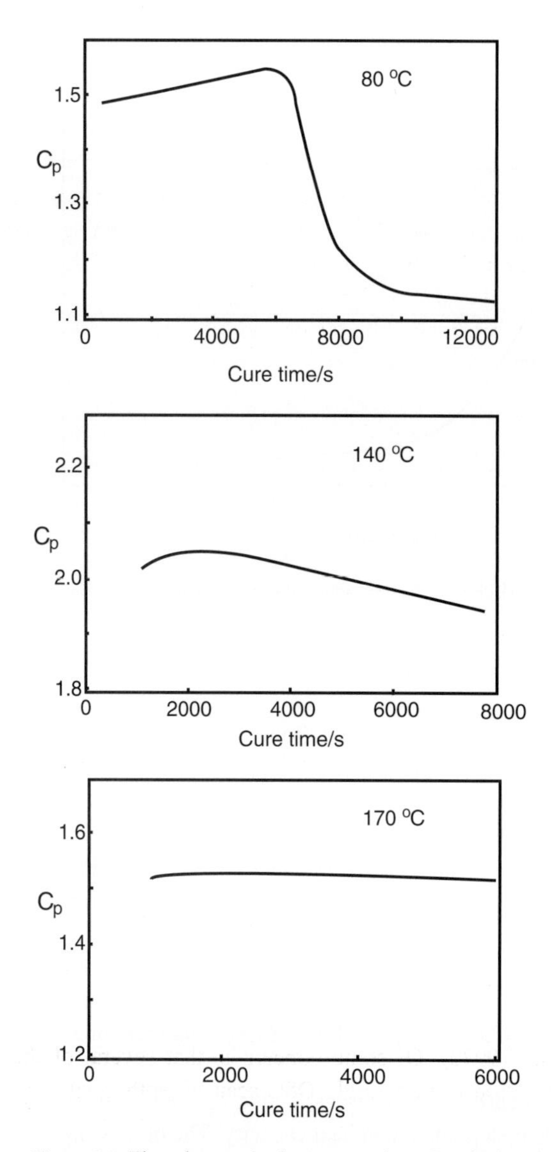

Figure 3.7 The change in heat capacity, C_p, during isothermal cure of an epoxy resin with an aromatic amine at temperatures below, near and above T_g. Adapted from Lange *et al.* (2000).

vulcanization of rubber (Brazier, 1981) the factors affecting ΔH_r were determined for various formulations of crosslinker (mostly sulfur) and accelerators (e.g. thiazoles) and fillers (mostly carbon black). Degradation reactions may also be studied so that overcure may be avoided in processing. While DSC is a favoured techniques for providing cure data on a wide range of elastomers (through comparison of ΔH_r values), it has been noted (Sircar, 1997) that the success of the enthalpy method depends on satisfaction of the following criteria.

1. The heat of vulcanization must be large enough to reduce error in any measurement of residual heat of vulcanization.

2. The residual reactants must not decompose after press cure.
3. The evolution of heat must be proportional to crosslinking.

Since the instantaneous heat flow is (with some assumptions) representative of the rate of reaction, there have been many attempts to extract kinetic and mechanistic information from scanning DSC studies. These have severe limitations (as discussed later) but at this stage it is useful to review the methods of analysis. The ultimate test of a scanning method is whether it is able to predict the same phenomena as observed in a comparable isothermal experiment.

The starting point for the use of scanning DSC for chemorheology of systems such as cure of an epoxy resin is the kinetic model which links the variables of temperature and reactant concentration to the rate of reaction. This is chosen to be a simple Arrhenius relation governed by the pre-exponential factor, A, and the activation energy, E_a,

$$dp/dt = A\exp[-E_a/(RT)]f(p), \tag{3.8}$$

where the form of $f(p)$ is chosen according to the chemistry of the system and the appropriate mechanistic scheme or experimental knowledge of the rate law. The most usual starting point is nth-order chemical kinetics, $f(p) = (1-p)^n$, so that in linear form Equation (3.8) becomes

$$\ln[(dp/dt)/(1-p)^n] = \ln A - E_a/(RT). \tag{3.9}$$

This is the basis of the Borchardt and Daniels method, which was originally developed for DTA (Borchardt and Daniels, 1957) and is now offered in almost all available commercial DSC software. The DSC heat flow, dH/dt, is treated as equivalent to the reaction rate dp/dt (as in an isothermal experiment) and plots of Equation (3.9) are made for various values of n, the so-called reaction order. The value yielding the best straight line is taken as the 'order' of the reaction and the values of E_a and $\ln A$ are then calculated from the slope and intercept of the plot (Barrett, 1967). The attractiveness of the method is that it yields apparent kinetic data from a single DSC trace, but it has been shown for many polymerizing systems that they have no kinetic meaning, and results that are at variance with those obtained by isothermal methods have been derived for many epoxy-resin and other reactive-network systems (Prime, 1997b). Of some concern is that several of the commercial packages offer the calculation from a single DSC scan of isothermal parameters such as gel time and the time to peak exotherm, both of which are valuable for reactive processing. It has been shown, and will be discussed later, that there are some limitations to this method (George and St John, 1993), but the most obvious is the over-simplification of the kinetics and mechanism of an often auto-accelerating reaction. In general it is inappropriate to apply the Borchardt and Daniels and the related Freeman and Carroll (Freeman and Carroll, 1958) methods to complex reacting systems since they assume simple nth-order kinetics.

In an extension of scanning techniques, heating methods that involve no assumptions regarding the order of reaction have been developed. The most widely used (and available also as standard software in commercial DSC systems) are the Kissinger and Ozawa methods of analysis. Both of these are based on the time taken to reach the same extent of conversion at various temperature-scanning rates and have an underlying assumption that the activation energy is independent of temperature. Since it has been found for many reacting systems that the extent of conversion, p_m, at the maximum of the cure exotherm,

T_m, is constant, this is used as the isoconversion point for the Kissinger analysis (Kissinger, 1957) of the DSC exotherms at various scan rates, β,

$$\mathrm{d}\ln(\beta/T_m^2)/\mathrm{d}(1/T_m) = -E_a/R. \tag{3.10}$$

The Ozawa method involves a refinement of this equation, namely

$$\mathrm{d}\ln\beta/\mathrm{d}(1/T_m) = -1.052E_a/R, \tag{3.11}$$

and is based on an empirical solution to the integral of $\exp[-E_a/(RT)]$ from the onset temperature (T_0) to the maximum (T_m). This has been shown to give apparent activation energies with an accuracy better than ±10% when compared with isothermal methods for the curing of DGEBA epoxy resin with a polyamide curing agent (Prime, 1997b). This has been adapted to ASTM method E698–79, although there are clear recommendations that any use of the data should be supported by an isothermal experiment (such as half-life measurement) to confirm the predicted behaviour. There are several concerns with these methods for complex reactions, especially when the DSC reaction profiles are compared with independent chemical information such as from NIR spectroscopy (Section 3.3.6). From these studies it has been seen that there is a significant scan-rate dependence of the coincidence of the exotherm maximum and the true reaction-rate maximum (George and St John, 1993). This heat-lag effect is most pronounced at and beyond T_m, where it distorts the DSC profile and, since the effect is non-linear with heating rate, will affect both the Kissinger method and the Ozawa method of analysis.

It has been noted (Richardson, 1989) that the reaction path may often change with temperature, and the parameters derived from the scanned DSC curve analysed by using either Equation (3.9) or Equation (3.10) represent some nonexistent 'average reaction'. So, even though the values of E_a and A will have no fundamental meaning, they may allow reasonable estimates to be made for quality-assurance purposes or for some baseline data for empirical modelling of the processing of resins as in the autoclave cure of composite materials (Ciriscioli and Springer, 1990). Of particular importance in the industrial process is the reaction which occurs during temperature profiles, such as ramping, and it is important to be able to transform the generated DSC data to a form that may be used for process control.

3.2.5 Process-control parameters from time–temperature superposition

The processing of epoxy-resin pre-impregnated fibre (prepreg) materials into a consolidated composite occurs in a series of programmed temperature profiles that are designed to control the development of the viscosity of the resin and then apply pressure to impregnate the fibres with resin. The cure temperature is chosen to optimize the exotherm and avoid thermal runaway. The ultimate T_g is achieved through a post-cure at an elevated temperature. The baseline data for this process are generated by DSC studies of the resins and prepreg. The starting point is the establishment of a thermochemical model, since the rate of cure depends on the rate of transfer of heat to the resin (Ciriscioli and Springer, 1990).

The principle of time–temperature superposition is that there is a temperature-shift factor that allows all data to be plotted on a master curve. This presupposes that there is no change in mechanism during the reaction, so that Equation (3.8) applies (Prime, 1997b). Such superposition processes are regularly used in rheology and the WLF equation is routinely applied when the system is above temperature T_g. When the system is controlled by chemical kinetics, the Arrhenius relation is applicable and the shift factor is the empirical activation energy, E_a. Of interest in prepreg cure in an autoclave that may have a large thermal lag is the equivalent isothermal time, τ, which may be attributed to a sample that has experienced a complex thermal profile. An example of importance is the time taken to reach gelation, t_{gel}, for the epoxy-resin prepreg, since this will decide the time at which to apply pressure in order to consolidate the composite. The calculation involves breaking up the profile into a series of isothermal steps of duration t_i at temperatures, T_i, with the accumulation up to the final temperature, T_r. Then it is possible to compute the accumulated cure time, τ, as

$$\tau = \sum_i \tau_i = \sum_i t_i \exp[E_a(T_i - T_r)/(RT_iT_r)]. \tag{3.12}$$

In this the way the effect of the rate of temperature ramping may be taken into account in the total cure profile. In such studies the empirical E_a data may be adequate, but for detailed kinetic modelling it is generally recognized that isothermal experiments are appropriate, and even they may have limitations, as discussed later.

3.2.6 Kinetic models for network formation from DSC

In any analysis of the kinetics of cure by DSC it is recognized that isothermal methods produce the more reliable data, for the reasons discussed in Section 3.2.4, which gave the limitations inherent in parameters determined from scanning DSC of polymerizing networks. In discussing the kinetic analysis of the DSC data generated by temperature-scanning experiments (Section 3.2.4) it was noted that one of the fundamental limitations was the assumption of simple *n*th-order kinetics of epoxy–amine reactions, even when there was evidence for autocatalytic behaviour. Mechanistic studies show that the reaction is catalysed by hydroxyl groups either present as impurities in the resin or generated by the ring-opening of the epoxy resin by the amine (Rozenberg, 1986). It is the latter process that results in auto-catalysis and requires the inclusion of the extent of reaction, p, explicitly in the rate law being used to fit the DSC data.

In one of the earliest analyses of cure kinetics by DSC, Horie *et al.* (1970) derived a kinetic relationship for epoxy–amine cure assuming near-equal reactivity of primary and secondary amines and no other reactions. This was later refined for a stoichiometric epoxy–amine system (Sourour and Kamal, 1976) and takes the form

$$\mathrm{d}p/\mathrm{d}t = (k_1 + k_2 p)(1 - p)^2, \tag{3.13}$$

where k_1 and k_2 are rate constants representing the uncatalysed and catalysed reactions, respectively. This has been generalized to the empirical equation (Barton, 1985)

$$\mathrm{d}p/\mathrm{d}t = (k_1 + k_2 p^m)(1 - p)^n, \tag{3.14}$$

where m and n are adjustable parameters representative of the order of the individual terms. Barton (1985) obtained a fit using $m = n = 1$ over the first 30% of conversion, implying that the dominant reaction mechanism is bimolecular. However, the small extent of conversion over which the equation applied limits the applicability in process control. As an illustration of the very empirical approach often used, the cure of the tetraglycidyl amine resin system, TGDDM, with 4,4′-diaminodiphenylsulfone (DDS) was shown (Mijovic and Wang, 1986) to fit Equation (3.14), but the values of m and n obtained were temperature-dependent.

Thus, while there is a fit, the equation can have no physical meaning regarding the kinetics of the system. This approach has been modified by introducing a separate exponent, l, for the uncatalysed reaction:

$$\mathrm{d}p/\mathrm{d}t = k_1(1-p)^l + k_2 p^m (1-p)^n. \tag{3.15}$$

For TGDDM/DDS the values $l=2$ and $m+n=2$ enabled a good fit for DSC data for up to 70% conversion, after which diffusion control of the reaction rate became dominant. The difficulty with this type of approach is that for each different resin it is necessary to determine not only new pseudo-rate constants but also new values for the exponents (Lee *et al.*, 1992, Mijovic and Wang, 1986). This supports the view that the exponents are more akin to curve-fitting parameters rather than being a true reflection of the underlying reaction mechanisms involved.

A different approach to using isothermal DSC data has been to derive rate expressions that encompass all the expected mechanistic features of an epoxy–amine cure, including a substitution effect on amine reactivity and etherification reactions (Chiao, 1990, Cole *et al.*, 1991) as discussed in Section 1.2.1 (Schemes 1.3 and 1.23). However, as more parameters are introduced, the possibility of achieving a fit increases, but often with a reduction in the mechanistic meaning. Also, as with other DSC studies, the assumption of equal reaction enthalpies for all epoxy reactions had to be made resulting in an underlying uncertainty in the conclusions. Nonetheless, the methodology described represented a major advance in the use of isothermal DSC in the study of cure kinetics, and it is only through the direct measurement of the absolute concentration of reacting species, such as the use of spectroscopic techniques, particularly NIR, that further advances in the understanding of reactive systems such as epoxy-resin networks has been made. These are described in the following sections.

3.3 Spectroscopic methods of analysis

3.3.1 Information from spectroscopic methods

In chemorheology there is a need to determine the chemical and physical changes to the polymer which are linked to the changes in rheological properties. In the area of reactive processing these will generally be accumulation of oxidation products (Xanthos, 1992, Zweifel, 1998), consumption of monomers (Xanthos, 1992) or formation of reactive intermediates (Sohma, 1989a, 1989b) and new functional groups during network formation (Dusek, 1986, St John and George, 1994). This requires the identification of the new functional groups formed or the consumption of the reactive species that lead to the changes in the rheological properties. These spectroscopic techniques then become complementary to thermal analysis as described in Section 3.2 and, if made quantitative, offer the opportunity for absolute kinetic information.

Since reactive processing changes the molar mass and its distribution, through chain scission, and/or crosslinking and/or branching, there is also the need to have complementary techniques that may measure properties that are sensitive to viscosity. These changes may span many orders of magnitude and include phase transitions. While many studies of composition have been performed on samples following reactive processing, an underlying consideration is that the methods should be adaptable **to real-time, on-line monitoring**. These issues will be addressed after the separate spectral techniques have been considered.

The chemical functional groups of organic compounds and the reactive intermediates have traditionally been measured by spectroscopic methods since the observables are the energy at which a particular spectroscopic transition occurs and the concentration (either absolute or relative) of the species which is undergoing the transition. Since the transitions are between quantized energy levels of the molecules, the spectrum contains, encoded, the chemical composition of the reactive intermediates and the groups which make up the polymer. The region of the spectrum in which these are observed depends on the energies involved in the transitions. In the following sections we will consider them in order of increasing energy.

3.3.2 Magnetic resonance spectroscopy

Electron-spin resonance (ESR) spectroscopy

Electron-spin resonance (ESR) has provided fundamental information regarding reactive intermediates, particularly free radicals formed during the process of polymer-chain scission (Sohma, 1989a, 1989b). Many studies have involved mechanochemical degradation, i.e. formation of free radicals and subsequent chain-scission and crosslinking reactions of these intermediates created by solid-state reactions when the polymer is sheared. It has been noted that the rate of radical formation is lower by a factor of about 100 in solution than that in the solid, but it is considered that the rate of formation of free radicals under shear in an extruder would lie between these extremes. The confirmation, by ESR, of free-radical formation during processing is the basis for the free-radical reaction scheme given in Chapter 1 for the oxidation of polymers during melt-processing.

The ESR technique has been used to identify and quantify free radicals in polymers, as well as to follow their reactions. There is a general requirement for the polymer to be sufficiently rigid that decay pathways are restricted in order to allow a significant steady-state concentration to form. The technique relies on the splitting of the electron-spin energy states in a magnetic field. There will be an absorption of energy corresponding to a transition from the lower to the upper energy level, provided that the incident energy in the microwave region matches the energy separation, ΔE.

The condition for energy absorption is

$$\Delta E = g\beta H, \tag{3.16}$$

where H is the applied magnetic field, β is the Bohr magneton and g is a constant, the spectroscopic splitting factor, which for a free electron has the value 2.0023. Since the population difference between the two states is small at normal temperatures (being governed by the Boltzmann distribution) spectra are collected by sweeping the applied field and noting the resonance condition. Structural information regarding the free radical is obtained by observing the splitting of electron-spin energy states due to hyperfine interaction with the nuclear spin of the protons or nitrogen atoms in the radical. (Since the backbone carbon atoms have no nuclear spin, the interactions are greatly simplified in ESR.) The g factor will shift from that for the free electron. Thus an alkyl macro-radical from the scission of a polymer with a methylene bridging unit will give a three-line spectrum with a relative intensity of 1 : 2 : 1 (Sohma, 1989b). In the presence of oxygen, the spectrum of polyolefins that have been oxidized collapses to a broad singlet characteristic of the peroxy radical which is the main chain carrier in the auto-oxidation of the polymers.

The radicals must be trapped in order to measure the spectrum, and a steady-state population of $\sim 10^{13}$ spins is required in the analytical volume, so it is impossible to observe free radicals by ESR during reactive processing. In principle it could be possible to trap the free radicals off-line by quenching the melt in liquid nitrogen without exposure to air, but this presents experimental challenges. A simple device that incorporates quenching by liquid nitrogen for near-real-time sampling of the polymer in the barrel of a reactive extruder has been described (Machado *et al.*, 1999). This could be applied to collecting and maintaining a sample for ESR study of the free-radical species present in the extruder at any stage of the process. Chemical trapping techniques offer greater opportunity for demonstrating the formation of free radicals due to the shear forces during extrusion (Sohma, 1989b). Spin traps such as nitroxyl radicals are frequently used in ESR spectroscopy to capture transient species, and a variation on this in which a hindered amine (the commercial HALS light stabilizer, Tinuvin 770, Figure 3.8 (Sohma, 1989a)) is included with PP during processing in an extruder has been described.

Following extrusion at 230 °C with a residence time of 30 s, the characteristic ESR spectrum of a triplet due to nitrogen coupling was observed from the sample (Sohma, 1989a). The results in Figure 3.8 show that the trapped free-radical concentration increased with the concentration of HALS and that the main contribution was the shear degradation of the polymer, not thermal scission since negligible radical formation occurred at 230 °C alone with in the same time frame.

In other reactive systems in which the network first gels and then vitrifies, free-radical species and radical cations may be stabilized physically as well as chemically and then analysed. An example is the vitrification of difunctional methacrylate monomers such as ethylene glycol dimethacrylate (EGDMA). The free radicals formed during the polymerization and crosslinking process are trapped in the glassy network and may be analysed by ESR. The difference in free-radical concentration between a system of methyl methacrylate (MMA) containing EGDMA and a neat MMA polymerization may be seen in Figure 3.9.

It is noted (Carswell *et al.*, 1996) that the concentration of radicals almost reaches millimolar levels in the crosslinking system, i.e. 40 times that for the uncrosslinked polymer. Species other than those responsible for network formation may also be observed, if they are stable, and then used to monitor crosslinking. The oxidation of amine-containing epoxy resins may occur during cure or on UV irradiation, and ESR has been used to identify the cation radical species formed (Figure 3.10) (St John, 1993, Fulton *et al.*, 1998).

By irradiating the curing system and measuring the concentration of this species by ESR it is possible to observe the considerable change in the network structure that occurs between gelation and vitrification and leads to an increasing population of trapped radical cations (Figure 3.10). In another use of ESR, the lifetime of the free radicals formed after γ-irradiation has been used to monitor the state of the network formed on cure. The occurrence of two distinct decay processes suggested that inhomogeneities in the network were being probed (Kent *et al.*, 1983).

The major limitations to the use of ESR other than for fundamental studies of the radical and other trapped species formed during reactive processing are the experimental requirements of the apparatus. There has been success in using ESR to monitor the concentration of the propagating free radicals during the emulsion batch polymerization of methyl methacrylate (Parker *et al.*, 1996) by using a time-sweep method for data acquisition and

HALS: Tinuvin 770

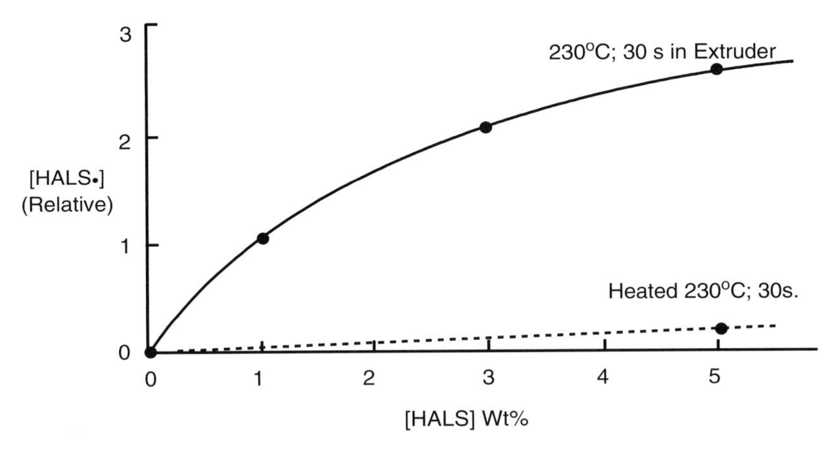

Figure 3.8. Radical formation due to mechanochemical degradation measured by using Tinuvin 770 as a spin trap and detecting the alkyl radicals via the change in nitrogen-centred radical concentration by ESR. Adapted from Sohma (1989a).

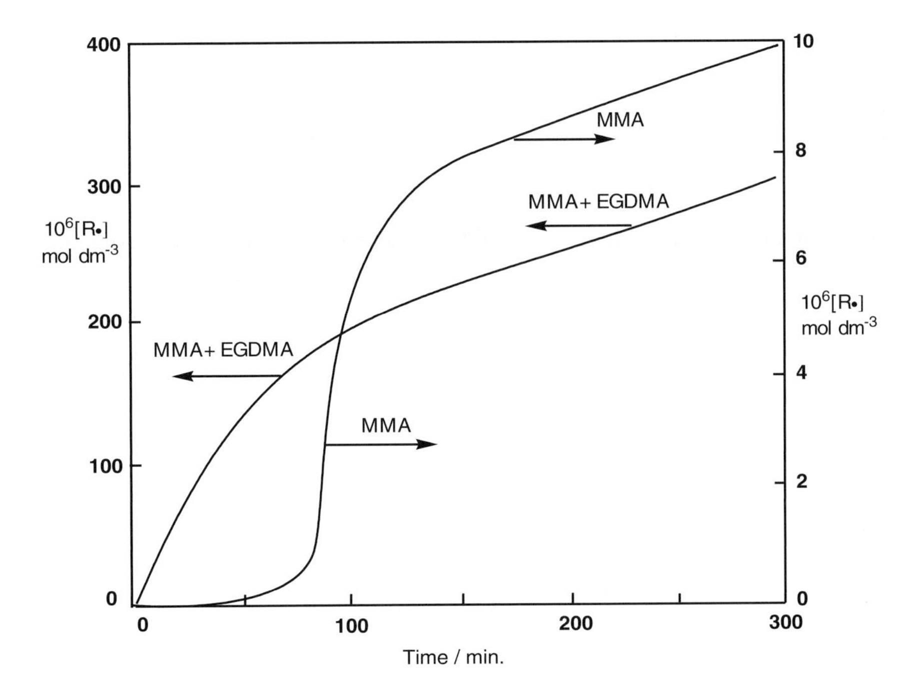

Figure 3.9. Changes in free-radical population [R·] during polymerization of MMA and after addition of the crosslinker EGDMA (note the different scales). Adapted from Carswell *et al.* (1996).

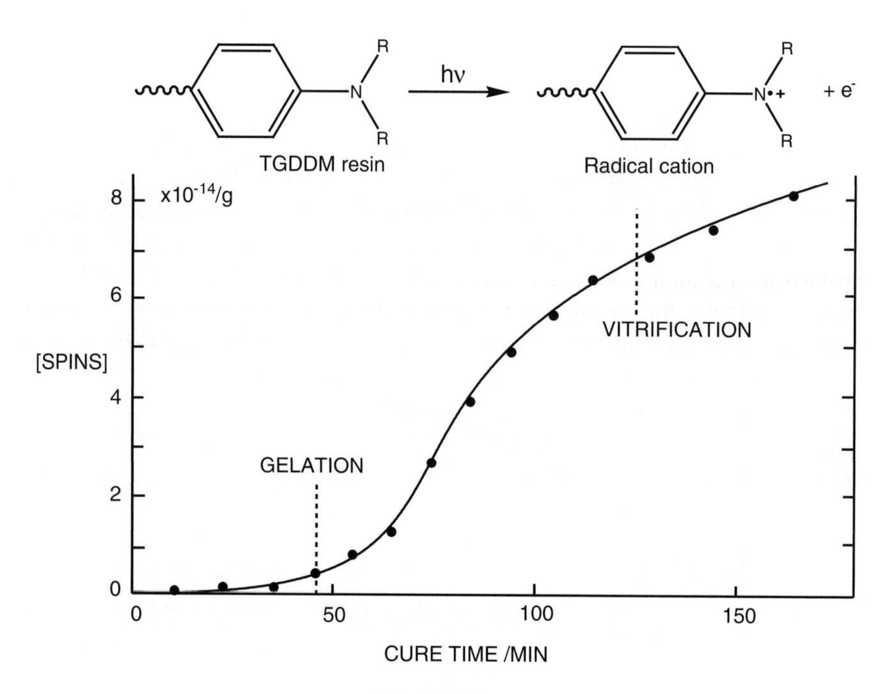

Figure 3.10. Change in spin population with time of cure of the TGDDM epoxy resin MY720 cured with 26.5% DDS at 160 °C. Brief UV irradiation is used to create the radical cation species analysed. Reproduced with permission from St John (1993).

a gravity-flow system leading into a collection reservoir. Although this provided an improvement over the previous field-sweep and closed-loop-flow system (Lau and Westmoreland, 1992) the technique still requires the sample to flow through the ESR cavity. For *in situ* study of the kinetics of reactive processes one of the simpler spectroscopic techniques for determining instantaneous concentration of reactant and/or product, as described in later sections, is preferable.

Nuclear magnetic resonance (NMR) spectroscopy

Nuclear magnetic resonance (NMR) has provided fundamental conformational and structural information on organic and inorganic polymers and has become the preferred spectroscopic technique for laboratory studies, including end-group analysis for determination of the number-average molar mass (M_n) (Tonelli and Srinivasarao, 2001). The quantized transitions occur in the microwave region of the spectrum and correspond to the spin energy states of certain nuclei. These states are degenerate unless in a magnetic field and, because of the small energy difference and the low population difference between these states, the sensitivity of the technique is poor unless resonance methods are employed. The sensitivity depends on the relative abundance of nuclei with the appropriate spin properties to display an NMR signal. The nuclei of interest in polymer science are ^{1}H, ^{13}C, ^{15}N, ^{19}F, ^{29}Si and ^{31}P. The low abundance of the NMR-active isotope of the most useful element, carbon, means that spectral acquisition times have to be long in order to obtain spectra with a good SNR. Analyses are therefore required off-line in order to determine the species formed in the process cycle. Unless the polymer may be dissolved in a suitable solvent, the NMR spectra are extremely broad.

It has been demonstrated that on-line monitoring in a laboratory environment of a copper-mediated living radical polymerization provides increased precision in kinetic analysis because of the greater number of data points obtained (Perrier and Haddleton, 2003). One major limitation in both ESR and NMR spectroscopy is the difficulty in applying the technique to either on-line or in-line analysis in real time in an industrial environment, although it can be expected that further advances in technology may lead to this possibility. An example of this is the high-temperature surface NMR probe that has been able to be adapted to operation at the die of an extruder (Gottwald and Scheler, 2005). While it was demonstrated that the system could operate in the aggressive environment, no interpretation of the relaxation time of PVC during extrusion was performed. The NMR technique is one of the most powerful off-line analytical tools since the signal is sensitive to the chemical, conformational and macroscopic environment and it has been used successfully for *in situ* study of Ziegler–Natta polymerization of propylene (Mori *et al.*, 1999), but much further development is needed in order to provide an on-line monitoring technique. Of some interest is the demonstration of rheo-NMR, in which magnetic resonance imaging is used to visualize polymer melts under shearing and extensional flows (Callaghan, 1999).

Since many products of reactive processing are solids, the advent of solid-state (magic-angle-spinning (MAS)) NMR spectroscopy has meant a greater understanding of networks after gelation and vitrification. An excellent example is with elastomers, and Koenig (1999) gives a detailed analysis of the information which solid-state NMR has provided regarding the mechanism of vulcanization, the effect of filler (e.g. silica and carbon black) on cure chemistry and also the thermal oxidation of the network. In relation to oxidation, Clough (1999) has demonstrated the extensive information which may be generated by using $^{17}O_2$ for oxidation studies and measuring the NMR spectrum of the resulting oxidation products. While this demonstrates the power of NMR, and there are further developments in magnetic resonance imaging that are having a major impact on polymer science, the direct application to chemorheology is not apparent except in assisting other techniques through establishing mechanisms, identifying the effects of impurities on the reaction route etc.

An example of this relevant to chemorheology of advanced epoxy-resin networks is the detailed studies of model compounds to identify the possible structural units from the reaction of *N*, *N*-diglycidylanilino groups with aromatic amines as models for the TGDDM/DDS epoxy-resin system (Laupretre, 1990) and then the use of ^{13}C CP/MAS NMR to characterize fully cured TGDDM/DDS networks (Attias *et al.*, 1990). From this it was concluded that the network structure of TGDDM/DDS had a considerably lower crosslink density than expected from the high functionality of the resin and hardener. Other researchers (Grenier-Loustalot *et al.*, 1990) found that the aliphatic-carbon peak could be resolved into primary- and secondary-alcohol, ether and epoxy bands, and that, for TGDDM and MY721, about 20% of epoxy groups formed ethers when cured at up to 185 °C. The final glass-transition temperature (T_g) decreased with increasing ether and hydroxyl group concentration. As is discussed later, these results are of value in interpreting data from NIR and other vibrational spectroscopic techniques that can be used for real-time reaction monitoring.

3.3.3 Vibrational spectroscopy overview – selection rules

The spectra corresponding to transitions between vibrational energy levels of the functional groups present in polymers occur in the **mid-infrared (MIR)** region of the spectrum, typically from 400 to 4000 cm^{-1}. Infrared spectroscopy generally involves the measurement of the wavelength (or energy) and the amount of radiation absorbed by the polymer from a beam of infrared radiation. The analysis of the spectrum therefore provides information about the composition of the system as well as the amount present. The vibrational spectra of small molecules such as water and carbon dioxide in the vapour state appear as many peaks in the MIR region due to the excitation of rotational states of the molecules, and these may be assigned to specific vibration and rotation modes of the molecules. This gives a characteristic band envelope for water and carbon dioxide, which is often seen in FT-IR transmission spectra because of the presence of water and carbon dioxide in the optical pathlength when polymer samples are analysed (Koenig, 1999). These rotations cannot be resolved in the condensed phase and only a broadened single band may be seen.

The formal requirement for these transitions to occur is that the molecular vibration should produce a change in dipole moment. Thus, in a molecule such as carbon dioxide, the symmetrical stretching vibration (ν_1) will be infrared **inactive** and the bending (ν_2) and asymmetric stretch (ν_3) modes will be **active**. The fundamental frequency at which a particular vibration will occur is given by the classical formula for a diatomic harmonic oscillator:

$$\nu = [1/(2\pi)](k/\mu)^{1/2} \tag{3.17}$$

where ν is the frequency of the fundamental vibration, k is the force constant for the bond undergoing vibration and μ is the reduced mass of the atoms in the molecule.

For a bending vibration, k will be lower than for stretching a bond, such that in CO_2 this frequency is lower (667 cm^{-1}) than that for a stretching vibration (2349 cm^{-1}).

The observed spectral bands correspond to allowed transitions between quantized energy levels of energy:

$$E_\nu = \left(\nu + \frac{1}{2}\right)h\nu, \tag{3.18}$$

where ν is the vibrational quantum number, which may take values 0, 1, 2 ...

The selection rule for allowed transitions is $\Delta\nu = \pm 1$, so the energy separation between the levels is $h\nu$, which is the observed spectral frequency. The fundamental band observed in the spectrum then corresponds to the transition from $\nu = 0$ to $\nu = 1$, since at normal temperatures most molecules will occupy the ground vibrational state.

These relationships are based on a pure harmonic oscillator, but real molecular vibrations are often anharmonic. Consequently the solution for the quantized harmonic oscillator (Equation (3.18)) is replaced by

$$E_\nu = \left(\nu + \frac{1}{2}\right)h\nu + \left(\nu + \frac{1}{2}\right)^2 h\nu\chi, \tag{3.19}$$

where χ is the anharmonicity constant, which is small and positive. The effect of anharmonicity is to relax the selection rule to $\Delta\nu = \pm 1, \pm 2, \pm 3, \ldots$ so allowing transitions corresponding to the first and second overtones as well as combination bands to be seen. Thus, in the spectrum of carbon dioxide, in addition to the two allowed fundamental vibrations, weaker bands corresponding to overtone bands (multiples of the fundamental) and combination bands (sums or differences of two fundamentals) are seen in the MIR. Both of these result from the anharmonicity of the vibrations of real molecules, which is the rule, rather than the exception, when studying polymers.

The combinations and overtones of many high-energy vibrations are at energies beyond the MIR spectral region and are therefore seen in the **near infrared (NIR)**. The extent to which this occurs depends on the anharmonicity of the particular vibration. The most anharmonic vibrations are those involving light atoms attached to heavy atoms, such as C–H, N–H and O–H (χ is large in Equation (3.19)), and it is found that these combinations and overtones are the most intense in the NIR spectra of organic molecules and polymers. As will be discussed in Section 3.3.6, these are still weaker than the fundamental modes by two orders of magnitude (Colthup *et al.*, 1990), so the pathlength needed to study the spectra at a reasonable absorbance value is of the order of millimetres rather than micrometres as in the MIR. One advantage of the NIR region of the spectrum is that the effects of atmospheric absorption bands are much less significant than in the MIR and the optical components may be of quartz or glass and are readily adapted to optical fibres for remote spectroscopy of reacting systems.

Classically, in a polyatomic molecule, a normal mode of vibration occurs when each nucleus undergoes simple harmonic motion with the same frequency as and in phase with the other nuclei. The possible normal modes of vibrational of a non-linear polyatomic molecule with N atoms are $3N - 6$ and are categorized as arising from stretching, bending, torsional and non-bonded interactions. Whether a particular mode is IR active depends on whether the symmetry of the vibration corresponds to one of the components of the dipole moment.

Information regarding the normal modes of a polyatomic molecule, which are not IR active, may often be obtained from the **Raman** spectrum. Raman spectroscopy is an inelastic-scattering technique rather than requiring the absorption or emission of radiation of a particular energy. The selection rule differs from the IR in that it is required that the incident electric field of the radiation can induce a changing dipole moment of the molecule. This results in a different symmetry requirement for the normal modes of vibration to be Raman active, since it now depends on the polarizability of the molecule.

The scattered radiation then contains information about the vibrational states of the molecule encoded as a band shift $\Delta\nu$, either to the high or low energy of the elastically scattered exciting laser radiation,

$$\Delta\nu = \nu' \pm \nu, \tag{3.20}$$

where ν' is the frequency of the incident radiation and ν is a vibrational frequency that satisfies the Raman selection rule.

The exciting laser radiation may be in the UV, visible or NIR region of the spectrum, so there is considerable flexibility in choosing the appropriate radiation source to minimize effects that mask the Raman spectrum, such as fluorescence from the polymer or impurities. However, since the Raman effect is inversely proportional to the fourth power of the wavelength of the exciting radiation, the scattering with NIR laser excitation will be weaker, although the interference from fluorescence is decreased.

As shown in Equation (3.20), analysis of the Raman spectrum may be performed to the lower (Stokes shift) or higher (anti-Stokes shift) energy of the exciting light, and the vibrational modes that may be resolved depend on the efficiency of removing the Rayleigh scattering. As with NIR, the technique may be used with little atmospheric interference and quartz or glass optical components. These and other experimental requirements are discussed in Section 3.3.7.

Table 3.1. Components of an interferometer (Figure 3.11) for optimum FT-IR spectral performance

Region	Source	Beam splitter	Detector
Mid-IR	Glowbar	KBr	MCT at 77 K or DTGS at room temperature
Near-IR	Tungsten–halogen	Quartz	InSb at 77 K

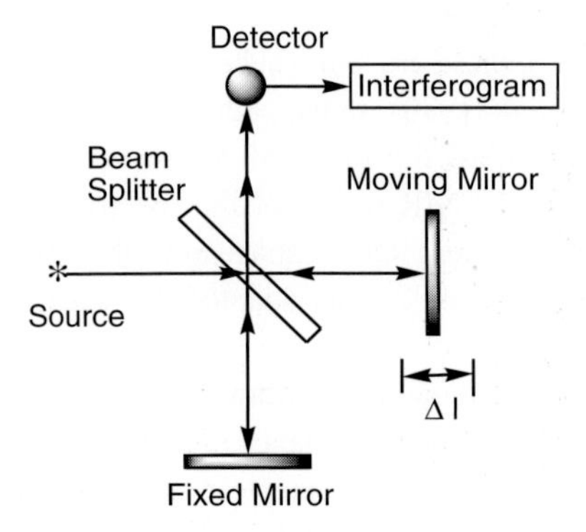

Figure 3.11. A schematic diagram of the operating principle of an FT-IR spectrometer, showing the Michelson interferometer.

3.3.4 Fourier-transform infrared (FT-IR) and sampling methods: transmission, reflection, emission, excitation

The spectral analysis of the chemical changes in reacting polymer systems requires reliable sampling of the polymer and calibration if quantification is required. The development of the FT-IR spectrometer has enabled digital spectroscopy with excellent signal-to-noise properties (SNR) from the MIR to the NIR, and the applications to polymers have continued to evolve (Koenig, 1984, 1999, Garton, 1992). The simple optical arrangement of the FT-IR spectrometer means that it may be readily adapted to on-line analysis using fibre-optics as described in Section 3.3.10. The following analytical methods are here described for application off-line.

The essentials of the FT-IR spectrometer are shown in Figure 3.11 (Koenig, 1984). Commercial instruments have the source, beam splitter and detector optimized for maximum SNR in the spectral region of interest (Table 3.1).

The beam splitter and moving and fixed mirrors constitute the interferometer, and the signal at the detector is an interferogram of the light source arising from the recombination of the waves that have travelled the different paths. If the moving mirror is oscillating at fixed frequency, and the source is monochromatic, the interferogram is a cosine function. The spectrum is transformed from an interferogram in the time domain to a spectrum in the frequency domain by taking the Fourier transform of the detector signal using the fast-Fourier-transform (FFT) algorithm. This computation is done in the background during spectral acquisition, which usually involves the co-addition of 16 or more interferograms, and the SNR increases as the square root of the number of scans co-added.

There is always a trade-off between the resolution (determined by the distance the moving mirror travels) and the time available for analysis. In off-line analysis of polymers, this is usually not a major factor, but in on-line analysis the SNR achieved may limit the detection limit of species and the analytical precision. A factor that also limits the resolution of the spectrometer is the finite distance the mirror travels in the interferometer. The FFT is taken between $\pm\infty$ so, the result is the presence of a spectral artefact (spectral side lobes) that is automatically removed by multiplication by an apodization function before the spectrum is displayed. This correction produces spectral line broadening, but this may be removed during Fourier self-deconvolution (with return of the side lobes) to assist in determining the presence of overlapping bands. This is just one of many data-analysis routines available to assist spectral interpretation, but there are limitations to quantitative analysis, which are discussed below.

Transmission spectra

In the simplest spectral-acquisition method, the sample to be analysed lies between the source and the interferometer. Since FT-IR is a single-beam spectral technique, the spectrum of the source is obtained in the absence of the sample and then obtained again with it in the beam. The ratio of the two generates a transmittance spectrum, which is then converted to an absorbance spectrum to enable spectral manipulation such as subtraction. An advantage of an FT-IR system is that spectra are acquired digitally with high wavenumber accuracy so that bi-component polymer films may be analysed by spectral subtraction if one of the components is known. An example is shown in Figure 3.12 for a sample containing

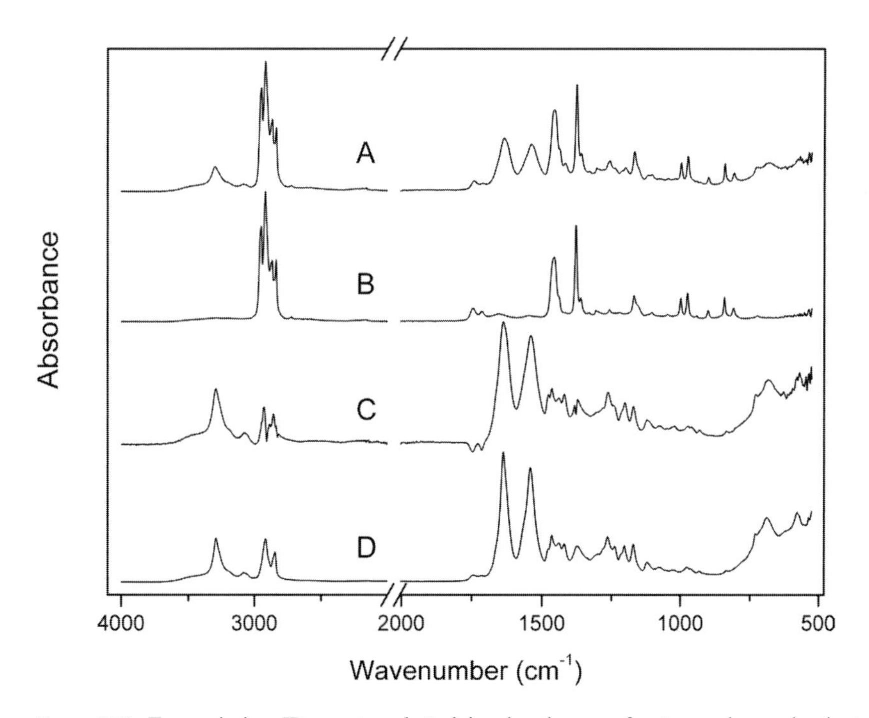

Figure 3.12. Transmission IR spectra plotted in absorbance of a two-polymer laminate (A), one component of which is able to be measured (B) whereas the other is unknown. Identification is possible by spectral subtraction to give spectrum (C), which may be compared with a library spectrum (D). Spectrum courtesy of Dr L. Rintoul.

two polymers. The first spectrum (A) is that of the bi-component system. The second spectrum (B) is that of polypropylene which was obtained by attenuated total reflection (ATR) on one surface of the system (see the next section). The third spectrum (C) is the result of subtracting the second from the first and shows features that are characteristic of a polyamide. This is confirmed by comparison with a spectrum of nylon-6 from the digital FT-IR library as shown in the fourth spectrum (D) in Figure 3.12. It may be seen by comparing this with the subtraction spectrum that there are minor differences due to artefacts introduced by the subtraction process, in which spectra may differ very slightly in wavenumber. This may produce 'derivative-like' spectra such as is seen around 2950 cm^{-1}. There are also negative bands at 1700–1750 cm^{-1} due to oxidation bands present in the polypropylene that are 'over-subtracted' out. Provided that these minor alterations are recognized, this method does offer analytical capabilities that may otherwise be difficult to obtain, such as when there is a laminate with one component surrounded top and bottom by another polymer so that the external polymer may be measured by ATR (see below) and the total laminate by transmission spectra.

Such analytical processes rely on the additivity of absorbance given in Beer's law:

$$A(v) = \log_{10}(I_0/I) = \varepsilon_1 c_1 b_1 + \varepsilon_2 c_2 b_2 + \cdots + \varepsilon_i c_i b_i, \tag{3.21}$$

where I_0 and I are the intensities of radiation without and with the sample present, respectively, $A(v)$ is the total sample absorbance at a frequency corresponding to absorption by all components and ε, c and b are the molar absorptivity (dm^3 mol^{-1} cm^{-1}), concentration (mol/dm^3) and pathlength (cm), respectively, of each component i at that frequency.

In many systems, such as homogeneous mixtures, b will be a constant for each component. In the MIR region of the spectrum, the molar absorptivities are typically 10–100 dm^3 mol^{-1} cm^{-1} so that, for solid polymers, a pathlength of about 10 μm is required. For this reason, sampling methods involving reflectance spectroscopy are employed for thick samples (Section 3.3.4). Beer's law will hold for systems in which there are no interactions between the components, such that spectral subtraction should enable quantitative analysis of the components, if all but one are known and standard spectra are available. These may be from one of the many libraries of polymer spectra available in digital form with commercial instrumentation. Where species are unknown, then methods such as least-squares regression analysis and principal-component analysis may be employed to determine the number of species present in a mixture as well as the principal components of each spectrum (Haaland *et al.*, 1985, Koenig, 1984, 1999). The interactions which may occur between components are important in blends, and small spectral differences are of importance in understanding the enthalpic contributions to miscibility and the origin of the intermolecular forces controlling the structure (Garton, 1992). Specific examples are considered in Section 3.3.5.

One of the features of transmission FT-IR spectra in the MIR is the appearance of atmospheric bands as either positive or negative spectra. This may often limit the analysis of oxidation products since the water-vapour bands lie in the same spectral region as those of some conjugated and carbonyl species. While atmospheric effects may be subtracted out, purging of the full optical path is the most reliable way of eliminating these interferences. It is also important to remember that analyses are more reliable at low absolute absorbance. The subtraction of two bands with high absorbance will heighten effects due to non-uniform sample thickness (Koenig, 1984). The reverse effect occurs when samples are of an extremely uniform thickness, so interference fringes are observed, which may swamp the spectral differences to be observed. One way of eliminating this is to measure spectra by a transmission–reflection technique so that the fringes are cancelled out (Spragg, 1984, Farrington *et al.*, 1990).

Polymer samples from reactive processing will often be in a form that does not permit simple transmission-spectrum measurement. Micro-sampling methods including IR microscopy are available for samples that are much smaller than the IR beam, but, if the material is opaque, then a reflection technique may be the most appropriate.

Reflection spectra

The transmission (T) of radiation by a sample is always lower due to reflection (R) losses, and, as shown in Figure 3.13(a), there is also a third component to be considered, namely emission (E), as discussed in the next section. In general $T+R+E=1$.

Internal-reflection spectroscopy (IRS), also known as attenuated total-reflection (**ATR**) spectroscopy, is one of the most widely used techniques to generate spectra from intensely absorbing samples. The principle of operation is shown in Figure 3.13(b).

The radiation from the IR source is transferred through mirrors to enter normal to the edge of a high-refractive-index optical element. This edge is cut at an angle to enable total internal reflection of the radiation through the element and then emergence through the opposite edge. Other mirrors gather the radiation and direct it to the interferometer. If a material of lower refractive index is in contact with the element along one or both sides, then, as the radiation reflects at or near the Brewster angle, there will be attenuation of the radiation due to absorption of the evanescent wave by the sample. This wave penetrates a distance, d, into the sample dependent on the refractive indices of the internal reflection element (IRE), n_1, and the sample, n_2, the angle of reflection, θ, and the incident wavelength, λ, as given by

$$d = \lambda / \left\{ 2\pi n_1 \left[\sin^2\theta - \left(n_2/n_1^2 \right) \right]^{1/2} \right\}. \tag{3.22}$$

The wavenumber dependence of the penetration depth for a KRS-5 optical element (a mixed salt of thallium bromide and iodide) ($n_1=2.5$) and a sample with $n_2=1.5$ is shown in Figure 3.13(c).

It may be seen that the total pathlength for a multi-reflection element with, say, six reflections will range from $\sim$13 μm at 800 cm^{-1} to $\sim$3 μm at 3000 cm^{-1}. Therefore ATR spectra are dependent on the experimental arrangements and apparatus. Other materials for ATR elements are ZnSe, Ge and diamond, with the choice being governed by the refractive index compared with those of the samples being studied.

As discussed later, in Section 3.4, this technique is particularly well adapted to coupling to optical fibres, so a probe that can be immersed in a polymer reaction stream may be fabricated to enable spectra to be collected in real time. The principles of total internal reflection and its application in fibre-optics are discussed in more detail in Section 3.4.1.

Another widely used reflection technique for polymer samples, particularly powders, is diffuse reflectance, which, when used with an FT-IR Spectrometer, is given the acronym **DRIFT**. The principle of operation requires a refinement of the scheme in Figure 3.13(a) to ensure that no specularly reflected radiation is collected. It has been shown that this may be achieved by ensuring that all radiation collected has passed through the particles rather than having been reflected from the surface by covering the material being analysed with a layer of finely ground potassium bromide. Under these circumstances the Kubelka–Monk

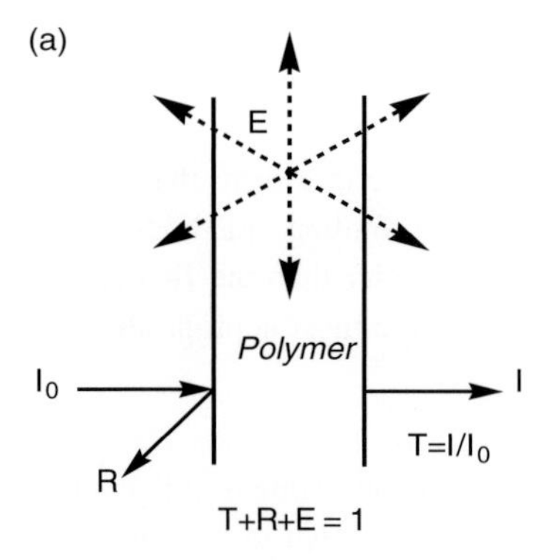

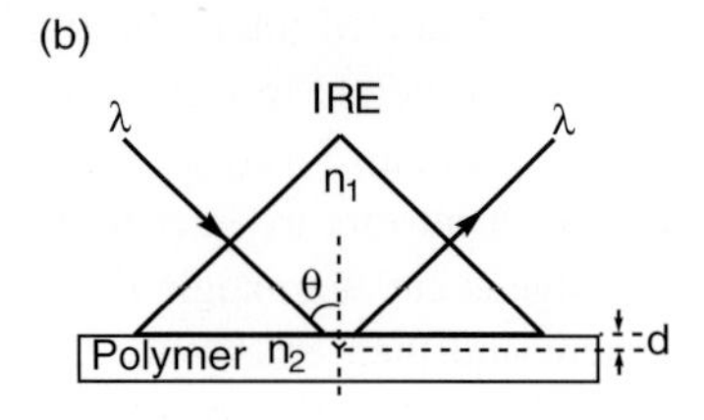

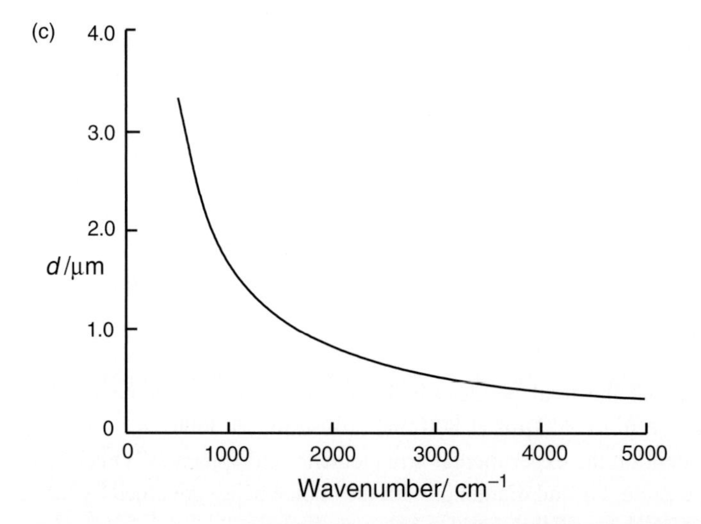

Figure 3.13 (a) A schematic diagram of light transmission (T), reflection (R) and emission (E) by a polymer film; I_0 and I are the incident and transmitted beam intensities. (b) A schematic diagram of single reflection with an internal reflection element (IRE) of refractive index, n_1, in contact with a polymer film, of refractive index n_2. The penetration depth, d, of the evanescent wave depends on the angle of incidence θ and wavelength, λ, of the radiation as given in (c). (c) A plot of Equation (3.22), showing the wavenumber ($1/\lambda$) dependence of penetration depth, d, of the evanescent wave for a KRS-5 IRE of refractive index 2.5 in contact with a polymer film of refractive index 1.5.

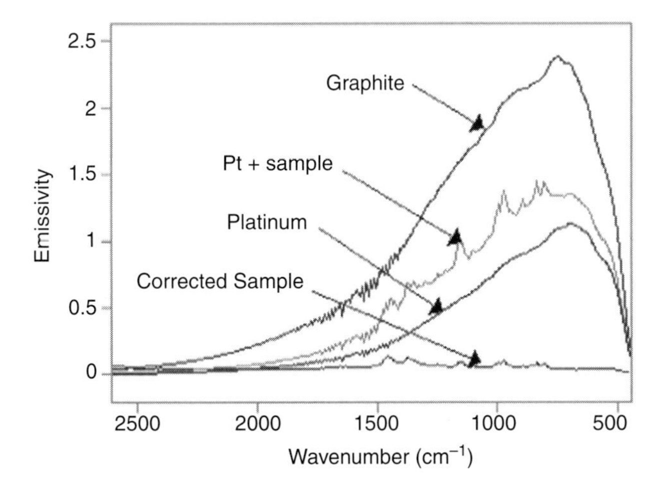

Figure 3.14. Emission spectra required for the generation of an equivalent absorbance spectrum from a heated sample (Equations (3.23) and (3.24)). Reproduced with permission from Blakey (2001).

correction (a routine built into FT-IR software for DRIFT) applies and quantitative analysis is feasible (Koenig, 1999).

Emission spectra

The contribution of emission to the infrared radiation detected by the FT-IR detector may be neglected under the normal conditions of measuring MIR spectra by any of the techniques described above. In most experiments, the sample is at room temperature, so there will be a negligible population of excited vibrational states created by processes other than absorption of the infrared radiation. There will be emission of radiation from those states populated by absorption, but this amounts to a negligible contribution since it is generated in all directions and only a small fraction will reach the interferometer, compared with the collimated beam from the light source.

However, if the sample is at an elevated temperature, above 150 °C, as in many processing operations, then significant emission in the MIR occurs. If the light source in the FT-IR system is replaced with a thin polymer film on a platinum hotplate, and the resulting single-beam spectrum referenced to a black body (graphite) at the same temperature, then an emissivity spectrum qualitatively equivalent to an absorbance spectrum is obtained. Figure 3.14 shows the sequence of measurements necessary to generate the emissivity spectrum.

Since the spectrometer will itself be emitting infrared radiation to the detector (at 77 K), determination of the emissivity requires the following measurements:

$$\varepsilon(v) = (I_{\mathrm{Sa}} - I_{\mathrm{Pt}})/\left(I_{\mathrm{bb}} - I_{\mathrm{Sp}}\right), \tag{3.23}$$

where I_{Sa}, I_{Pt}, I_{bb} and I_{Sp} represent the intensities of infrared emission at each frequency, v, from the sample, platinum sample holder, black body and spectrometer, respectively. The final corrected emissivity spectrum of the sample as a function of wavenumber is shown in Figure 3.14 and may be scale-expanded for detailed analysis.

A simple analysis (George *et al.*, 1995) gives the relation between absorbance $A(\nu)$ and emissivity $\varepsilon(\nu)$ as

$$A(\nu) = -\frac{1}{2}\ln[1 - \varepsilon(\nu)]. \tag{3.24}$$

This relation holds only for low values of emissivity, and, in order to ensure that there is no loss of spectral definition due to reabsorption and to prevent the creation of thermal gradients in the polymer, a thickness less than 10 μm is required. The technique, like all emission methods, is extremely sensitive (Chase, 1981), and it has been applied to study the cure chemistry of thermoset resins on metal substrates, and thus is appropriate as a technique for monitoring reactive processing such as coating deposition (Pell *et al.*, 1991) and epoxy-resin prepreg crosslinking (George *et al.*, 1996). An example of the use of emissivity to study the oxidation of polypropylene in real time is shown later in Figure 3.22.

Excitation spectra

Excitation spectra are generated when the wavelength of radiation absorbed by a sample is varied and the intensity of emitted radiation is measured. The resulting spectrum of emission intensity against excitation wavelength is analogous to the absorption spectrum of the sample. While such measurements are usually associated with luminescence processes (Section 3.3.8), the analogous technique in the MIR is photo-acoustic spectroscopy (PAS). In this technique, the extent to which the sample is heated by the incident radiation is measured by the pressure change to gas contained in an enclosed chamber. Since the FT-IR radiation is modulated at the frequency of the moving mirror, this will appear in the audio range of 1–2 kHz. The pressure sensor is a microphone that detects a photo-acoustic equivalent of the interferogram, which therefore contains the absorption information for the sample. An advantage of the PAS technique is that reflected or scattered radiation does not affect the modulated signal, so it is particularly effective with highly filled samples, such as those with carbon black (Koenig, 1984). Depth profiling to one-half the thermal diffusion length from the surface may be performed by varying the modulation frequency of the incident radiation by altering the mirror velocity. However, PAS spectra are generally of low SNR, so a large number of scans needs to be co-added and thick samples are prone to saturation effects

It has been reported that PAS can be used to monitor reactive processing of network polymers (Taramae *et al.*, 1982), but it is noted that at high temperatures it may be necessary to account for emission from the sample during the PAS experiment (McGovern *et al.*, 1985).

3.3.5 Mid-infrared (MIR) analysis of polymer reactions

Since the MIR provides identification of many of the chemical groups in the polymer with good sensitivity and the possibility of quantification, it has frequently been employed to follow the reactions occurring during reactive processing of polymers and thus assist in determining the link between chemical and rheological changes. The sampling methods discussed in the previous section have been applied with varying success. The following examples are intended to be representative of the approaches employed and are not exhaustive.

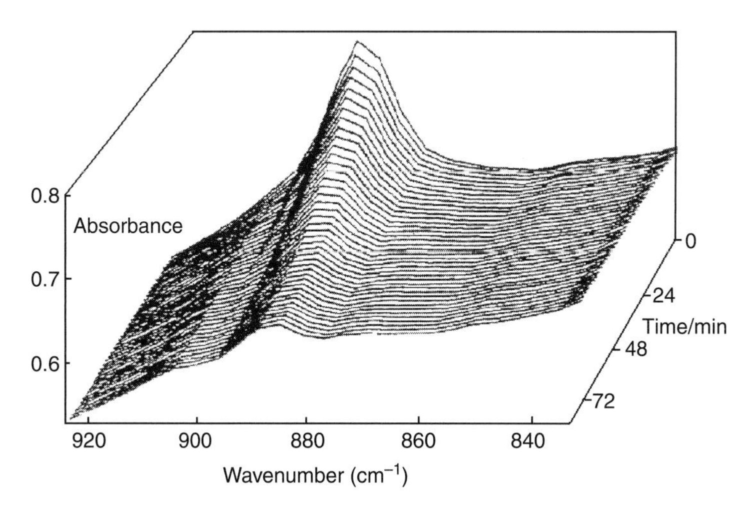

Figure 3.15. A real-time FT-IR 'waterfall' plot of the change in the absorbance of the $=CH_2$ wag (at 887 cm^{-1}) during the polymerization of isobutylene. Reprinted with permission from Storey (1998). Copyright 1998 American Chemical Society.

Addition polymerization

Vinyl polymerization reactions may be studied by transmission FT-IR, but, because of the short pathlength (e.g. 50 μm) required in the MIR and the need for uniform heating of the monomer in the absence of air, alternative approaches such as ATR are employed. The traditional, laborious gravimetric approach to the study of kinetics of the living cationic polymerization of isobutylene has been replaced with a real-time MIR analysis of the consumption of monomer by monitoring the disappearance of the $=CH_2$ wagging band at 887 cm^{-1} (Storey *et al.*, 1998). In Figure 3.15, a stack plot of successive spectra obtained by ATR using a composite diamond insertion probe is shown (after subtraction of solvent) for a reaction time of over one hour.

Kinetic analysis of peak height and peak area gave identical profiles and enabled temperature fluctuations due to the high reaction exotherm to be highlighted. In a further study with this system, which uses a light-conduit pipe so that the reaction is remote from the spectrometer, the block copolymerization of styrene to living polyisobutylene chain ends was monitored (Storey *et al.*, 1999). It was shown that the real-time analysis for styrene content agreed with the off-line analysis by ^{1}H NMR. A similar study of the kinetics of carbocationic polymerization of styrene and isobutylene using an ATR fibre-optic probe produced similar agreement between the MIR method and gravimetry (Puskas *et al.*, 1998). Fibre-optic probes are available with MIR transmitting optics as accessories for all brands of FT-IR spectrometers for laboratory studies of polymerization, oxidation and functionalization reactions. The availability of commercial laboratory-scale polymerization reaction systems incorporating ATR-IR probes (e.g. Mettler-Toledo React-IR™) has allowed qualitative and mechanistic studies of metallocene-catalysed copolymerization, polycondensation and polyaddition reactions (Sahre *et al.*, 2006) as well as high-output polymer screening (HOPS) to determine catalyst productivities for copolymerization by using kinetic measurements in real time (Tuchbreiter, 2004). These systems and further applications are discussed in the section on remote spectroscopy.

An industrial system based on transmission that is able to analyse molten polymer from a reactor stream by using a variable-pathlength cell and a heated transfer line has been reported (Fidler *et al.*, 1991). The system is designed to tolerate pressures up to 300 bar and temperatures up to 400 °C, this requires optical components in the sample area to be made of diamond or ZnSe. The application reported involved the determination of vinyl acetate concentration in polyethylene, and good quantification was obtained by using the overtone of the ester band rather than the fundamental. This enabled a longer pathlength to be used and so avoid the high back-pressure from the narrow pathlength necessary to obtain quantification of an intense band such as that of the ester. As will be discussed later, the use of overtone and combination bands and longer pathlengths is a feature of NIR spectroscopy, which is the technique of choice for applications such as reactive extrusion (Dumoulin *et al.*, 1996).

Photopolymerization of acrylic monomers and related systems is a challenging area for the application of real-time MIR spectroscopy since the reactions are often completed within seconds, so there needs to be an excellent SNR in the spectra if reliable kinetic data are to be generated. The temporal resolution of an FT-IR spectrometer at a resolution of 16 cm^{-1} is 11 ms, so 100 spectra can be gathered in the time frame of the fastest reaction, but without multiplexing to improve the SNR. Using a diamond ATR element of area 4 mm^2 that could also be heated to 200 °C, uniform irradiation was possible and the cure reactions of films of thicknesses ranging from 1 to 20 μm could be followed (Scherzer and Decker, 1999).

Network polymerization

The determination of the chemical changes in the development of a three-dimensional network as the curing polymer moves from a liquid through a gel to a glass may be readily studied by IR techniques. Systems that have been well studied include polyurethanes (Koenig, 1999), cyanate esters (Bauer and Bauer, 1994) and epoxy resins (St John and George, 1994). Because it attains a high T_g on curing with aromatic amines such as diaminodiphenylsulfone (DDS) at temperatures from 160 to 180 °C, tetraglycidyl diamino diphenyl methane (TGDDM) has widely been used as an aerospace epoxy resin for composite materials. It has therefore been studied widely by a range of techniques (St John and George, 1994) including MIR spectroscopy. While IR cannot directly provide rheological information, it may enable the factors which control the kinetics of the network formation to be determined. In the study of reacting systems it is important, because of the overlapping bands in the MIR, to have a detailed assignment of all peaks in the spectrum so that the changes which occur on mixing and then subsequent reaction may be understood.

Figures 3.16(a) and (b) show the MIR spectra in two spectral regions for the TGDDM resin (Ciba MY721) and the aromatic amine hardener DDS as well as the mixture containing 27% DDS, which is typical of commercial formulations used in composite materials. The assignments of the significant bands in the spectra are given in Tables 3.2 for the TGDDM (MY721 resin) and 3.3 for DDS.

The changes in the MIR spectra on curing may be seen in Figure 3.17, which shows spectra of the TGDDM + 27% DDS mixture before and after curing for three hours and the difference spectra (uncured – cured). The bands which have become negative represent the functional groups which have participated in the reaction.

From a comparison of the first spectrum (Figure 3.17(a)) with the assignment tables, it is seen that the intensities of all N–H bands have decreased, as have those of the two epoxy C–H stretching bands at 3049 and 2993 cm^{-1}.

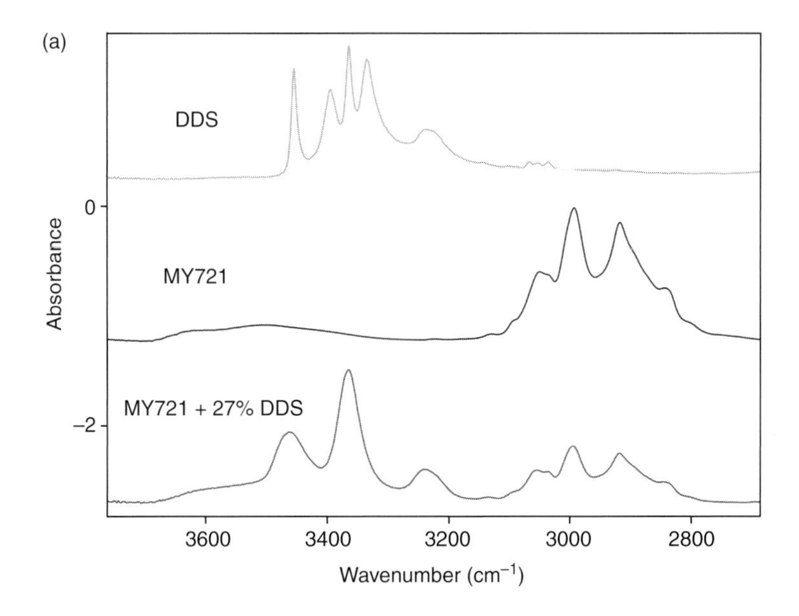

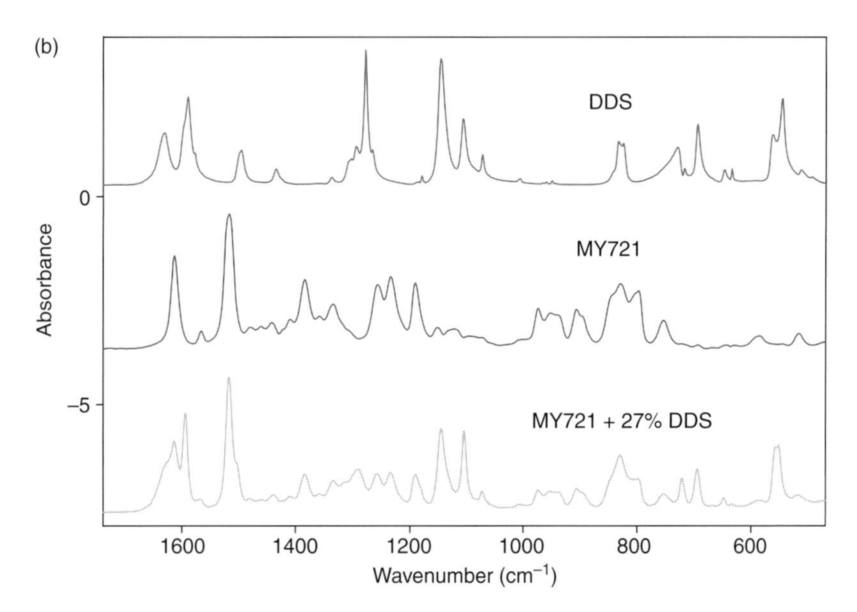

Figure 3.16 Mid-IR spectra for resin, hardener and their mixture, (a) in the hydrogenic region and (b) in the fingerprint region. Reproduced with permission from St John (1993).

At the same time there is an increase in intensity for the methylene groups at 2918 and 2844 cm^{-1} and a broad, hydrogen-bonded O–H band. All of these observations are a result of the ring-opening reaction of the epoxide with primary amine, the major chain-extension reaction occurring up to gelation of the resin.

The changes in the fingerprint region, Figure 3.17(b), are more complex due to the overlapping bands, some of which broaden and shift duc to thc increase in polarity of the network as curing takes place. In spite of this it is possible to discern two amine-related

Table 3.2. Mid-infrared band assignments for TGDDM (MY721 resin)

Peak position (cm^{-1})	Vibrational mode assignment
3623, 3508	O–H stretch; impurity hydroxyl groups
3049, 2993	C–H stretch; terminal epoxy group
2918, 2844	C–H stretch; CH_2 groups
1613	Phenyl-ring quadrant stretch
1516	Phenyl-ring semicircle stretch
1384	Aromatic C–N stretch; tertiary amine
1257	Symmetric ring breathing; epoxy group
1191	C–H wag
906	Asymmetric ring bend; epoxy group
829	C–H bend; epoxy ring
797	In-plane quadrant bend; benzene ring
585	Out-of-plane quadrant bend; benzene ring

Table 3.3. Mid-infrared band assignments for DDS hardener

Peak position (cm^{-1})	Vibrational mode assignment
3462 (3454, 3365)	N–H asymmetric stretch; primary amine (split in crystal)
3366 (3395, 3335)	N–H symmetric stretch; primary amine (split in crystal)
3238	N–H deformation overtone; primary amine
1630	N–H deformation; primary amine
1589	Phenyl-ring quadrant stretch
1495	Phenyl-ring semicircle stretch
1303	Sulfone asymmetric stretch
1278	Aromatic C–N stretch; primary amine
1146	Sulfone symmetric stretch
1107, 1072	In-plane bend; *para*-substituted phenyl
833	C–H wag; *para*-substituted phenyl
729	Phenyl-ring in-plane quadrant bend
694	N–H wag; primary amine
562	Sulfone wag
544	Phenyl-ring out-of-plane quadrant bend

bands (at 1630 and 544 cm^{-1}), which are suitable for monitoring in addition to the main epoxide band at 906 cm^{-1}.

Such detailed spectral analysis was first reported by Morgan and Mones (1987), who showed, through transmission MIR, the difference between the rates of the primary- and secondary-amine reactions with epoxy and the etherification reaction, which requires higher cure temperatures. A detailed knowledge of the mechanism of cure can enable, in certain circumstances, a rationalization of the differences among gel times for resins of various purities. In a study of the cure of tetraglycidyl diamino diphenyl methane (TGDDM) epoxy resins of various purities (St John *et al.*, 1993) it was shown that the major factor controlling the gel time was the catalysis of the cure reaction by residual hydroxyl groups present as impurities (e.g. glycols) in the resin.

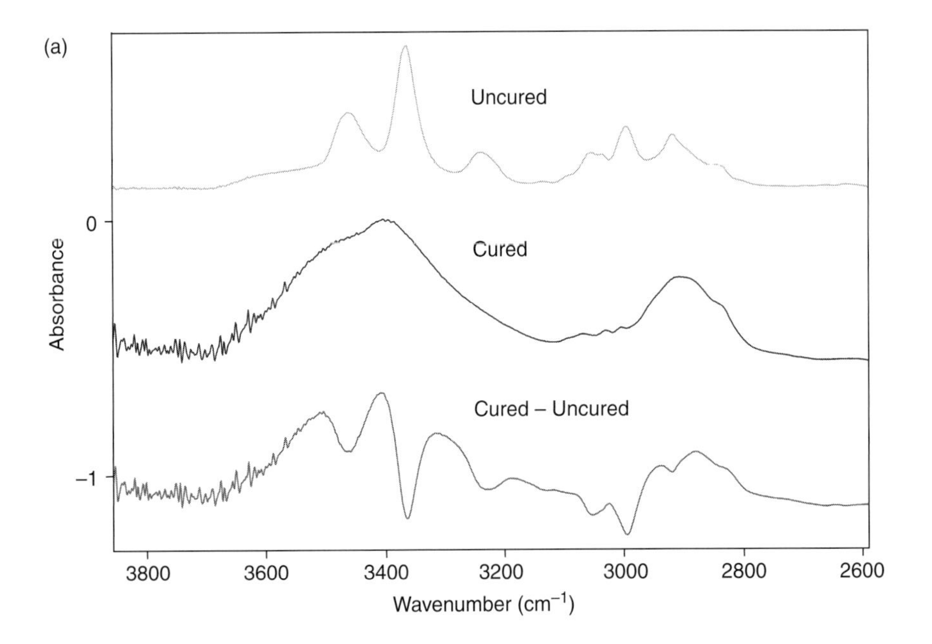

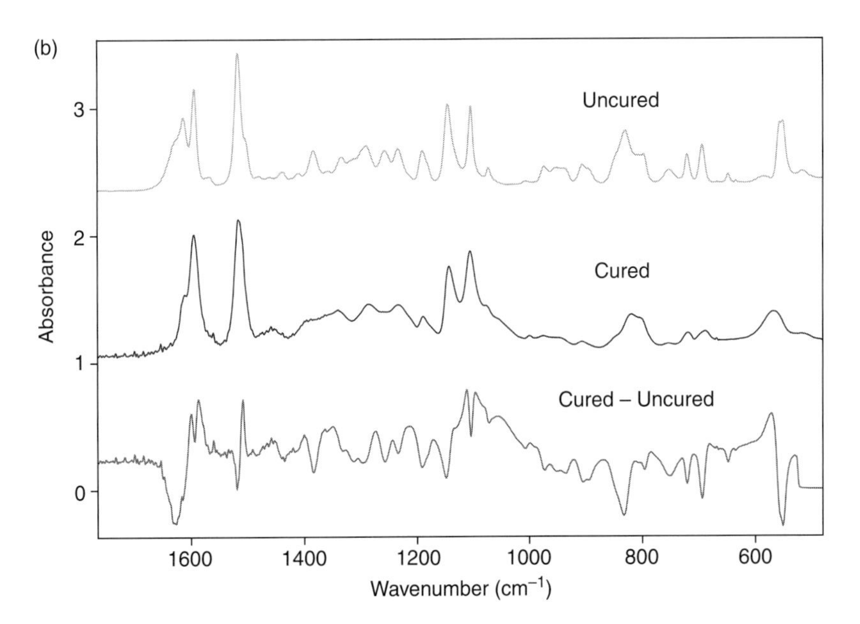

Figure 3.17. Changes in the MIR spectra in (a) the hydrogenic region and (b) the fingerprint region of MY721 cured with 27% DDS for 3 hours at 160 °C. Spectral subtraction (cured – uncured) shows the (negative) bands that change on cure. Reproduced with permission from St John (1993).

Figure 3.18(a) shows the MIR spectrum in the –OH stretch region of three TGDDM resins, namely commercial resins MY720 and MY721 and a purified resin from the flash chromatography of MY721 (labelled TGDDM). It can be seen that there is a progressive decrease in the total band area with increasing purity of the resin.

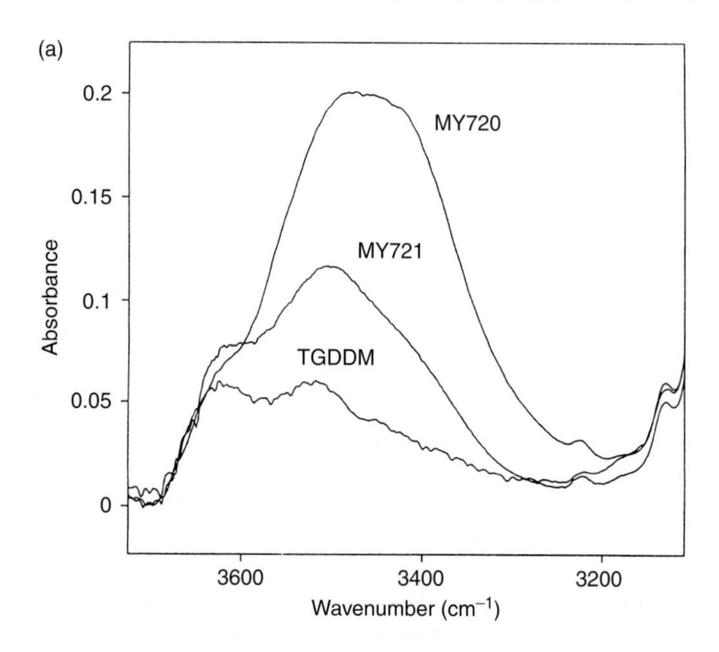

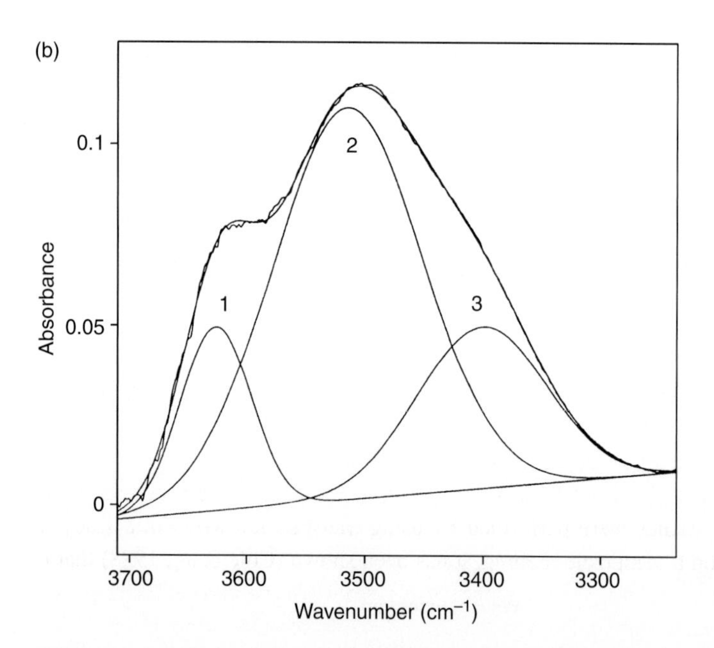

Figure 3.18. (a) Mid-IR spectra in the —OH stretch region of MY720, MY721 and TGDDM resins. Note the decrease in absorbance and band shift with increasing resin purity. (b) Curve fitting of the —OH band for MY721 from (a) as three components. Note the close fit of the sum of the bands to the original spectrum. (St John *et al.*, 1993).

The overlapping bands may be resolved and assigned to free and hydrogen-bonded hydroxyl groups of impurities in the resin as shown in Figure 3.18(b) for the MY721 commercial TGDDM. The bands are assigned to **1**, free O–H (3650–3590 cm^{-1}); **2**, intra-molecularly hydrogen-bonded O–H (3570–3450 cm^{-1}); and **3**, a 'polymeric' strongly bound

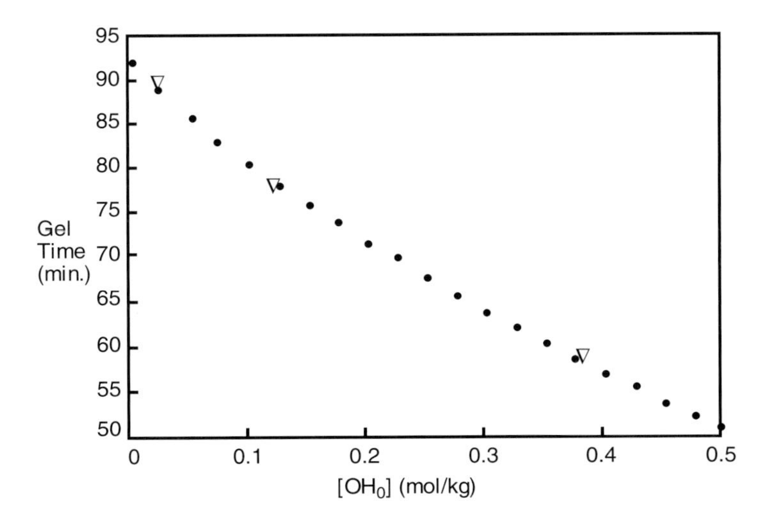

Figure 3.19. A plot of the gel time of TGDDM resins of various levels of purity as a function of $[OH]_0$ as determined by MIR. The curve is calculated from a model of —OH group catalysis of epoxy ring-opening (St John *et al.*, 1993).

type of hydrogen-bonding (below 3450 cm^{-1}). The most highly purified sample still contains hydroxyl groups, but there are equal amounts of the three types. As the resin increases in hydroxyl content there is an increase in the fraction of intramolecular and strongly bound hydroxyl groups. This has the effect of increasing the catalysis of the ring-opening of the epoxy by the amine nucleophile, with a resultant shortening of the gel time of the resin for the same concentration of amine.

This is shown in Figure 3.19 through the relation between the gel time of the resin and the initial –OH concentration in the resin (Oh *et al.*, 2006, St John *et al.*, 1993). The experimentally measured values are shown and the effect is an increase of up to 50% in gel time as the resin is purified.

The reasons behind the observed differences in rheological behaviour of these TGDDM resins could be determined by analysis of the kinetics of the system and the quantitative analysis of the hydroxyl content by MIR or other methods of chemical analysis. This is a clear example of the effect of chemical purity of the resin on rheological properties.

The above studies were performed by using transmission MIR. In a study of a similar resin system on a composite prepreg, it has been shown (Cole *et al.*, 1988) that *in situ* ATR using a Ge optical element was quantitatively more reliable than extracting the resin from the prepreg and analysing by transmission IR. The error arose due to variability in sample thickness of the cast film and thus uncertainty in the absorbance values which were used for calculation of the concentration of components. The applicability of DRIFT was very limited for the *in situ* sample due to the high specular component from the glossy carbon-fibre prepreg surface. As was noted in Section 3.3.4, the Kubelka–Monk function will generate reliable absorbance spectra only if all components are diffusely reflected. The KBr-overlayer method of Koenig (1999) would avoid this problem and should produce reliable DRIFT spectra.

The problem of reliably analysing the epoxy-resin component of a carbon-fibre epoxy prepreg has also been addressed by using FT-IR emission spectroscopy (FTIES)

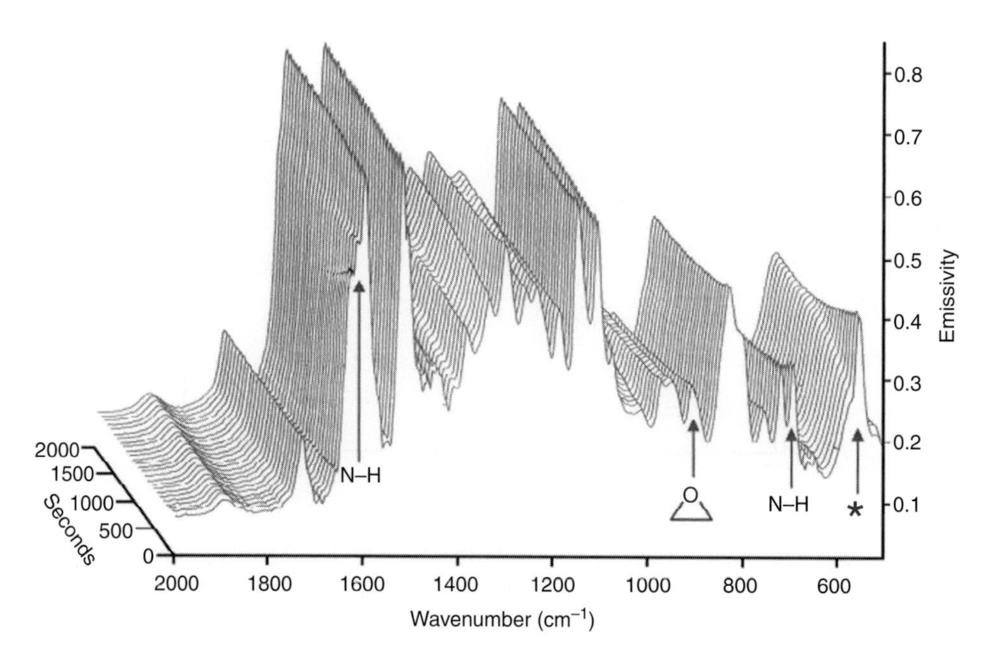

Figure 3.20. Successive FT-IR emission spectra from a thin film of resin from a carbon–epoxy-resin prepreg during cure. Bands that change during cure are marked (George *et al.*, 2006).

(George *et al.*, 1996) as described above in Section 3.3.3. Instead of removing the resin from the fibres by extraction, a small amount was transferred to the hotplate of the emission oven by contact, and the spectra were collected as the thin film of epoxy resin and curing agent was heated through its cure cycle. High-quality spectra were generated (Figure 3.20) and the amine and epoxy bands identified in MIR transmission (Figure 3.16 and 3.17) were identified and found to change quantitatively as cure proceeded.

One band for amine and one for epoxy suitable for such analysis are marked. Figure 3.21 shows an expansion of the region highlighted with a * in Figure 3.20, which corresponds to the phenyl-ring quadrant bend mode of DDS at 552 cm^{-1} (George *et al.*, 2006).

There is a small shift of this band to higher wavenumber in emission compared with MIR absorption (Table 3.3), but of most importance is that it is seen to decrease with cure, and this decrease is accompanied by an increase in intensity of the band that is initially a shoulder at 575 cm^{-1}. There is an isobestic point (marked) that is invariant in emissivity as the two spectral regions change in intensity. This is an indication that there is a smooth inter-conversion of the reactant to product and that Beer's law is obeyed throughout cure. This means that these bands may be used for quantitative analysis of the cure reaction. The FTIES experiment, carried out on a vanishingly small amount of resin transferred from the surface of the prepreg, was thus able to generate the cure kinetics of the resin and identify the components. Further, FTIES was found to be able to follow the ageing of the prepreg and so provide a rapid quality-assurance technique. This technique has also been applied to the thermal cure kinetics of photocatalysed dicyanate esters and directly compared with MIR transmission spectra (Liu and George, 2000). It has been shown to be reliable to high extents of conversion for quantitative analysis by applying Equation (3.24) to the determination of absorbance, A, calculated for each wavelength from the emissivity, $\varepsilon(\nu)$,

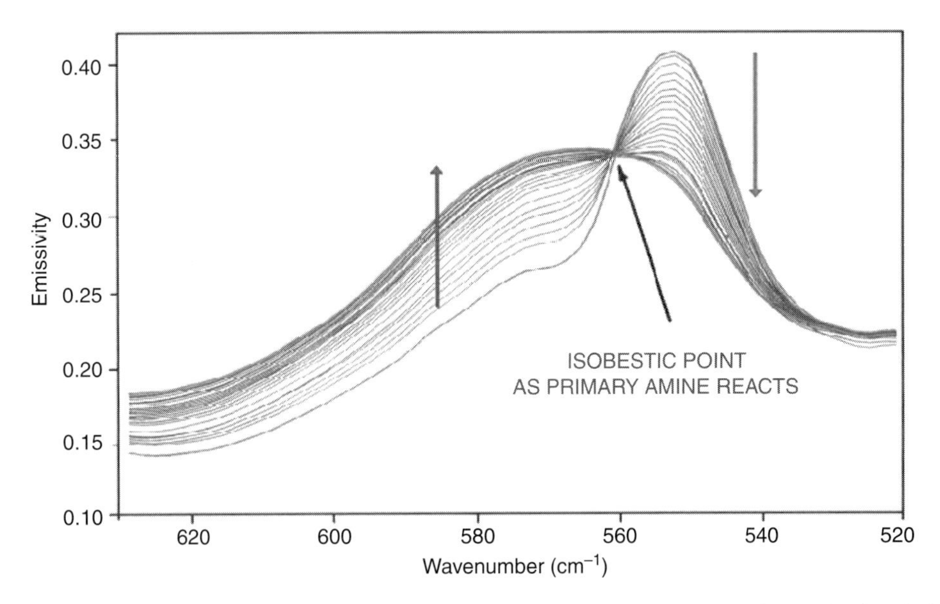

Figure 3.21. Detail from Figure 3.20 marked with * showing the decrease and growth of emission bands during cure of thin film of epoxy-resin prepreg.

determined from separate emission spectra using Equation (3.23). Emission spectra offer the prospect of being able to monitor the reactive processing of thin films of thermosetting resins on metallic substrates (Pell *et al.*, 1991).

Oxidation reactions

Among the earliest uses of MIR transmission and ATR spectroscopy has been the study of the oxidation of polymers. The principal oxidation products from polyolefins, polyamides and elastomers such as EPDM are carbonyl-containing species including aldehydes, ketones, esters, lactones, acids and peracids. The reaction scheme for their formation has been determined from studies of model compounds and polymers over the past 20 years, and a typical scheme is shown in Scheme 3.1 (Blakey, 2001).

All of these functional groups have a distinctive absorption band in the region, and a Fourier self-deconvolution routine is often used to attempt to resolve the often broad resulting absorbance band to give the best fit of the various species to the composite spectrum.

In determining this fit, model compounds as well as specific reactions of the separate components to produce adducts with distinctive spectra are used to identify the individual spectra of the components. The latter strategy is also used when, under the processing conditions, the components have short lifetimes and are reactive intermediates such as hydroperoxides.

An example of this is the range of derivatization reactions devised by several authors (Lacoste *et al.*, 1991, Carlsson *et al.*, 1987, Delor *et al.*, 1998). This has enabled the identification of p-, s- and t- hydroperoxides and also peracids in the oxidation of polypropylene. In the study of oxidation reactions during processing it has been shown that, in the absence of deliberate addition of oxygen or pro-oxidants such as transition-metal salts, chain scission is not accompanied by oxidation (Gonzalez-Gonzalez *et al.*, 1998). It has

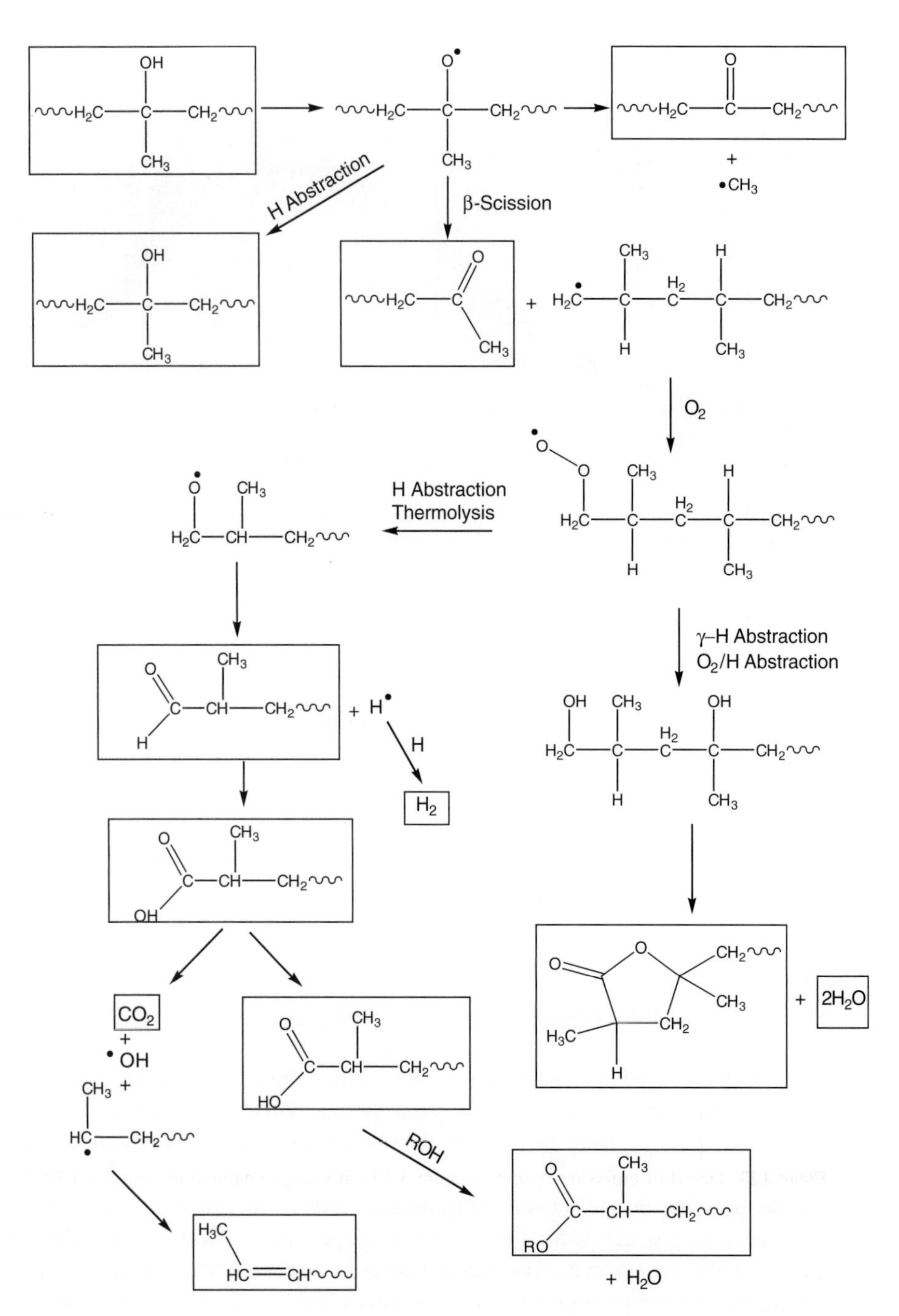

Scheme 3.1. Formation of oxidation products in the degradation of polypropylene through the tertiary hydroperoxide intermediate that may be identified by MIR spectroscopy (often combined with derivatization). After Blakey (2001).

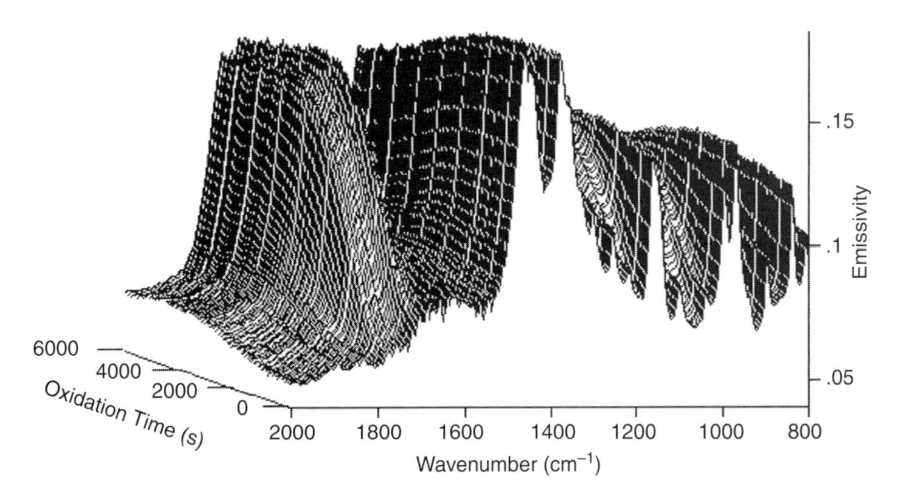

Figure 3.22. Time-resolved FT-IR emission spectra (every 13 s) showing the increase in emissivity in the region 1650–1800 cm^{-1} due to formation of oxidation products. Reproduced with permission from Blakey (2001).

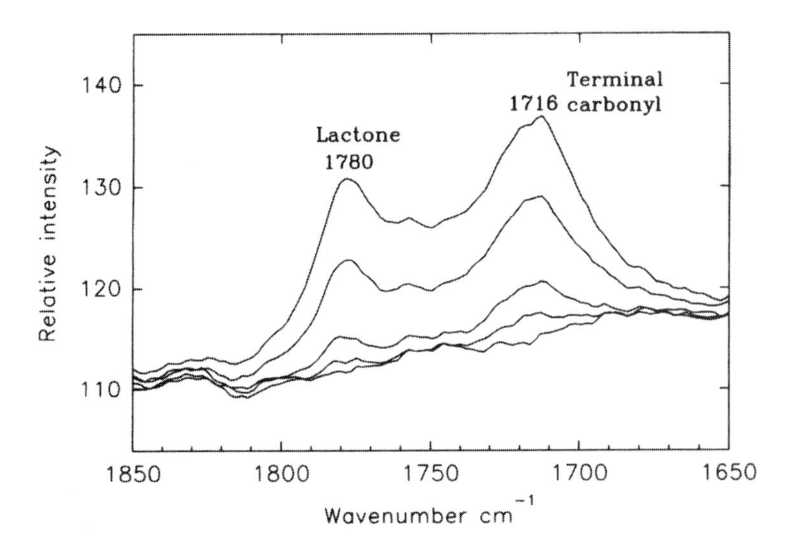

Figure 3.23. Detail of emission spectra (Figure 3.22) showing expanded the region 1650–1850 cm^{-1} that enables identification of oxidation products as marked (George *et al.*, 1995).

been usual to determine the products off-line, so real-time monitoring has proved difficult, but IR emission offers an alternative approach.

Laboratory studies of FT-IR emission (George *et al.*, 1995, Celina *et al.*, 1997) have demonstrated that it may provide real-time oxidation profiles. Figure 3.22 shows successive emission spectra from a single particle of polypropylene at 150 °C in the presence of oxygen.

These were collected every 13 s and the development of the oxidation bands with time is apparent, while the remaining bands are unaltered. Figure 3.23 shows a detail of five of these spectra over the carbonyl spectral region (1850 cm^{-1} to 1650 cm^{-1}) and the formation of the range of oxidation products as predicted by the scheme in Figure 3.20 may be seen, of which the lactone and terminal carbonyl are highlighted.

This should be adaptable to on-line analysis by using the transient IR emission methodology developed by (Jones and McClelland (1990, 1993) for other applications. In this a hot gas jet is used to bring about transient heating of a thin layer of the sample so that reabsorption effects from thick samples (Section 3.3.4) are avoided. Having a moving stream of sample means that only a thin layer of emitting polymer is seen by the spectrometer, so a continuous monitor of composition is obtained during reactive processing. This may be adapted to real-time monitoring using an appropriate MIR-transmitting optical fibre (George *et al.*, 2006) as discussed later.

Molar-mass changes by end-group analysis

One of the critical property changes during reactive processing is to the polymer's molar mass, and a fundamental goal of chemorheology is to understand how chemical change and molar mass relate to rheological properties. The measurement of molar mass usually requires the dissolution of the polymer and its analysis by a physical technique sensitive to either the numbers of molecules present in solution (e.g. osmometry) to measure the number-average molar mass (M_n) or the radius of gyration of the molecule (e.g. light scattering) to measure the weight-average molar mass (M_w). Determination of the full distribution of molar masses in a polydisperse sample requires a chromatographic technique such as size-exclusion chromatography. A method for the measurement of accurate values of M_n in condensation polymers such as polyesters and, in particular, polyamides has been end-group analysis of the polymer in solution. For polyamides, separate potentiometric titration of the carboxylic acid and amine end-group concentration (g/mol) gives the simple relation (Kamide *et al.*, 1993)

$$M_n = 2/([\mathrm{COOH}] + [\mathrm{NH_2}]). \tag{3.25}$$

While this works well for polymers of moderate molar mass ($<5\times10^4$ mol/g) and well-defined end-group composition, it does require the use of difficult-to-handle solvent mixtures. It would be simpler if the titration could be performed spectroscopically and in real time so that a continuous record could be obtained during the course of reaction without the need to dissolve the polymer. Nuclear magnetic resonance (NMR) provides one method, provided that the terminal protons may be resolved and integrated as discussed in Section 3.3.2. An infrared method has been illustrated for polyesters, for which a similar relation to Equation (3.25) is obtained, where hydroxyl end groups replace the amine end groups of the polyamide. The analysis of poly(butylene terephthalate) end groups in the MIR (Kosky *et al.*, 1985) to measure [COOH] was achieved at 3290 $\mathrm{cm^{-1}}$ and at 3535 $\mathrm{cm^{-1}}$ (Oh *et al.*, 2006). Because the end groups are isolated, model compounds may be used to determine the molar absorptivities necessary to convert absorbance to concentration, Equation (3.21). The analysis of hydroxyl groups is complicated spectroscopically because of the strong hydrogen bonding (shown above in Figure 3.18(a) for TGDDM-based epoxy resins) which affects spectral line width as well as molar absorptivity, so careful calibration against titration data is necessary to establish quantification.

There is a further complication in the analysis of polyesters (and several other polymers) and this is the presence of cyclic oligomers during the polymerization process, which are not removed from the commercial polymer. These oligomers will affect the rheological and mechanical properties but, since they have no end groups, cannot be detected by end-group

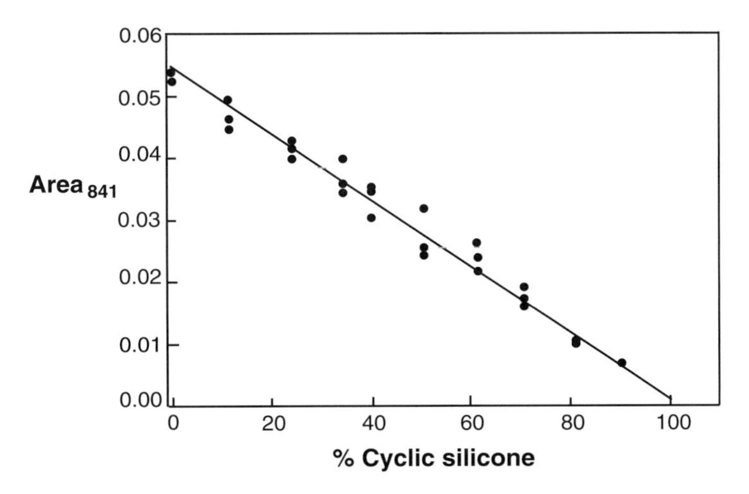

Figure 3.24. Relative area of the band at 841 cm^{-1} (due to trimethylsilyl end groups) in the MIR spectrum of a mixture containing linear and cyclic silicones.

analysis (either spectral or chemical). This effect is also seen with polydimethylsiloxanes (PDMSs), which are polymerized by the ring-opening reaction of a cyclic precursor so that a typical level of residual cyclic material is 10%. The end-group analysis of PDMS in the MIR region may be most readily performed at 841 cm^{-1}. Study of model compounds (Noll, 1968) enables assignment of the absorption of the dimethyl (in-chain) group [–$(Me)_2$Si–O–] to the strong bands between 790 and 830 cm^{-1}, compared with the trimethyl end groups [$(Me)_3$Si–O–] as two weak bands at 841 and 750 cm^{-1}. The effect of the presence of cyclic oligomers on the absorbance of the band at 841 cm^{-1} (and thus the apparent molar mass of the sample) may be seen in Figure 3.24, which shows the decrease in the band area (A_{841}) as a function of cyclic-silicone content. At low levels, the effect on calculated M_n is about 15%.

In spite of the limitations, in suitable polymers, MIR analysis offers a rapid technique for studying changes in molar mass that occur during processing. The greatest limitation is in the adaptation of the technique to on-line analysis and the requirement for thin samples due to the high molar absorptivities. The near-infrared region offers many experimental advantages, as discussed below, but there are severe restrictions on the functional groups which may be analysed at the levels encountered in chemorheological applications, particularly those involving controlled oxidation.

3.3.6 Near-infrared (NIR) analysis of polymer reactions

It was noted in Section 3.3.3 that the overtone and combination bands of many high-energy molecular vibrations (e.g. stretching modes) occur in the near-infrared (NIR) region of the spectrum from 4000 cm^{-1} to 14 000 cm^{-1}. This is a region that is spectrally accessible to either FT-IR or grating spectrophotometers and has the added attraction of being a region where glass and quartz are transmitting and atmospheric interferences are minimal. The intensity carried by NIR absorption bands depends on the anharmonicity of the functional group being measured, as given by the value of χ in Equation (3.19). Thus, while N–H, O–H and C–H modes may be seen in the NIR because of the large anharmonicity introduced by the light H atom, bands such as –C=O modes that are strong in the MIR are very difficult to observe in the NIR, since the anharmonicity is small. The molar absorptivities of the overtones and combination bands in N–H, C–H and O–H are lower by one to two orders of magnitude than those of the fundamental vibrations, so pathlengths of millimetres are typically used, again simplifying sample handling. Consequently, this spectral region has been the most heavily studied for process control, often involving the use of fibre-optics (as discussed later).

Addition and condensation polymerization

Most studies involving the use of NIR to monitor the polymerization process have been focussed on monitoring the reactant profiles as a function of time (Miller, 1991, Santos, 2005) but there are many applications in which NIR can detect changes in the polymer structure, moisture content or changes to reaction conditions that may affect the final product (Lachenal, 1995). For example, the polymerization of ethylene (Buback and Lendle, 1983) was followed by monitoring the decrease in intensity of the first overtone of the vinyl C–H stretch bands at 6140 and 5920 cm^{-1}, and the corresponding increase in intensity of the polyethylene bands at 5790 and 5680 cm^{-1}. The use of the NIR for quantitative analysis is illustrated by the determination of the absolute rate constants for propagation and termination in the free-radical polymerization of cyclohexyl acrylate (Yamada *et al.*, 2000). An obvious advantage of the NIR region is seen in enabling aqueous polymerization systems to be analysed, such as the emulsion polymerization of styrene (Wu *et al.*, 1996). One of the major limitations identified by them in NIR studies of an emulsion system was light scattering, which could be minimized by using second-derivative spectra. In an extension of this, it has been shown that NIR can detect the onset of monomer-droplet formation, which is important in avoiding thermal runaway or specification drift of the product (Vieira *et al.*, 2001). It has been shown that process variables in emulsion copolymerization (conversion, composition and particle diameter) could be determined with some success (although aggregation and temperature variations affected the particle measurements). Of some interest is the use of on-line NIR-probe fault detection for bubble formation and polymer-film deposition in the optical path, which would otherwise limit the reliability of any dip-probe technique (Reis *et al.*, 2004).

Bulk polymerization of methyl methacrylate within a mould was monitored by using a diode-array spectrometer and a sampling design that allowed the light beam to traverse the sampling thickness twice and so minimize optical artefacts (Aldridge *et al.*, 1993). In the solution polymerization of methyl methacrylate in a semi-continuous lab-scale reactor, an *in situ* NIR transmission probe was used to generate conversion data and assess the robustness of calibration when the solvent content was varied (Cherfi *et al.*, 2002). In polyurethane synthesis (Brimmer *et al.*, 1992) it was shown that, from an understanding of the peak assignments for reactants and products and calibration against isocyanate standards, it was possible to monitor the reaction profile for both isocyanate consumption and polyurethane formation by using second-derivative NIR spectra. This enabled process optimization since the completion of the reaction occurred after a much shorter time than had previously been determined by intermittent analysis. The optimum time for the addition of a polyol chain extender was also determined. The use of the NIR data to feed back into control of the process conditions has been demonstrated in the solution polymerization of styrene to control both monomer conversion and polymer molar mass by using a Kalman-filter state estimator and first-principles model for the control loop (Fontoura *et al.*, 2003).

In condensation polymerization, the use of NIR to follow the reactant concentrations at elevated temperatures (200–300 °C) has been applied to the synthesis of aromatic and aliphatic polyesters (Dallin, 1997). This provides an accurate alternative to the routine measurement of acid values, hydroxyl number and viscosity, with the added advantage of providing continuous data throughout the polyesterification, which allows optimum conversion. Related examples of the use of NIR include studies of esterification of low-molar-mass analogues (Blanco and Serrano, 2000) or transesterification of an existing polymer or copolymer (Sasic *et al.*, 2000). The NIR method allowed quantitative determination of rate constants, end points and yield and equilibrium constants as well as mechanistic information, which allowed the esterification to be optimized through the use of higher temperatures and an excess of acetic acid (Blanco and Serrano, 2000).

Network polymerization

The analysis of the cure reactions of networks has been facilitated by the use of NIR spectroscopy because of the rapid development of the crosslinked system which renders solution spectroscopy impossible. The advantage of use of the NIR region over the MIR studies, previously discussed in Section 3.3.3, is the ability to analyse thicker samples (typically of thickness 1–5 mm) and thus obtain results more representative of the bulk curing reactions. The quantitative features of NIR spectral analysis were recognized with one of the earliest analyses of the epoxide and hydroxyl-group content of a commercial DGEBA epoxy resin before and during cure (Dannenberg, 1963). The advent of high-performance composite materials, and the need for a deep understanding of the chemistry of cure and post-cure of networks and the correlation with mechanical and environmental performance, meant that there was a need for spectral techniques that would complement the DSC and DMA methods being used for cure analysis (Gupta *et al.*, 1983, Cole *et al.*, 1988). Most of these studies were in the MIR and the limitations of this spectral region for kinetic analysis were discussed earlier (Section 3.3.5). There have been several studies of DGEBA-based epoxy resins cured with a range of hardeners in which the NIR studies provide quantification and the MIR studies, which are less reliable quantitatively, are able to assist in mechanistic understanding (Poisson *et al.*, 1996, Strehmel and Scherzer, 1994). For example, in the case of DGEBA cured with dicyandiamide and accelerated by diphenyl dimethyl urea (Diuron) it was found that, while the heterogeneous system resulted in light scattering, this did not affect quantification in the NIR, but the MIR results showed poor correlation with NIR results and those from other methods such as SEC and chemical titration (Poisson *et al.*, 1996). The use of simultaneous analytical methods together with NIR offers the advantage that artefacts can be determined and a deeper understanding of subtle changes in analytical results can be gained, for example by NIR–dielectric methods (Kortaberria, 2003).

The use of NIR for monitoring the reactive processing of high-temperature-curing, high-performance resins such as TGDDM (e.g. the Ciba resin MY721) cured with DDS was first reported in an application that used fibre-optics to couple the sample to the spectrometer (George *et al.*, 1990, 1991). In Figure 3.25 is shown a NIR spectrum for a sample of TGDDM (MY721) containing 27% DDS.

The assignment of the bands in the spectrum to combination and overtone bands of the fundamental vibrations assigned earlier for both TGDDM and DDS (Tables 3.2 and 3.3) is shown in Table 3.4 for both the resin and the amine.

Table 3.4. Near-IR band assignments for TGDDM + DDS

Peak position (cm^{-1})	Band assignment (and fundamental bands)
4519	Epoxy combination (C–H stretch + ring breathing)
4557	Primary-amine combination (N–H symmetric stretch + C–N stretch)
4804	Primary-amine combination (N–H symmetric stretch + deformation)
5072	Primary-amine combination (N–H asymmetric stretch + deformation)
5645, 5774	Methylene first overtones (C–H stretch)
5887, 5987	Aromatic first overtones (C–H stretch)
5873, 6057	Terminal-epoxy first overtones, (C–H stretch)
6575	Primary-amine first overtone (N–H symmetric stretch)
6684	Primary-amine first overtone (N–H asymmetric stretch)
7000	Hydroxyl first overtone (O–H stretch)
8627, 8837	Terminal-epoxy second overtones (C–H stretch)
9801	Primary-amine second overtone (N–H asymmetric stretch)
10191	Hydroxyl second overtone (O–H stretch)

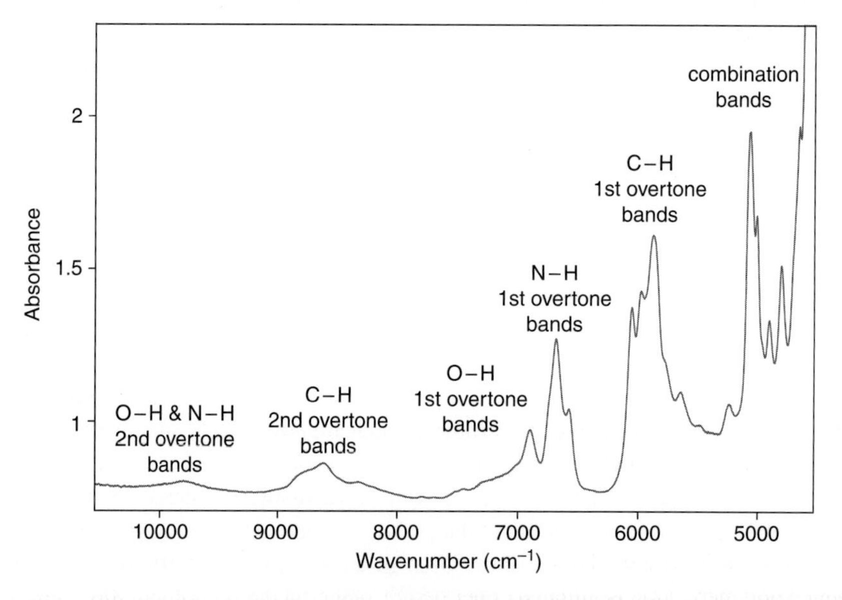

Figure 3.25. Near-IR spectra and region assignments for the epoxy resin MY721 with 27% DDS hardener before cure reaction at 160 °C. Reproduced with permission from St John (1993).

It should be noted immediately that there are many fewer bands in the NIR; also they are broader and overlapping of bands is common. In some instances, Fourier self-deconvolution (Section 3.3.4) is required in order to assign bands, but the original bands should be used when possible for quantitative analysis since Beer's law is generally obeyed and quantification is straightforward. When the sample in Figure 3.25 was heated at 160 °C for three hours for the cure reaction to occur, the spectral changes shown in Figure 3.26 were observed.

The bands that have become negative correspond to consumption of the amine and epoxide, and it is the detailed analysis of these bands which permits a quantitative analysis of the cure kinetics, which may then be applied to real-time monitoring of reactive processing.

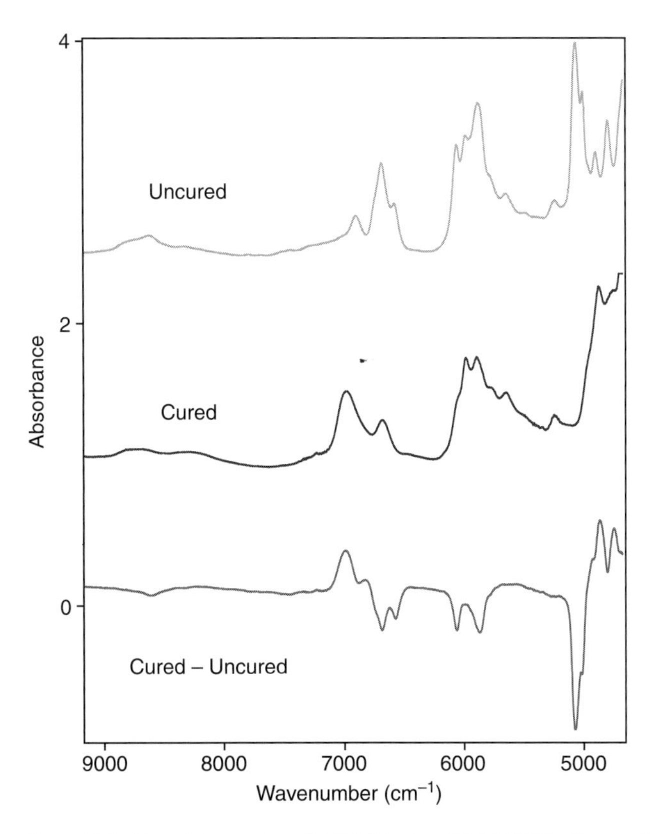

Figure 3.26. Near-IR spectra of MY721 plus 27% DDS before and after cure at 160 °C for 3 hours together with the difference spectrum to highlight bands that change during cure. Reproduced with permission from St John (1993).

There are many commercially significant networks that are based on free-radical polymerization of multifunctional monomers. Obvious examples include crosslinked surface coatings (Paul, 1989), styrene-crosslinked unsaturated polyesters (Melby and Castro, 1989) and crosslinked acrylates and related systems for optical transparencies (Carswell *et al.*, 1996), all of which require process control as the network is formed in order to achieve optimum performance. These are often copolymers with monofunctional monomers in which polymerization is initiated either by UV radiation and a photoinitiator or by a thermal free-radical initiator such as a peroxide, so the polymerization is often rapid and results in a heterogeneous network (Dusek, 1996). The kinetics for network formation are those for free-radical polymerization, so, in contrast to the stepwise polymerization of epoxy resins, during which the molar-mass increases slowly, there is the immediate formation of high-molar mass polymer, which is initially soluble in the monomer, but, because of the multifunctionality, will soon crosslink and become gel particles. This gel may form at extents of conversion as low as 10%–15%, so subsequent reaction will occur in the gel particles, but macroscopic gelation may be delayed due to the formation of intramolecular crosslinks and cyclization (Dusek, 1996). Near-infrared spectroscopy may be applied in these systems in the way that it is applied to other free-radical polymerizations (Yamada *et al.*, 2000), and it has been shown (Rey *et al.*, 2000) that, by establishing a molar absorption coefficient for each of the reacting species, both methacrylate and vinyl conversions could be obtained after spectral deconvolution. The effects of inhibition by oxygen and onset of diffusion control were demonstrated and are of vital importance in monitoring of reactive processing.

3.3.7 Raman-spectral analysis of polymer reactions

Raman spectroscopy is attractive for monitoring polymer reactions in real time because it is conceptually simple, requiring only a single wavelength of exciting light and a monochromator and detector, and hence is potentially non-invasive. However, as noted in Section 3.3.3, the low intensity of the scattered Raman radiation (the scattering efficiency being typically 10^{-10}) and the tendency of many polymers and reactants to fluoresce limit the sensitivity of the technique. Quantification also needs special consideration since the analytical volume may often be imperfectly defined.

The intensity, I_ν, of Raman scattering is given (Jawhari *et al.*, 1990) by

$$I_\nu = \alpha BVF_{\text{instr}}, \tag{3.26}$$

where α represents the sample-based parameters to be determined (e.g. concentration of species analysed), B is the brightness of illumination, V is the volume of scattering material viewed by the spectrometer and F_{instr} takes account of all the instrumental factors which need to be kept constant for quantification (e.g. detector sensitivity, monochromator efficiency, refractive index of the sample and excitation wavelength). The instrument-dependent terms may be eliminated by adding an internal standard, but in many process applications this is not practicable and a band known to be invariant throughout the analysis is required. The challenges in quantitative Raman spectroscopy have been reviewed (Pelletier, 2003), and complex samples may be accessed through the use of spectral pre-processing, noise analysis and multivariate methods (as discussed later).

The minimization of fluorescence from Raman spectra has been in part achieved by moving the exciting radiation from the visible to the NIR region of the spectrum. This opens up the possibility of using an FT-IR spectrometer as the spectral device and gaining the FT advantages of wavelength precision and excellent SNR.

Addition and network polymerization

One of the early applications of FT-Raman was for the curing reaction of an epoxy resin at room temperature over several days (Agbenyega *et al.*, 1990), although there was no quantitative analysis of the spectra to determine the cure kinetics. Earlier studies had shown (O'Donnell and O'Sullivan, 1981) that with even conventional monochromator-based Raman spectrometers the changes in concentration of monomer (diethylene glycol *bis*(allyl carbonate)) at temperatures from 70 to 85 °C could be detected from the strong (–C=C–) stretching vibration at 1643 cm^{-1} as network formation proceeded. It was noted that Raman was superior to MIR analysis of this band in terms of sensitivity. The use of the same band has been reported for microemulsion polymerization (Feng and Ng, 1990) and, while there are some concerns regarding quantification, it opens up the possibility of using Raman for reactive-process monitoring in aqueous systems. The emulsion polymerization of vinyl acetate has been monitored by FT-Raman (Ozpozan *et al.*, 1997), but it was necessary to increase the concentration above normal production levels in order to achieve sufficient Raman scattering. The issue of SNR is raised as an area of concern in most applications of Raman spectroscopy in on-line processing. However, as discussed later, in-line FT-Raman

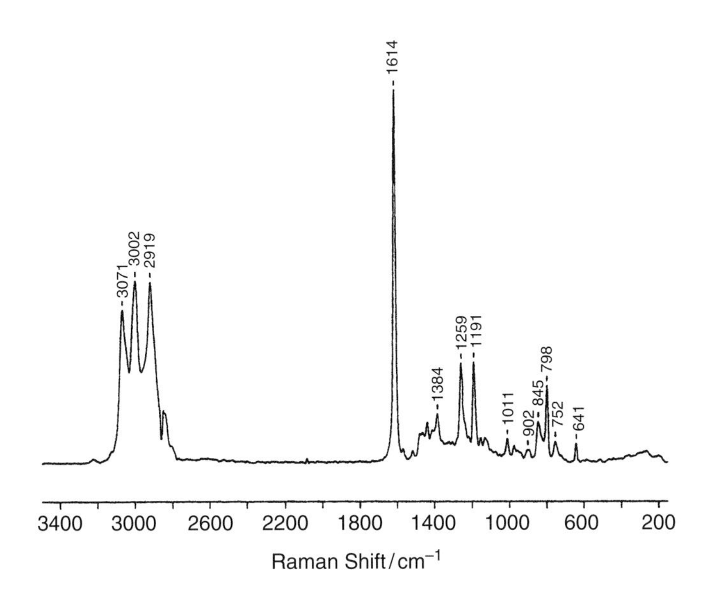

Figure 3.27. The FT-Raman spectrum of the TGDDM epoxy resin MY721. The prominent bands may be contrasted with those from the MIR (Figure 3.16) reflecting the different selection rules for the two vibrational spectroscopic techniques (de Bakker *et al.*, 1993a).

monitoring of emulsion polymerization (Charmot *et al.*, 1999, Bauer *et al.*, 2000) has successfully been demonstrated using a ruggedized version of the laboratory spectrometer.

In a study of ATRP of halogen-free amino-functionalized poly(methyl methacrylate), the solvent (anisole) was used as the internal standard that enabled the on-line consumption of monomer as given by the decrease in the Raman band at 1641 cm^{-1} to be followed with a time resolution of 2 min. Close agreement was found with off-line analysis by ^{1}H NMR, but it was noted that the immersion probe detects a thin layer (thickness ~500 μm) on the surface, which may adhere at high conversion when the viscosity is high, and then material exchange with the polymerizing bulk might not occur (Fischer *et al.*, 2006). For these reasons, non-contact analysis is preferred, but might not always be practical.

In the area of monitoring of the cure of high-performance epoxy resins for composite materials the relative merits of FT-Raman have been compared with those of FT-IR in the NIR region (as discussed in Section 3.6.2) (de Bakker *et al.*, 1993a).

Figure 3.27 shows the FT-Raman spectrum of TGDDM (MY721), and the assignment of the prominent bands is made in Table 3.5.

The Raman-spectral data from the aromatic amine curing agent DDS are also shown in Figure 3.28, and assignments are given in Table 3.6.

It is of interest to compare these with the MIR spectra discussed earlier (Figures 3.16 and 3.17 and Tables 3.2 and 3.3) and note the differences in the spectra and, in particular, band intensities arising from the different selection rules. Of particular interest are the differences in the analytical band used in the MIR for epoxide analysis (904 ± 2 cm^{-1}) due to an asymmetric ring vibration, which is very weak in the Raman, where the preferred analytical band is the symmetrical ring vibration (at 845 cm^{-1}) that is not active in the MIR.

For complete kinetic analysis and to give a cross-check when using spectral methods for monitoring reactive processing, it is important to monitor several bands due to groups that participate in the chemical reactions for network formation and at least one band that does

Table 3.5. Band assignments for Raman spectrum of TGDDM (MY721)

Peak position (cm^{-1})	Vibrational mode assignment
3071	CH_2 epoxy asymmetric stretch and C–H phenyl stretch
3002	CH_2 epoxy symmetric stretch
2919	CH_2 alkyl asymmetric stretch
1614	Phenyl quadrant stretch
1440, 1385	CH_2 deformation
1260	Epoxy-ring breathing
1191, 1010	*Para*-substituted phenyl ring
902	Epoxy, asymmetric ring vibration
845	Epoxy, symmetric ring vibration
798	*Para*-substituted phenyl ring: adjacent C–H wag
642	*Para*-substituted phenyl ring

Table 3.6. Band assignments for Raman spectrum of DDS

Peak position (cm^{-1})	Vibrational mode assignment
3367	NH_2 asymmetric stretch
3330	NH_2 symmetric stretch
3065	Phenyl C–H stretch
1629	NH_2 symmetric deformation
1597	Phenyl quadrant stretch
1500	*Para*-substituted phenyl-ring semicircle stretch
1280	Sulfone asymmetric stretch
1140	Sulfone symmetric stretch
1107	C–S–C asymmetric stretch
1073	C–S–C symmetric stretch; sulfone rock
822	*Para*-substituted phenyl ring: adjacent C–H wag
634	*Para*-substituted phenyl ring
482	Sulfone deformation
289	C–S–C deformation

not change (an internal standard). For the TGDDM/DDS system, the primary amine is a sensitive index of the early reaction up to gelation, so the vibrations of $-NH_2$ have been monitored both in MIR (Morgan and Mones, 1987) and in NIR (St John and George, 1992) analyses of cure. Unfortunately, in the Raman, all of the bands that are attributable to this group are weak, so this immediately limits the power of the Raman analysis compared with the NIR, where the $-NH_2$ combination band at 5072 cm^{-1} (Table 3.4) has been shown to give an excellent SNR and is ideal for monitoring the epoxide–primary-amine reaction (St. John and George, 1992). A comparison of the potential application of the two methods to monitoring the cure of TGDDM/DDS may be seen in Figure 3.29.

In this plot the Raman data are referenced to the peak at 2919 cm^{-1} which is due to a stretching mode of the methylene group in TGDDM and therefore should not vary during cure because this group does not participate in the reaction. It may be seen that the poorer SNR for the Raman spectra results in a high uncertainty in the degree of conversion

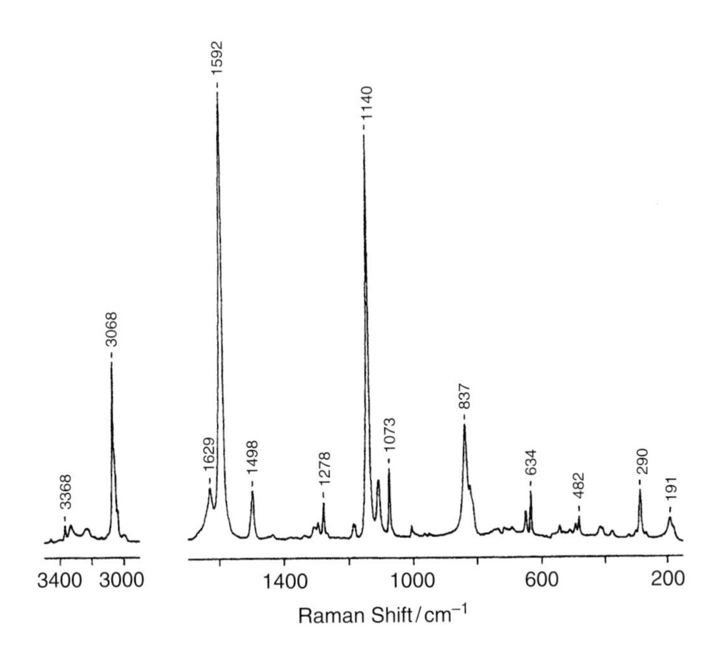

Figure 3.28. The FT-Raman spectrum of the aromatic hardener DDS (de Bakker *et al.*, 1993a).

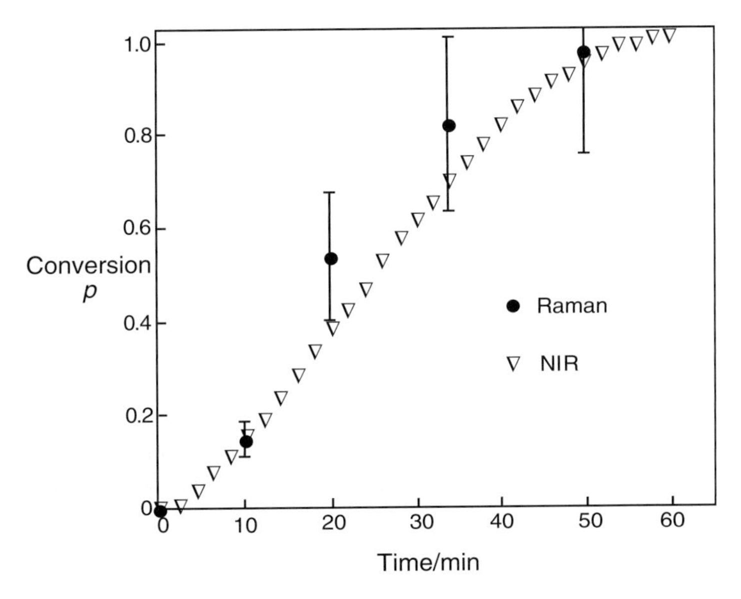

Figure 3.29. Comparison of the degree of conversion, p, as a function of cure time for TGDDM cured with DDS as measured by NIR and Raman methods. Adapted from de Bakker *et al.* (1993a).

of $-NH_2$ at any time, which is emphasized by the long time between data points due to the large number of interferograms which need to be co-added to generate the spectrum. This is contrasted with the large number of data points that establish the cure profile from the NIR curve. It was concluded that NIR was superior to Raman spectroscopy for real-time monitoring of the resin cure, but an interesting insight into a competing side reaction involving cyclization of a chlorohydrin impurity in the commercial resin was obtained from an analysis of the Raman spectrum (de Bakker *et al.*, 1993a). This could not be detected by other spectral techniques. The strong fluorescence of these resin systems prevents the use of grating systems, and FT-Raman systems will not have the superior SNR and time resolution seen in charge-coupled-device (CCD)-based detector systems with which spectra may be collected every 50 s (Fischer *et al.*, 2006).

Degradation reactions

There have been several studies of the degradation of polymers, which may be adapted to the study of reactions leading to changes in properties (including viscosity during processing) and thus are relevant to chemorheology. One of the most obvious changes which may be monitored by FT-Raman is the dehydrochlorination of poly(vinyl chloride) (PVC). The mechanism of dehydrochlorination and stabilization of PVC has been studied for some time (Braun, 1981, Starnes, 1981), and the main reason for the rapid production of hydrogen chloride is known to be the facile unzipping reaction which is auto-catalysed. This results in the formation of extended sequences of polyenes (as shown below), which may be readily detected by Raman spectroscopy:

$$-(\mathrm{CH_2CHCl})_n- \rightarrow -(\mathrm{CH{=}CH})_m-(\mathrm{CH_2CHCl})_{n-m}- + m\mathrm{HCl}. \tag{3.27}$$

dramatic changes in the FT-Raman spectrum occur on heating PVC and its copolymer with poly(*N*-vinyl-2-pyrrolidone) (PVP) (Dong *et al.*, 1997) due to polyene formation. Figure 3.30 shows intense bands around 1500, 1100, 610–690 and 1430 $\mathrm{cm^{-1}}$ corresponding to $(-C{=}C-)_n$, $(=C-C=)_n$ and C–Cl stretching vibrations and CH_2 deformation bands, respectively. The extent of dehydrochlorination and polyene formation of 90% PVC/10% PVP blends with respect to time at 120 °C can be seen from 13 consecutive spectra presented in stacked form. Each spectrum was recorded at 5-minute intervals over 50 scans, while keeping the temperature constant at 120 °C (Kaynak *et al.*, 2001). The PVP is able to undergo hydrogen bonding to PVC, and increasing the amount of PVP in the PVC/PVP blends decreased the dehydrochlorination temperature and time and resulted in an earlier decrease in the mean polyene length. The sensitivity of Raman Spectroscopy for detecting polyenes means that as little as 10^{-4}% dehydrochlorination of PVC and its copolymers can be detected (Hillemans *et al.*, 1993).

While successful in monitoring the formation of highly conjugated products on degradation, Raman spectroscopy is much less sensitive than IR for the detection of oxidation products such as carbonyls, so that any measurements made to monitor changes during oxidation involve secondary effects such as the loss of aliphatic C–H compared with aromatic C–H during the oxidation of an acrylonitrile–EPDM–styrene terpolymer as a way of locating the site of degradation (Mailhot and Gardette, 1996) rather than direct measurement of oxidation bands.

3.3.8 UV–visible spectroscopy and fluorescence analysis of polymer reactions

The use of electronic spectroscopy to follow polymer reactions has been limited because of the broad spectra resulting from electronic transitions and the lack of molecular structural information in the spectra. The only exception to this is in polymer photochemistry, where information regarding the excited states and the pathways for energy loss may be exploited

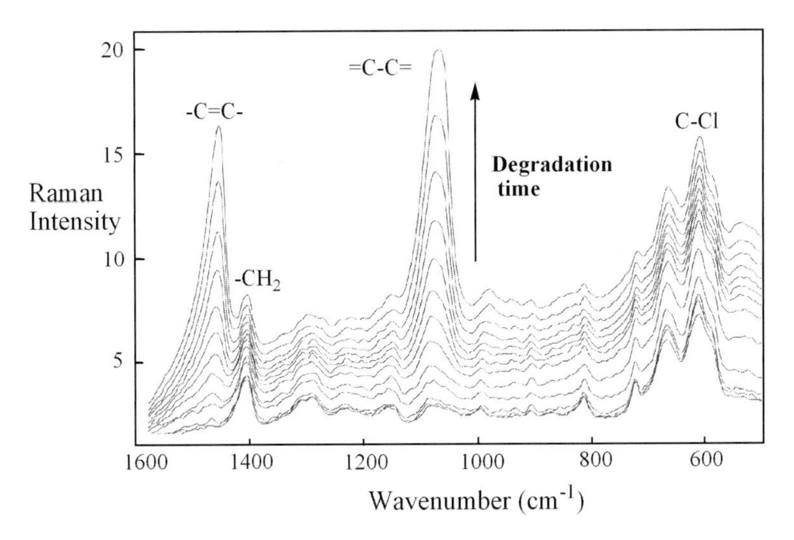

Figure 3.30. Successive Raman spectra collected during the thermal degradation of a PVC : PVP blend (90 : 10) at 120 °C, showing the rapid increase in the bands at 1100 and 1500 cm^{-1} due to polyene formation. Adapted from Kaynak *et al.* (2001).

to determine the reaction pathway and develop strategies for controlling the reaction rate. Thus, for the chemorheology of photoinitiated reactions, detailed knowledge of the excited states and energy-transfer routes to enable sensitization of the reaction and/or the possible quenching of excited states is invaluable (Guillet, 1987). By exploiting this fundamental knowledge and identifying particular spectral parameters that are sensitive to viscosity changes, it may be possible to achieve non-invasive monitoring of the viscosity and phase changes which accompany network formation.

Figure 3.31 and Table 3.7 describe the processes of emission and energy relaxation which may take place following absorption of radiation of intensity I_a einsteins dm^{-3} s^{-1} expressed in terms of the rates of the process (George, 1989b). These relaxation processes may involve either the singlet excited states, leading to fluorescence from the first excited singlet state, S_1, or radiationless processes of internal conversion (shown as wavy lines); or the triplet states which may be populated by inter-system crossing and then decay by emission of phosphorescence from the first excited triplet state, T_1, or radiationless inter-system crossing to the ground (singlet) state, S_0.

In an isolated molecule these processes will be unimolecular (as shown by the rate expressions in Table 3.7), but in solution or the molten state there will be processes of energy transfer that may take the form of quenching of either the singlet or the triplet state, leading to a decrease in the intensity of emitted fluorescence or phosphorescence. These are bimolecular processes as shown in Table 3.7. Because direct population of the triplet state by absorption from the ground singlet state ($T_1 \leftarrow S_0$) is formally forbidden, the lifetime of this state is long (from milliseconds to several seconds in the absence of bimolecular processes). This long lifetime is the reason why the triplet state is photochemically active and also why phosphorescence may be detected only in the absence of quenching molecules such as ground-state oxygen. In contrast, the transition to the first singlet state ($S_1 \leftarrow S_0$) is allowed, so the lifetime of the first singlet state is very short (nanoseconds in the absence of

Table 3.7. Photophysical processes and their rates after absorption of light, intensity I_0, in terms of the rate coefficients and populations of excited states shown in Figure 3.31

	Transition or reaction	Rate
Unimolecular process		
Absorption	$S_1^v \leftarrow S_0$	$I_a = I_0(1-10^{-\varepsilon cd})$
Internal conversion	$S_1^v \rightarrow S_1$	$k_{IC}[S_1^v]$
Internal conversion	$S_1 \rightarrow S_0$	$k_{IC}[S_1]$
Inter-system crossing	$S_1 \rightarrow T_1^v$	$k_{ISC}[S_1]$
Fluorescence	$S_1 \rightarrow S_0\ (+\ h\nu)$	$k_F[S_1]$
Phosphorescence	$T_1 \rightarrow S_0\ (+\ h\nu')$	$k_P[T_1]$
Inter-system crossing	$T_1 \rightarrow S_0$	$k'_{ISC}[T_1]$
Bimolecular process		
Singlet quenching	$S_1 + Q \rightarrow S_0 + Q^*$	$k_Q[S_1][Q]$
Triplet quenching	$T_1 + Q \rightarrow T_0 + Q^*$	$k'_Q[T_1][Q]$
Excimer formation	$S_1 + S_0 \rightarrow (S_1S_0)^*$	$k_{DM}[S_1][S_0]$
Excimer emission	$(S_1S_0)^* \rightarrow 2S_0\ (+\ h\nu'')$	$k_{FD}[(S_1S_0)^*]$

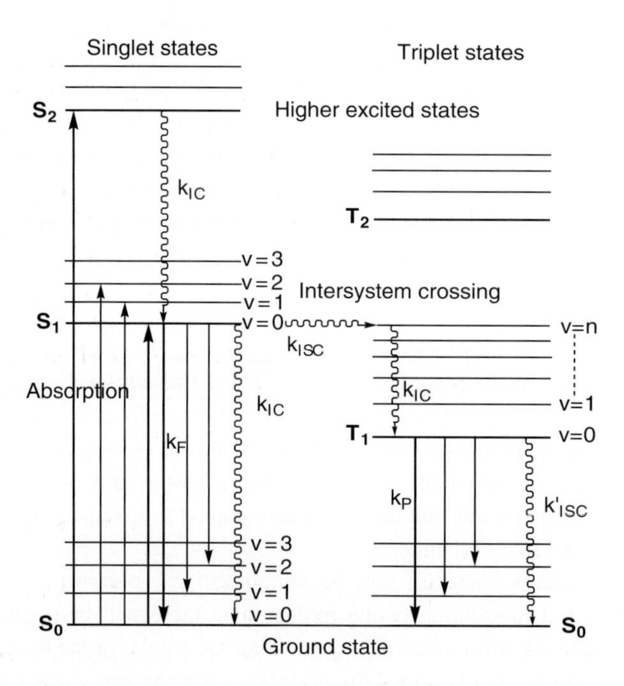

Figure 3.31. An energy-level diagram for an aromatic molecule or similar chromophore that may absorb radiation in the visible or UV region of the electromagnetic spectrum (George, 1989).

quenching) and the more intense the absorption, the shorter the lifetime. However, bimolecular processes do occur and an interesting example is excimer emission through the formation of an excited state complex $(S_1S_0)^*$ between an excited molecule (S_1) and a ground state molecule (S_0). These bimolecular processes are relevant in this particular application since the rate constants for quenching (k_Q) and excimer formation (k_{DM}) as shown in Table 3.7 will be dependent on the encounter probability of molecules in solution and therefore related to the viscosity.

In the situation described as Stern–Volmer quenching, the fluorescence quantum yield Φ_F^o is reduced to Φ_F in the presence of Q, a quenching molecule of concentration [Q],

$$\Phi_F^0/\Phi_F = 1 + k_Q\tau_F[Q], \tag{3.28}$$

where τ_F is the lifetime of the first singlet excited state in the absence of any quenching, i.e. $\tau_F = 1/(k_F + k_{IC} + k_{ISC})$. The rate coefficients for these competing processes are shown in Table 3.7.

In solution the quenching rate coefficient will be the diffusion-controlled rate coefficient for encounter, k_d, which is related to the viscosity η by the Debye equation:

$$k_Q = k_d = 8RT/(3\eta). \tag{3.29}$$

Thus there is the potential for the quenching of excited states to be used as a monitor of the viscosity of the system containing a fluorescent molecule with an excited-state lifetime appropriate for a measurable change in fluorescence intensity (i.e. from Φ_F^o to Φ_F) to occur over the range of viscosity of interest. The quenching molecule also must meet the energetic requirements of having an energy level lower than the singlet excited state S_1 of the chromophore and also having a rapid loss of the excitation energy through radiationless decay (k_{IC} is large) so that emission is not observed from the quencher.

Excited-state quenching is one example of excitation energy transfer from a donor (excited) molecule, D*, to a previously unexcited acceptor molecule, A, having a lower-lying singlet energy level at which the acceptor does not itself fluoresce. In many cases, A* may emit radiation and then there will be an emission spectrum, which, depending on the concentration of A, will feature the spectra of both D and A and, in the case of total energy transfer, emission from A alone. In certain circumstances the rate of dipole–dipole energy transfer can be so efficient that it may occur at greater than encounter distances (e.g. 30–100 Å) such that the rate of energy transfer exceeds the diffusion controlled rate and becomes independent of viscosity. In this regime, the fluorescence is unsuitable as a monitor for reactive processing. However, there has been a particular example of these systems in which the D and A are electron-donor and -acceptor species and are located on the one molecule to form intramolecular charge-transfer fluorescence probes (Song and Neckers, 1996). These exhibit spectral shifts during polymerization or network formation that are due to the effect of microviscosity on the accessibility of the twisted charge-transfer state.

Excimer emission is a broad structureless emission red-shifted from the normal fluorescence and is observed at high chromophore concentration because it requires close encounter of the excited and the unexcited molecule to form the encounter complex $(S_1S_0)^*$ with a separation of ~3 Å. This will depend on the encounter occurring during the lifetime of the excited state so that, as the viscosity increases, there will be a decrease in the intensity of excimer emission. This is discussed and an example shown in Section 3.3.8.

Monitoring polymerization by UV–visible absorption spectroscopy

This approach requires that the polymer or an additive contains a chromophore that changes the wavelength of its absorption maximum as the reactive processing procceds. In one well-studied example of the addition of a low concentration of a probe molecule, p,p'-diamino azobenzene (DAA – Scheme 3.2), to an epoxy resin, the probe actually participates in the cure chemistry without affecting the overall course of network formation (Sung *et al.*, 1986, Yu and Sung, 1990). The reaction of the probe mimics that of the curing agent and, through studies of a series of five model compounds corresponding to progressive reaction of the amine with the epoxide, to give secondary or tertiary adducts that are either monofunctional or difunctional, the corresponding red shifts in the $\pi^* \leftarrow \pi$ absorption spectra were recorded. These were then deconvoluted as cure proceeded to yield the relative contributions to the different reactions of the amine. This molecule is thus not a viscosity probe but a reactivity probe, and its use requires that the reactivity of the probe with the resin has the same kinetic parameters as for the hardener. The reactivity of aromatic amines varies with the base strength of the amine, which is influenced by the bridging group, e.g. DDS (with a sulfone bridging group) is a weaker base than DDM (which has a methylene bridging group) and requires a higher reaction temperature. Indeed, the authors note that DAA reacts faster than DDS and hence corrections need to be made (Sung *et al.*, 1986). This is a severe restriction to the general use of this approach in real-time monitoring of the reaction, but it can provide

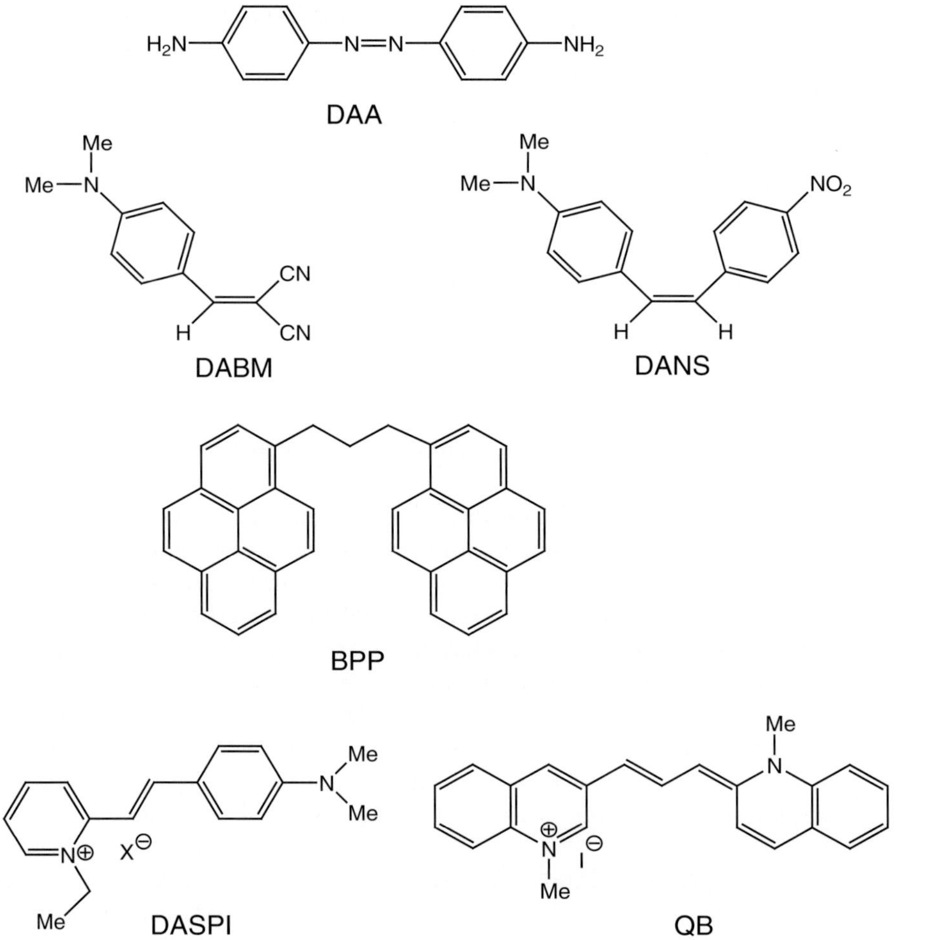

Scheme 3.2. Molecular probes used for UV absorption and fluorescence monitoring of the reactions for network formation.

interesting fundamental information that can complement that obtained from the other approaches such as NIR spectroscopy, which was described above in Section 3.3.6. However, the advantage of NIR and MIR spectroscopies is that they can follow the actual reactive species, so no assumptions need to be made regarding the relative reactivity of the probe and the resin components. It is also an advantage if the spectral properties of the probe are sensitive to the viscosity of the medium rather than the reactivity.

It has been suggested that the DDS itself will show a progressive red shift of up to 24 nm in the $\pi^* \leftarrow \pi$ absorption spectra with cure, and this is most readily seen from the fluorescence excitation spectra (Paik and Sung, 1994), which are equivalent to absorption spectra. It was shown that three techniques, namely UV reflection, UV transmission and fluorescence excitation, gave similar spectral shifts as the cure reaction took place at 160 °C (Yu and Sung, 1995). Excitation spectra are generally more sensitive at low concentrations of analyte but may saturate and become unrepresentative of the absorption properties at high concentrations. This was illustrated with a DGEBA-based epoxy resin and, while there was a linear relation between the spectral shift and the extent of reaction as determined by IR, it was noted that there was a temperature dependence of the shift for the DDS. This was accommodated by using a rhodamine dye as an internal standard, which also allowed for any variations in lamp intensity, alignment etc. Thus even this method needed to have a probe added, which does result in some limitations in a processing environment.

Monitoring polymerization and network formation by time-resolved and steady-state fluorescence

Fluorescence probes are widely used in chemistry, biology, physics and materials science to detect specific functional groups present in a system and to monitor relaxations occurring on the time scale of the lifetime of the excited state. It has been noted (Bajorek *et al.*, 2002) that most probes used in monitoring polymer relaxations and thus applicable in reactive processing fall into the following categories:

- excimer-forming probes
- intramolecular charge-transfer probes
- twisted intramolecular charge-transfer or TICT probes
- Fluorescent salts

In many cases the differences among the mechanisms of action of the last three categories are subtle and are related to whether there is a shift in the fluorescence spectrum with changing polarity of the medium and the actual nature of the excited-state relaxation that is environment-sensitive. Since there are now many hundreds of organic molecules available as fluorescence probes (Song and Neckers, 1996, Vatanparast, 2001, Bosch *et al.*, 2001, Bosch, 2005, Loutfy, 1981, 1982, 1986, Wróblewski *et al.*, 1999, Rettig, 1986, Quirin and Torkelson, 2003), only a few representative examples are considered. In choosing a probe issues to be considered include the likely reactivity of the probe under the conditions of reactive processing and the ease of extracting the data of interest (Mikes *et al.*, 2002a, 2002b).

The p,p'-diamino azobenzene (DAA) reactivity probe, described in the previous section, has been demonstrated to form reaction products with DGEBA-based epoxy resins that have a greater fluorescence intensity than does DAA itself. By examining model adducts with phenyl glycidyl ether, the intensity enhancement varied from 1400 times to 9 times depending on whether the DAA reacted at a crosslink or a chain end, respectively

(Sung *et al.*, 1985). Part of this enhancement is due to a greater molar absorptivity, but there is also a contribution from an enhanced fluorescence quantum yield. When incorporated into a curing epoxy-resin system at a level of 0.1–0.3 wt.%, the increase in intensity with time is sigmoidal, with the point of inflection identified with the gelation of the resin. The levelling off in intensity was identified with vitrification of the resin, but it was recognized that this was due not to the rapid increase in viscosity, but rather to a loss of reactivity consistent with DAA being a *reactivity probe*, not a viscosity monitor.

The possibility of using a probe molecule possessing intrinsic fluorescence properties that change with the viscosity of the medium has been considered for some time (Tredwell and Osborne, 1980). This change in the intrinsic *unimolecular* rate processes for fluorescence is to be contrasted with the bimolecular quenching of fluorescence which may follow Stern–Volmer kinetics, as discussed above in Section 3.3.8. By using a solvent series that increased in viscosity, it was shown that the fluorescence lifetime (τ_F) of the rhodamine dye *N*-(2-tolyl)-*N*′-(2-tolyl-5-sulfonate) rhodamine, Fast Acid Violet 2R (or FAV) increased from 10 to 160 ps as a result of viscosity-dependent ultrafast internal conversions (k_{IC}, Table 3.7). The relation followed was

$$\tau_F = C\eta^{2/3}. \tag{3.30}$$

However, it was the microviscosity, rather than the bulk viscosity which was probed by using picosecond time-resolved fluorescence methods, since the value of C was constant within a homologous series of solvents but was different for monofunctional and difunctional alcohols. The probe molecule has a natural (radiative) lifetime, τ_F^o of ~4.6 ns, so the reduction in lifetime results from the competition between fluorescence ($k_F \sim 2\times10^8\ s^{-1}$) and the high rate of internal conversion ($k_{IC} \sim 10^{11}\ s^{-1}$). This was attributed to rotation of the N-tolyl group of the dye, which then has to sweep through the solvent and hence probes the local environment.

This concept was extended by Loutfy (Law and Loutfy, 1981, Loutfy, 1982, 1986), who studied a series of *p*-(*N,N*-dialkyl amino)benzylidene malononitriles. These ‘molecular rotors’ have a large side group attached to the double bond of the malononitrile, as shown for the structure DABM in Scheme 3.2. Since the molecule loses excitation energy by rotation about the double bond, if this is inhibited, there will be a lowering of k_{IC} and thus an increase in the fluorescence intensity. When these probes were incorporated into a free radical polymerization of styrene, acrylates or methacrylates, there was a sigmoidal increase in emission intensity with time. When this was converted to viscosity, Equation (3.30) was followed over the range from 0.2 to 100 cP, indicating that the macroscopic and microscopic viscosities are both probed by the molecular rotor. However, as the macroscopic viscosity increased even further as the glassy state was approached, the microviscosity remained constant and there was no change in fluorescence intensity. *This probe therefore is sensitive to gelation, not to vitrification.*

The relative sensitivity of reactive and non-reactive monosubstituted nitro-stilbene probe molecules for the photopolymerization of diacrylates and dimethacrylates has been studied by real-time fluorescence (Jager *et al.*, 2001). It was found that the acrylate-terminated nitrostilbenes were more sensitive than the non-reactive probes at the early stages of the reaction. This was attributed to the higher rate of incorporation of the reactive probes into the highly crosslinked regions which form during early stages of the polymerization reaction (Jager and Vanden Berg, 2001). There has been detailed study of the non-reactive DANS

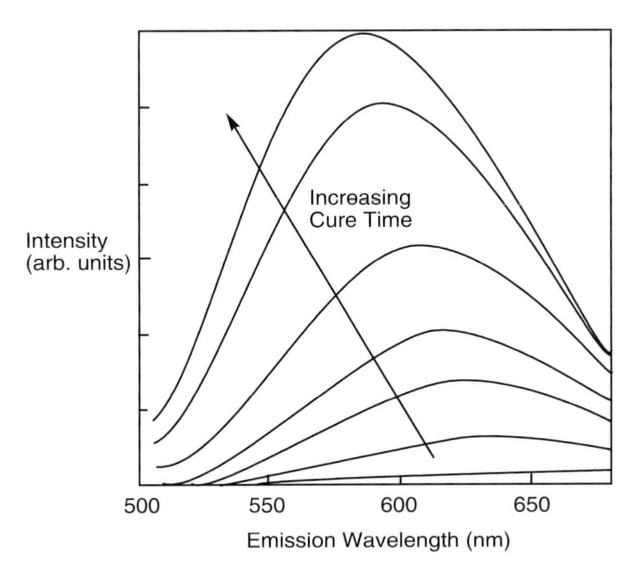

Figure 3.32. The change in the emission spectrum of the molecular probe DANS (Scheme 3.2) with cure time of epoxy resin at 80 °C. Adapted from Quirin and Torkelson (2003).

probe, 4-(*N*,*N*-dimethylamino)-4′-nitrostilbene (Scheme 3.2) which shows polarity sensitivity and thus a shift in the fluorescence spectrum as cure of an epoxy resin takes place (Quirin and Torkelson, 2003) (Figure 3.32). This sensitivity arises since the molecule contains electron-donor and -acceptor moieties, which results in significant charge separation in the excited state. The probe is self-referencing and insensitive, to temperature, so non-isothermal cure reactions could be followed (Quirin and Torkelson, 2003).

Figure 3.33 shows the effect of conversion of epoxy (i.e. increasing viscosity) on the wavelength of the maximum emission for excitation at 475 nm and the ratio of emission intensity at 588 and 564 nm. There is a sharp increase in emission intensity but no spectral shift at the conversion corresponding to the T_g at the cure temperature, indicating that the *probe is sensitive to both conversion and vitrification*.

It has been noted (Strehmel *et al.*, 1999) that the fluorescent dyes 1,1′-dimethyl-2,2′-carbocyanine iodide (QB) and *p*-*N*,*N*-dimethylamino-styryl-2-ethylpyridinium salt (DASPI) may be used both to monitor the crosslinking process and to detect T_g. These dyes have twisted intramolecular charge-transfer (TICT) states that are accessible from the first excited singlet state. Such TICT states may be considered to arise from intramolecular donor–acceptor electron transfer whereby the new state is stabilized in a geometry perpendicular to the original conformation (Figure 3.34).

Rotation occurs about the double bond in a manner similar to what occurs in the malononitrile derivatives described by Loutfy (1982, 1986). However, instead of measuring the steady-state emission intensity, the rate coefficient for the formation of the TICT state was obtained by picosecond laser fluorescence lifetime measurements (Strehmel *et al.*, 1992, 1999, Younes *et al.*, 1994).

If k_{AB} is the rate coefficient for formation of the TICT state, the measured lifetime is

$$\tau_F = 1/(k_F + k_{IC} + k_{AB}). \tag{3.31}$$

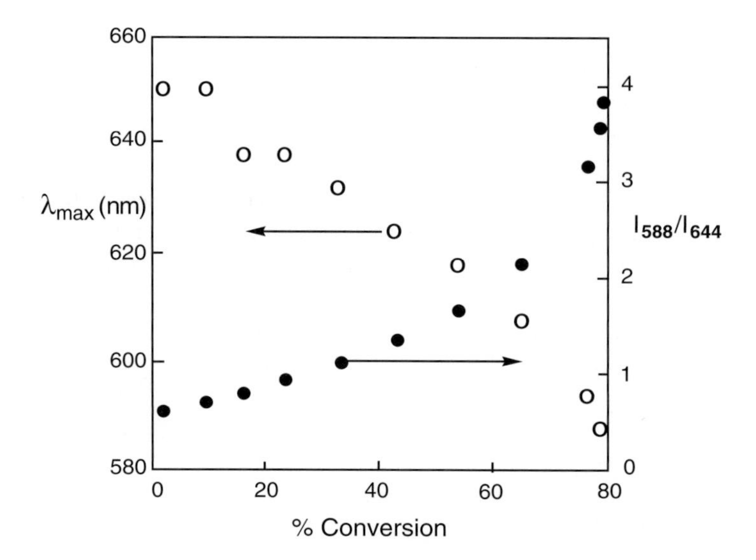

Figure 3.33. The fluorescence emission intensity ratio and peak wavelength of DANS (see Figure 3.32) as a function of epoxy-resin conversion at 80 °C. Adapted from Quirin and Torkelson (2003).

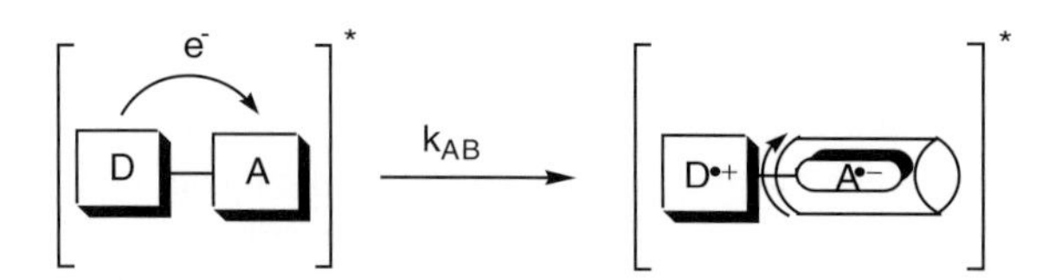

Figure 3.34. A schematic diagram showing movement occurring in the case of a TICT mechanism in a molecular probe such as DASPI and the reaction volume probed. After Strehmel *et al.* (1999).

The reference lifetime in the absence of TICT formation, τ_F', is obtained by measuring the fluorescence decay in a rigid solvent at 77 K, at which temperature it may be assumed that $k_{AB} \sim 0$. Then,

$$k_{AB} = \left[1/\tau_F - 1/\tau_F'\right] = 1/\tau_{AB}. \tag{3.32}$$

Plots of ln τ_{AB} against $1/T$ (K) of DASPI in a DGEBA resin show that the probe is able to detect T_g of the resin as a change of slope. Interestingly, when the cure of the resin was studied with these two probes, different profiles were obtained, with DASPI having an onset for the increase of τ_{AB} at the gel point, whereas for QB the gel point is the inflection in the τ_{AB} versus time curve. *Thus QB behaves exactly as did the molecular rotor in sensing gelation, whereas DASPI monitors the network formation beyond gelation and into the glass*. From a study of the cure of a range of resins and curing agents with DASPI, these results are rationalized in terms of the available free volume that the probe is able to sense (Strehmel *et al.*, 1999).

In a study of seven different dyes containing electron-donor and -acceptor groups, it was noted that the TICT states are sensitive to the polarity of the matrix as well as to the microviscosity (Vatanparast *et al.*, 2001). From a study of the curing of a DGEBA epoxy resin with methyl nadic anhydride, significant blue-shifts in fluorescence spectra were observed during cure. These shifts were linked to polarity changes, so they were not reliable indicators of the degree of cure. It was also shown that various probes exhibited different behaviours in the same curing environment, with fluorscence some of them even decreasing in intensity.

The choice of viscosity probes for polymerization is not restricted to dye molecules, and certain organometallic complexes show changes in luminescence during thermal and photochemical polymerization (Lees, 1998b). In particular, it was noted that metal carbonyl derivatives containing lowest-energy metal-to-ligand charge-transfer transitions are sensitive to viscosity changes of over five orders of magnitude when incorporated into epoxy and cyanate ester thermosets and acrylate and epoxy photo-curing resins (Lees, 1998a).

It was noted in Section 3.3 that the bimolecular processes, which compete with fluorescence, such as energy transfer, quenching and excimer formation, should offer a way to monitor the changes in the viscosity of the medium. In contrast to the probes discussed before, the intensity of fluorescence generally will **decrease** with an increase in viscosity as the resin cures. In a number of studies, Wang *et al.* (1985, 1986a, 1986b) investigated the changes in the fluorescence properties of excimer-forming dyes dissolved in epoxy resins and other polymerizing systems. The fluorescence of the probe 1,3-(*bis*-1-pyrene) propane (BPP in Scheme 3.2) shows all of the features of excimer emission outlined at the end of Section 3.3.8, viz. a structured free-molecule or 'monomer' emission and a broad red-shifted excimer spectrum. Pyrene has been the most-studied molecule that readily forms a sandwich-type excimer as the planes of the molecule come together (Guillet, 1987). In solution, the intensity of excimer emission (I_E) of pyrene increases compared with the monomer emission, I_M, with increasing concentration, as expected for a bimolecular process, but for BPP the excimer may form intramolecularly, regardless of the concentration. The sensitivity of the intensity of excimer emission to change in viscosity arises since the excited pyrene group may adopt a large number of conformations, only a few of which will allow it to overlap sufficiently with an unexcited group on the same molecule to achieve the minimum in energy to stabilize the excimer. Thus, as viscosity increases, the probability of achieving this conformation within the lifetime of the excited state *decreases*, so decreasing the excimer emission intensity (Fanconi *et al.*, 1986). This has been shown for the amine cure of DGEBA (Wang *et al.*, 1985), for which, after an initial increase in I_E/I_M, due to the decrease in viscosity as the temperature was increased from 22 to 60 °C, there was an eight-fold decrease as the resin gelled. Thus a plot of the inverse intensity ratio, I_M/I_E, allows a comparison with the known viscosity profile on curing of an epoxy resin (Apicella, 1986).

There have been studies in which pyrene alone has been used as the fluorescent probe for *in situ* polymerization (Pekcan *et al.*, 1997, 2001, Pekcan and Kaya, 2001). It is assumed that the observed enhancement of fluorescence intensity or increase in singlet-state lifetime of the 'monomeric' pyrene as the network forms arises from a decrease in bimolecular quenching events as the viscosity increases and k_Q decreases (Equations (3.28) and (3.29)). It is noted that the spectra showed no evidence of pyrene excimer formation, suggesting that the pyrene is isolated in the network at the concentrations employed.

An alternative approach to obtaining a photophysical measurement of viscosity that may be applied to monitoring polymer cure is to measure the polarization properties of the fluorescence of an added probe (Scarlata and Ors, 1986). This again measures a property of the molecule that depends on its rotation in the polymerizing medium. The fluorescence is excited with polarized light and the extent of depolarization of the emission will depend on the rotation of the probe following excitation, in addition to the depolarization due to the angle between the absorbing and emitting dipoles of the molecule. The emission anisotropy, r, is given by

$$r = (I - I_{\perp})/(I + 2I_{\perp}), \tag{3.33}$$

where I and $I_{\perp}$ are the luminescence intensities when measured through analysing polarizers which are oriented parallel and perpendicular, respectively, to the plane of polarization of the exciting light. The limiting anisotropy, r_0, applies to the value obtained if the probe is motionless for the excited-state lifetime, τ. The link between anisotropy and viscosity is given by the Perrin equation:

$$r_0/r = 1 + (k_B\tau/V)(T/\eta), \tag{3.34}$$

where k_B is the Boltzmann constant and V the molecular volume.

It was shown (Scarlata and Ors, 1986) that there was an increase in polarization of the probe molecule (a 4-amino naphthalimide derivative) during the photoinitiated cure of an acrylate-based coating formulation as the viscosity increased.

Again it is noted that this method, in common with many others discussed above, requires the addition of a well-characterized probe molecule. Only a few studies have exploited the intrinsic luminescence properties of the polymer. The authors of some chemiluminescence studies described later attempted to use the changes in intrinsic photophysical and photochemical properties of the polymers as they cure to probe the viscosity and phase changes in the system.

In monitoring the cure of advanced epoxy resins by fluorescence, it was noted that the TGDDM molecule (as in the commercial epoxy resin MY721) had an available route for internal conversion that involved torsional rotation, analogous to the molecular rotors. It was demonstrated that the fluorescence of the resin itself could provide a self-probe of the viscosity changes as cure proceeded (Levy, 1984). As the network developed, the four available epoxy groups on each molecule progressively reacted, resulting in a decrease in the rate of internal conversion, so the fluorescence intensified. It was recognized that the simplicity of the measurement meant that it could be adapted to on-line processing by using fibre-optics as discussed in Chapter 6. However, these studies also showed that the fluorescence may arise from an unidentified impurity that is acting as the probe (Levy and Schwab, 1987). A different thermosetting system that exhibits self-probe fluorescence is a resole-based phenolic resin (Vatanparast, 2002). Unlike the epoxy system, the emission intensity decreases almost linearly with the degree of cure, which has been attributed to the consumption of the phenolic groups in the resin on cure. There is also a red-shift in the spectrum that is consistent with a decrease in polarity as the cure proceeds.

Luminescence studies of degradation reactions

The extreme sensitivity of emission spectroscopy in comparison with absorption spectroscopy has meant that there have been many studies of the effect of degradation of the polymer on the fluorescence and phosphorescence properties (Allen and Owen, 1989a, 1989b). The quenching of phosphorescence by oxygen and other molecules, especially at elevated temperatures, means that, for on-line monitoring of degradation during processing and similar chemorheological processes, fluorescence or total luminescence is more appropriately monitored. In some cases the emission will arise from several species, although the effectiveness of energy transfer in fluid systems very often results in the species with the lowest-lying energy levels dominating the emission spectrum. Commercial polymers will also contain pigments and additives such as antioxidants and UV absorbers, so these will also add their spectral features.

It has been noted in Section 3.3.7 that the rapid formation of polyenes was a feature of thermal degradation of PVC and that these could be most readily detected in the Raman spectrum. The polyenes also have a characteristic fluorescence spectrum that may be used to study the spatial as well as temporal development of degradation (Remillard *et al.*, 1998). Other polymers for which the detection of conjugated oxidation products or a change in the extent of conjugation of the backbone or side groups is most readily performed by luminescence include polyethylene terephthalate (Edge *et al.*, 1995), polycarbonate (Chipalkatti and Laski, 1991) and the polyolefins (Allen and Owen, 1989a, 1989b, Jacques and Poller, 1993). In PET the appearance of excimer fluorescence at the expense of molecular ('monomer') emission on heating at melt temperatures is interpreted as due to main-chain breakdown and aggregation of the terephthalate fragments, while the longer-wavelength species are attributed to hydroxylation of the aromatic ring. More extensive oxidation results in quinone and stilbene quinone species, which are responsible for the yellowing of the polymer. Bisphenol-A polycarbonate at processing temperatures forms phenyl *o*-phenoxy benzoate, which is fluorescent. The progress of degradation was followed by monitoring emission intensities at 330 and 347 nm (Chipalkatti and Laski, 1991).

The changes in the luminescence properties of polyolefins with degradation have been attributed to the formation of α, β-unsaturated ketones (or ene-ones) (Jacques and Poller, 1993, Allen and Owen, 1989b), which may have a pro-oxidant effect. While these species emit strong phosphorescence, the detection of these and other conjugated sequences by fluorescence has been suggested (Jacques and Poller, 1993, Allen and Owen, 1989b).

3.3.9 Chemiluminescence and charge-recombination luminescence

The oxidative-degradation processes of polymers as discussed in Chapter 1 are frequently accompanied by the emission of weak visible luminescence, which, because the energy necessary to populate the excited states has been provided by a chemical reaction, is generally termed chemiluminescence (CL) (George, 1989a). The low quantum yield ($\Phi_{CL} \sim 10^{-9}$) often limits spectral analysis to broad wavelength bands, so the identification of the emitting chromophores and also elucidation of the precise chemistry of the light-producing reactions are often subject to speculation (Blakey and George, 2001, Blakey *et al.*, 2001). In analytical chemistry, CL has proven to be a powerful technique for both organic (Rakicioglu *et al.*, 2001) and inorganic (Zhang *et al.*, 2001) analysis in solution. In those applications, the CL reactions are chosen to have a high quantum yield (Φ_{CL} ranging from 10^{-2} to 0.9) so that the limits of detection are sub-ng/ml. Since CL is a measure of the rate of the light-producing reaction, the intensity of emission, I_{CL}, is given by

$$I_{CL} = \Phi_{CL} R, \tag{3.35}$$

where R is the rate of the reaction which produces the emitting species in the excited state. In spite of the low value of Φ_{CL} in polymer oxidation, reactions with rates as low as 10^{-12} mol dm^{-3} s^{-1} may be studied, which correspond to free-radical concentrations of 10^{13} radicals dm^{-3}, which in an analytical volume of 1 ml gives a sensitivity ~1000 times that of ESR (Section 3.3.2).

In oxidizing polyolefins such as polypropylene, the oxidation in the melt may be treated as a homogeneous system kinetically, in contrast to the situation in the solid state (George and Celina, 2000), such that the rate of oxidation is dependent on the rate of reaction of the chain-carrying alkyl peroxy radical with the polymer. In the Russell mechanism for CL, the energy for the excitation of CL is given by the termination reaction of two peroxy radicals (Blakey *et al.*, 2006). Formally this requires one of the radicals to be either primary or secondary, so that the chemistry occurring in the light-producing reaction(s) may change with the extent of oxidation (George, 1989a). However, this may be simplified kinetically if steady-state conditions prevail, and the rates of initiation (by hydroperoxide thermolysis) and termination (by peroxy-radical recombination) are equal:

$$I_{CL} = \Phi_{CL} k_t [RO_2\cdot]^2 = \Phi_{CL} k_d [ROOH]^n, \tag{3.36}$$

where k_t is the rate coefficient for bimolecular termination of the alkyl peroxy radicals of steady-state concentration $[RO_2\cdot]$, k_d is the rate coefficient for the decomposition of the initiating hydroperoxides, ROOH, and n is the order of this reaction. In an auto-oxidation at elevated temperatures, [ROOH] will change continuously, passing through a maximum and then decreasing. This is most clearly seen in the oxidation of PVC during processing (Scott, 1993). At lower temperatures and in the absence of transition-metal salts, which catalyse the decomposition through a redox cycle, there may be an accumulation of hydroperoxide as an oxidation product, but in the temperature regime for processing this will not occur.

In principle, the CL from polyolefins could be used as a monitor of the extent of oxidation during reactive extrusion in the presence of peroxides to decrease the molar mass (Brown, 1992), but this has not been reported. Attention has been paid to the effect that change in viscosity of a polymerizing network has on the CL profile, as described in the following section.

Monitoring network polymerization with CL

There have been several applications in the curing reactions of epoxy resins that exploit the CL mechanism as shown in Equation (3.36), i.e. the CL arises from the recombination of peroxy radicals (George, 1986, George and Schweinsberg, 1987, Kozielski *et al.*, 1995). In these experiments the resin, either in the presence or in the absence of the amine hardener, was heated through the normal cure cycle in the presence of air so that peroxy radicals could form. The resulting profile of CL over the time for normal cure was measured so that I_{CL} would reflect the changes to the three terms: Φ_{CL}, k_t and $[RO_2\cdot]$ in Equation (3.36) as the resin passed through the gel point and into the glassy state as the network developed.

Initial studies focussed on TGDDM and DDS, and it was found that there was a change in CL profile as the viscosity increased, with the CL decreasing with time of cure in the presence of hardener (George and Schweinsberg, 1987, George, 1986). In contrast, when the resin alone was heated for the same length of time in air, the CL increased, as expected for the increase in hydroperoxide concentration with time of oxidation. When this relationship was plotted on a semi-logarithmic scale there was a clear change in slope, which was correlated with the onset of gelation. This was interpreted as arising from k_t becoming diffusion-controlled so that Equation (3.29) applied with $k_t = k_d$. This would rationalize the observed result of a decrease in I_{CL} with time of cure. However, it would also be expected that Φ_{CL} would become sensitive to viscosity, given that it had been demonstrated that, in

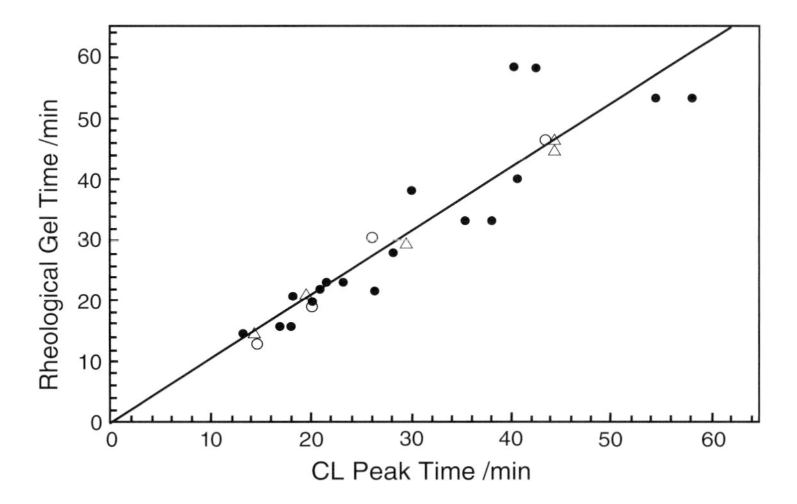

Figure 3.35. A correlation plot of rheological gel time (from crossover of G' and G″) and the maximum in chemiluminescence (CL) intensity over the temperature range 140–200 °C for the following epoxy-resin/hardener systems: ○, Tactix 742/27% DDS; Δ, Shell 1153/22% DDS; and • Shell 1071/27% DDS. Adapted from Kozielski *et al.* (1995).

TGDDM, the fluorescence intensity and lifetime increased with viscosity (Equation (3.30)) since it has a structure too similar to that of the molecular rotors (Levy, 1984). Thus, on combining the two relations, it would be expected that

$$I_{\mathrm{CL}} = C'RT\eta^{-0.33}[\mathrm{RO_2\cdot}]^2, \tag{3.37}$$

so that the intensity of CL should decrease with time as the viscosity increases. In principle, the change in $[RO_2\cdot]$ with time should follow the profile for the oxidation of the neat resin since the viscosity should be constant, but this will hold only if the oxidation reaction does not become diffusion-controlled in the curing resin. The complexity of the CL profile from epoxy resins was demonstrated when systems other than TGDDM were studied (Kozielski *et al.*, 1995). It was found that all systems exhibited a maximum or deflection in CL with time of cure that correlated with the gel times as measured by the crossover of G' and G'' in separate measurements (Figure 3.35).

However, it was also shown that the CL–time profile was sensitive to the purity of the resin and the amine, DDS, which was found to contain a quenching agent, so the quantum yield of emission was reduced in the fluid state (Equation (3.28)). It is therefore concluded that simple CL experiments are not generally applicable as a method for monitoring the cure of epoxy resins. In addition to the self-probe fluorescence of TGDDM, the CL measurements rely on absolute emission intensities, which, in the absence of an internal standard for calibration, is unreliable.

It was noted above that measurements of the lifetime of emission provide an alternative way of accessing the photophysical parameters of a polymer. The analogous approach in CL is to subject the oxidizing polymer to an external perturbation and then observe the change in CL as the system returns to the steady state (George, 1981). This will generally take the form of a short period of UV irradiation after which there is a burst of CL followed by decay, which may be analysed according to an assumed kinetic model, such as second-order kinetics for the CL arising from the bimolecular termination of peroxy radicals. Such an analysis leads to the following relation for the change in CL intensity, I, with time:

$$I^{-1/2} - I_0^{-1/2} = (k_t/\Phi_{CL})^{1/2}t. \quad (3.38)$$

Thus a plot of $I^{-1/2}$ against time should be linear and have a slope of $(k_t/\Phi_{CL})^{1/2}$, while the lifetime of the decay will depend on the initial peroxy radical concentration:

$$\tau_{CL} = (\Phi_{CL}/k_t)^{1/2} I_0^{-1/2}. \quad (3.39)$$

It was shown that these equations could be used to analyse the CL from thin films of epoxy resins during cure (Schweinsberg and George, 1986) and there was a change in the parameter $(k_t/\Phi_{CL})^{1/2}$ as cure passed through gelation. This was also shown to be applicable to monitoring the cure of a carbon-fibre–epoxy-resin prepreg (George, 1986), but the change in the above parameter as the resin transformed to a gel was only ten-fold and this, together with the time taken to measure the decay properties, was considered a severe limitation to the usefulness of the CL techniques for studying the early stages of network formation. An interesting observation was that the emission intensity increased rapidly as the resin vitrified, becaming insensitive to the presence of oxygen. The lifetime of the emission also increased and it was concluded that the emission arose from charge-recombination luminescence (Billingham *et al.*, 1989).

Charge-recombination luminescence

The origin of charge-recombination luminescence (CRL) during the cure of epoxy resins has been studied in detail (St John, 1993, Sewell *et al.*, 2000, Billingham *et al.*, 1989). From the ESR and UV–visible spectra it was found that the photo-irradiation of glassy TGDDM produced a radical cation centred on the tertiary amine site (Figure 3.10), analogous to the Wurster's Blue cation (St John, 1993). While photo-ionization may occur in the fluid state, the photo-electron must be trapped before electron–ion recombination can occur, so the system must both have a trapping site and be sufficiently rigid to separate the ion pair for a sufficient time to allow the decay to be resolved over times ranging from seconds to hours. This has been exploited in studying the vitrification during cure of TGDDM, and it has been determined that it is the hydroxyl groups which are more important than the amine sites in determining the stability of the ion pair (St John, 1993, Sewell *et al.*, 2000). The rapid increase in CRL was found to occur always at the same extent of reaction regardless of the cure temperature. In a study of various epoxy-resin and diacrylate networks, it was found that, although the CRL was weaker and shorter-lived than that from TGDDM, the vitrification of these resins too could be monitored (Lange *et al.*, 1998, 1999). In the case of the diacrylate network, there was evidence presented for micro-vitrification since the CRL was detected at low extents of conversion, which is a further example of the extreme heterogeneity of these networks (Dusek, 1996).

Experimentally the technique is simple since the sample may be irradiated from time to time and the emission intensity measured by a photomultiplier over a period of tens of seconds (Lange *et al.*, 1998). However, these studies confirmed the limited usefulness of CRL alone for real-time monitoring since the emission is significant only in the glassy state, so this technique cannot detect the gel point for the majority of networks. When combined with other luminescence methods it has potential use because it is independent of the atmosphere and is readily adaptable to fibre-optics.

3.4 Remote spectroscopy

The spectroscopic techniques described in Section 3.3 have all been developed for application in an analytical laboratory and generally require polymer samples to be taken to the spectrometer. For the monitoring of chemorheological processes and reactive processing, it is necessary to take the spectrometer to the reacting system or the rheometer. For systems such as NMR that require large magnetic fields to operate, this is clearly impractical, so magnetic-resonance spectroscopy has not been assessed for real-time remote monitoring in-line other than in the preliminary studies such as of the high-temperature surface NMR probe that has been adapted to operation at the die of an extruder (Gottwald and Scheler, 2005). There is greater potential for NMR and MRI techniques in off-line monitoring and also for providing fundamental information such as imaging of flow (Callaghan, 1999). In contrast, vibrational and electronic spectroscopy, both in absorption mode and in emission mode, lend themselves to remote sampling by coupling with fibre-optics. Some examples of laboratory-based optical-fibre systems were given in Section 3.3.5. In the following, specific examples are presented from the viewpoint of sampling and spectroscopic considerations such as analytical volume. The use of remote sensors for process control has been researched for the past 30 years, and the expansion of fibre-optic telecommunication networks and the decreasing cost of low-loss fibre-optic cable have provided an impetus for research in this area (Krohn, 1988, Boisde and Harmer, 1996). Developments in opto-electronics and photonics have been responsible for sophisticated fibres ranging from single-mode ultrathin monofilament fibres and bundles transmitting a single wavelength to large-diameter multimode fibres that transmit a wide range of wavelengths and are therefore suitable for remote spectroscopy. A detailed review of the application of optical-fibre sensors to the process monitoring of reinforced organic matrix composites (Fernando and Degamber, 2006) includes a comprehensive survey of the fundamental principles of optical fibres and experimental configurations for monitoring of thermosetting reactions. Hardware available for monitoring of conversion, copolymer composition and particle size by techniques in addition to fibre-optic spectroscopy has been reviewed comprehensively (Kammona *et al.*, 1999).

3.4.1 Principles of fibre-optics

The fundamental requirement of all radiation-transfer techniques in remote spectroscopy is that the radiation be transferred from the spectrometer to the sample, probe the reactions or transformations of interest and then return the modified beam of radiation to the spectrometer for the measurement of intensity at each wavelength in the spectral region of interest, all without any contributions from the transfer medium and with little loss in energy. While fibre-optics meets these requirements under favourable circumstances, there are potential artefacts and limitations, which may be understood from the principles of operation.

The key principle in the operation of fibre-optics, and also in several of the sampling methods, is total internal reflection. This is illustrated in Figure 3.36, which shows three cases for a ray of light travelling in a medium of refractive index n_1, incident at three different angles, θ, to the interface with a medium of lower refractive index, n_2.

It is noted that only for values of θ greater than θ_c will there be no refracted ray such that light is totally internally reflected at the interface and propagates within the medium of

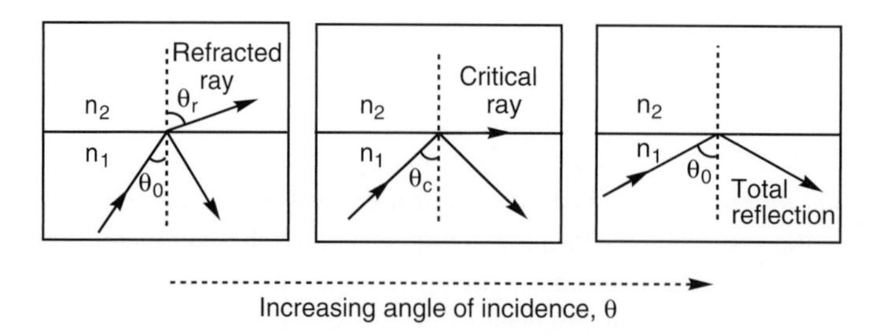

Figure 3.36. Refraction and reflection at an interface between media of refractive indices n_1 and n_2 below, at and above the critical angle, θ_c ($n_1 > n_2$).

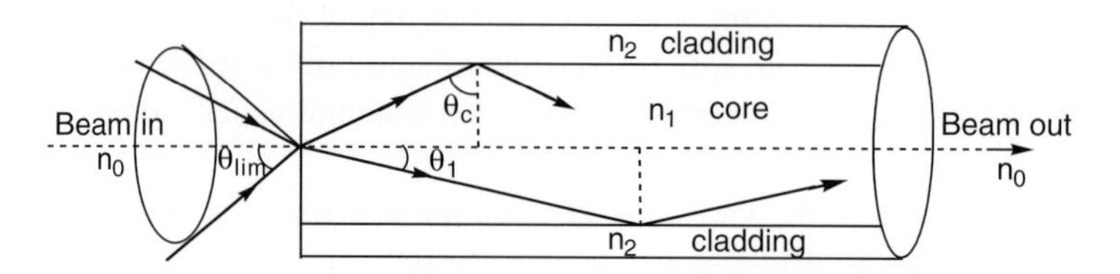

Figure 3.37. A schematic diagram of an optical fibre showing propagation of light beams defined by the numerical aperture of the fibre and the refractive indices n_1 and n_2 of the core and cladding materials; n_0 is the refractive index of air. Adapted from Boisde and Harmer (1996).

refractive index n_2. Thus, if the light is launched into a fibre at angles less than $\sin^{-1}(n_2/n_1)$, there is propagation along the length of the fibre and the attenuation of the energy depends on the following.

1. The intrinsic absorption properties of the fibre material at the wavelength, λ, of the radiation. This is usually expressed in optical power loss, a_λ, in dB/m, per unit length of fibre, L, which is related to the spectrophotometric absorbance of the material, A (Equation (3.21)), by the relation

$$a_\lambda = 10A/L. \tag{3.40}$$

2. The numerical aperture (NA) of the fibre. This is shown in Figure 3.37, which is a simple ray diagram for the entry of radiation into the fibre under the condition for total internal reflection (i.e. $\theta < \theta_c$) and thus propagation along the length. This defines the limiting angle of incidence, θ_{lim}, at the end of the fibre in contact with the air (refractive index n_0), and then

$$\mathrm{NA} = n_0 \sin\theta_{lim} = \left(n_1^2 - n_2^2\right)^{1/2}. \tag{3.41}$$

   Thus the numerical aperture, NA, depends on the materials of construction of the fibre and the larger the difference between the refractive indices of the cladding and the fibre core, the greater the range of angles for rays that may propagate along the fibre. This is relevant for matching the numerical apertures of spectroscopic and other optical components for analysis of the beam in order to optimize the SNR.
3. The absorption of the evanescent wave. The evanescent wave is the component of the totally internally reflected wave that penetrates into the medium of lower refractive index (i.e. the cladding) by a distance that depends on the angle of incidence. This is illustrated in

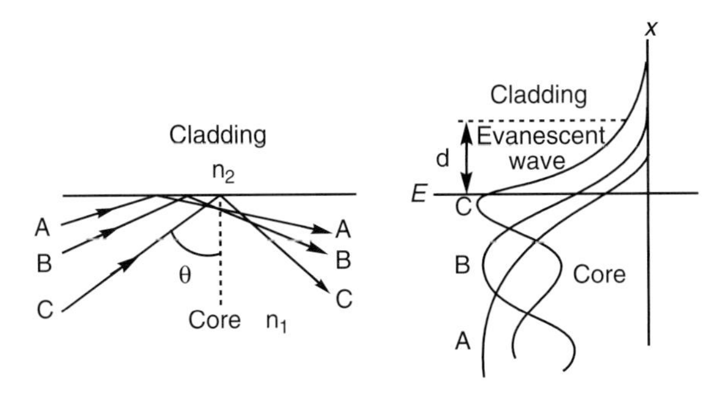

Figure 3.38. The penetration depth, *d*, for the evanescent wave from the core into the cladding for three beams A, B and C; *E* is the electric field strength.

Figure 3.38, which schematically illustrates three beams A, B and C at different angles of incidence to the cladding and the depth of penetration of the evanescent wave for each. *E* is the electric-field intensity and it can be seen that this decays exponentially into the cladding.

The wavelength and angle dependences of the depth of penetration, *d*, were given earlier (Equation (3.22)), and this principle is used in the MIR for the sampling technique of ATR spectroscopy. Figure 3.13(c) illustrated this, and, for typical refractive indices of fibre core and sample of 2.5 and 1.5, respectively, and an angle of incidence of 45°, the depth of penetration of the evanescent wave is about 0.15λ. The effect of the evanescent wave on fibre-optic spectroscopy may be illustrated by the example of plastic-clad silica (PCS) fibre optics for remote spectroscopy in the NIR spectral region, as discussed below.

Examples of the effect of these factors on remote spectroscopy may be seen by considering the use of PCS fibre in a simple capillary transmission cell for the measurement of NIR absorption spectra (George *et al.*, 1991). The energy from a source such as a tungsten lamp or glow-bar emitting in the NIR is focussed on a polished end of the PCS fibre by using a quartz lens or a parabolic mirror. The optics are designed to optimize transfer to the fibre by matching the NA of optics and fibre. The various methods of coupling the fibre to the reacting system are discussed in the next section, but the simplest method is to cut and polish the fibre and place the sample between the two ends at a distance as short as possible. To maximize transfer from the launching fibre to the gathering fibre, collimating lenses (again matched to the NA) may be used, or, if pathlengths are short, the energy losses for simply mounting the sample in a metal capillary cell may be tolerable. The gathering fibre is then returned to the detector of the spectrometer and the signal recorded. For high-quality spectra it is necessary to subtract the spectrum of the fibre itself from that of the sample plus fibre. Figure 3.39 gives the NIR spectrum of a 2-m length of silicone-clad silica fibre-optic monofilament (George *et al.*, 1991) and several features may be noted.

(i) The absorbance of ~3 for the fibre material at 7100 cm^{-1}, which would correspond to an attenuation of 15 dB/m at this wavelength. This arises from the first overtone of the –OH vibration of silica at 3540 cm^{-1} and, if spectral information is required in this region, it is necessary to use a fibre core with a low silica –OH concentration. Other bands attributed to –OH are at 8000 and 10 500 cm^{-1}. When used in a capillary cell, this results in a dead band (Figure 3.40) with a total fibre length of only 2 m. Another dead band corresponding

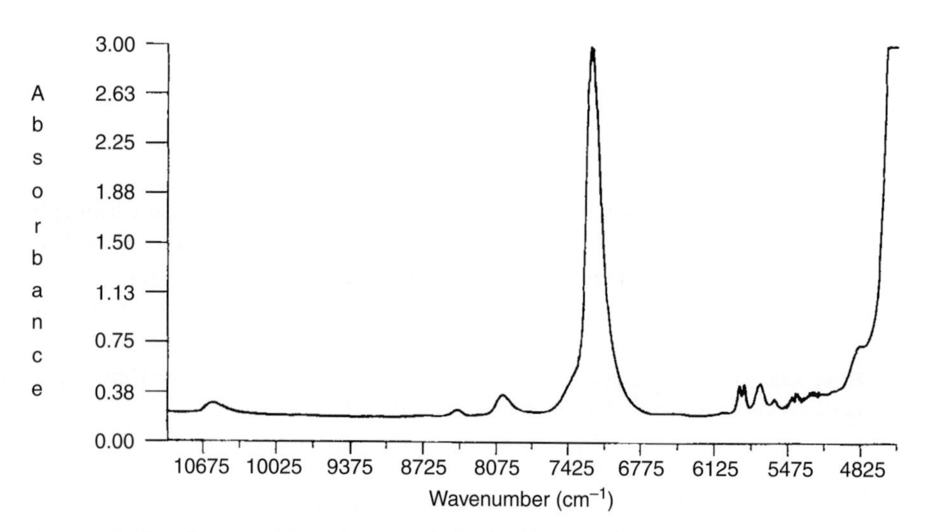

Figure 3.39. The NIR absorbance of a 2-m length of PCS optical fibre, showing the spectral artefacts that may be introduced by species present in core and cladding (George *et al.*, 1991).

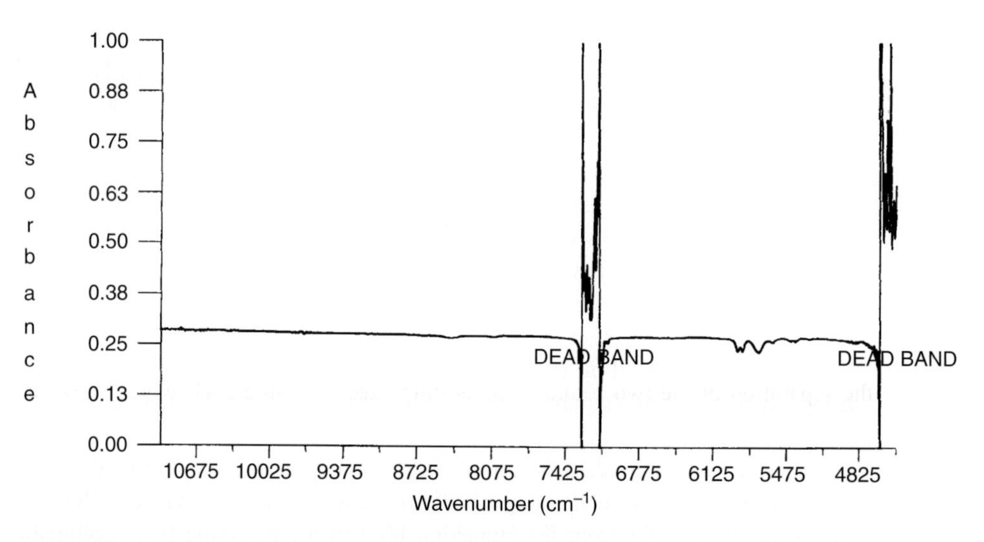

Figure 3.40. Dead bands and negative bands arising in a transmission cell used for remote spectroscopy due to the core and cladding bands shown in Figure 3.39 (George *et al.*, 1991).

to the cut-off of the transmission of silica in the IR is seen at 4800 cm^{-1}. At wavelengths away from the –OH bands the attenuation is of the order of 3–10 dB/km.

(ii) The presence of weaker, but sharper, bands in the range 5400–6000 cm^{-1}. These correspond to the first and second overtone and combination bands of the methyl group of poly(dimethyl siloxane) used for the cladding. That these arise from the evanescent wave may be calculated from the measured absorbance of the methyl band of 0.15 at 5950 cm^{-1}. Using a reported molar absorptivity of 0.1 dm^3 cm^{-1} mol^{-1} (Goddu and Delker, 1960), an equivalent pathlength of 500 μm may be calculated from Beer's law (Equation (3.21)). By using the attenuated total reflection relation (Equation (3.22)) it is found that this pathlength corresponds to the attenuation of the evanescent wave for a launch angle of 75.1°, which is close to the limiting angle of 76° (Equation (3.40)) of the fibre-optics, which has a numerical aperture of 0.4. The effect of these cladding bands on the use of the capillary cell for spectroscopy may be seen in Figure 3.40. Negative bands may arise due to a mismatch between the sample and reference pathlengths. Bending of the fibre may contribute, since this will change the angle at the interface and thus the depth of penetration into the silicone cladding. The effect of these cladding bands on quantitative fibre-optic spectroscopy is severe with certain polymers such as epoxy resins, and significant interference of analytical bands for the monitoring of the reactive processing has been reported (George *et al.*, 1991). This may be overcome by moving to silica-clad silica, which has a low –OH content to minimize attenuation. In this way it is possible to obtain spectra in a remote capillary cell that are of good SNR and free of interfering negative bands.

3.4.2 Coupling of fibre-optics to reacting systems

The fibre-optics provides the conduit for transferring the analysing radiation to the reacting polymer and returning it to the spectrometer for spectral analysis. The coupling of the spectrometer to the fibre-optics is now routine, and producers of most commercial vibrational-spectroscopy instruments (FT-IR in the MIR and NIR and Raman spectroscopy) offer accessories for launching radiation into fibre-optics tailored for the spectral region of the instrument as well as a range of sampling options. Since the materials of construction and techniques employed are dependent on the wavelength range of the radiation, these are best considered separately.

Remote visible and ultraviolet absorption and emission spectroscopy

Operation in the near-UV and visible regions is inexpensive since quartz or glass optics may be used. One major concern is the intrinsic fluorescence properties of the fibre core material which, because of the long pathlength of fibre, may interfere with the analytical signal. The same fibre may be used for excitation and emission with an external beam splitter to enable the separation of the two beams. This is illustrated in Figure 3.41, which shows a system

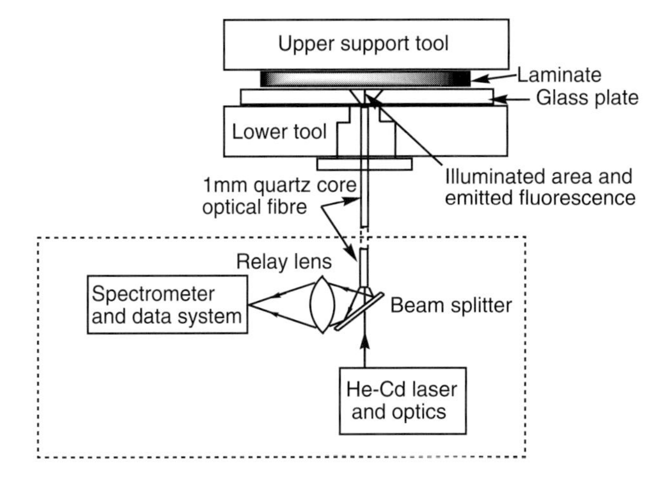

Figure 3.41. A system for remote fibre-optic fluorescence monitoring of viscosity changes during processing of a carbon-fibre–epoxy-resin laminate. Adapted from Levy and Schwab (1987).

developed for monitoring the viscosity-dependent laser-excited self-fluorescence of an epoxy resin (Section 3.3.8) during the processing of a multiple-ply fibre-reinforced composite material.

In this case a single quartz-core fibre (of diameter 70–1000 μm) is mounted flush with a glass plate, which is in contact with the reacting resin. The output fluorescence may be separated from scattered exciting light either by cut-off filters (Levy, 1986) or by a spectrometer (Levy and Schwab, 1987, Paik and Sung, 1994) before the detector. When filters are used, only the intensity of emission is monitored, whereas with a spectrometer the changes in wavelength of the emission as cure proceeds may be obtained. The latter was found to be of value (Levy and Schwab, 1987) since the run-to-run reproducibility of the absolute intensity of the emission was poor. This was attributed to the differences in resin thickness viewed by the fibre-optics as the resin flowed during the early stages of cure. Embedding the fibre will minimize such energy losses (as discussed later). The change in the maximum wavelength of emission as cure proceeded gave a distinctive and reproducible profile that revealed the chemorheological events of a minimum in viscosity and the gel point of the resin. It was also insensitive to the changes in absolute intensity that occurred during the processing cycle. The reasons for the wavelength shift may be linked to the change in the DDS hardener since a progressive red-shift of the fluorescence excitation spectra by up to 24 nm was reported to occur as the primary amine is converted into the secondary amine (Paik and Sung, 1994). This is the dominant reaction up to gelation (St John and George, 1992). A similar system of a single fibre for gathering fluorescence has been used for following the solidification of a polyolefin after injection moulding (Bur and Thomas, 1994). The probe was inserted through the ejector-pin channel and mounted flush with the mould using a sapphire window. A molecular-rotor dye (Loutfy, 1982, 1986) was used as the fluorophor (Section 3.3.8) so that the emission intensity increased as the viscosity increased. It was found that there was a significant contribution of reflected light, which, as noted above, should be removed by effective filtering.

Another reason for the change both in emission intensity and in the transmission properties of the thermosetting resin is the systematic change in refractive index as the network polymerizes. This change is typically from 1.56 (uncured) to 1.615 (cured). While this has the effect of changing the efficiency of energy transfer from the sample to the fibre due to the change in the numerical aperture of the fibre when it is embedded in the resin and so altering the absolute intensity, it has been exploited in a number of sensor applications for monitoring the polymerization of networks (Krohn, 1988, Powell *et al.*, 1995, 1998). In a simple demonstration of the use of a single mode rather than the usual multimode fibres, an optical fibre with its cladding intact is immersed into the curing resin and illuminated by light from a laser diode. The radiation reflected along the same fibre bundle is picked off at a Y-junction and measured by a photodiode. In order to improve the SNR, the light source is modulated and lock-in detection is used (Cusano *et al.*, 2000). It was shown that the intensity of the reflected radiation, R, should vary with the refractive indices, n_2, of the resin and n_1 of the fibre according to

$$R = [(n_1 - n_2)/(n_1 - n_2)]^2. \tag{3.42}$$

This analysis would predict that the reflection coefficient, R, would decrease with cure, since n_2 will progressively increase, whereas R was found to increase systematically and produce a profile that followed that of a DSC conversion–time curve for an epoxy resin.

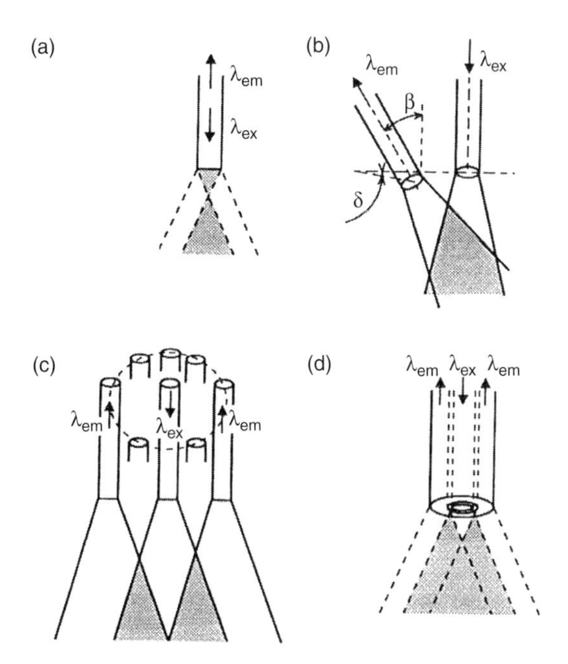

Figure 3.42. The arrangement of excitation (λ_{ex}) and emission (λ_{em}) beams from optical fibres for remote fluorescence or Raman spectroscopy. The resulting analytical volume is shaded. Reproduced with permission from Boisde and Harmer (1996). Copyright 1996, Artech House.

This suggests that the process of energy transfer to the fibre is more complex and that changes to the NA of the fibre may be important.

An alternative to the illumination system of the type shown in Figure 3.41 is to use separate excitation and gathering fibres, either as separate bundles, held side by side, or as a mixed bundle with a central excitation core surrounded by gathering fibres. This avoids the use of a beam splitter and efficiently transfers the emitted radiation to the detector in the second arm of the bifurcated bundle. This geometry is equivalent to the front-face illumination mode of fluorescence spectroscopy and has been used by several researchers when studying the monitoring of reactive processing using molecular probes that are sensitive to viscosity (Vatanparast *et al.*, 2001) or using energy-transfer efficiency as a measure of blend morphology when the fibre-optics is coupled to a rheometer (Vorobyova and Winnik, 2001). The analytical volume in this and the single-fibre mode has been considered for several different possibilities shown in Figure 3.42 (Boisde and Harmer, 1996).

These considerations apply to remote Raman spectroscopy using fibre-optics (Section 3.4.2) as well as fluorescence. While a single-fibre system offers convenience, especially in confined spaces and with small samples as may be encountered in an on-line sampling port or a reactor, the double-fibre system has superior SNR (Louch and Ingle, 1988). This is especially so when the fluorescence intensity from the sample is low and high excitation intensities must be employed. In this case the fluorescence from the single fibre will also be detected, severely limiting the length of fibre that can be used. As may be seen qualitatively from Figure 3.42, the volume probed and thus sensitivity increase if the excitation and emission fibres are at an angle of 15°–30° to one another. In general the excitation and gathering efficiencies are governed by the NA of the fibres. Commercial systems are available with configurations

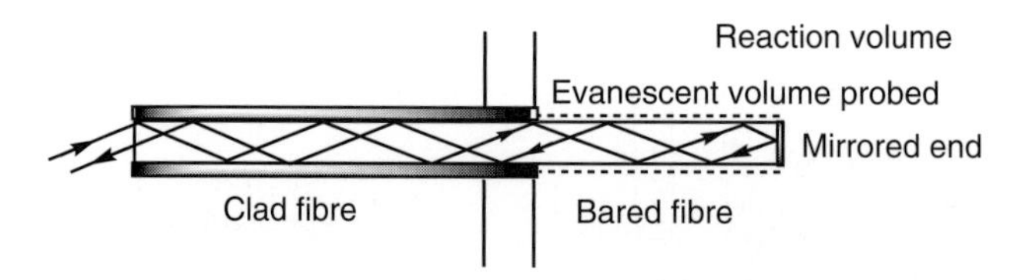

Figure 3.43. Use of a bared optical fibre as a reaction probe by using the evanescent wave to sample a volume controlled by the refractive index of the fibre and reactant. Adapted from Park and Sung (1994).

(c) and (d), but such geometries are expensive and are not always suited to the rugged environment of reactive processing. They do, however, overcome the limitations of the fluorescence of single fibres, which can become dominant when the optical fibre is bent (Myrick *et al.*, 1990). An alternative geometry is the 'optical-fibre-facing' configuration in which the excitation and gathering fibres are in-line, and filters are used to remove spurious radiation, but this requires a sampling arrangement similar to that for transmission spectroscopy, which might not always be compatible with the available space in the reaction zone. The transfer efficiency may be increased by using graded-refractive-index lenses bonded to the ends of the fibres as well as collimating optics such as Winston cones (Myrick *et al.*, 1990). The use of fibre-optics in conjunction with other cure-monitoring techniques to obtain simultaneous spectral and physical information on reactive processing is discussed later in this chapter.

In one common method used with multimode fibres (Paik and Sung, 1994, Powell *et al.*, 1995, 1998), the cladding is stripped from the optical fibre and the polymerizing network is sensed by the evanescent wave (Figure 3.43).

As the refractive index of the resin approaches that of the fibre, the energy is lost since the condition for total internal reflection (Section 3.4.1) is no longer met. In a variation on this method, a short section of the optical fibre is replaced with fully cured resin of the same dimensions, before immersion in the curing resin system. Initially, the energy will be transmitted by total internal reflection along the resin section, but, when the refractive index of the curing resin equals that of the resin section, energy will again be lost. The system is also able to monitor the progressive change in refractive index as well as the fully cured state, since there will be a change in the critical angle, $\sin^{-1}(n_{\text{uncured}}/n_{\text{cured}})$, as the network develops, and the detector will see a progressive change in signal (Afromowitz, 1988, Lam and Afromowitz, 1995a, 1995b). One advantage of using a length of fibre with the same refractive index as that of the fully cured resin as the sensor is that the sensitivity can be controlled by merely increasing the length of the sensing fibre, but the penetration of the evanescent wave is limited at shorter wavelengths, so the sensor length may need to become prohibitively large in order to achieve an acceptable SNR in the visible region. However, this small depth of penetration can enable the interphase region in composite materials to be studied (Paik and Sung, 1994, Woerdeman and Parnas, 2001). (The interphase is the region of resin adjacent to the reinforcing fibre in which the properties of the resin differ from the bulk due to the reactions between the resin and the coupling agents deposited on the fibre surface and the effect of the activity of the surface on the local concentrations of both resin and hardener.) This may extend several tens of nanometres from the surface. When properties closer to those of the bulk polymerization are to be monitored, the limited penetration depth of the evanescent wave may be overcome by operating in the NIR and MIR regions. In this case, the penetration depth increases from $\sim$50 nm in the visible region to $\sim$0.1 μm in

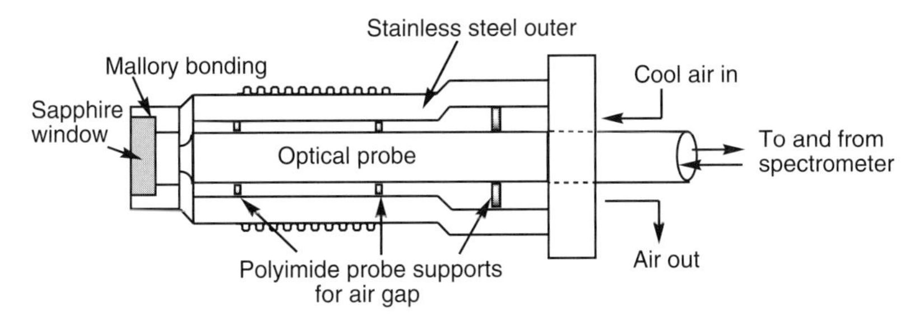

Figure 3.44. A fibre-optic probe with air cooling and a sapphire window suitable for use in a reactive extruder or other rugged environment. After Hansen and Khettry (1994).

the NIR and ~1 μm in the MIR. The precise depth of penetration for a particular wavelength may be calculated from Equation (3.22).

Remote NIR absorption and reflection spectroscopy

Most attention has been paid to applications of absorption and reflection spectroscopy in the NIR region for industrial process control (Dallin, 1997, Kammona *et al.*, 1999, Fernando and Degamber, 2006). This arises from the robust nature of the optical components and the availability of commercial systems that are based on gratings as well as interferometers. Transmission and reflectance probes that are designed to interface with extruders, and in particular reactive extruders, are commercially available, and their performance has been assessed quantitatively (Hansen and Khettry, 1994). The interface for a rugged NIR probe to monitor the molten flow of an extruder is shown in Figure 3.44.

The essential features are the protection of the optical components by appropriate windows, brazed to the housing with a special Mallory alloy to minimize thermal stress, and the flow of cooling air to remove the transferred heat. The pathlength for transmission spectra may be varied from 1 to 9 mm (Hansen and Khettry, 1994). The efficiency of transfer of radiation in the transmission mode may be increased by incorporating spherical quartz lenses at the ends of the optical fibres, but there must be no interference with the flow properties of the extruder, and the differences of coefficient of thermal expansion between the quartz and the housing must be accommodated by the high-temperature potting compounds used.

By mounting two such probes between 1 and 10 mm apart, a transmission efficiency of 12% was obtained (Rohe *et al.*, 1998). In a consideration of the problems associated with the process/monitor interface, a multifunctional interface was designed for an extruder (Reshadat *et al.*, 1999), which enabled successive monitoring of the melt in the barrel and at the die and the solid extruded strand. The melt-in-barrel monitoring relied on diffuse reflectance and for the other interfaces a reflecting mirror was used opposite the probe (the 'transflectance' mode of operation), which had a central illumination bundle and an outer return bundle. The diffuse-reflectance probe had lower efficiency than that of the transflectance probe. It was noted that measurement in the strand position for a highly scattering semi-crystalline polyolefin probably produced a diffuse-reflectance rather than a transmission spectrum, in spite of the presence of the mirror. The differences among the spectra from the three positions meant that each required a separate calibration routine. A major feature that emerges for all practical designs is that the ports should be designed to enable interchange of probes without affecting the process stream. An analysis of the materials of construction for windows in the high-pressure zones (Reshadat *et al.*, 1999) shows that sapphire has the best combination of thermal and mechanical properties and has a reflection loss of between 7.0% and 7.5% in the NIR, depending on wavelength.

The limitation of performance in the NIR due to the materials of construction of the fibre has been noted (Dallin, 1997), and the small residual –OH functionality in most commercial fibres will absorb energy and result in an upper spectral range of 2250 nm. This has been shown in the example in Section 3.4.1 (Figure 3.39), and the dynamic range is generally reduced from an upper absorbance limit of about 3.0–3.5 in NIR spectroscopy in the instrument to about 1.5–2.0 when monitoring is performed using fibre-optics (Dallin, 1997). However, the major contribution to this limit is the actual loss of energy at each interface in the optical path rather than the intrinsic energy losses in the fibre alone. One way of minimizing fibre energy losses is to shorten the path or, in transmission measurements, place either the light source (Calvert *et al.*, 1996) or the detector (Dallin, 1997) adjacent to the sample, with the main function of the fibre being to enable the spectrometer to be operated remotely from the process line, which is usually a hostile environment for the optical system of a grating or interferometer. It has been shown that a total optical-fibre length of 2 km may be used with a transmission cell for the real-time analysis of a process stream in the 1–2 μm spectral range by using a pair of multimode silica fibres that have very low –OH absorption (Mackison *et al.*, 1992).

Quartz fibres generally have a refractive index that is too low for evanescent-wave spectroscopy of curing epoxy resins (although, as shown earlier, in Figure 3.39, the NIR absorption spectrum of the lower-refractive-index silicone cladding is recorded in the spectrum of the PCS quartz fibre-optics). However, if a section is fabricated with a material of higher refractive index (e.g. tungsten oxide or sapphire) then ATR spectra of the substrate in the NIR region may be obtained. A dip probe may be constructed by coating the end of the element with gold so that the effective pathlength is doubled because also the return beam probes the substrate. The length of the element is governed by the molar absorptivity of the bands to be analysed, but, for the curing of composite materials in an autoclave, sufficient attenuation of the evanescent wave is obtained for one to be able to monitor the chemical changes as the network polymerizes (Druy *et al.*, 1988b). In the particular application, the fibre must be surrounded by resin in the article being fabricated and the effect that the permanently embedded fibre (which will be at least ten times the diameter of a reinforcing fibre in the composite) will have on the mechanical properties and integrity of the product must be considered. Alternatively, the sensing fibre may be located in a test piece accompanying the part in the autoclave. In spite of the limitations, the interphase and in particular the reactions occurring between the fibre surface, the adsorbed silane coupling agent and either epoxy resins or unsaturated polyester resins have been monitored using evanescent-wave spectroscopy in the NIR (Johnson *et al.*, 2000). In this application, a bundle of specially fabricated fibres having a flint-glass core of diameter 25 μm and a soda-lime glass cladding of thickness 1 μm approximated the thickness of typical E-glass reinforcing fibres. A further advantage of the system was that the cladding composition approximated that of E-glass so that the interphase around the special fibre bundle was considered to approach that around the reinforcing fibres. The length of the sensing fibre is a definite limitation to the use of ATR in the NIR, so more applications have been researched in the MIR spectral region.

Remote MIR absorption and reflection spectroscopy

Remote MIR spectroscopy is limited by the lack of inexpensive fibre-optics operating in the wavelength range where the fundamental vibrational modes of molecules participating in the chemorheological changes are observed. This has not prevented several fibre-optic systems that are fabricated from chalcogenide glasses (e.g. arsenic germanium selenide) (Young *et al.*, 1989, Compton *et al.*, 1988) and silver halide crystals (Kupper *et al.*, 2001) being made available.

The longer wavelength of the MIR means that ATR spectroscopy may be readily performed by removing cladding to bare a short section of the fibre, which then becomes the ATR element. In addition to the high cost, chalcogenide fibres do not permit transmission to wavenumber lower than $\sim$1200 cm^{-1}, so detailed analysis in the fingerprint region of the MIR is prevented. These fibres have a refractive index of 2.4 and are available with core diameters of 120 μm and a silicone-coating thickness of 90 μm so that, with a service temperature of 250 °C, they are suitable for *in situ* monitoring of composites through both the cure cycle and post-cure cycles. A length of 3–6 cm of cladding could be removed by soaking in chloroform to swell the silicone, which could then be stripped off with a decladding tool. This section was then immersed in the resin undergoing crosslinking and the interferogram obtained in a conventional FT-IR instrument with an identical length of fibre was used to obtain the background reference spectrum. The spectral quality enabled the imide bands at 1780, 1710 and 1350 cm^{-1} to be monitored (Compton *et al.*, 1988). The attenuation loss of 10–15 dB/m is high, restricting use in this application to a total fibre length of about 3 m. The attenuation may be reduced and spectral range increased by purification of the precursor materials. For example, a chalcogenide with a composition of $Ge_{30}As_{10}Se_{30}Te_{30}$ has a large absorption band between 1400 and 1100 cm^{-1} that limits the cut-off, but on purification this is totally removed, so the fibre may transmit out to 900 cm^{-1} (Sanghera and Aggarwal, 1999). An alternative, narrower-diameter, fibre (70 μm, plus a 10-μm-thick cladding) with a lower refractive index of 1.5 and a lower loss of $<$0.5 dB/m made from a heavy-metal fluoride glass was also assessed (Druy *et al.*, 1988a), but had a cut-off of 2300 cm^{-1} and hence was restricted in the application of epoxy cure monitoring to following amine functionality at 3367 cm^{-1}.

Polycrystalline silver halide fibres of composition $AgBr_{1-x}Cl_x$ with a refractive index of 2.2 have a much wider spectral range (4000 to 750 cm^{-1}) and are available with optical losses of less than 0.1 dB/m at 1000 cm^{-1}(Kupper *et al.*, 2001). The strength of the fibres depends on the value of x, with the highest strength and strain at break corresponding to $x = 0.5$ (Barkay and Katzir, 1993). The fibres are available with a square cross-section (0.75 mm) to enable coupling with a high aperture (NA = 0.55) to FT-IR spectrometers, and there is a range of ATR-based probes that allow dip analysis of reacting systems. The fibres do have a limited continuous operating temperature and are moisture sensitive after prolonged exposure, so they cannot be used for high-temperature process monitoring without special precautions to protect the fibres.

Remote Raman spectroscopy

Raman spectroscopy lends itself to remote monitoring of reactive processing using optical fibres because the excitation and scattered radiation will be in the visible or NIR spectral regions and thus with in the bandwidth of quartz fibres. Thus all the principles described earlier in Section 3.4.2 apply. It was initially believed that remote Raman spectroscopy was limited in application (Harmer and Scheggi, 1989) because of

- the poor efficiency of Raman scattering
- the poor collection efficiency (low NA) of the receiving fibre and matching to the analysing monochromator
- the high fluorescence and Raman scattering of the fibre itself
- the low excitation intensity in the analytical volume

At about the same time as these concerns were expressed, the optimization of an optical-fibre system for use in remote Raman spectroscopy was described (Hendra *et al.*, 1988). The areas explored are similar to those for remote emission spectroscopy discussed earlier in Section 3.4.2, and techniques for achieving the efficient transfer of the weak scattered radiation to the gathering probe include the use of

- six collection fibres around the excitation fibre
- excitation normal to the surface and collection at an angle of 17°
- reflective surfaces around the fibres to increase collection efficiency (such as an internally reflecting capillary to contain the sample)

The overall efficiency of excitation with the optimized fibre-optic probe compared with excitation with the same samples in an actual Raman spectrometer ranged from 50% for liquids to 37% for solids. The biggest single limitation was identified as fluorescence of the sample or components of the fibre-optics. The effects of fluorescence are largely removed by operating in the NIR region of the spectrum as in Fourier transform raman spectroscopy, for which excitation is traditionally with a Nd:YAG laser operating at 1.064 μm (9393 cm^{-1}). (YAG is yttrium aluminium garnet.)

Grating-based spectrometers may also operate with NIR lasers, and a compact fibre-optic system (Shim and Wilson, 1997) using either a titanium : sapphire laser (785 nm) or a tunable high-power diode laser (778 ± 10nm) providing up to 500 mW with a spectral bandwidth of less than 0.02 nm has been described. The probe used six collection fibres around the excitation fibre. A study of different fibre types resulted in silica-clad silica being the fibre of choice (over sapphire and liquid light guides) because it gave the lowest background. The detection system was chosen to be a liquid-nitrogen-cooled charge-coupled device (CCD) behind an *f*/1.8 axial transmissive imaging spectrograph, which is compact and has a high optical transmission of >50%. The high optical speed resulted from the use of holographic components both for the notch filter (for removing the excitation radiation) and for the grating (for spectral dispersion). In spite of the use of NIR excitation, with its lower Raman scattering as well as lesser fluorescence, there was still underlying sample fluorescence, which was removed by fitting the baseline with a fifth-order polynomial and then subtracting. This can cause some artefacts in weak spectral regions. The value of SNR for the grating system compared with the FT-Raman system was determined both theoretically and experimentally and found to be between 2.0 and 2.5 in favour of the grating system. This was believed to be due to the greater Raman cross-section at the shorter excitation wavelength and the higher transmission (50%) versus 10% for the FT-Raman system, which has many more optical elements, so offsetting the intrinsic advantage of high throughput of an interferometer.

The efficiency of a system is strongly dependent on the fibre-optic components, and any increase in the active area and NA of the bundle as well as the optimum geometry for the type of sample will greatly enhance sensitivity and reduce spectral collection time. In the processing environment, the Raman probe must be rugged, and in one such system (Niemela and Suhonen, 2001) an absorptive CdTe filter replaced the usual holographic notch filter for laser line rejection at 830-nm excitation with no loss in efficiency. Another factor affecting overall sensitivity is the volume of the sample from which the scattered radiation is gathered. An analysis of the sampling depth for an FT-Raman spectrometer revealed that this was generally limited by the scattering properties of the sample rather than a limitation in depth of field of the spectrometer and optics system (Barrera and Sommer, 1998). In some low-absorbing and optically transparent samples, the depth of analysis approached 10 mm. When applied to the real-time monitoring of processing this ensures that Raman spectra are measuring the bulk reactivity of a system and are not sensitive to surface effects and reactions at the interphase as may occur with evanescent-wave techniques (Section 3.4.2).

A feature of Raman spectra is the temperature sensitivity of the ratio of anti-Stokes scattering to Stokes scattering. As noted in Section 3.3.3 the anti-Stokes scattering lies to higher energy than the Rayleigh line and so arises from the molecules which are in the higher-lying vibrational state. The population in this state and thus the intensity will depend on the Boltzmann distribution, so the ratio of intensities, I_{AS}/I_S is given by

$$I_{AS}/I_S = (\lambda_S/\lambda_{AS})^4 \exp[-hv/(k_B T)], \qquad (3.43)$$

where λ_S and λ_{AS} are the Stokes- and anti-Stokes-band wavelengths, respectively, and v is the frequency of the exciting laser radiation. If either the sample or the fibre shows a Raman band that does not change with the reaction occurring, then the ratio I_{AS}/I_S may be used to determine the temperature (Dakin, 1989). In the case of the fibre being the sensor, by using a pulsed laser and gating the Raman spectra, it is possible to determine temperatures at different points (e.g. 10 m apart) along lengths of fibre. This is not expected to be applicable in real-time processing because of the shorter distances involved, in particular since the capability of this method is considered to be 1 K or less per metre of fibre (Dakin, 1989). If, however, the sample has a band that is invariant with conversion and also shows strong Raman scattering, then the technique has an obvious advantage for measuring temperature at the same point as that at which the concentration of reactive species is measured. This has been demonstrated for the measurement of both local temperature and conversion of a commercial cyanate ester polymer (AroCy L-10) (Cooper *et al.*, 1996).

A remote monitoring probe for *in situ* Raman spectroscopy based on a single fibre terminating in a Teflon capillary, so enhancing the Raman signal 15-fold compared with the fibre-optics alone, has been described (Aust *et al.*, 1996). When immersed in an epoxy resin, the lower refractive index of the tube ensures that it acts as a waveguide, so enhancing the signal. A further feature of this simple system is that the multivariate data-analysis routine (see Section 3.5) not only allowed the epoxy-resin conversion to be followed in the C–H region of the spectrum, but also showed the effect of temperature fluctuations. This was achieved by using a peak at 3069 cm^{-1} that remained unchanged during cure, but was temperature-sensitive. Unlike the earlier system (Cooper *et al.*, 1996) which ratios the anti-Stokes and Stokes shifts (Equation (3.43)), this approach is based solely on the Stokes-shifted band and so does not yield a direct measurement of temperature without calibration.

3.5 Chemometrics and statistical analysis of spectral data

The description of the various spectral techniques to obtain information regarding the chemical changes occurring in the reacting polymer during processing has focussed on the methods for obtaining compositional information by analysis of the absorption, reflection or emission spectra of the polymer. Particular attention has been paid to the transfer of radiation to and from the reacting system by fibre-optics, so that spectra with maximum SNR are obtained. The level of analysis employed has been to use Beer's law or an analogous calibration to determine the concentration of reacting species from the changes in the spectral absorbance (Equation (3.21)). It is assumed that the bands to be analysed are well resolved from the remainder of the spectrum and that a particular wavelength of absorption may be associated with the species of interest. This is a univariate method since only a single absorption band is used.

When spectra are collected from complex systems containing many components or the changes in spectra might not be readily discerned, then statistical analysis of the spectral data will often be required. The starting point for statistical analysis of spectral data is to have a digitized set of spectral parameters for the process being studied. This may be a set of absorbance spectra consisting of an ordered set of N pairs of values of the independent variable, wavenumber (cm^{-1}), and the dependent variable absorbance. This is then related to the variable of interest to be determined, viz. concentration, through a linear relation such as Beer's law involving the molar absorptivity, ε, as in Equation (3.21), or a non-linear relation that must be independently determined through a calibration routine. Fourier-transform-IR and other spectrometers that are computerized provide digitized data that are immediately available for analysis, so chemometric routines may be applied either during or after acquisition.

The questions that may be asked in analysing the data measured by, say, a fibre-optic NIR or Raman spectral analysis system as described in Section 3.4.2 above are as follows.

(i) How many components are present in the system and which of these are responsible for the spectral changes observed?
(ii) What are the changes in concentration or other property of the polymer and reactants that are leading to the change?
(iii) What is the relation of these to the reactive-processing operation, and can these changes be linked to the chemorheological properties of the system?
(iv) Can a reliable calibration routine be established to enable on-line monitoring of the process variables?

Each of these questions can be answered provided that the data are collected with the appropriate level of precision. There are many specialized texts (Adams, 1995, Jolliffe, 1986, Malinowski and Howery, 1980, Pelikan *et al.*, 1994) and review articles (Lavine, 2000) on chemometrics, and in the following section it is intended to address only those methods most commonly used in spectral analysis of chemorheological data using remote spectroscopy.

3.5.1 Multivariate curve resolution

The purpose of the techniques grouped under this heading, e.g. factor analysis and principal-component analysis, is to determine the number and, if possible, the relative amounts of the various components which, taken together, can recreate the full spectral data set collected over a processing operation. The analysis is multivariate since all the data over the spectral region are used in the computation, in contrast to the univariate approach of using a single band, usually at the wavelength of maximum absorbance.

Consider the set of spectra generated from a polymer sample as it undergoes chemical reaction during processing. At each time interval, a new set of spectra is collected and averaged to give a spectrum for that particular time. These are digitized and stored. As the reaction takes place, some spectral bands will remain unchanged and others will change in absorbance at a particular wavelength due to the chemical reaction changing either the concentration of the species or the chemical nature of the species. If there is a change in chemical species then there may be changes to all of the variables in the spectrum, viz. frequency, bandwidth and absorbance.

The challenge is to determine which components are changing and their concentration at any point in time. Collecting a large number of spectra corresponding to the various compositions generates a large data set that can allow this to be achieved. Modern FT-IR spectrometers come with chemometric routines for this data analysis, and stand-alone systems that can perform these computations in real time as spectra are collected are also available. The main issue is the quality of the spectral data when quantitative analysis is required.

Principal-component analysis (PCA)

This is a statistical analysis of spectral data in order to reduce the data set to the *minimum* number of principal components which, with an appropriate weighting, can explain all of the variations in the spectra that have occurred throughout the process.

The total spectral output from the reaction sequence may be expressed as a series of linear combinations of these components, so the outputs from a PCA analysis of a data set are

- the *score vector* which indicates how closely each sample relates to the principal component
- the *loading vector* which contains the spectral information in the principal component.

For example, the first principal component (PC1) may be the average spectrum generated from all the spectra of all the sample sets, since this should account for the greatest variance in the data set. Subsequently, principal components PC2 etc. are obtained by accounting for as much as possible of the remaining data after PC1 has been subtracted. In this way the principal components are generated, with each contributing to the total spectrum at any point in time by a varying amount as the composition changes through the reaction. Since, in PCA, the first principal component mirrors the average spectrum, mean centring is often performed as the starting point.

Thus the spectrum at each time (or composition profile) will be a linear combination of principal components given by the product *score* × *loading* for that particular sample. It is often valuable to follow the change in a principal component throughout the reaction cycle. This can be seen in Figure 3.45, which is the Raman spectrum before and after the cross-linking reaction of an amine-cured epoxy resin as well as the PC1 *loading* vector after mean centring (Stellman *et al.*, 1995).

This is a relatively simple application since, after background subtraction and normalization (an important part of data manipulation, which will be discussed in detail later) the PCA of the 25 spectral scans collected over the reaction time of 2.5 minutes showed that only one principal component was required in order to explain 86.4% of the spectral variance. The remaining principal components were found to contain mostly noise, so it is concluded that the plot of PC1 against time will capture the important spectral changes occurring during cure. The *score* plot of PC1 against cure time is shown in Figure 3.46;

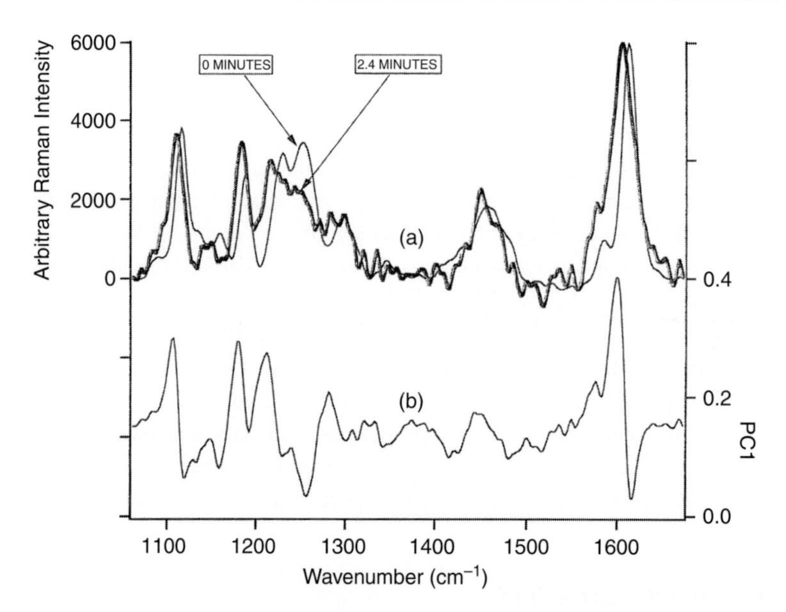

Figure 3.45. (a) Raman spectra from an epoxy resin (DER332/T403) before and after 2.4 minutes of microwave curing and (b) a loading plot of PC1 versus Raman shift. From Stellman *et al.* (1995). Copyright 1995 by Society for Applied Spectroscopy; reproduced with permission of Society for Applied Spectroscopy.

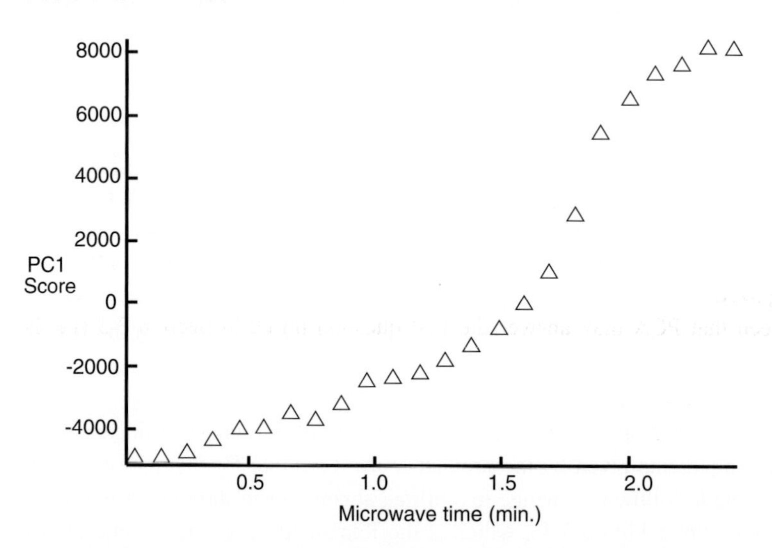

Figure 3.46. A score plot of PC1 versus cure time for the system shown in Figure 3.45. From Stellman *et al.* (1995). Copyright 1995 by Society for Applied Spectroscopy; reproduced with permission of Society for Applied Spectroscopy.

multiplication of the *score* at each time by the *loading* should regenerate the spectrum at that time.

This loading plot shows features similar to the sigmoidal curve generated by the integrated heat flow during a DSC run for the analysis of epoxy-resin cure (Figure 3.5, **Section 3.2.2**) that measures the extent of reaction as a function of cure time.

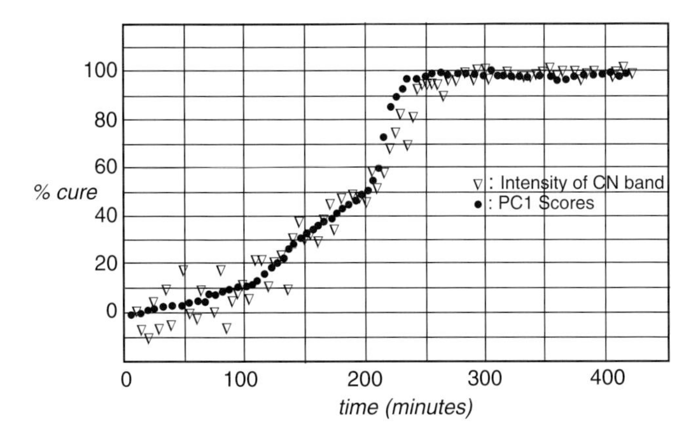

Figure 3.47. A comparison of cure reaction monitoring by a univariate method (CN band intensities) and a multivariate method (PC1 scores) for a cyanate ester resin. Adapted from Cooper (1999).

The system has thus been reduced to a univariate problem and conventional analysis may be employed for subsequent quantification if required. In a closely related study of the curing of a cyanate ester rather than an epoxy resin (Cooper, 1999) a similar profile to Figure 3.46 was obtained after PCA (Figure 3.47), and a comparison with the univariate analysis of single peak intensities that are the major peaks in the first principal component after mean centring was made.

This provides an interesting comparison of the univariate and multivariate approaches. By virtue of using all of the spectral data in PCA, the SNR achieved is much higher than for any single-wavelength analysis, providing a more precise plot of extent of reaction over time. However, there is limited mechanistic information since the subtle changes due to reaction intermediates may be lost in lower-order principal components that, because they do not contribute greatly to the overall variance from the average spectrum, are ignored due to their low statistical weight. A detailed analysis of a single band enables these subtleties to be recognized.

It is seen that PCA may answer the first question posed in Section 3.5 (i.e. how many spectral components are changing during the processing reaction?) as well as, in part, question (iii) regarding the relation to the chemorheological changes on cure. However, the technique does not address the remaining, more quantitative, questions since there has been no calibration of the spectral data. This would require regression of the data against a calibration set that has been constructed under similar conditions.

3.5.2 Multivariate calibration

The data generated from a NIR or Raman spectrum do not immediately provide the concentrations of the species at any time, so there is no predictive capability. Construction of a calibration set requires an independent measure of the property, e.g. by HPLC or by NIR of known mixtures of the components. Two such methods are principal-component regression (PCR) and partial least squares (PLS). As soon as quantitative analysis is considered, the question of noise and reproducibility of the data set becomes important. It is therefore necessary to treat the raw data to remove the drift in baseline etc. which will occur over a long period of spectral acquisition.

Pre-treatment of spectral data for quantitative analysis

In all quantitative analysis the drift in the spectra is commonly addressed by normalization of all peaks against an invariant peak. This requires knowledge of the system as well as assignments for the major bands. Where possible, normalization is the first step in pre-treatment of the data. If Raman spectra are strongly influenced by fluorescence then this is removed (by a polynomial fit to the broad, underlying profile) prior to normalization (Stellman *et al.*, 1995).

There have been many studies of the most effective way to address a common feature in NIR and Raman spectra of reacting systems, viz. the change in spectral baseline. Figure 3.48(a) shows a typical output from a NIR diffuse-reflectance spectral measurement of 12 kinds of ethylene–vinyl acetate (EVA) copolymers differing in vinyl acetate content (Shimoyama *et al.*, 1998).

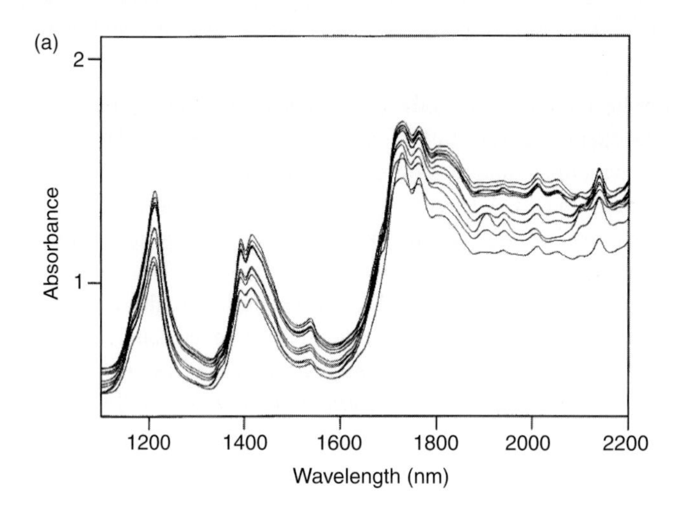

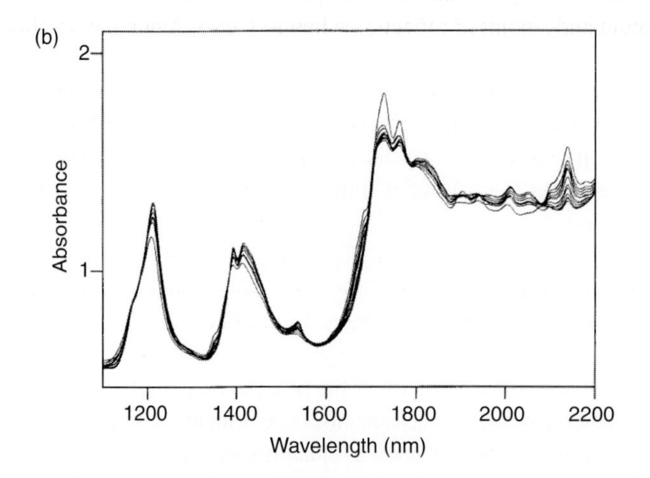

Figure 3.48. (a) The NIR spectra of 12 kinds of EVA copolymers before Multiplicative Scatter Correction (MSC) treatment (Shimoyama *et al.*, 1998). (b) The effect of MSC on the spectra shown in (a). Copyright 1998 John Wiley and Sons, Inc.; reproduced with permission.

It is apparent that there has been significant offset of the baseline between samples and hence that the data are unsuitable for quantitative analysis in the raw form. Three different methods of correction of these data have been compared (Shimoyama *et al.*, 1998):

- multiplicative scatter correction (MSC)
- taking the first derivative
- taking the second derivative

The first approach, MSC, reduces the differences due to light scattering by using the assumption that the wavelength dependence of light scattering is different from that of the NIR absorption (Fischer *et al.*, 1997). An average spectrum of the set is used, and all spectra are then regressed against this average spectrum and the slope and intercept determined. After ratioing the slope to the ideal spectrum, an offset value is generated, which is then subtracted from the original spectrum to give the MSC-corrected spectrum. This processing retains the chemical information in the spectra while removing the non-chemical differences such as scattering.

The effect of MSC on the spectra of EVA copolymers in Figure 3.48(a) is shown in Figure 3.48(b) (Shimoyama *et al.*, 1998). The changes occurring in the region of 1200 nm and, more particularly, between 2100 and 2200 nm, are now apparent.

However, there are many instances where regression against a physical property such as viscosity is required and it is difficult to make assumptions about the nature of the correction. Derivative spectra have the advantage that they are sensing the change in slope of the spectrum so that, at the baseline, the derivative is zero and thus the offset in the intensity or absorbance scale due to baseline drift is eliminated. There is also an enhancement of shoulders in the spectrum – an obvious advantage when dealing with heavily overlapped spectra as in the NIR. However, as in all pre-processing there is an increase in noise in the system. Figure 3.49 shows the original and first-derivative spectra from another study of EVA copolymers (Hansen and Vedula, 1998), and the systematic change in the derivative spectra with change in composition may readily be seen.

The second derivative enables the location of the peaks in the spectra and thus assists in assignment, since it will be negative if the original spectrum has a local maximum, positive if it has a local minimum and a point of inflection when it is zero. Again the baseline drift is eliminated (Shimoyama *et al.*, 1998) and either the first or the second derivative may be used in subsequent data analysis by one of the regression methods. In a comparison of the three correction methods specified above, it was found that MSC was best for the discrimination of the EVA copolymers due to their chemical similarity (Shimoyama *et al.*, 1998), while in a comparison of rheological properties (Vedula and Hansen, 1998) the first-derivative spectra were satisfactory.

In spectral data there will be noise from both multiplicative scatter (MS) and additive scatter (AS). Normalization works well to limit the effects of MS and the derivative spectra remove AS, but only the MSC method can minimize the effects of both (Shimoyama *et al.*, 1997).

Principal-component regression (PCR)

This involves an inverse-squares regression method to provide quantification of the system in which the score and loadings of the principal components have been determined by PCA. Since these contain all of the spectral variations in the system, applying a regression analysis to a calibration set of concentrations from independent measurements creates a

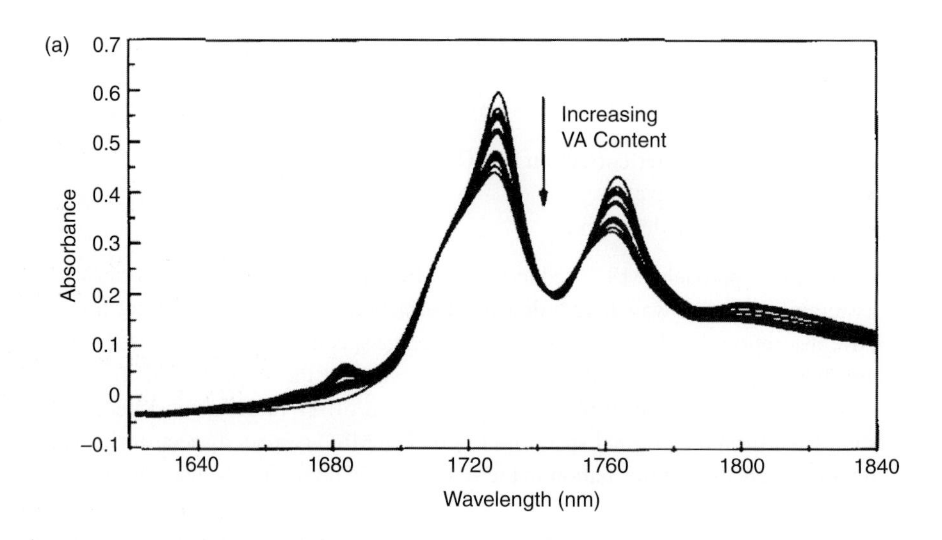

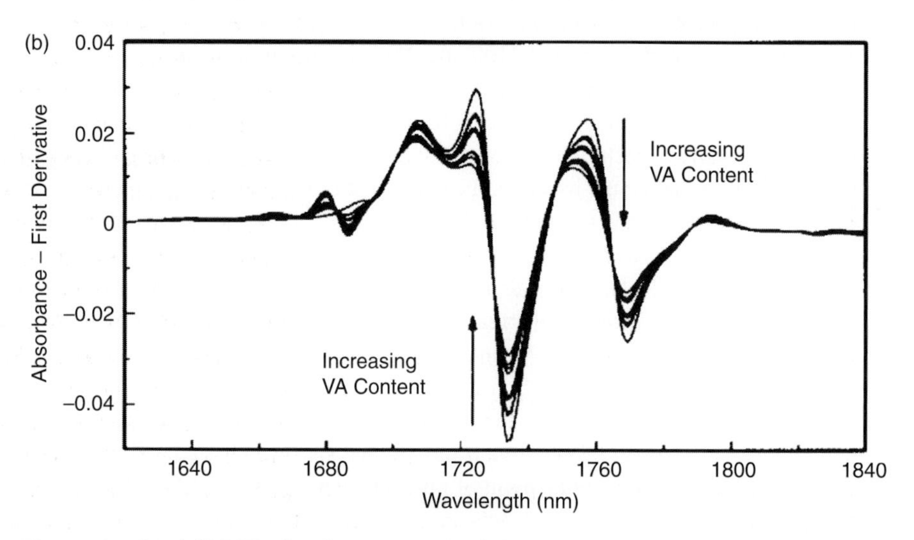

Figure 3.49. Overlaid NIR absorbance spectra of EVA copolymers and the first-derivative plots (lower traces) that result in better spectral resolution and remove baseline offsets (Hansen and Vedula, 1998). Copyright 1998 John Wiley and Sons, Inc.; reproduced with permission.

robust data set against which unknown samples may be quantitated. An example of this is the extension of the PCA analysis shown earlier in Figures 3.45 and 3.46 to a precise determination of the percentage cure of an epoxy resin as a function of time by calibrating the Raman data against DSC data. The available data for both Raman Spectroscopy and DSC were divided into two groups to give calibration and validation sets. Attention was focussed on high extents of cure, and Figure 3.50 shows the relationship between the percentage cure and PC1 score from 93% to 99% cure at two different concentrations of hardener. This study also highlighted the problems of operating with fibre-optics when a stress develops in the fibre. The multivariate analysis immediately shows this as an increase in standard error of prediction (SEP), indicating that spectral artefacts have been introduced, rendering the calibration invalid (Stellman *et al.*, 1995).

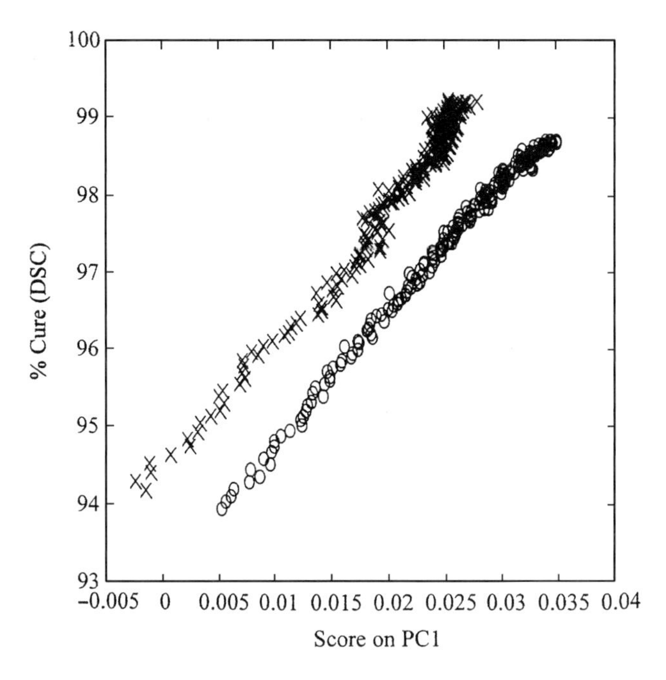

Figure 3.50. The relationship between PC1 score and percentage cure as determined by DSC for the epoxy-resin/hardener system DER 332/Jeffamine T-403 in the following mixing ratios: ○, 100/45; and X, 100/55. Aust *et al.* (1997). Copyright 1997 by Society for Applied Spectroscopy; reproduced with permission of Society for Applied Spectroscopy.

Partial least squares (PLS)

This approach is closely related to PCR but actually uses the concentration data generated for the calibration set in the process of determining the principal components and their score and loading vectors. This results in principal components that are quite different from those in PCR since there is an intrinsic weighting for concentration, so the first few loading vectors contain the majority of the concentration information. This results in two sets of vectors: one for the spectral data and one for the concentration data. These will be interrelated through a regression, which may be a simple linear relation (as in PCR), but this is not assumed and the decomposition of the two sets of data is performed simultaneously.

The two methods have been compared in a study of in-line monitoring of vinyl acetate content in flowing molten EVA copolymers (Khettry and Hansen, 1996), and it was concluded that PLS performed fractionally better than PCR. This has generally been observed, particularly in the presence of interference (Lavine, 2000), but it was noted that overfitting of the data by PLS needed to be avoided. This is achieved by using a statistical analysis such as predicted residual sum of squares (PRESS) to determine the residuals after each principal component has been assigned. The optimum number of factors is defined by a minimum in PRESS. This was illustrated in a study of the prediction of rheological properties of a series of EVA copolymers with various vinyl acetate (VA) contents and melt-flow indices (MIs) (Hansen and Vedula, 1998, Vedula and Hansen, 1998). The PLS model used a singular-value decomposition (SVD) method in order to determine the number of principal components. In this study, PC1 reflected the compositional changes (i.e. [VA]) and PC2 and PC3 the MI changes. The calibration set was developed using a cross-validation method in

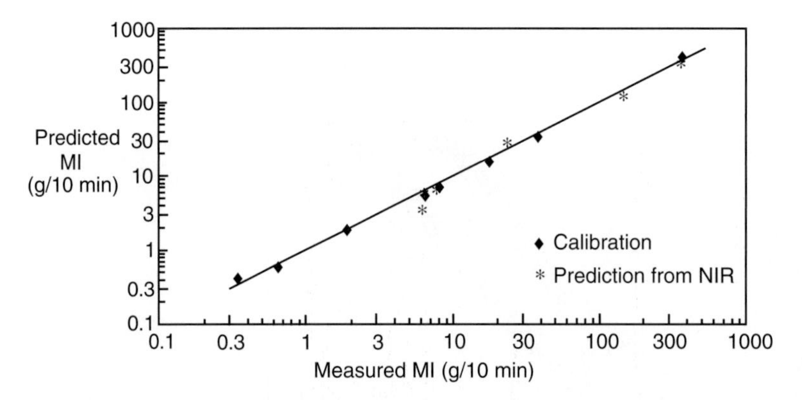

Figure 3.51. Use of NIR to predict the melt index (MI) of a series of EVA copolymers with differing VA contents. A three-factor model was used to predict ln(MI) using the PRESS and the F-statistic criterion (Hansen and Vedula, 1998). Copyright 1998 John Wiley and Sons, Inc.; reproduced with permission.

which one sample was left out of the set and then used to predict the value from the calibration based on the remaining samples. In addition to PRESS, the standard errors of calibration (SEC) and of prediction (SEP) were calculated. This leave-a-sample (LAS) method was then used for each of the samples in the set, in turn. Figure 3.51 shows the plot which links the NIR predictions for MI using the three principal components and the PRESS criterion.

The values of SEC and SEP are 0.14 and 0.46, respectively. The validity of a calibration set when working with materials that may be variable in properties must be continually questioned and, if the results deviate by greater than $\pm$ SEP the calibration might no longer be valid. In a study of the in-line monitoring of the composition of PE/PP blends by NIR spectra, the effect of outliers on the PLS analysis was seen by noting prediction results that consistently exceeded the measured data (Rohe *et al.*, 1998). In a refinement of the data prior to PLS analysis it was found that spectral averaging, smoothing and data reduction (Rohe *et al.*, 1999) allowed the SEP to be reduced to 0.25%, in comparison with 2% for the untreated data. This reiterates the importance of pre-treatment of data before multivariate regression.

3.5.3 Other curve-resolution and calibration methods

It has been noted that there are now more than 20 different multivariate curve-resolution methods in addition to those discussed and some of these have been applied to remote monitoring of chemical reactions (Quinn *et al.*, 1999). Examples of interest include those such as self-modelling curve resolution (SMCR) which can be used to resolve composition profiles of reactants and products (as well as reactive intermediates) from knowledge of the spectra, without a basis set of known concentration. A further study of SMCR in reactive processing has highlighted some of the limitations of the approach in fibre-optic NIR (Sasic *et al.*, 2000), since it is necessary to have at least a knowledge of the spectra of the reactants and products. The drawback for on-line analysis is that the concentration profiles are all relative unless there is one known concentration. While not all spectra can be resolved due

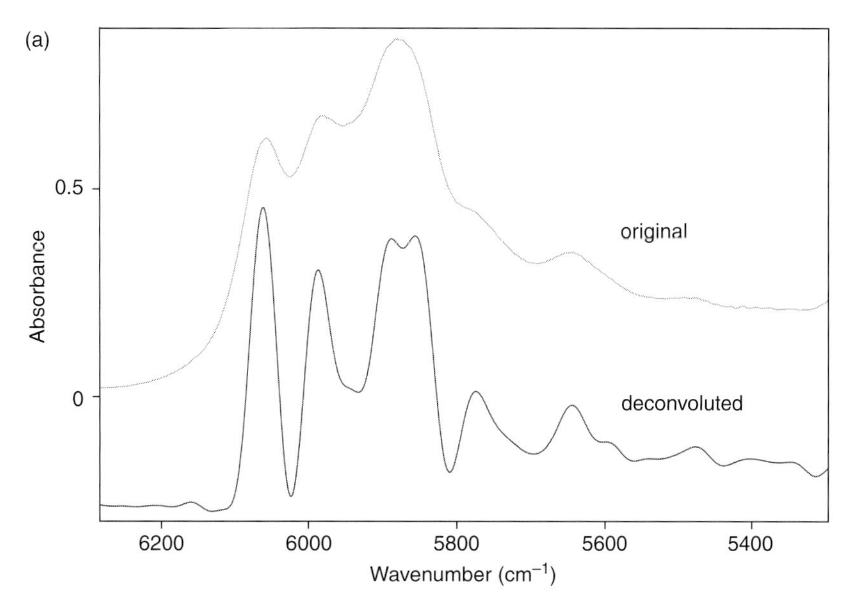

Figure 3.52. The NIR spectrum of MY721 + 27% DDS (a) before and (b) after Fourier self-deconvolution ($K = 2$). This shows the complete resolution of the epoxy band at 6057 cm^{-1} from the other C—H overtone bands. Reproduced with permission from St John (1993).

to the overlap in the NIR, a good estimate of the number of components may be made. As for other multivariate techniques, pre-treatment of the data is required, and in this application MSC (Section 3.5.2) was effective at generating spectral data that, when analysed over narrow wavelength regions, gave meaningful reaction profiles.

It is important to note that among the techniques for pre-treatment of data there are information-recovery processes including Fourier self-deconvolution (FSD). This is particularly relevant in FT-IR since the spectra as obtained have been apodized. This involves the removal of an artefact appearing in the Fourier transformation due to the finite travel of an interferometer so that the limits of the transformation are not infinite as required mathematically. The effect on the spectrum is to produce side-lobe artefacts, which, in order to aid spectral interpretation, are removed by multiplication by an apodization function. This improvement in cosmetic quality of the spectrum comes at the price of band broadening, so the purpose of FSD is to reduce the spectral bandwidth by removing the apodization and further narrowing the spectrum by multiplying it by the appropriate function. Since this will introduce noise into the system, the spectra must have a high SNR before FSD is performed. The resulting improvement in resolution often assists peak assignment, but quantification must now involve band areas instead of peak heights (Koenig, 1999).

Figure 3.52 shows the effect of Fourier self-deconvolution (FSD) on NIR spectra from an epoxy-resin system (St John, 1993), providing the resolution of the epoxy band and so enabling quantitation (through band area) of the consumption kinetics during reactive processing. This would not have been possible before the application of FSD. A chemometric analysis will also produce the same result for the cure profile and may be regarded as more accurate, since the entire spectrum of the resin is involved, but, if absolute concentrations are required (for example to determine rate coefficients), then a band-area analysis after FSD is unambiguous.

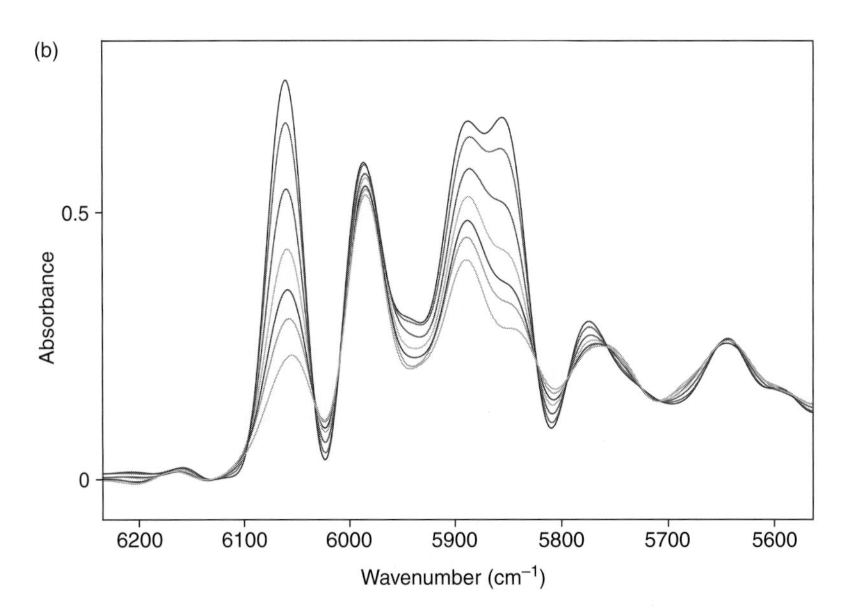

Figure 3.52. (cont.)

3.6 Experimental techniques for determining physical properties during cure

3.6.1 Torsional braid analysis

As noted by Palmese and Gillham (1987), torsional braid analysis (TBA) was an important measurement technique for monitoring changes in physical properties during cure. In TBA specimens are prepared by impregnating the thermoset material into heat-cleaned glass braids. The pendulum is set into motion to produce freely damped waves with frequencies ranging from 0.02 to 2.0 Hz. Measurements of frequency and decay constants characteristic of each wave provide two dynamic mechanical properties, namely relative rigidity (elastic) and logarithmic decrement (viscous). Maxima in logarithmic decrement curves are used to identify transitions. Both isothermal and non-isothermal testing can be conducted to analyse transitions and ageing effects. Typical data for an epoxy-resin system are shown in Figure 3.53.

Note that the gelation and vitrification are identified by the maxima in the logarithmic decrement. Note, however, that Stutz and Mertes (1989) found that for their epoxy-resin system the first peak was not caused by gelation of the resin, but occurred at approximately the same viscosity for each resin and additionally was observed only when a fibre braid was used as support in a TBA. Wetton and Duncan (1993) also found that glass fibres or wire meshes have a stronger influence on the storage/elastic modulus than they do on the loss modulus, thus disrupting decrement measurements. In general it is proposed that the braided supports can have an influence on the results due to interfacial, flow phenomena or internal stresses.

There are other supports used in torsional braid (or perhaps, more strictly termed, supported dynamic mechanical) measurements. These include solid substrates (solids that are free of mechanical transitions under the testing conditions), which have been shown to improve the precision of results over those obtained using conventional glass braids (Wetton, 1986). However, these solid substrates generally require thick coatings. Also wire-mesh

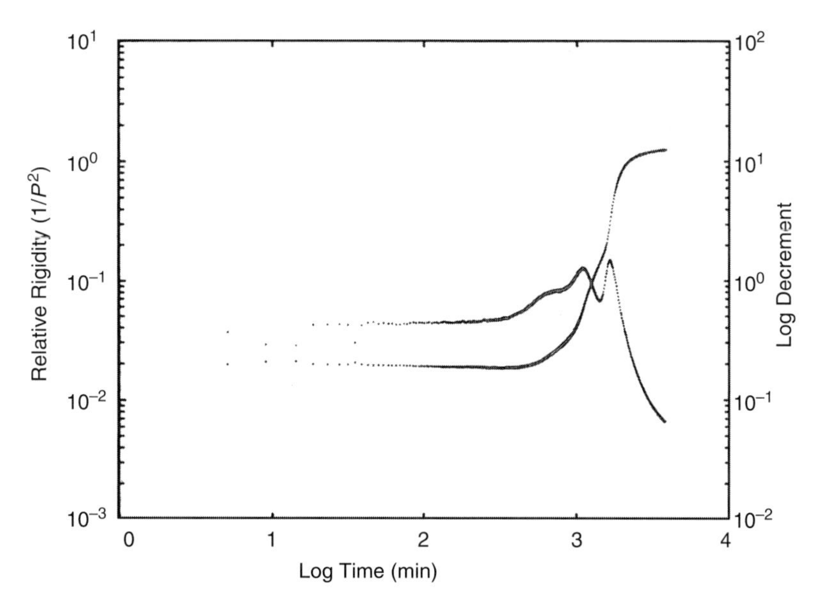

Figure 3.53. Isothermal cure of an epoxy-resin using torsional braid analysis (TBA). Reprinted from Figure 57 (Prime, 1997). Copyright (1997), with permission from Elsevier.

substrates have been used, which offer the advantages of the solid substrate, without the experimental difficulties of applying a thick, even coating. Prime *et al.*, (1988) showed that measurements obtained using the wire-mesh supports were significantly more reproducible than were those from tests using glass braids. Semi-quantitative measurements of G' and G'' have also been developed for the wire-mesh tests (Dillman *et al.*, 1987), which provide a more in-depth analysis of viscoelasticity changes for the curing systems.

3.6.2 Mechanical properties

Mechanical analysis is a common approach for evaluating polymer properties. For reactive systems we shall limit ourselves to the thermal mechanical analyser (TMA) and the dynamic mechanical thermal analyser (DMTA).

The TMA essentially operates by measuring the dimensional change of, or the penetration into, a material at a specified time, temperature and (typically non-oscillating) load stress. The important variables to measure for reactive systems are the glass-transition temperature, softening temperature and heat-deflection temperature, the linear coefficient of thermal expansion and physical ageing effects on these variables. For example, Figure 3.54 shows the use of a knife-edge probe to determine the softening points both of the inner and of the outer coatings of a thermoset-coated motor winding wire (Prime, 1997a).

Much more detailed procedures are described in Prime (1997a). Ramis *et al.*, (2003) use shrinkage measurements on a TMA to determine the gelation point of polyester-based thermoset coatings. As shown in Figure 3.55, the gel point is defined as the point at which the rate of change of shrinkage abruptly slows down.

Alternatively, Ramis and Salla (1997) define gelation of a polyester resin via Figure 3.56 as the point at which a submerged TMA probe becomes embedded ('stuck') in the resin.

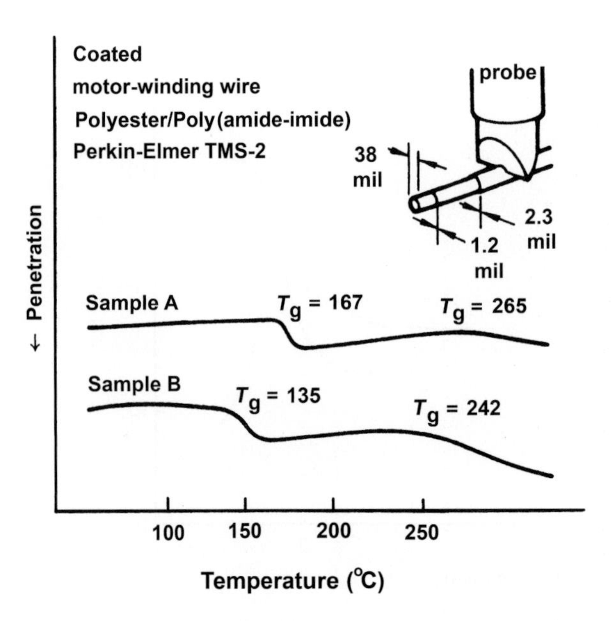

Figure 3.54. Typical TMA curves for a motor-winding wire. Reprinted from Figure 42 (Prime, 1997). Copyright (1997), with permission from Elsevier.

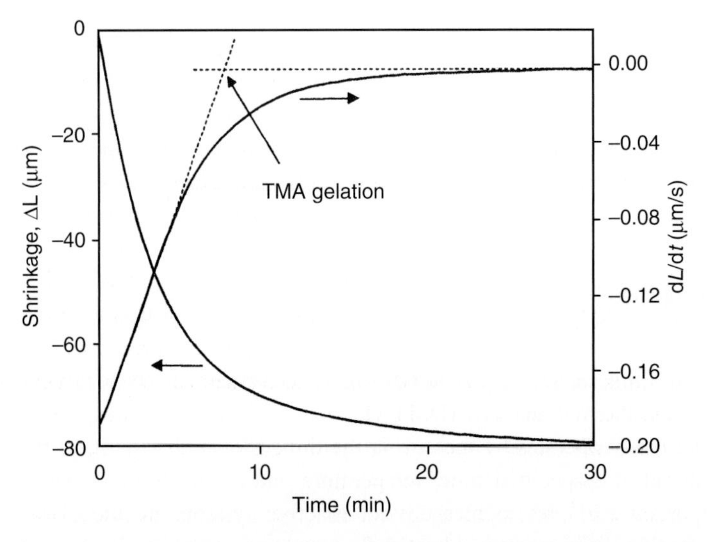

Figure 3.55. Gelation determined from shrinkage and shrinkage rate versus time data from TMA. Reprinted from Figure 3 (Ramis *et al.* 2003). Copyright (2003), with permission from Elsevier.

Garcia and Pizzi (1998) presented a detailed paper on the use of TMA for monitoring the gelation of phenol–formaldehye resins. This paper noted that there were two peaks of the modulus in TMA curves during cure; the first peak is related to critical entanglements during cure and the second peak is related to the crosslinking network. They suggest that the initial rise of the second peak should be considered as the gel point in these resin systems wherein entanglements are important. Cadenato *et al.*, (1997) highlighted differences when measuring gel and vitrification times via TMA and DMTA of an epoxy–amine system.

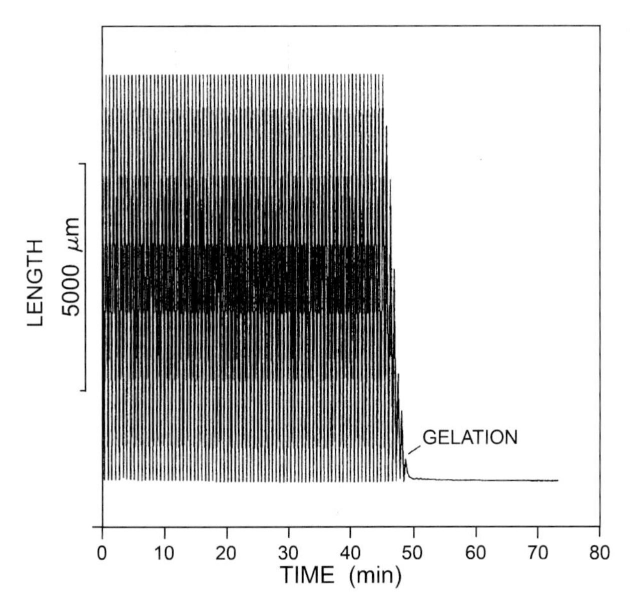

Figure 3.56. Gelation determined by damping of oscillations from TMA. Reprinted from Figure 1 (Ramis and Salla 1997). Copyright (1997), with permission from Elsevier.

These differences were removed when one presented the data as a function of conversion, instead of time, indicating that there are differences in curing between DMTA and TMA.

The DMTA measures the complex modulus, compliance and viscosity in various modes of deformation (shear, flexural, tensile, bending, compression), typically at a range of applied oscillatory deformation frequencies. The glass-transition temperature, gelation and vitrification, secondary transitions, creep, mechanical properties and stress relaxation may all be determined using this technique. Also isothermal and non-isothermal temperature profiles may be used. The DMTA is relatively insensitive to events prior to gelation, but becomes an essential tool for monitoring transitions and mechanical properties after gelation.

In dynamic mechanical analysis for thermosetting systems the principal use is to monitor the glass-transition temperature (T_g). In a DMTA T_g is defined as the maximum in loss modulus, loss compliance or loss tangent. For example, Figure 3.57 (Prime, 1997a) shows the various transitions for a cured magnetic ink coating that had previously been cured under air and N_2.

Figure 3.58 (Labana, 1985) generically shows the effects of increased crosslinking on the T_g of curing systems.

Gelation and vitrification may also be observed using a DMTA. Generally gelation is described by a peak in loss modulus, whereas vitrification is described as a peak in loss modulus or loss tangent. Figure 3.59 (Prime, 1997a) shows the progression of the storage and loss moduli (under shear) of a curing DGEBA–DDS system. This work shows the gelation and vitrification as peaks in the loss modulus.

The DMTA has also been used to monitor the changes in dynamic viscoelastic properties during cure as highlighted by Dillman and Seferis (1989) for TGDDM/DDS epoxy-resin-based systems. Hofmann and Glasser (1990) developed a DMTA system for DGEBA/m-PDA epoxies impregnated onto glass braids. They define gelation and vitrification by

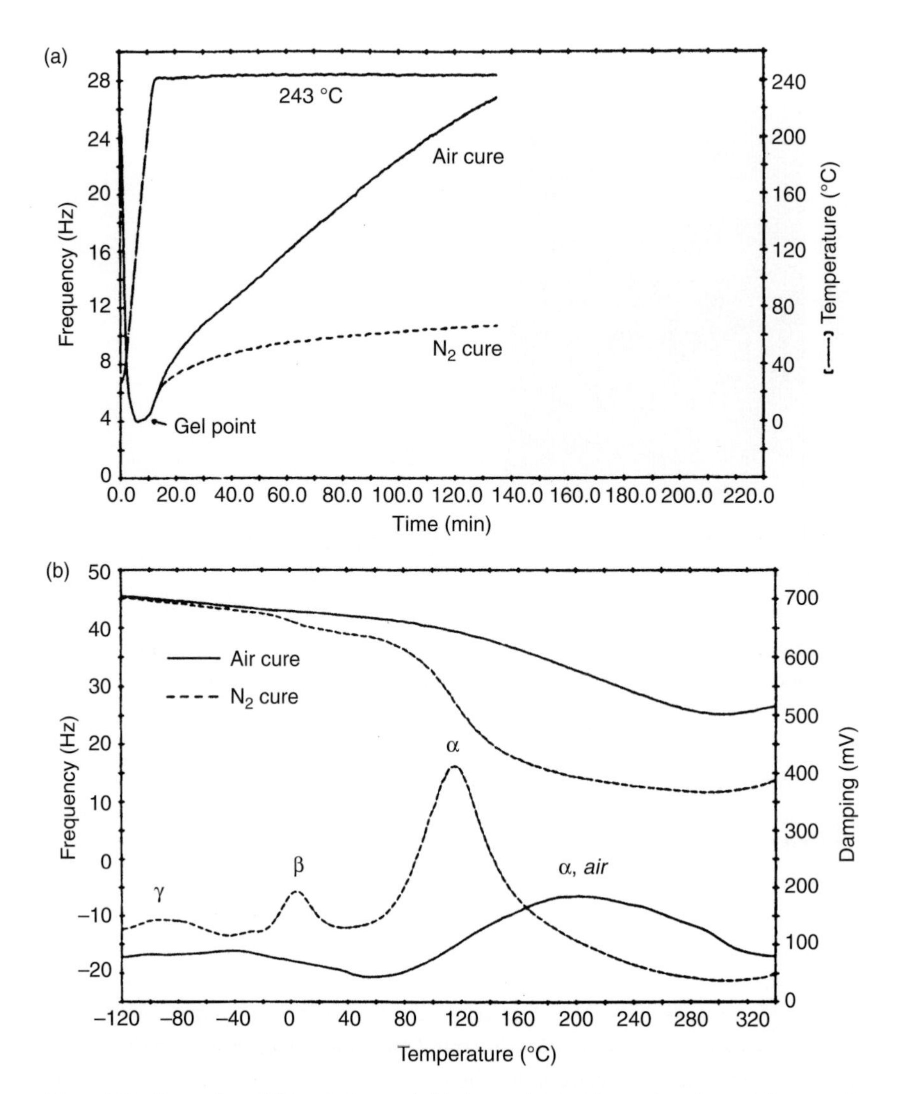

Figure 3.57. Examples of (a) gelation and (b) thermal transitions of curing coatings as determined by DMTA. Reprinted from Figure 46 (Prime, 1997). Copyright (1997), with permission from Elsevier.

maxima in the loss tangent and also monitor the progression of cure conversion by DSC. They compile a TTT diagram from the DMTA and DSC data, Figure 3.60, clearly showing gelation, vitrification, maximum-conversion-rate and end-of-conversion points at a range of isothermal curing temperatures.

A fundamental property that determines the state of a reacting system is its extent of cure or chemical conversion (α). Several papers have shown that there is a unique relationship between the glass-transition temperature (T_g) and α that is independent of cure temperature and thermal history. This may imply that molecular structures of materials cured with different histories are the same or that the changes in molecular structure do not affect T_g. There are generally accepted to be two approaches to modelling glass-transition–conversion relationships, namely thermodynamic and viscoelastic approaches. These are summarized in Table 3.8.

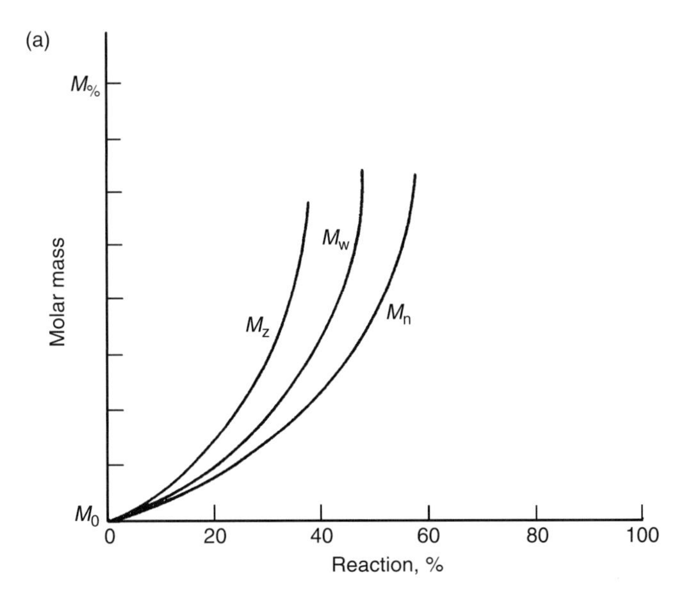

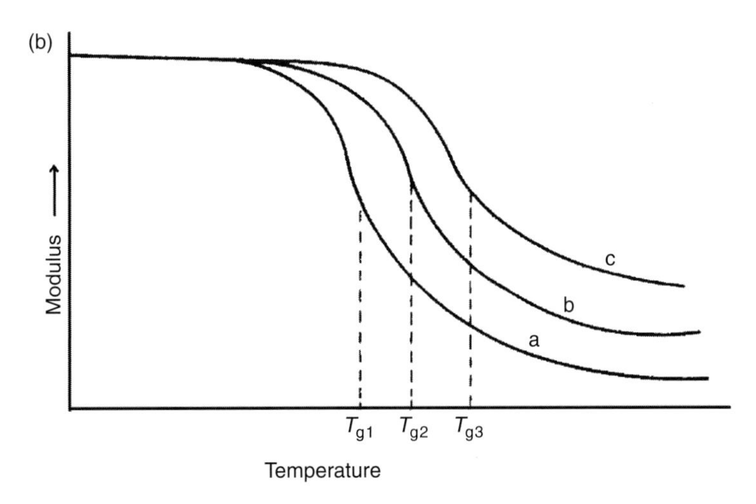

Figure 3.58. An example showing (a) the variation in molar mass with extent of reaction and (b) the variation of modulus versus temperature for polymers with three different crosslink densities, as determined by DMTA. Adapted from Figures 1 and 4 (Labana, 1985). Copyright (1985). Reprinted with permission of John Wiley and Sons, Inc.

3.6.3 Dielectric properties

Measurements of dielectric properties of reactive materials constitute an attractive method for determining physical properties during curing, especially because of their easy application on-line to processing equipment. Dielectric analysis involves the measurement of electrical polarization and conduction properties of a sample subject to a time-varying electric field. Dielectric properties such as the permittivity, loss factor and conductivity may be measured and related to the glass transition, secondary transitions, rheological phenomena, vitrification, segmental mobility and dipolar relaxations.

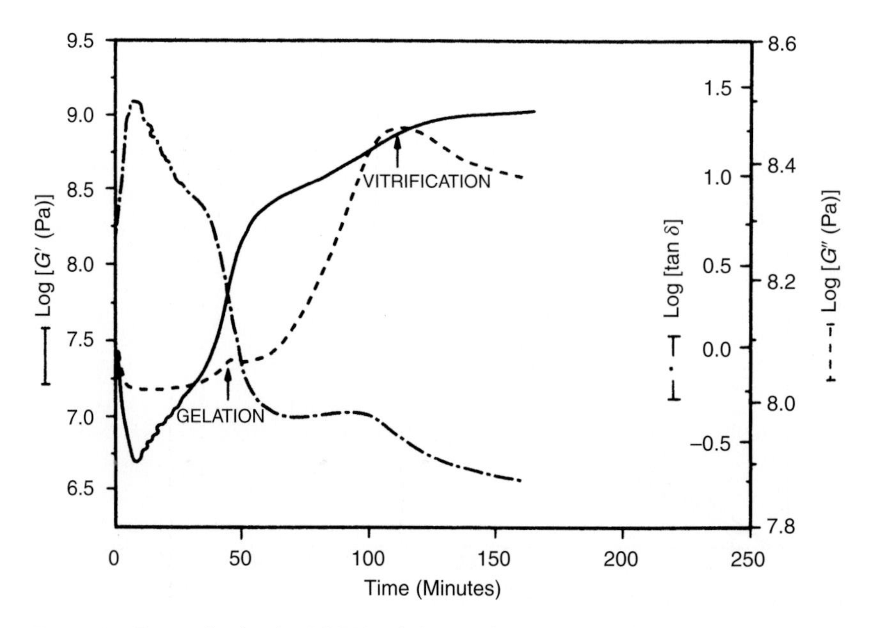

Figure 3.59. Determination by DMTA of changes in storage and loss moduli during cure. Reprinted from Figure 59 (Prime, 1997). Copyright (1997), with permission from Elsevier.

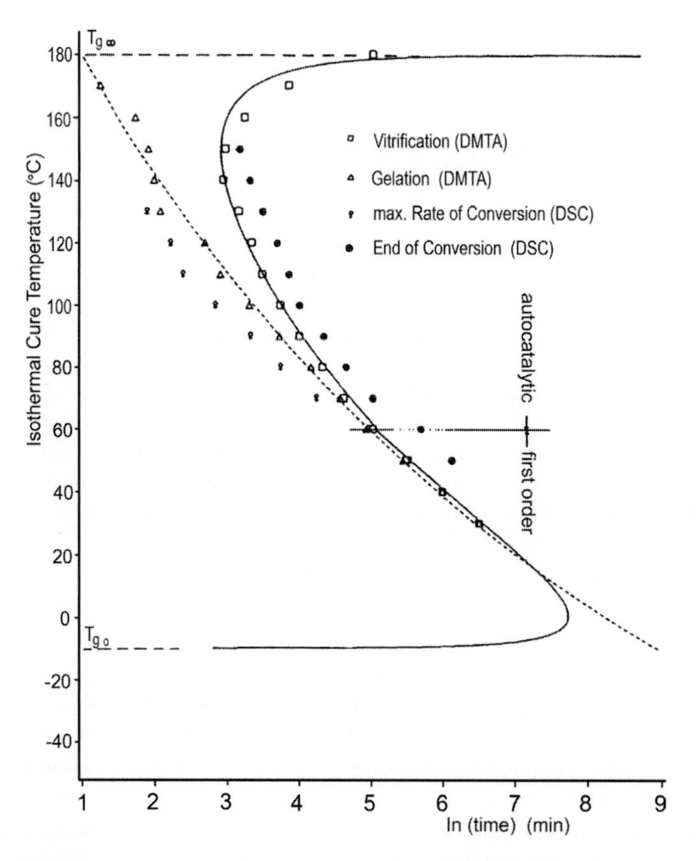

Figure 3.60. A DMTA-derived TTT diagram. Reprinted from Figure 6 (Hofmann and Glasser, 1990) Copyright (1990), with permission from Elsevier.

Table 3.8. Models for T_g

Model	Equation	Features
Thermodynamic Hale (Hale *et al.*, 1991)	$\frac{1}{T_{gu}} = \frac{1}{T_{g0}} - ka$	$T_{gu} = T_g$ of uncrosslinked polymer; k is a constant and α is the conversion. This equation is for pre-gel in the absence of substantial crosslinking.
Fox (Fox and Loshaek, 1955)	$T_g = kX$	X is the degree of crosslinking.
Dimarzio (Dimarzio, 1964)	$\frac{1}{T_g} = kX$	
Dibennedetto (Dibennedetto, 1987)	$\frac{T_g - T_{gu}}{T_{gu}} = \frac{(E_\infty/E_0 - C_\infty/C_0)X}{1-(1-C_\infty/C_0)X}$	X is the crosslink density or fraction of all segments crosslinked, E is the lattice energy, C is the segment mobility. This gives a good fit for many systems.
Modified Dibennedetto, (Adabbo and Williams, 1982)	$\frac{T_g - T_{g0}}{T_{g0}} = \frac{(E_\infty/E_0 - C_\infty/C_0)a}{1-(1-C_\infty/C_0)a}$	T_{g0} is the T_g of polymer at $\alpha = 0$. This is good for many systems except for highly crosslinked multifunctional epoxy-resin moulding compounds (Hale 1991).
Couchman (Couchman, 1978)	$\frac{T_g - T_{g0}}{T_{g\infty} - T_{g0}} = \frac{\lambda a}{1-(1-\lambda)a}$	λ is an adjustable structure-dependent parameter, $\lambda = T_{g0}/T_{g\infty} = \Delta C_{p\infty}/\Delta C_{p0}$, where $\Delta C_{p\infty}$ is the difference in heat capacity for glassy and rubbery and liquid states at T_g for 100% conversion and ΔC_{p0} is the difference in heat capacity for glassy and rubbery and liquid states at T_g for 0% conversion. Pascault and Williams (1990) had a modified ΔC_p relationship, $\Delta C_p \sim 1/T$.
Hale (Hale *et al.*, 1991)	$T_g = \frac{1/(1/T_{g0} - ka)}{1 - k_2X/(1-\mu X^2)}$	k is a constant, k_2 is a constant for crosslink effects, X is the crosslink density (mol/mol) and μ is a constant for non-ideal effects at high conversion. This equation is good for highly functional systems.
Vendetti and Gillham (Vendetti and Gillham, 1993)	$\ln T_g = (1-a)\ln T_{g0} + \frac{(\Delta C_{p\infty}/\Delta C_{p0})a \ln T_{g\infty}}{1 - a + (\Delta C_{p\infty}/\Delta C_{p0})a}$	Where ΔC_p is independent of T.
Viscoelastic Gan (Gan *et al.*, 1991)	$T_g = \frac{E_r}{R\ln[C_1(1-a)^{\varphi} + C_2]}$	E_r is the activation eneregy of transition from the glassy to the rubbery state, R is the gas consant, $C_1 = \exp[E_r/(RT_{g0})] - \exp[E_r/(RT_{g\infty})]$, $C_2 = \exp[E_r/(RT_{g\infty})]$ and ϕ is a chain-entanglement parameter. Good fits have been obtained (T_g changes due to change in relaxation time).

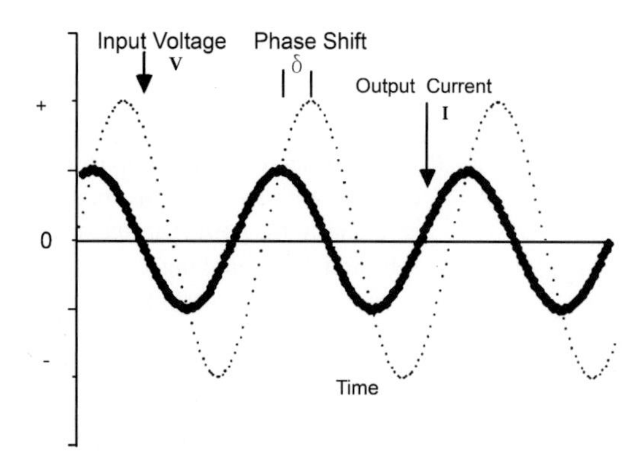

Figure 3.61. The voltage response for dielectric materials. Reprinted from (Prime, 1997). Copyright (1997), with permission from Elsevier.

In dielectric analysis a sample is placed between two electrodes and a sinusoidal voltage (at specific frequencies of interest) is applied to one electrode. This applied voltage establishes an electric field in the samples. In response to this the sample can become electrically polarized (dielectric polarization) and may also conduct a net charge from one electrode to the other (ionic conduction). Both the dielectric polarization and ionic conduction give rise to a current, the amplitude of which is dependent on the frequency of the applied voltage, the temperature and the structural properties of the sample. Figure 3.61 (Prime, 1997a) shows input voltage and output current waveforms.

Note that the amplitude and phase shift are recorded and may be related to the dielectric properties of the medium (as long as plate geometries are also included in the assessment).

The dielectric permittivity of a medium (relative to the permittivity of free space, $\varepsilon_0 = 8.85 \times 10^{-12}$ F/m) is given by ε' and measures the polarization of the medium per unit applied electric field. The dielectric loss factor arises from energy loss during time-dependent polarization and bulk conduction. The loss factor is written as ε''. The loss tangent or dissipation of the medium, $\tan\delta$ is defined by $\varepsilon''/\varepsilon'$. The orientation of molecular dipoles has a characteristic time τ_d. Typically τ_d is short early in the cure but grows large at the end of the cure.

The simplest model of dipole orientation is that given by Debye (1912), in which one assumes a single relaxation time for all species. This leads to the following expressions for ε' and ε''

$$\varepsilon' = \varepsilon_u + (\varepsilon_r - \varepsilon_u)\Big/\left[1 + (\omega\tau_d)^2\right], \tag{3.44}$$

$$\varepsilon'' = \sigma/(\omega\varepsilon_0) + (\varepsilon_r - \varepsilon_u)\omega\tau_d/\left[1 + \left(\omega\tau_\delta^2\right)\right], \tag{3.45}$$

where ε_u is the unrelaxed permittivity (or baseline permittivity) and ε_r is the relaxed permittivity equal to the bulk permittivity when molecular dipoles align with the electric field to the maximum extent.

Reiterating, there are generally two effects considered in curing systems, namely conduction of ionic resin impurities and molecular-dipole orientation. The ionic conductivity $\sigma = \Sigma_i q_i N_i \mu_i$, where μ_i is the ion mobility, N_i is the number of ions per unit volume and q_i is the charge, for species i. Qualitative relationships between conductivity and properties of

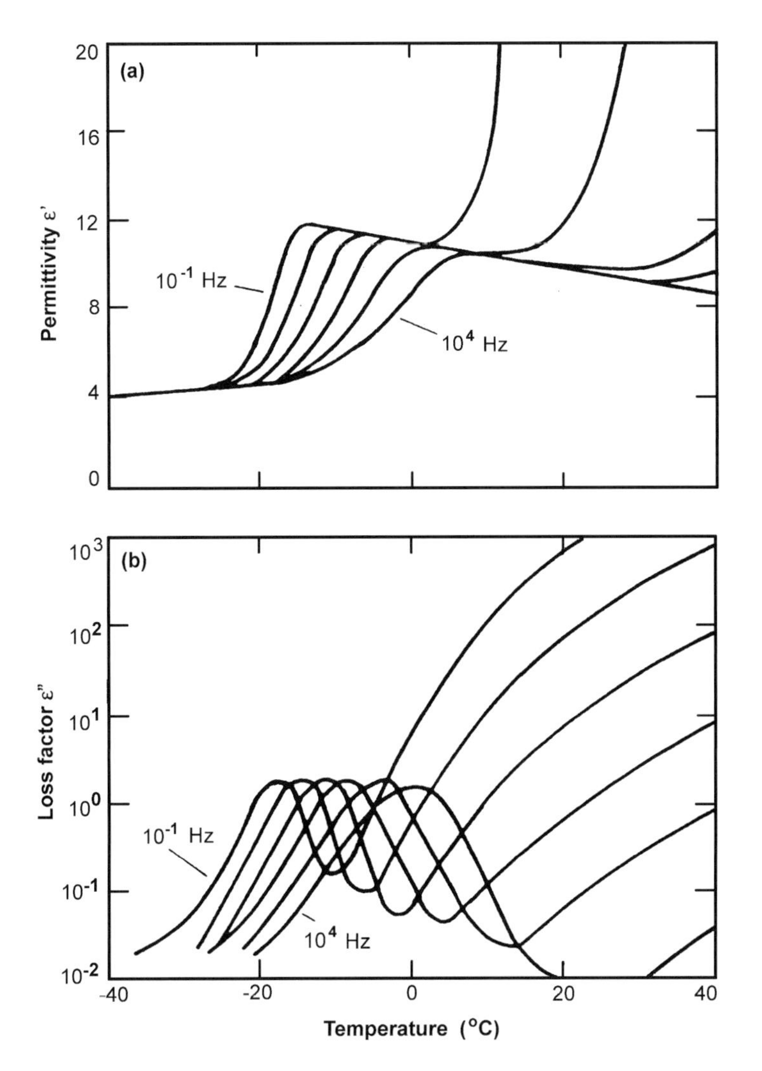

Figure 3.62. Peaks in permittivity and loss factor denoting T_g for a dielectric test. Reprinted from Figure 73 (Prime, 1997). Copyright (1997), with permission from Elsevier.

the resin can be established with reference to simplifications. For example, Stokes' law states that $\mu_i = q_i/(6\pi\eta r_i)$ where r_i is the sphere radius and η is the viscosity of the medium. In this simple model the mobility and hence conductivity varies with $1/\eta$.

Both the dipole-relaxation time and the ionic conductivity are related to the glass-transition temperature T_g. As a material is heated through its glass-transition temperature, static dipoles gain mobility and start to oscillate in an electric field. This causes an increase in permittivity and a loss-factor peak is noted. Obviously this motion is affected by frequency (lower frequencies have greater effects). This effect is shown in Figure 3.62 (Prime, 1997a), which shows the peaks in permittivity and loss factor at T_g.

Note the frequency dependence of the peaks. The low-frequency peaks are known to correspond well to other thermal measures of T_g.

There are differing opinions on whether gelation can be measured by dielectric analysis. Ionic conductivity is related to viscosity, in that conductivity decreases with increasing viscosity (increasingly restricted mobility). Some authors attribute a rapid decrease (Acitelli *et al.*, 1971) or an inflection (Boiteux *et al.*, 1993) in conductivity with gelation. However, Senturia and Sheppard (1986) argue that auto-catalysis and the proximity to vitrification can also contribute to large decreases in conductivity.

There is also a good relationship between dielectric properties and chemical kinetics. The relaxed permittivity is related to chemical composition because it is a strong function of the concentration of polar molecules. Also dipolar mobilities and ionic mobilities depend on the extent of reaction and the changing value of T_g during cure. The relaxed permittivity (maximum dipolar alignment at a given temperature and chemical composition) can be qualitatively related to chemical changes, but is generally not a useful quantitative measure due to the occurrence of complex inter-relationships.

The average dipole mobility at a given temperature and conversion (as given by the frequency of the maximum in loss factor, or 1/frequency $= \tau_d$) and the distribution of relaxation times (as evaluated from measurements of the frequency dependences of the permittivity and loss factor) are also important in determining transition properties. Early in cure, the dipole relaxation time correlates well with viscosity. The dipole loss peak has also been shown to have a relationship to the vitrification peak. Viscosity and resistivity (1/conductivity) are correlated before gelation but diverge at gelation. Attempts to correlate conductivity, chemical conversion and T_g have been made. The sensitivity of conductivity to T_g changes means that conductivity can be used as a good measure for T_g.

Bidstrup (Bidstrup, 1986, Bidstrup and Senturia, 1987, Bidstrup *et al.*, 1989) highlights microdilatometry (combined dielectric and rheological measurements on a small parallel-plate sample) as a useful technique with potential for on-line and combined dielectric–rheological data generation. Lane (Lane *et al.*, 1986), Kranbuehl (Kranbuehl *et al.*, 1987, 1990), Gorto (Gorto and Yandrastis, 1987, 1989) and Boiteux (Boiteux *et al.*, 1993) all highlight in detail the benefits of using combined dielectric and rheological methods to determine the gelation transition and properties of epoxy-resin systems from micro-dilatometry. Kenny *et al.* (1991), Sanjana (1986) and Hsieh and Ho (1999) extend the application of dielectrics to thermoset polymer composite materials.

3.6.4 Rheology

A brief introduction to rheology

Rheology is the study of flow and deformation of matter. It is used for many applications, including

- classification of materials,
- characterization of flow properties,
- troubleshooting processing problems,
- modelling, controlling and optimizing processing and
- understanding flow phenomena in complex fluids.

From these uses it is evident that rheology is both an excellent diagnostic for flow behaviour and extremely useful in assisting optimization and prediction of processing. There are many excellent and complete texts on rheology (Macosko, 1994, Larson, 1999); however, there are relatively few that describe experimental rheology in sufficient depth to allow one to begin experiments. Thus it is the aim of the next section to provide a background to experimental rheology in order to understand chemorheology better.

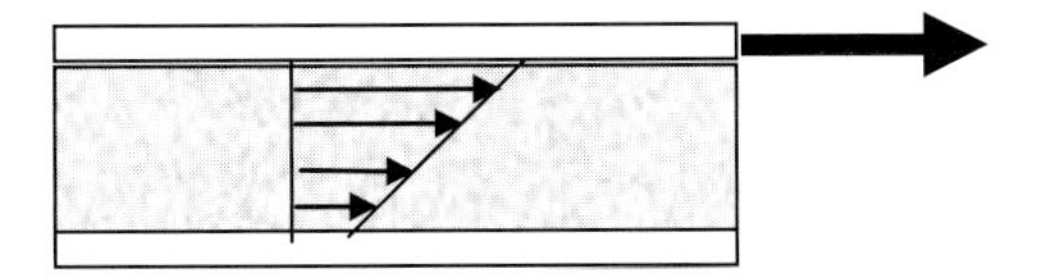

Figure 3.63 Simple shear deformation.

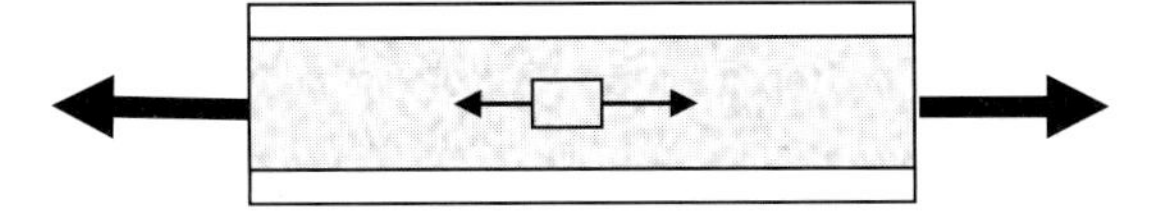

Figure 3.64 Simple elongational deformation.

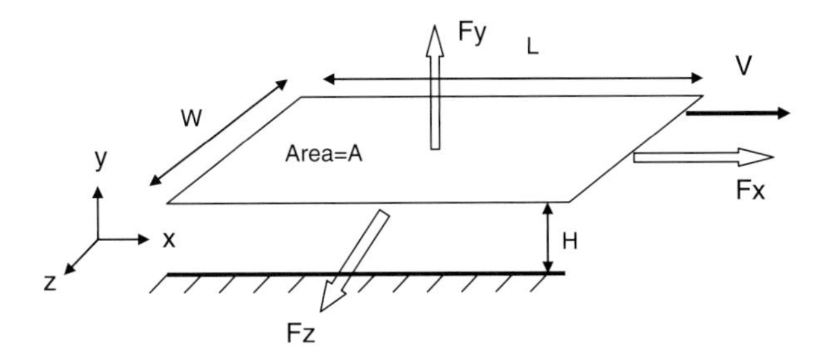

Figure 3.65 Ideal steady-shear flow.

Shear, steady and dynamic, transient; elongational rheology

There are various broad classifications of rheology. In this text we shall use the broad classifications of shear and elongation rheology.

Shear rheology occurs when simple shearing deformation is applied to the material and shear rheological properties are measured, Figure 3.63.

In elongational rheology the material is subject to stretching (elongational) deformation and elongational rheological properties are determined, Figure 3.64.

Shear rheology itself can be further subdivided into cases of steady shear, dynamic shear and transient shear.

Steady-shear rheology

In steady-shear rheology continuous steady shearing deformation is applied to the material, and steady-shear properties are measured. For the simplest case of an infinite parallel-plate system, Figure 3.65, one can define relevant properties as follows;

The shear rate (γ') is defined as the change in fluid velocity (V) over the gap between the infinitely parallel plates (H) which is perpendicular to flow,

$$\gamma' = V/H. \tag{3.46}$$

The shear stress (σ) is defined as the shear force induced by the fluid on the plate in the direction of flow (F) divided by the area normal to flow (A) upon which the force is acting,

$$\sigma = F/A. \tag{3.47}$$

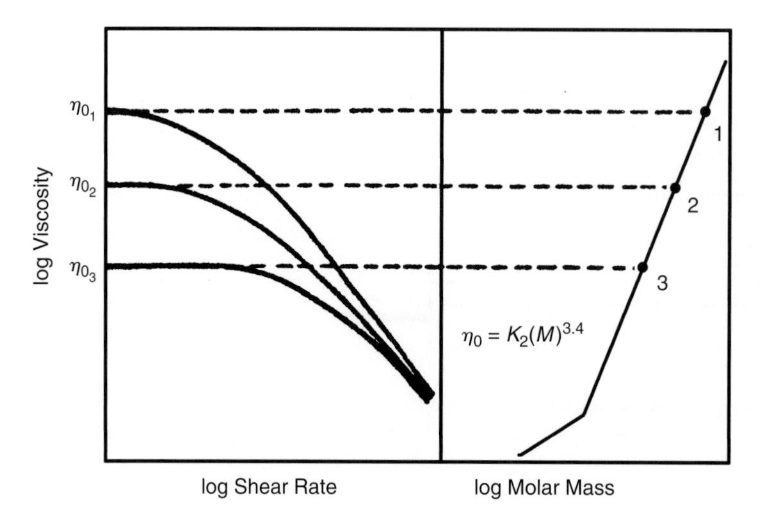

Figure 3.66 Steady-shear viscosity for a polymer system at three different molar masses.

The viscosity may then be defined as the stress divided by the shear rate,

$$\eta = \sigma/\gamma' = FH/(AV). \tag{3.48}$$

The normal stresses (N_{1s} or N_{2s}) (which are induced in flow of viscoelastic fluids) are defined as the force normal (F_x or F_z) to flow divided by the area in that direction,

$$N_{1s} = F_x/A, \tag{3.49a}$$

$$N_{2s} = F_z/A. \tag{3.49b}$$

The normal-stress difference (NSD) may also be defined as the difference between the two normal stresses perpendicular to flow,

$$\text{NSD} = N_{1s} - N_{2s} \tag{3.50}$$

More detailed analyses have been given for analysis of more complex geometries that are used in practical rheometers and for extensions to three-dimensional flow for general modelling applications, and the reader is recommended to refer to Macosko (1994) for those applications.

Figures 3.66 and 3.67 show typical steady shear flow results for polymer systems.

Figure 3.66 shows the steady-shear viscosity for a polymer system at three molar masses. Note the plateau in viscosity at low shear rates (or the zero-shear viscosity). Also note how the zero-shear viscosity scales with M_w to the power 3.4. (This is predicted by Rouse theory (Rouse, 1953).) Figure 3.67 shows the viscosity and first normal-stress difference for a high-density polyethylene at 200 °C. Note the decrease in steady-shear viscosity with increasing shear rate. This is termed shear-thinning behaviour and is typical of polymer-melt flow, in which it is believed to be due to the polymer chain orientation and non-affine motion of polymer chains. Note also that the normal-stress difference increases with shear rate. This is also common for polymer melts, and is related to an increase in elasticity as the polymer chain motion becomes more restricted normal to flow at higher shearing rates.

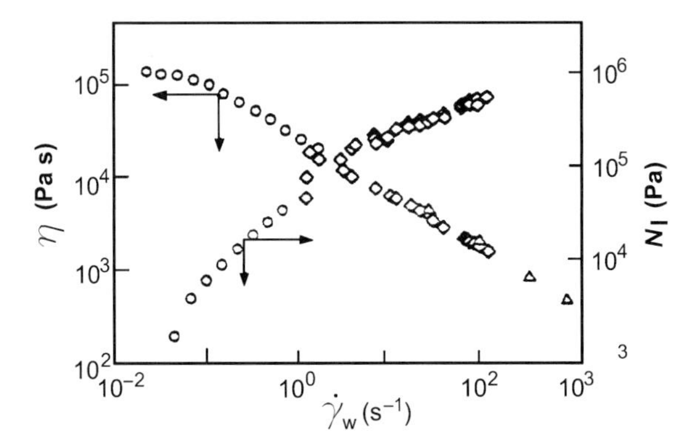

Figure 3.67. Viscosity and first normal-stress difference versus shear rate for a HDPE melt. Adapted from Figure 6.3.4 (Macosko, 1994). Copyright (1994). Reprinted with permission of John Wiley and Sons, Inc.

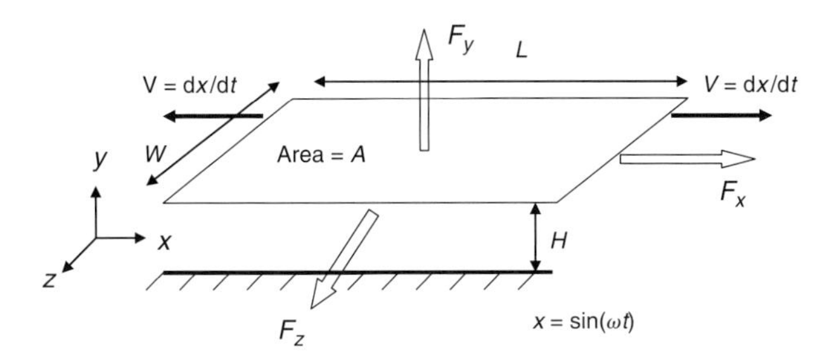

Figure 3.68. Ideal dynamic-shear flow.

Dynamic-shear rheology

In dynamic-shear rheology an oscillating or dynamic shearing deformation is applied to the material, and dynamic-shear properties are measured. For the simplest case of an infinite parallel-plate system, Figure 3.68, one can define relevant properties as follows.

The shear rate (γ') is defined as the change in the fluid's oscillatory velocity ($V = a\omega \cos(\omega t)$) over the gap between infinite parallel plates that is normal to the flow,

$$\gamma' = v/H = (a\omega/H)\cos(\omega t) = \gamma_0 \omega \cos(\omega t), \tag{3.51}$$

where the strain amplitude (γ_0) is defined as $\gamma_0 = a/H$.

The shear strain (γ) is defined by the fluid oscillatory displacement ($x = a \sin(\omega t)$) divided by the gap,

$$\gamma = x/H = (a/H)\sin(\omega t) = \gamma_0 \sin(\omega t). \tag{3.52}$$

Now, for a solid material (entirely elastic), one would expect the force (F) or stress (σ) to be in phase with or 'follow' the displacement (or strain, γ). That is,

$$F = F_m \sin(\omega t) \tag{3.53}$$

or

$$\sigma = (F_m/A)\sin(\omega t) = \sigma_0 \sin(\omega t), \tag{3.54}$$

where $\sigma_0 = F_m/A$.

For a Newtonian liquid (entirely viscous) one would expect the force or stress to follow the rate of motion,

$$\sigma = \eta\gamma' = \eta\gamma_0\,\omega\,\cos(\omega t). \tag{3.55}$$

Also you can reach this conclusion if you expect the force to lag relative to the displacement by 90°,

$$F = F_\mathrm{m}\sin(\omega t - 90) = F_\mathrm{m}\cos(\omega t) \tag{3.56}$$

or

$$\sigma = (F_\mathrm{m}/A)\cos(\omega t) = \sigma_0\cos(\omega t). \tag{3.57}$$

However, for a viscoelastic fluid the stress must arise from a contribution of in-phase and out-of-phase components,

$$\sigma = \sigma_0[A\sin(\omega t) + B\cos(\omega t)], \tag{3.58}$$

where $\sigma_0 A\sin(\omega t)$ represents the in-phase or solid-like stress response and $(\sigma_0 B\cos(\omega t)$ represents the out-of-phase or liquid-like stress response.

Rewriting this as

$$\sigma = \gamma_0[G'\sin(\omega t) + G''\cos(\omega t)], \tag{3.59}$$

we define G' as a storage (or solid-like) modulus and G' as the loss (or liquid-like) modulus.

One may also define the loss tangent ($\tan\delta$) by

$$\tan\delta = G''/G', \tag{3.60}$$

where $\delta = 0$ for a pure solid and $\delta = 90°$ for a pure liquid.

Other parameters that may be defined in dynamic shear rheology are the viscous dynamic viscosity, $\eta' = \mathrm{G}''/\omega$, and the solid dynamic viscosity, $\eta'' = G'/\omega$.

The complex viscosity, $\eta*$, may be defined as

$$\eta* = \sqrt{\left(\eta'^2 + \eta''^2\right)}. \tag{3.61}$$

Interestingly, it has been shown experimentally for many systems (Cox, 1958) that the profile of the complex viscosity ($\eta*$) as a function of the dynamic frequency (ω) is equivalent to the profile of the steady-shear viscosity (η) with respect to the shear rate (γ') for the same system. That is,

$$\eta*(\omega) = \eta(\gamma') \tag{3.62}$$

for a range of polymer systems. Milner (1996) has begun to develop a theoretical rationale behind this empirical rule.

Typical dynamic-shear profiles are shown in Figures 3.69 and 3.70. Figure 3.69 shows a typical set of profiles of elastic modulus, loss modulus and dynamic viscosity versus applied frequency for a polymer melt. This example (Rheometrics-Scientific, 1990) is a polyethylene melt at 160°C. Initially the loss modulus is higher than the storage modulus because the melt behaves more like a viscous fluid. This can generally be said to be due to the fact that the imposed frequency is low enough that most of the molecules in the system have enough time to relax after the perturbation, and the fluid dissipates the stresses due to the imposed shear. At higher frequencies the storage modulus grows larger than the loss

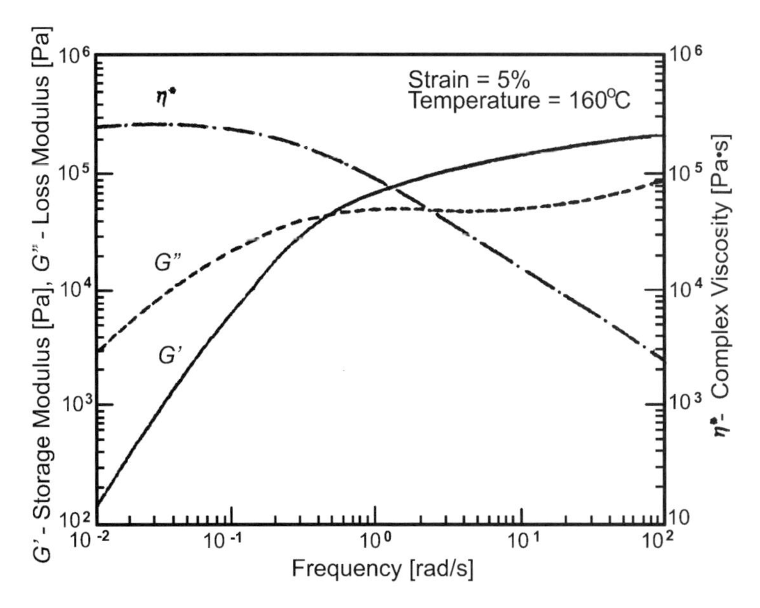

Figure 3.69. Typical profiles of elastic modulus, loss modulus and dynamic viscosity versus applied frequency for a polymer melt.

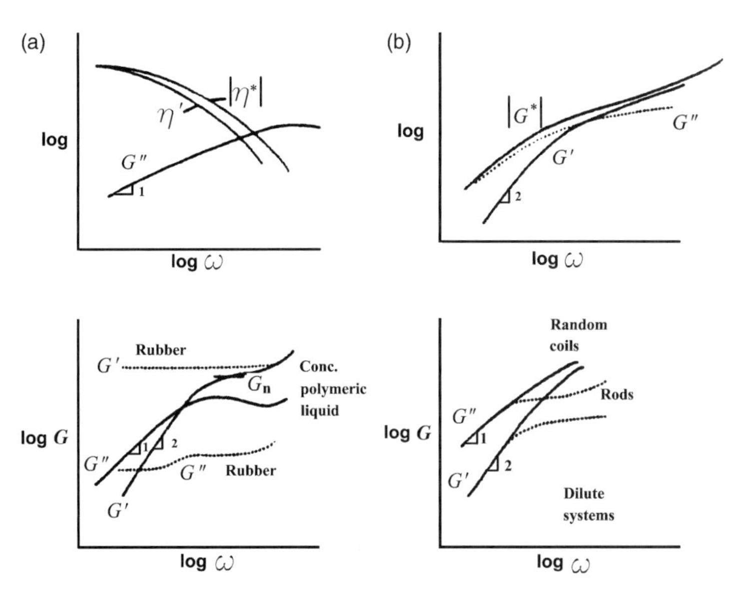

Figure 3.70. Dynamic profiles for a range of polymer systems from (a) concentrated to (b) dilute systems. Adapted from Figure 3.3.5 (Macosko, 1994). Copyright (1994). Reprinted with permission of John Wiley and Sons, Inc.

modulus, and the system is said to behave more like a solid. This may be explained by the fact that larger molecules become trapped by the higher frequencies of imposed shear and do not have enough time to relax. The viscosity profile also starts from a plateau at low frequencies where the behavior is pseudo-Newtonian (or fluid-like) and then decreases with increasing frequency (or becomes more pseudo-plastic or shear-thinning). Figure 3.70, adapted from Macosko 1994, shows deviations from a concentrated polymer solution. Note in Figure 3.70 how the profile of a concentrated polymer solution is similar to that for the polymer-melt system described in Figure 3.69. Here note that initially at low frequencies the storage modulus scales with frequency to the power 2 (since $G' \sim \omega^2$) and the loss modulus scales with frequency to the power 1 (since $G'' \sim \omega^1$). This can be predicted for Newtonian systems, and further describes the initial low-frequency response as Newtonian or fluid-like. Next note that Figure 3.70 shows the response of a rubber system and note how the storage modulus now dominates the loss modulus ($G' \gg G''$; since the system is more solid-like) and the storage modulus is independent of frequency ($G' \sim \omega^0$; which is also typical of a pure solid).

Time–temperature superposition

An interesting extension of dynamic rheology is time–temperature superposition (TTS) (Ferry, 1980). Essentially TTS recognizes for some systems (known as thermorheologically simple systems) the equivalence of the effects of time (or imposed frequency) and temperature throughout certain regions of behaviour. A corollary of this is to consider that, for a thermorheologically simple fluid, the response to a low frequency at a low temperature would be equivalent to the response to a high frequency imposed at a high temperature. In effect this allows one to determine the response of a system, at a standard (reference) temperature, over a larger range of frequencies than could be determined experimentally.

Dependences of dynamic properties are given by Ferry (1980):

$$\ln[\eta T_r \rho_r / (\eta_r T \rho)] = \ln(a_T), \tag{3.63}$$

where η is the viscosity, T is the temperature, ρ is the density, with subscript r denoting the reference state, and a_T is the shift factor.

One can obtain responses reduced to a specific reference temperature by use of the following expressions:

$$G'_r = G' T_r \rho_r / (T\rho) \; \omega_r = \omega a_T; \tag{3.64}$$

$$G''_r = G'' T_r \rho_r / (T\rho) \omega_r = \omega a_T; \tag{3.65}$$

$$\eta_r = \eta T_r \rho_r / (a_T T \rho); \quad \omega_r = \omega a_T. \tag{3.66}$$

Figure 3.71 shows how dynamic profiles for the loss modulus versus frequency may be reduced to a single reference temperature.

The overlapping of curves from a range of temperatures to a single curve at the reference temperature defines its thermorheological nature.

Transient shear

Transient shear is defined as when a material is subject to an instantaneous change in deformation and the response as a function of time is measured. For example, Figure 3.72 shows an ‘instantaneously’ applied step-strain test.

The imposed strain is denoted in Figure 3.72(a) and the measured stress relaxation is shown in Figure 3.72(b) (where G, the relaxation modulus, is defined as $\sigma(t)/\gamma_0$; where $\sigma(t)$ is the stress at time t after the step strain and γ_0 is the magnitude of the step strain.)

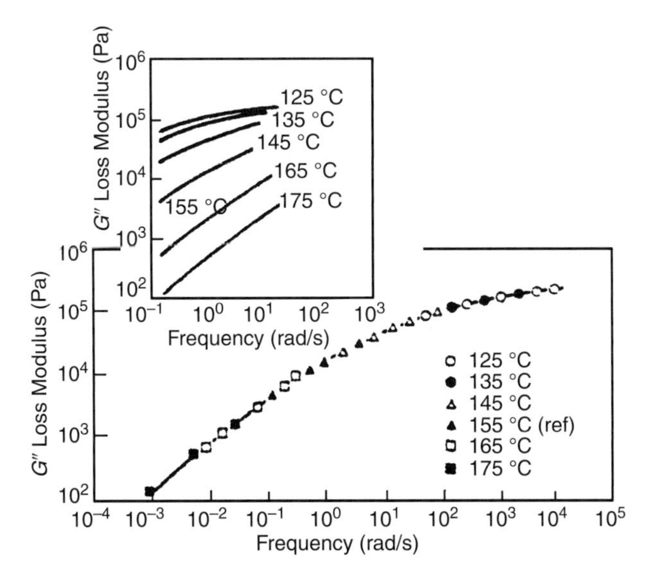

Figure 3.71. An example of time–temperature superposition (TTS) of rheological data. Adapted from Figure 11.1 (Ferry, 1980). Copyright (1980). Reprinted with permission of John Wiley and Sons, Inc.

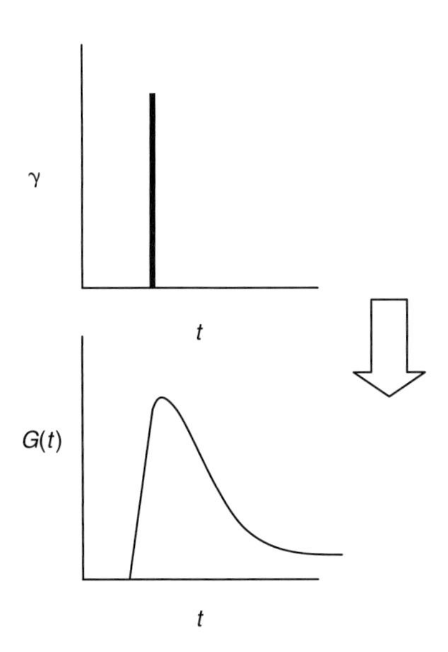

Figure 3.72. A transient rheological test showing (a) instantaneously applied strain and (b) the associated stress response.

Another example of a transient measurement is given in Figure 3.73. Figure 3.73(a) shows the start up of an imposed strain rate and the subsequent measurement of the stress build-up ($\sigma^+(t)$) over time. Figure 3.73(b) shows the relaxation of an applied strain rate and the subsequent measurement of the stress decay ($\sigma^-(t)$) over time. An example of a typical stress build-up for a polymer melt is shown in Figure 3.74.

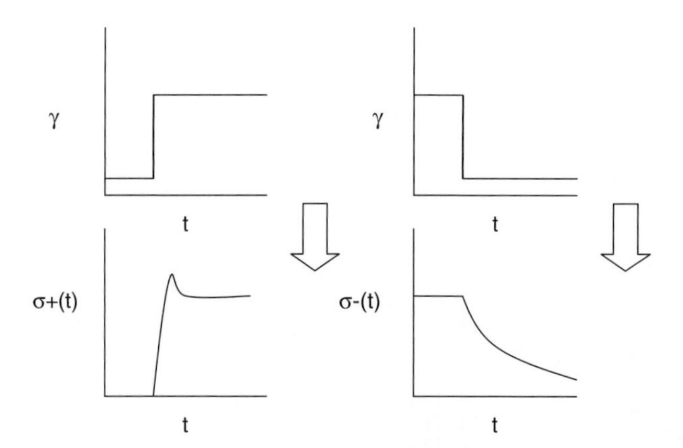

Figure 3.73. A transient rheological test showing (a) the start up of an imposed strain rate and the subsequent measurement of the stress build-up, and (b) the cessation of an imposed shear rate and the subsequent stress decay.

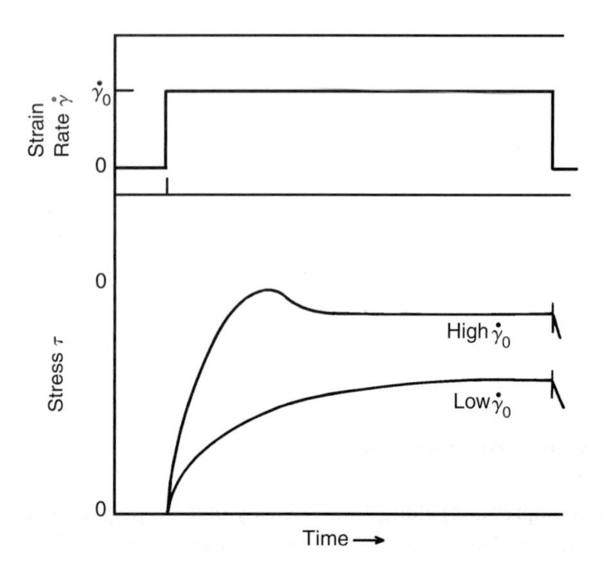

Figure 3.74. An example of a typical stress build-up during start up of shearing for a polymer melt.

Note that, in the higher-shear-rate test, the polymer shows a large stress overshoot. This overshoot is common for viscoelastic fluids and is believed to be related to hindrance to chain disentanglement and conformational alterations.

Elongation rheology

Elongation rheology is defined as when the material is subject to stretching (elongational) deformation and elongation-rheology properties are measured. Figure 3.75 shows ideal elongational flow.

The elongation rate (γ_e') may be defined as the rate of deformation per unit length (dv/dx) in the direction of flow,

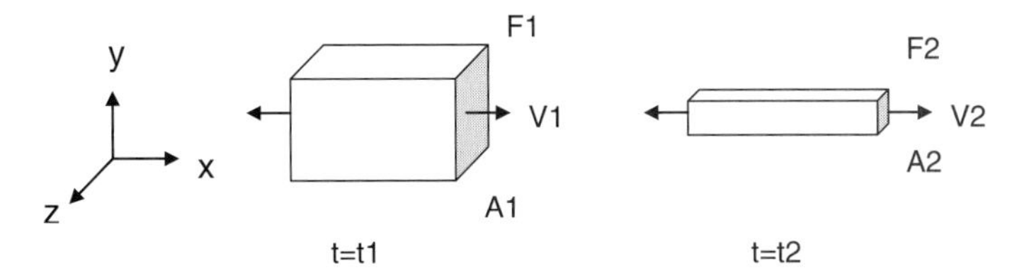

Figure 3.75. Ideal elongational flow.

$$\gamma'_{\mathrm{e}} = \mathrm{d}v/\mathrm{d}x. \tag{3.67}$$

The normal stress (σ_{n}) may be defined as the force ($F(t)$) divided by the area ($A(t)$) in the direction of flow,

$$\sigma_{\mathrm{n}} = F(t)A(t). \tag{3.68}$$

Note that the area may decrease as a function of time.

The elongation viscosity (η_{e}) is defined as the normal stress (σ_{n}) divided by the rate of deformation (γ'_{e}),

$$\eta_{\mathrm{e}} = \sigma_{\mathrm{n}}/\gamma'_{\mathrm{e}}. \tag{3.69}$$

The Trouton ratio (Tr) may also be defined as the ratio of the elongation viscosity (η_{e}) to the steady-shear viscosity (η),

$$\mathrm{Tr} = \eta_{\mathrm{e}}/\eta. \tag{3.70}$$

Typically this value ranges from 3.0 for Newtonian systems to up to 1000 for non-Newtonian systems.

There are many types of elongation flow, which are shown in Figure 3.76. Simple or uniaxial extension, equibiaxial extension and planar extension will provide different elongational properties specific to their type of deformation. The type of elongation flow selected will depend on the type of elongation flow that is encountered during the specific polymer process to be investigated. An example of the effect of the type of elongation flow on the properties of a polymer solution is shown in Figure 3.77.

Here uniaxial and biaxial (compression) elongation viscosity of a 1% polyacrylamide in a glycerine–water system is shown. Note that the material shows a reasonably constant resistance to biaxial (compression) elongation, but quite an increase in uniaxial extension with increasing extension rate. This has implications for processing of this fluid using predominantly uniaxial or biaxial flows.

Rheological classifications

Simple classifications of fluids can be made on the basis of their rheological profiles. Figure 3.78 shows the (a) shear stress and (b) viscosity profiles for various systems. From Figure 3.78 one may define the following systems. *Newtonian* systems have a constant viscosity with respect to shear rate. *Dilatant* (or shear-thickening) systems have a viscosity that increases with respect to shear rate. *Pseudo-plastic* (or shear-thinning) systems have a viscosity that decreases with respect to shear rate. *Yield-stress* materials are materials that have an initial structure that requires a finite stress before deformation can occur. The stress that initiates deformation is defined as the yield stress.

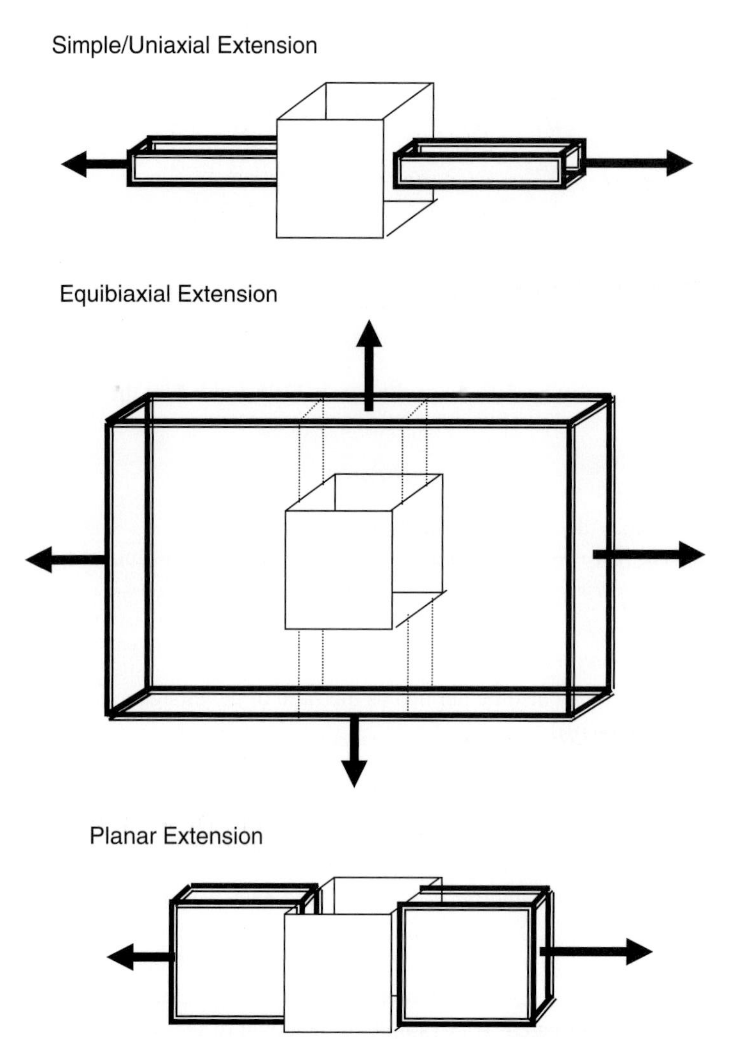

Figure 3.76. Types of elongational flow.

Figure 3.79 shows the viscosity profiles for time-dependent fluids. From Figure 3.79 one may define the following systems. *Time-independent* fluids obviously undergo no change in viscosity with respect to time. *Rheopectic* fluids show an increase in viscosity with respect to time. *Thixotropic* fluids show a decrease in viscosity with respect to time.

Table 3.9 shows many various rheological models used to categorize and model fluid systems. It is written in terms of the three-dimensional form (where terms are discussed in Macosko (1994)) and the simple two-dimensional shear-flow relationship (where terms have been defined here).

Typical fluid viscosities for various materials and typical shear-rate ranges for their processing operations are shown in Tables 3.10 and 3.11 (Macosko, 1994).

Rheometers

Many commercial rheometers are available both for shear and for elongation rheology. For an intensive review the reader is referred to Macosko (1994). In summary, Figure 3.80

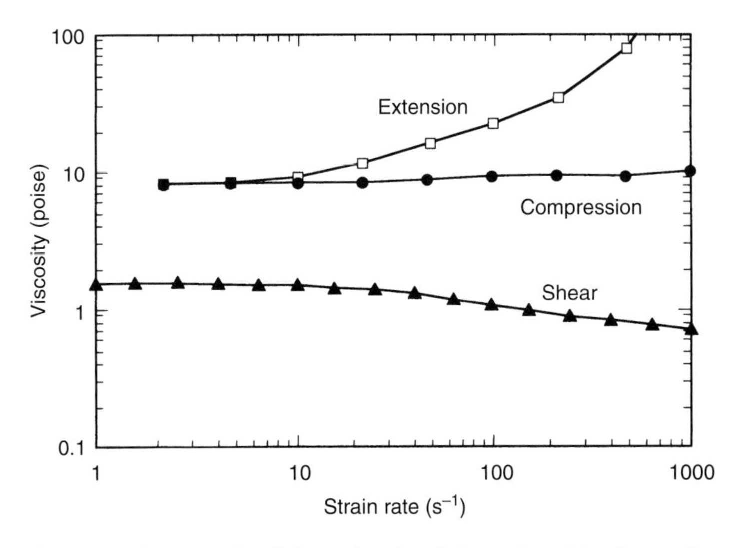

Figure 3.77. An example of elongational and shear viscosities for a polyacrylamide solution. From Figure 7.7.9 (Macosko, 1994). Copyright (1994). Reprinted with permission of John Wiley and Sons, Inc.

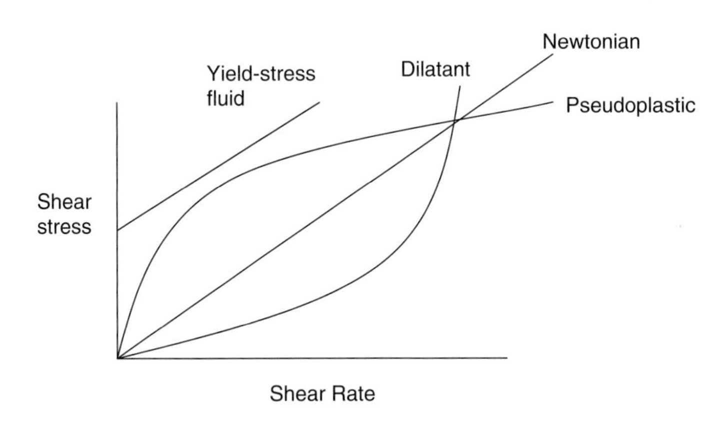

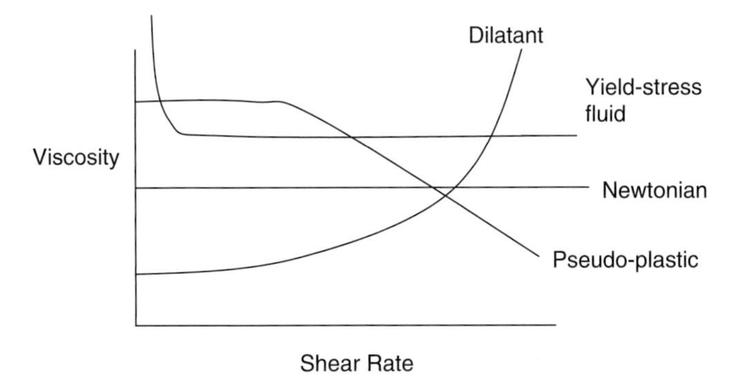

Figure 3.78. Classifications of fluids on the basis of dependences of shear stress and shear viscosity on shear rate.

Table 3.9. Constitutive models for viscosity

Model	Form	In simple shear
Viscous models		
Newtonian	$\tau = \eta(2D)$	$\tau_{12} = \eta\gamma'_{12}$ $\eta = \text{constant}$
General viscous fluid	$\tau = f(2D)$	$\tau_{12} = \eta\gamma'_{12}$ $\eta = \eta(\gamma')$
Power law	$\tau = m\Pi_{2D}^{(n-1)/2}(2D)$	$\tau_{12} = m\gamma'^{n}_{12}$ $\eta = m\gamma'^{n-1}_{12}$
Cross		$(\eta - \eta_\infty)/(\eta - \eta_0) = 1/[1 + (K\gamma')^{(1-n)/2}]$
Ellis		$\eta/(\eta - \eta_0) = 1/[1 + (K\gamma')^{(1-n)/2}]$
Plastic models		
Bingham	$\tau = GB$ for $\Pi\tau < \tau_y^2$ $\tau = (\eta_0 + \tau_y/\Pi_{2D})^{1/2}2D$ for $\Pi\tau > \tau_y^2$	$\tau = G\gamma$ or $\gamma = 0$ for $\tau < \tau_y$ $\tau = \eta\gamma' + \tau_y$ for $\tau > \tau_y$
Linear viscoelastic models		
General linear viscoelastic model	$\tau + \lambda\ \mathrm{d}\tau/\mathrm{d}t = \eta\gamma'$	$\eta = \sum_i G_i\lambda_i = \eta_0$
Non-linear viscoelastic models		
Upper convected Maxwell	$\tau + \lambda(\tau)^\nabla = 2\eta_1 D$	$\eta = \eta_0$ and $\varphi_1 = 2\eta_0\lambda$
Oldroyd B	$\tau = \tau_{UCM} + 2\eta_s D$	
KBKZ		$\tau_{12} = \gamma \int M(t)h(\gamma)\mathrm{d}t$
Maxwell type	$(\tau)^\nabla + f_c(\tau, D) + 1/\lambda(\tau) + f_d(\tau) = 2GD$	(Various models based on different expressions for f_c and f_d)
Molecular constitutive equations		
Elastic dumbell	$\tau + \lambda(\tau)^\nabla = -n_0 k_B T\lambda^2 D$	$\eta = \eta_s + n_0 k_B T\lambda$ and $\varphi_1 = 2n_0 k_B T\lambda^2$
Zimm model	Finite-extensibility non-linear elastic (FENE) extension of dumbell model	$G(t) = \frac{1}{3} n_0 k_B T \exp(-t/\lambda)$
Rouse model	Multimode Zimm model	$G(t) = \frac{1}{3} n_0 k_B T \sum_i \exp(-t/\lambda_i)$
Pom-pom model		$G(t) = G_c(1 + a)\int (1 - x)^a \exp(-t/\tau(x))\mathrm{d}x$

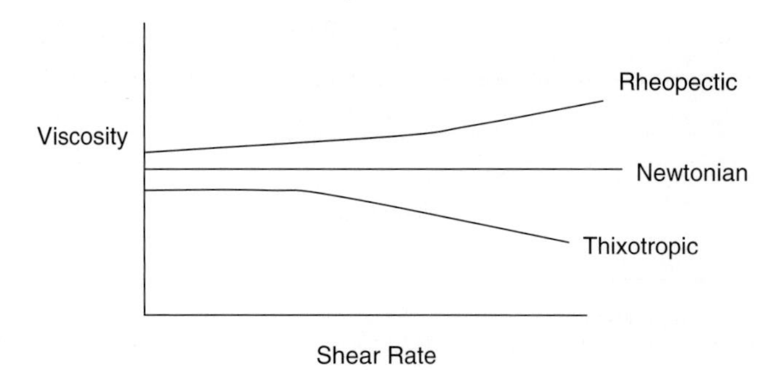

Figure 3.79. Classification of fluids on the basis of the time dependence of shear viscosity.

Table 3.10. Viscosity of common materials at room temperature

Liquid	Approximate viscosity (Pas)
Glass	10^{40}
Molten glass (500 °C)	10^{12}
Asphalt	10^{8}
Molten polymers	10^{3}
Heavy syrup	10^{2}
Honey	10^{1}
Glycerine	10^{0}
Olive oil	10^{-1}
Light oil	10^{-2}
Water	10^{-3}
Air	10^{-5}

Adapted from Macosko (1994, Table 2.3.1).

Table 3.11. Shear rates of typical processes

Process	Typical range of shear rates (s^{-1})	Application
Sedimentation of fine powders in a suspending liquid	10^{-6}–10^{-4}	Medicines, paints
Levelling due to surface tension	10^{-2}–10^{-1}	Paints, printing inks
Draining under gravity	10^{-1}–10^{1}	Painting and coating; emptying tanks
Screw extruders	10^{0}–10^{2}	Polymer melts, dough
Chewing and swallowing	10^{1}–10^{2}	Foods
Dip coating	10^{1}–10^{2}	Paints, confectionery
Mixing and stirring	10^{1}–10^{3}	Manufacturing liquids
Pipe flow	10^{0}–10^{3}	Pumping, blood flow
Spraying and brushing	10^{3}–10^{4}	Spray-drying, painting, fuel atomization
Rubbing	10^{4}–10^{5}	Application of creams and lotions to the skin
Injection-moulding gate	10^{4}–10^{5}	Polymer melts
Milling pigments in fluid bases	10^{3}–10^{5}	Paints, printing inks
Blade coating	10^{5}–10^{6}	Paper
Lubrication	10^{3}–10^{7}	Gasoline engines

Adapted from Macosko (1994, Table 2.3.2).

(shear) and Figure 3.81(elongation) give a good review of various commercial rheometers. Figure 3.82 shows the appropriate shear- and elongation-rate ranges for these rheometers.

The advantages of various types of shear and elongation rheometers are discussed in Table 3.12 (shear) and Table 3.13 (elongation).

3.6.5 Other techniques

SAXS, XRD

X-ray diffraction tests are based on the fact that incident X-ray radiation is diffracted at certain angles by regularly spaced inclusions. The two main techniques used in reactive polymer systems are wide-angle X-ray diffraction (WAXD) and small-angle x-ray

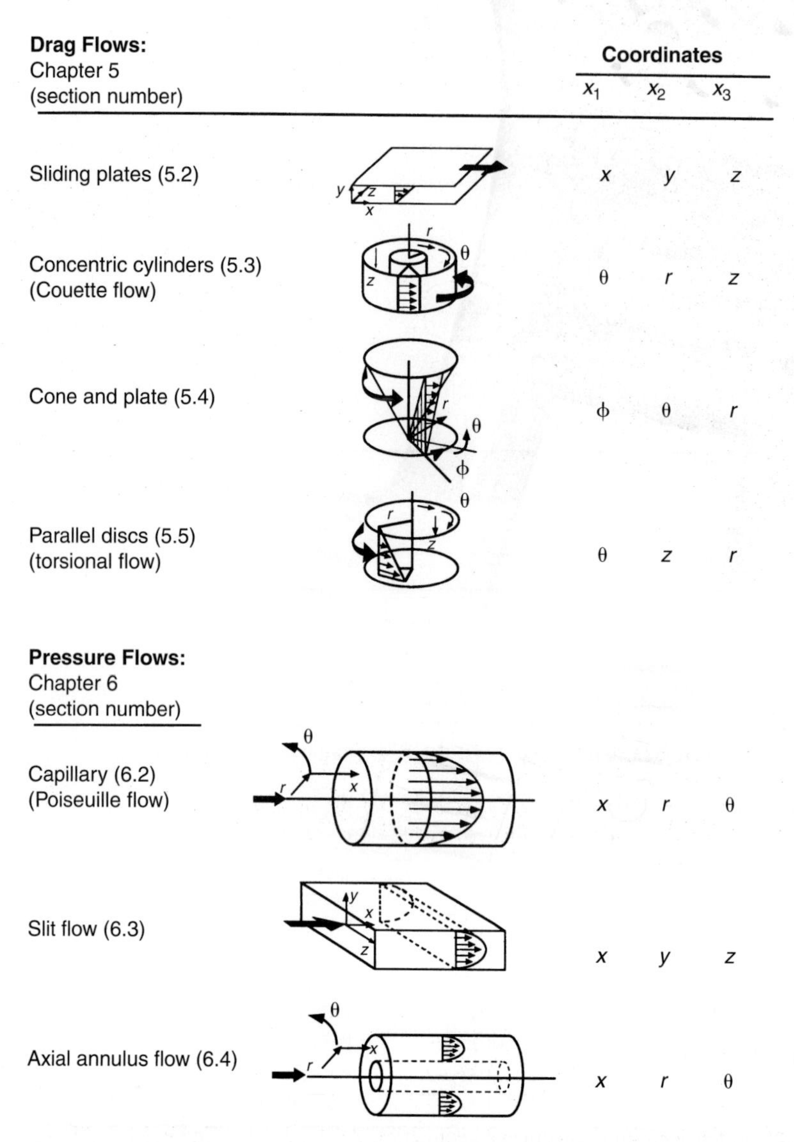

Figure 3.80. Shear-flow rheometer geometries. (Adapted from Figure 5.1.2 (Macosko, 1994)). Copyright (1994). Reprinted with permission of John Wiley and Sons, Inc.

scattering (SAXS); WAXD (using angles from 5° to 120°) is able to monitor scales of order from about 1 to 50Å, whereas SAXS (using angles from 1° to 5°) is able to measure structures of the order of 70–500Å.

A large amount of literature recently has focussed on the use of XRD to examine structural order in epoxy-resin nanocomposite systems. Lan *et al.* (1996) examined the initial separation (intercalation) of clay layers in self-polymerization of Epon 828 epoxy resin with onium-ion-exchanged clays. Much XRD work on epoxy-resin and thermoset nanocomposites has continued since this work, up until and including recent work by Becker *et al.* (2003a, 2003b) examining the clay structure of a range of epoxy-resin systems with a range of different functionalities and curing conditions.

Geometry (section number)	Co-ordinates x_1	x_2	x_3
Extension (7.2)	x or x	r or y	θ or z
Compression (7.3)	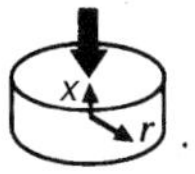 r	θ	x
Sheet stretching (7.4)	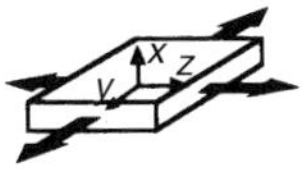 x	y	z
Fibre spinning (7.5)	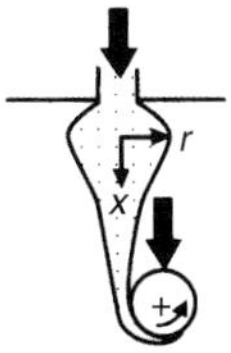 x	r	θ
Bubble collapse (7.6)	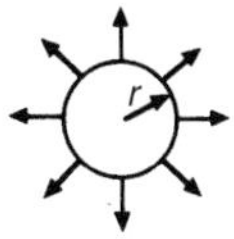 r	θ	ϕ
Stagnation flows (7.7)	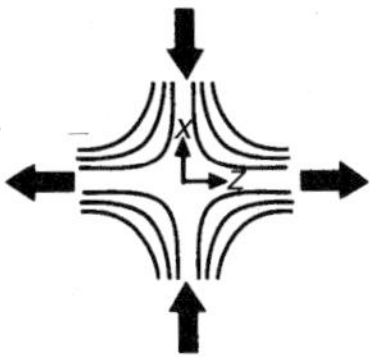 x	y	z
Entrance flows (7.8)	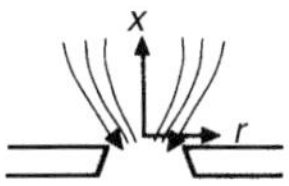 x	r	θ

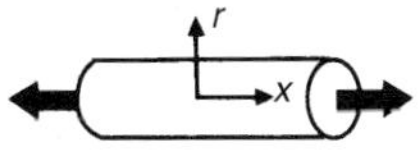

Figure 3.81. Extensional-flow rheometer geometries. Adapted from Figure 7.1.2 (Macosko, 1994). Copyright (1994). Reprinted with permission of John Wiley and Sons, Inc.

Much work has been reported on studying the structure of thermoset resins via SAXS, especially focussing on interpenetrating network polymers (IPNs), thermoset nanocomposites, rubber-modified thermosets and thermoset–thermoplastic blends. Most recently Guo *et al.*, (2003) have examined the use of SAXS to monitor the nanostructure and crystalline phase structure of epoxy–poly(ethylene–ethylene oxide) thermoset–thermoplastic blends. This work proposes novel controlled crystallization due to nanoscale confinements.

SANS

Small-angle neutron scattering (SANS) is also used for characterizing structure in reactive systems. With SANS (using angles $<2°$) one can examine polymer morphology and chain dimensions, typically enhanced by deuteratation (which enhances differences in

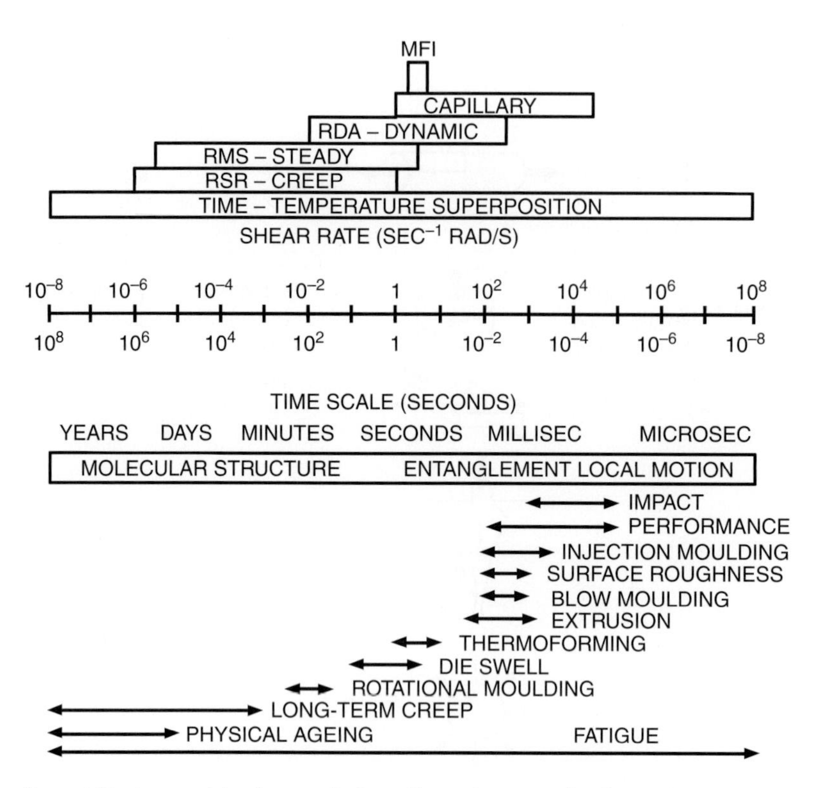

Figure 3.82. Appropriate shear and elongation-rate ranges for rheometers.

neutron-scattering length versus undeuterated chains). Once again the most common systems investigated in recent literature are thermoset–thermoplastic blends (due to the ability to control structure during cure of these systems). Elliniadis *et al.* (1997) examined the structure of a thermoset–thermoplastic blend of epoxy–polyarylsulfone. They were able to use SANS to determine an effective Flory–Huggins interaction parameter and incorporated this into a recent model for linear–branched polymer blends to predict the evolution of phase diagrams under isothermal cure. Carter *et al.* (2003) examined the phase structure during cure of low-temperature-cured epoxy–thermoplastic blends for aerospace composites using SANS.

PALS

Positron-annihilation spectroscopy (PALS) provides a measure of the free volume of polymers via the lifetime of orthopositroniums (oPs). The typical lifetime of 1–3 ns corresponds to free-volume cavities of 0.3–0.7 nm. For reactive polymers, PALS can be linked to T_g, gas diffusion rates, crosslink density, network formation and ageing and post-cure effects. There have been some very instructive reviews (Hill, 1995, Simon, 1997) on PALS that also incorporate examples involving reactive polymers. Suziki (2001) used PALS to monitor T_g (in terms of the temperature dependence of the positronium (Ps) lifetime) and gelation (in terms of an increase in the intensity of Ps and a sharp decrease in the lifetime) of a curing epoxy cresol/phenol novolac system. Wang *et al.* (2003) also examined physical ageing of epoxy-resin systems using PALS. They found that the side group plays an important role in determining free-volume properties of epoxy cresol resins, i.e. the stronger the crosslinking of curing agent, the smaller the free volume. However, during ageing the

Table 3.12. Advantages and limitations of various types of shear rheometers

Method	Advantages	Disadvantages
Sliding plates	Simple design Homogeneous Linear motion high η, $G(t, \gamma)$ $t \geq 10^{-3}$ s	Edges limit $\gamma < 10$ Gap control Loading
Falling ball	Very simple Needle better Sealed rheometer High T, p	Not very useful for viscoelastic fluids Non-homogeneous Transparent fluid Need p
Concentric cylinders (Couette flow)	Low η, high γ' Homogeneous if $R/R \geq 0.95$ Good for suspension settling	End correction High-η fluids are difficult to clean N_1 impractical
Cone and plate	Homogeneous $\beta \geq 0.1$ rad Best for N_1 Best for $G(t, \gamma)$	High η: γ' low, edge failure, loading difficult Low η: inertia Evaporation Need good alignment
Parallel discs (torsional flow)	Easy to load viscous samples Best for G' and G'' of melts, curin Vary γ' by h and Ω $(N_1 - N_2)(\gamma')$	Non-homogeneous: not good for $G(t, \gamma)$. OK for $G(t)$ and $\eta(\gamma)$ Edge failure Evaporation
Contained bobs (Brabender, Mooney)	Sealed Process simulator	Indexers Friction limits range
Capillary (Poiseuille flow)	High γ' Sealed Process simulation η_{ext} from Δp_{ent} Wide range with L	Corrections for Δp_{ent} time-consuming Non-homogeneous no $G(t, \gamma)$ Bad for time dependence Extrudate swell only qualitative for N_1
Slit flow	No Δp_{ent} with wall-mounted pressure transients $\eta(p)$ p_{cx}, p_h give N_1	Edge effects with $W/B > 5$ Similar to capillary Difficult to clean
Axial annular flow	Slit with no edges Δp can give N_2	Difficult construction and cleaning
Squeeze flow	Simple Process simulation $\eta(\gamma')$ at long times	Index flow: mixed shear rates and shear transients

Adapted from Macosko (1994, Table 6.5.1).

free volume diminished as a result of the formation of the local order structure due to the chain-segmental rearrangement.

GPC

The molecular weight or, more rigorously, molar mass (M) and molecular-weight distribution (MWD) of a reactive system (shown in Figure 1.6 and discussed in Section 1.1.3) can be monitored via gel permeation chromatography (GPC). The sample is introduced to the

Table 3.13. Advantages and limitations of various types of elongational rheometers

Method	Advantages	Disadvantages
Tension	Homogeneous 'Clean' data Wind-up is easiest to generate	Requires high viscosity Sample gripping Low ε' Sample history, preparation Need bath
Lubricated compression	Simple sample preparation, grip Easy to generate small displacement	Need lubricant High η $\eta_b < \eta_e$ $\varepsilon \leq 2$
Fibre spinning, ductless siphon	Low η Process simulator Sample preparation easy	Entrance condition Corrections; g, F_D Photo
Bubble collapse	Simple sample preparation Process simululation	Transparent for photo $\eta_b < \eta_e$ Not homogeneous
Stagnation	Large strain centre but non-homogeneous Birefringence Opposed nozzles low η convenience of a wide range of ε	Wall effects (e.g. shear) Lubricant Stability of flow
Entrance flows	Simplest Wide η range Process simulation	Complex flow

Adapted from Macosko (1994, Table 7.9.1).

system in a solvent. As the material is passed through the GPC column, the material is selectively retained on the basis of on its size (or M). From the time the material fractions spend in the GPC column, a distribution of molecular weights can be obtained. For network systems this method is limited as the M approaches infinity, the MWD becomes infinitely broad and the material becomes insoluble at gelation. However, for reactively modified polymers GPC is widely used to examine changes in M of modified polymers. Japon *et al.* (2001) highlight the use of GPC in determining molecular structural changes in reactively modified epoxidized-PET. Chen and Chiu (2001) examine the use of GPC to validate a Monte Carlo model of the increase in M of the sol part of a curing epoxy-resin network.

SEM/TEM

Scanning electron microscopy (SEM) involves scanning an electron beam (5–10 nm) across a surface and then detecting the scattered electrons. Literature abounds, with work focussing on the use of SEM in the fracture and failure of epoxy resins and other thermoset polymers. Also work on multiphase thermosets (thermoset–thermoplastic blends, thermoset nanocomposites, interpenetrating network (IPN) polymers) is abundant.

AFM

Atomic-force microscopy (AFM) is a technique that examines the surface topography of polymer samples. The probe moves across the surface and a piezo-electric detector can monitor the topography in tapping or contact modes. Shaffer *et al.* (1995) examined the interphase regions in rubber-modified epoxies, finding that AFM assessed interphase dimensions more accurately than did SEM. Karger-Kocsis *et al.* (2003) examined the morphology of an epoxy–vinyl ester IPN using AFM, finding interesting two-phase structures, and could also monitor post-cure modifications of this structure.

3.6.6 Dual physicochemical analysis

There has been recent interest in developing dual spectroscopic–rheological techniques for characterizing polymer solutions and melts. Primarily this is because of the advantage of being able to obtain chemical and physical information simultaneously from the system to be monitored. Obviously, this is of extra relevance for curing systems that have time-dependent chemical and physical properties.

Spectroscopy–rheology

A review of rheo-optical techniques by Sherman *et al.* (1996) notes that there has been an increase in the use of rheo-optic set-ups both for FT-IR dichroism and for dynamic IR dichroism spectroscopies for polymer melts and polymer blends. Skytt *et al.* (1996) highlight the use of simultaneous measurement of the transient or steady-state rheological properties and IR dichroism to characterize orientation in polymer melts. However, there is little reference to dual spectroscopic–rheological techniques for reactive polymer systems in the literature.

The authors have also previously proposed (George and Halley, 2000) a novel system in which a NIR spectrometer is coupled through fibre-optics to a Rheometrics ARES rheometer to provide a unique capability for thermosetting systems. The particular advantages of working in the NIR region of the spectrum have been discussed in Section 3.3.6, including the adaptation of fibre-optics for real-time reaction monitoring (Sections 3.4.1 and 3.4.2). Two approaches can be employed for coupling the NIR fibre-optic system to a rheometer. The first uses single-filament fibres and micro-positioners above and below the rheometer plates in order to transfer and collect the NIR radiation. The second uses a dedicated NIR source above the plates and an optical fibre for return of the radiation to the spectrometer. The first is expected to have a poorer SNR than the second, but has a better-defined pathlength, which is essential for quantitative analysis. Kinetic data acquisition routines can be developed to synchronize rheological and spectral data. In principle the two methods should provide the absolute viscosity at any point for comparison with the absolute rate coefficient (since NIR measures concentration in absolute units). At the first level this is able to calibrate the fibre-optic probe as a pseudo-rheometer for this system. In addition to the above, the availability of rheological and conversion data as well as the absolute rate parameters provides an opportunity to examine the theories for the onset of diffusion control as the crosslinking network develops and so unambiguously connect the chemical and rheological changes.

Chemoviscosity profiles may also be correlated to chemical rate parameters. Simultaneous acquisition of the NIR spectrum and low-shear dynamic rheology may also be undertaken from prior to gelation to beyond vitrification for various epoxy-resin systems for selected isothermal profiles to provide network information from rheological power-law indices, fractal dimensions and network rheological structure parameters.

Recently Benali *et al.* (2004) reported the development of a new laboratory-made system that allows the combination of rheometric (Alessi *et al.*, 2005) and spectroscopic (Fourier-transform-near infrared (FT-NIR) spectroscopy) measurements for reactive polyurethane materials along the lines of these principles.

Rheo-dielectrics

The development of simultaneous rheometry and dielectric spectroscopy is more mature than is the rheo-optics of reactive systems. Initial work by Senturia *et al.* (1982, Senturia and Sheppard, 1986) highlighted the use of microdielectrometry to monitor the complex dielectric constant of reactive polymer materials during or after cure. Gorto and Yandrastis (1989) extended this work by examining the simultaneous use of microdielectrometry and rheometry to monitor the dielectric loss factor and the dynamic viscosity of thermosetting bismaleimides and epoxies, by mounting a microdielectric sensor on the bottom plate of a parallel-plate rheometer. Dipole peaks were correlated to vitrification phenomena, and the maximum in the loss factor followed the same heating rate dependence as the minimum in the viscosity. Further work on rheo-dielectrics for reactive thermosets was carried out by Maffezzoli *et al.* (1994), Simpson and Bidstrup (1995) and Mijovic *et al.*, (1995), who obtained detailed correlations between dielectric-spectroscopic data and rheological properties.

Other dual methods

There are also more recent developments of other dual physio-chemical experimental methods. For example Durand *et al.* (2006) presented a laboratory-made system that allows the coupling of dielectric analysis and Fourier-transform near-infrared spectroscopy (FT-NIR) to follow the cure of polyepoxy reactive systems. Complementary data are provided by the simultaneous dielectric analysis (the vitrification phenomenon) and near-infrared spectroscopic analysis (the extent of the reaction).

Callaghan (2006) recently reviewed work on rheo-NMR spectroscopy that is useful in determining simultaneous molecular ordering and molecular dynamics in relation to complex fluid rheology. Even though at present systems are limited to polymer solutions and low-temperature melts, extension of this technique to reactive polymer systems would be interesting.

References

Acitelli, M., Prime, R. & Sacher, E. (1971) *Polymer*, **12**, 333–343.

Adams, M. (1995) *Chemometrics in Analytical Spectroscopy*, Cambridge: Royal Society of Chemistry.

Afromowitz, M. A. (1988) *J. Lightwave Technol.*, **6**, 1591–1594.

Agbenyega, J. K., Ellis, G., Hendra, P. J. *et al.* (1990) *Spectrochim. Acta A*, **46**, 197–216.

Aldridge, P., Kelly, J. J., Callis, J. B. & Burns, D. H. (1993) *Anal. Chem.*, **65**, 3581–3585.

Alessi, S., Calderaro, E., Fuochi, P. *et al.* (2005) *Nucl. Instrum. Methods Phys. Res.* B, **236**, 55–60.

Allen, N. S. & Owen, E. D. (1989a) Luminescence studies of the photooxidation of polymers, in Zlatkevitch, L. (Ed.) *Luminescence Techniques in Solid State Polymer Research*, New York: Marcel Dekker.

Allen, N. S. & Owen, E. D. (1989b) Polymer analysis by luminescence spectroscopy, in Zlatkevitch, L. (Ed.) *Luminescence Techniques in Solid State Polymer Research*, New York: Marcel Dekker.

Apicella, A. (1986) Effect of chemorheology on epoxy resin properties, in Pritchard, G. (Ed.) *Developments in Reinforced Plastics – 5*, Essex: Elsevier Applied Science.
Attias, A. J., Bloch, B. & Laupretre, F. J. (1990) *J. Polym. Sci., Polym. Chem.*, **28**, 3445–3466.
Aust, J. F., Booksh, K. S. & Myrick, M. L. (1996) *Appl. Spectrosc.*, **50**, 382–387.
Bair, H. E. (1997) Thermal analysis of additives in polymers, in Turi, E. A. (Ed.) *Thermal Characterization of Polymeric Materials*, San Diego, CA: Academic.
Bajorek, A., Ciepluch, M. & Paczkowski, J. (2002) *J. Polym. Sci. A: Polym. Chem.*, **40**, 3481–3488.
Barkay, N. & Katzir, A. (1993) *J. Appl. Phys.*, **74**, 2980–2982.
Barrera, B. A. & Sommer, A. J. (1998) *Appl. Spectrosc.*, **52**, 1483–1487.
Barrett, K. E. J. (1967) *J. Appl. Polym. Sci.*, **11**, 1617–1626.
Barton, J. M. (1985) The application of DSC to the study of epoxy resins curing reactions, in Dusek, K. (Ed.) *Applied Polymer Science – Epoxy Resins and Composites I*, Berlin: Springer-Verlag.
Bauer, C., Amram, B., Agnely, M. *et al.* (2000) *Appl. Spectrosc.*, **54**, 528–535.
Bauer, M. & Bauer, J. (1994) Aspects of the kinetics, modelling and simulation of network build-up during cyanate ester cure, in Hamerton, I. (Ed.) *Chemistry and Technology of Cyanate Ester Resins*, London: Chapman and Hall.
Becker, O., Cheng, Y., Varley, R. & Simon, G. (2003b) *Macromolecules*, **36**, 1616–1625.
Becker, O., Simon, G., Varley, R. & Halley, P. (2003a) *Polym. Eng. and Sci.*, **43**, 850–862.
Benali, S., Bouchet, J. & Lachenal, G. (2004) J. *Near Infrared Spectrosc.*, **12**, 5–13.
Bidstrup, S. (1986), Microdielectrics, PhD thesis, University of Minnesota.
Bidstrup, S., Sheppard, N. & Senturia, S. (1989) *Polym. Eng. Sci.*, **29**, 325–328.
Bidstrup, W. & Senturia, S. (1987) *Society of Plastics Engineers ANTEC '87*, pp. 1035–1038.
Billingham, N. C., Bott, D. C. & Manke, A. S. (1981) Application of thermal analysis methods to oxidation and stabilization of polymers, in Grassie, N. (Ed.) *Developments in Polymer Degradation – 3*, London: Applied Science.
Billingham, N. C., Burdon, J. W., Kozielski, K. A. & George, G. A. (1989) *Makromol. Chem.*, **190**, 3285–3294.
Blakey, I. (2001) Aspects of the oxidative degradation of polypropylene, Thesis, School of Physical Sciences, Brisbane, Queensland University of Technology.
Blakey, I. & George, G. A. (2001) *Macromolecules*, **34**, 1873–1880.
Blakey, I., George, G. A. & Billingham, N. C. (2001) *Macromolecules*, **34**, 9130–9138.
Blakey, I., Goss, B. & George, G. (2006) *Aust. J. Chem.*, **59**, 485–498.
Blanco, M. & Serrano, D. (2000) *Analyst*, **125**, 2059–2064.
Boisde, G. & Harmer, A. (1996) *Chemical and Biochemical Sensing with Optical Fibers and Waveguides*, Boston, MA: Artech House.
Boiteux, G., Dublineau, P., Feve, M. *et al.* (1993) *Polym. Bull.*, **30**, 441–447.
Borchardt, H. J. & Daniels, F. J. (1957) *J. Amer. Chem. Soc.*, **79**, 41–46.
Bosch, P., Catalina, F., Corrales, T., Peinado, C. (2005) *Chem. Eur. J.*, **11**.
Bosch, P., Fernando-Arizpe, A. & Mateo, J. L. (2001) *Macromol. Chem. Phys.*, **202**, 1961–1969.
Braun, D. (1981) Thermal degradation of poly(vinyl chloride), in Grassie, N. (Ed.) *Developments in Polymer Degradation – 3*, London: Applied Science.
Brazier, D. W. (1981) Calorimetric studies of rubber vulcanization and rubber vulcanisates, in Grassie, N. (Ed.) *Developments in Polymer Degradation – 3*, London: Applied Science.
Brimmer, P. J., Monfre, S. L. & Dethomas, F. A. (1992) Real-time monitoring for the production of polyurethane, in Murray, I. & Cowe, I. A. (Eds.) *Making Light Work: Advances in Near Infrared Spectroscopy*, Weinheim: VCH.
Brown, S. B. (1992) Reactive Extrusion: A survey of chemical reactions of monomers and polymers during extrusion processing, in Xanthos, M. (Ed.) *Reactive Extrusion: Principles and Practice*, Munich: Hanser-Verlag.

Buback, M. & Lendle, H. (1983) *Makromol. Chem.*, **184**, 193.
Bur, A. J. & Thomas, C. L. (1994) *Polym. Mater. Sci. Eng.*, **71**, 415–416.
Cadenato, A., Salla, J., Ramis, X. *et al.* (1997) *Thermal Analysis*, **49**, 269–279.
Callaghan, P. (2006) *Current Opinion Colloid Interface Sci.* **11**, 13–18.
Callaghan, P. T. (1999) *Rep. Prog. Phys.*, **62**, 599–668.
Calvert, P., George, G. A. & Rintoul, L. (1996) *Mater. Chem.*, **8**, 1298–1301.
Carlsson, D. J., Brousseau, R., Zhang, C. & Wiles, D. M. (1987) *Polym. Deg. Stab.*, **17**, 303–318.
Carswell, T. G., Garrett, R. W., Hill, D. *et al.* (1996) The use of ESR spectroscopy for studying polymerization and polymer degradation reactions, in Fawcett, A. H. (Ed.) *Polymer Spectroscopy*, Chichester: John Wiley.
Carter, J., Emmerson, G., Lo Faro, C., McGrail, P. & Moore, D. (2003) *Composites* A, **34**, 83–91.
Celina, M., Ottesen, D. K., Gillen, K. T. & Clough, R. L. (1997) *Polym. Deg. Stab.*, **58**, 15–31.
Charmot, D., Amram, B., Agnely, M. (1999) *Bruker Report*, **147**, 32–33.
Chase, D. B. (1981) *Appl. Spectrosc.*, **35**, 77–81.
Chen, Y. & Chiu, W. (2001) *Polymer*, **42**, 5439–5448.
Cherfi, A., Fevotte, G. & Novat, C. (2002) *J. Appl. Polym. Sci.*, **85**, 2510–2520.
Chiao, L. (1990) *Macromolecules*, **23**, 1286–1290.
Chiou, B.-S. & Khan, S. A. (1997) *Macromolecules*, **30**, 7322–7328.
Chipalkatti, M. H. & Laski, J. J. (1991) *Polym. Mater. Sci. Eng.*, **64**, 131–132.
Ciriscioli, P. R. & Springer, G. S. (1990) *Smart Autoclave Cure of Composites*, Lancaster, PA: Technomic.
Clough, R. L. (1999) Oxidation of elastomers, in Mallinson, L. (Ed.) *Lifetime Prediction of Materials*, Oxford: Kluwer.
Cole, K. C., Hechler, A. A. & Noel, D. (1991) *Macromolecules*, **24**, 3098–3110.
Cole, K. C., Pilon, A., Noel, D. *et al.* (1988) *Appl. Specrosc.*, **42**, 761–769.
Colthup, N. B., Daly, L. H. & Wiberley, S. B. (1990) *Introduction to Infrared and Raman Spectroscopy*, London: Academic.
Compton, D. A. C., Hill, S. L., Wright, N. A. *et al.* (1988) *Appl. Spectrosc.*, **42**, 972–979.
Cooper, J. B. (1999) *Chemom. Intell. Lab. Syst.*, **46**, 231–247.
Cooper, J. B., Vess, T. M., Campbell, L. A. & Jensen, B. J. (1996) *J. Appl. Polym. Sci.*, **62**, 135–144.
Cox, W. (1958) *J. Polym. Sci.*, **28**, 619.
Cusano, A., Breglio, G., Giordano, M. *et al.* (2000) *Sensors Actuators*, **84**, 270–275.
Dakin, J. (1989) Distributed optical fiber sensor sytems, in Culshaw, B. & Dakin, J. (Eds.) *Optical Fiber Sensors Systems and Applications*, Norwood, MA: Artech House.
Dallin, P. (1997) *Process Control and Quality*, **9**, 167–172.
Dannenberg, H. (1963) *SPE Trans.*, 78–88.
Debakker, C. J., George, G. A., St. John, N. A. & Fredericks, P. M. (1993a) *Spectrochim. Acta. A*, **49**, 739–752.
Debakker, C. J., St John, N. A. & George, G. A. (1993b) *Polymer*, **34**, 716–726.
Debye, P. (1912) *Phys. Z*, **13**, 97.
Delor, F., Teissedre, G., Baba, M. & Lacoste, J. (1998) *Polym. Deg. Stab.*, **60**, 321–331.
Dillman, S. & Seferis, J. (1989) *J. Macromol. Sci. – Chem. A*, **26**, 227–247.
Dillman, S., Seferis, J. & Prime, R. (1987) *Proceedings of the 16th North American Thermal Analysis Society Conference*, pp. 429–435.
Dong, J., Fredericks, P. M. & George, G. A. (1997) *Polym. Deg. Stab.*, **58**, 159–169.
Druy, M. A., Elandjian, L. & Stevenson, W. A. (1988a) *SPIE Int. Soc. Opt. Eng.*, **986**, 130–134.
Druy, M. A., Elandjian, L., Stevenson, W. A. *et al.* (1988b) *SPIE Int. Soc. Opt. Eng.*, **1170**, 150–9.
Dumoulin, M. M., Gendron, R. & Cole, K. C. (1996) *Trends Polym. Sci.*, **4**, 109–114.

Durand, A., Hassi, L., Lachenal, G. *et al.* (2006) *J. Near Infrared Spectrosc,* **14**, 161–166.

Dusek, K. (1986) Network formation in curing of epoxy resins, in Dusek, K. (Ed.) *Advances in Polymer Science 78: Epoxy Resins and Composites III*, Berlin: Springer-Verlag.

Dusek, K. (1996) *Angew. Makromol. Chem.*, **240**, 1–15.

Edge, M., Allen, N. S., Wiles, R., McDonald, W. & Mortlock, S. V. (1995) *Polymer*, **36**, 227–234.

Elliniadis, S., Higgins, J., Clarke, N. *et al.* (1997) *Polymer*, **38**, 4855–4862.

Fanconi, B. M., Wang, F. W., Hunston, D. & Mopsik, F. (1986) Material characterization for systems performance and reliability, In McCauley, J. W. & Weiss, V. (Eds.) *Materials Characterization for Systems Performance and Reliability*, New York: Plenum.

Farrington, P. J., Hill, D., O'Donnell, J. H. & Pomery, P. J. (1990) *Appl. Spectrosc.*, **44**, 2543–2545.

Feng, L. & Ng, K. Y. S. (1990) *Macromolecules*, **23**, 1048–1053.

Fernando, G. F. & Degamber, B. (2006) *Int. Mater. Rev.*, **51**, 65–106.

Ferry, J. (1980) *Viscoelasticity of Polymers*, New York: Wiley.

Fidler, R. A., Rowe, D. & Weis, G. (1991) On-line infrared process control of molten polymers via a high pressure, high temperature flow cell, in *Proceedings of SPE ANTEC '91*, pp. 850–855.

Fischer, D., Bayer, T., Eichhorn, K.-J. & Otto, M. (1997) *Fresenius J. Anal. Chem.*, **359**, 74–77.

Fischer, D., Sahre, K., Abdelrhim, M. *et al.* (2006) *C. R. Chim.*, **9**, 1419–1424.

Fontoura, J. M. R., Santos, A. F. *et al.* (2003) *J. Appl. Polym. Sci.*, **90**, 1273–1289.

Freeman, E. S. & Carroll, B. (1958) *J. Phys. Chem.*, **62**, 394–397.

Fulton, M. I., Pomery, P. J., St. John, N. A. & George, G. A. (1998) *Polym. Adv. Technol.*, **9**, 75–83.

Gallagher, P. K. (1997) Thermoanalytical instrumentation, techniques and methodology, in Turi, E. A. (Ed.) *Thermal Characterization of Polymeric Materials*, San Diego, CA: Academic Press.

Garcia, R. & Pizzi, A. (1998) *J. Appl. Polym. Sci.*, **70**, 1111–1119.

Garton, A. (1992) *Infrared Spectroscopy of Polymer Blends, Composites and Surfaces*, Munich: Hanser-Verlag.

George, G., Hynard, N., Cash, G., Rintoul, L. & O'Shea, M. (2006) *C. R. Chim.*, **9**, 1433–1443.

George, G. A. (1981) Use of chemiluminescence to study the kinetics of oxidation of solid polymers, in Grassie, N. (Ed.) *Developments in Polymer Degradation – 3*. London: Applied Science.

George, G. A. (1986) *Mater. Forum*, **9**, 224–236.

George, G. A. (1989a) Chemiluminescence of polymers at nearly ambient conditions, in Zlatkevitch, L. (Ed.) *Luminescence Techniques in Solid State Polymer Research*, New York: Marcel Dekker.

George, G. A. (1989b) Luminescence in the solid state: general requirements and mechanisms, in Zlatkevitch, L. (Ed.) *Luminescence Techniques in Solid State Polymer Research*, New York, Marcel Dekker.

George, G. A., Cash, G. A. & Rintoul, L. (1996) *Polym. Int.*, **41**, 169–182.

George, G. A. & Celina, M. (2000) Homogeneous and heterogeneous oxidation of polypropylene, in Halim Hamid, S. (Ed.) *Handbook of Polymer Degradation, 2nd edn*, New York: Marcel Dekker.

George, G. A., Celina, M., Vassallo, A. M. & Cole-Clarke, P. A. (1995) *Polym. Deg. Stab*, **48**, 199–210.

George, G. A., Cole-Clarke, P. A. & St John, N. A. (1990) *Mater. Forum*, **14**, 203–209.

George, G. A., Cole-Clarke, P. A., St John, N. A. & Friend, G. (1991) *J. Appl. Polym. Sci.*, **42**, 643–657.

George, G. A. & Schweinsberg, D. P. (1987) *J. Appl. Polym. Sci.*, **33**, 2281–2292.

George, G. A. & St John, N. A. (1993) *Chem. Australia*, **60**, 654.

Gmelin, E. (1997) *Thermochim Acta*, 304/**5**, 1–26.

Goddu, R. F. & Delker, D. A. (1960) *Anal. Chem.*, **32**, 140.

Gonzalez-Gonzalez, V. A., Neira-Velazquez, G. & Angulo-Sanchez, J. L. (1998) *Polym. Deg. Stab.*, **60**, 33–42.

Gorto, J. & Yandrastis, M. (1987) *Proceedings of Society of Plastics Engineers ANTEC* '**87**, pp. 1039–1042.

Gorto, J. & Yandrastis, M. (1989) *Polym. Eng. Sci.*, **29**, 278–284.
Gottwald, A. & Scheler, U. (2005) *Macromol. Mater. Eng.*, **290**, 438–442.
Grenier-Loustalot, M., Metras, F., Grenier, P., Chenard, J. & Horny, P. (1990) *Eur. Polym. J.*, **26**, 83.
Guillet, J. (1987) *Polymer Photophysics and Photochemistry*: Cambridge, Cambridge University Press.
Guo, Q., Thomann, R., Gronski, W. *et al.* (2003) *Macromolecules*, **36**, 3635–3645.
Gupta, A., Cizmecioglu, M., Coulter, D. *et al.*, (1983) *J. Appl. Polym. Sci.*, **28**, 1011–1024.
Haaland, D. M., Easterling, R. G. & Vopicka, D. A. (1985) *Appl. Spectrosc.*, **39**, 73–84.
Hansen, M. G. & Khettry, A. (1994) *Polym. Eng. Sci.*, **34**, 1758–1766.
Hansen, M. G. & Vedula, S. (1998) *J. Appl. Polym. Sci.*, **68**, 859–872.
Harmer, A. & Scheggi, A. (1989) Chemical, biochemical and medical sensors, in Culshaw, B. & Dakin, J. (Eds.) *Optical Fiber Sensors: Systems and Applications*, Norwood, MA: Artech House.
Hendra, P. J., Ellis, G. & Cutler, D. J. (1988) *J. Raman Spectrosc.*, **19**, 413–418.
Hill, A. (1995) *ACS Symp.*, **603**, 63–80.
Hillemans, J. P., Colemonts, C. M., Meier, R. J. & Kip, B. J. (1993) *Polym. Deg. Stab.*, **42**, 323–333.
Hofmann, K. & Glasser, W. (1990) *Thermochim. Acta*, **166**, p169–184.
Horie, K., Hiura, M., Sawada, M., Mita, I. & Kambe, H. (1970) *J. Polym. Sci.* A-1, **8**, 1357–1372.
Hsieh, T. & Ho, K. (1999) *Polym. Eng. Sci.*, **39**, 1335–1343.
Jacques, P. P. L. & Poller, R. C. (1993) *Eur. Polym. J.*, **29**, 83–89.
Jager, W. F. & Vanden Berg, O. (2001) *Polym. Prepr.*, **42**, 807–808.
Jager, W. F., Wallin, M. & Fernandez, M. V. (2001) *Polym. Prepr.*, **42**, 751–752.
Japon, S., Luciani, A., Nguyen, Q., Leterrier, Y. & Manson, J. (2001) *Polym. Eng. Sci.*, **41**, 1299–1309.
Jawhari, T., Hendra, P. J., Willis, H. A. & Judkins, M. (1990) *Spectrochim. Acta A.*, **46**, 161–170.
Johnson, F. J., Cross, W. M., Boyles, D. A. & Kellar, J. J. (2000) *Composites* A, **31**, 959–968.
Jolliffe, I. T. (1986) *Principal Component Analysis*, New York: Springer-Verlag.
Jones, K. J., Kinshott, I., Reading, M. *et al.* (1997) *Thermochim. Acta*, **304/5**, 187–199.
Jones, R. W. & McClelland, J. F. (1990) *Anal. Chem.*, **62**, 2074–2079.
Jones, R. W. & McClelland, J. F. (1993) *Process Control and Quality*, **4**, 253–260.
Kamide, K., Saito, M. & Miyazaki, Y. (1993) Molecular weight determination, in Hunt, B. J. A. J. (Ed.) *Polymer Characterization*, London: Blackie Academic.
Kammona, O., Chatzi, E. G. & Kiparissides, C. (1999) *J. Macromol. Sci. – Rev. Macromol. Chem. Phys.* C, **39**, 57–134.
Karger-Kocsis, J., Gryshchuk, O. & Schmitt, S. (2003) *J. Mater. Sci.*, **38**, 413–420.
Kaynak, A., Bartley, J. P. & George, G. A. (2001) *J. Macromol. Sci. – Pure Appl. Chem.* A, **38**, 1033–1048.
Kenny, J., Trivisano, A. & Berglund, L. (1991) *SAMPE J.*, **27**, 39–46.
Kent, G. M., Memory, J. M., Gilbert, R. D. & Fornes, R. E. (1983) *J. Appl. Polym. Sci.*, **28**, 3301–3307.
Khettry, A. & Hansen, M. G. (1996) *Polym. Eng. Sci.*, **36**, 1232–1243.
Kissinger, H. E. (1957) *Anal. Chem.*, **29**, 1702–1706.
Koenig, J. L. (1984) Fourier transform infrared spectroscopy of polymers, in *Advances in Polymer Science 54: Spectroscopy: NMR, Fluorescence, FT-IR*, Berlin: Springer-Verlag.
Koenig, J. L. (1999) *Spectroscopy of Polymers*, Amsterdam: Elsevier Science.
Kortaberria, G., Arruti, P., Mondragon, I. (2003) *Macromol. Symp.*, **198**, 389–398.
Kosky, P. G., Mcdonald, R. S. & Guggenheim, E. A. (1985) *Polym. Eng. Sci.*, **25**, 389–394.
Kozielski, K. A., Billingham, N. C., George, G. A., Greenfield, D. C. L. & Barton, J. M. (1995) *High Perform. Polym.*, **7**, 219–236.
Kranbuehl, D., Delos, S., Hoff, M., *et al.* (1987) *Proceedings of Society of Plastics Engineers ANTEC '87*, pp. 1031–1034.

Kranbuehl, D., Hood, D., Wang, Y. *et al.* (1990) *Polym. Adv. Technol.*, **8**, 93–99.
Krohn, D. A. (1988) *Fiber Optic Sensors – Fundamentals and Applications*, Research Triangle Park, NC: Instrument Society of America.
Kupper, L., Heise, H. M. & Butvina, L. N. (2001) *J. Molec. Struct.*, **563–4**, 173–181.
Labana, S. (1985) Cross-linking, in Mark, H. (Ed.) *Encyclopedia of Polymer Science and Engineering*, 2nd edn, New York: Wiley.
Lachenal, G. (1995) *Vib. Spectrosc.*, **9**, 93–100.
Lacoste, J., Carlsson, D. J., Falicki, S. & Wiles, D. M. (1991) *Polym. Deg. Stab.*, **55**, 34.
Lam, K.-Y. & Afromowitz, M. A. (1995a) *Appl. Opt.*, **34**, 5635–5638.
Lam, K.-Y. & Afromowitz, M. A. (1995b) *Appl. Opt.*, **34**, 5639–5643.
Lan, T., Kaviratna, D. & Pinnavaia, T. (1996) *J. Phys. Chem. Solids*, **57**, 1005–1010.
Lane, J., Seferis, J. & Bachmann, M. (1986) *Polym. Eng. Sci.*, **26**, 346–353.
Lange, J., Altmann, N., Kelly, C. T. & Halley, P. J. (2000) *Polymer*, **41**, 4949–5955.
Lange, J., Ekelof, R. & George, G. A. (1998) *Polymer*, **40**, 149–155.
Lange, J., Ekelof, R. & George, G. A. (1999) *Polymer*, **40**, 3595–3598.
Larson, R. (1999) *The Structure and Properties of Complex Fluids*, New York: Oxford University Press.
Lau, W. & Westmoreland, D. G. (1992) *Macromolecules*, **25**, 4448–4449.
Laupretre, F. J. (1990) *Prog. Polym. Sci.*, **15**, 425–474.
Lavine, B. K. (2000) Chemometrics. *Anal. Chem.*, **72**, 91R–97R.
Law, K. Y. & Loutfy, R. O. (1981) *Macromolecules*, **14**, 587–591.
Lee, S.-N., Chiu, M. & Lin, H.-S. (1992) *Polym. Mater. Sci. Eng*, **32**, 1037–1046.
Lees, A. J. (1998a) *Polym. Polym. Composites*, **6**, 121–131.
Lees, A. J. (1998b) *Coord. Chem. Rev.*, **177**, 3–35.
Levy, R. L. (1984) *Polym. Mater. Sci. Eng.*, **50**, 124–126.
Levy, R. L. (1986) *Polym. Mater. Sci. Eng.*, **54**, 321–324.
Levy, R. L. & Schwab, S. D. (1987) *Polym. Mater. Sci. Eng.*, **56**, 169–174.
Liu, H. & George, G. A. (2000) *Polym. Int.*, **49**, 1501–1512.
Louch, J. & Ingle, J. D. (1988) *Anal. Chem.*, **60**, 2637–2640.
Loutfy, R. O. (1981) *Macromolecules*, **14**, 270–275.
Loutfy, R. O. (1982) *J. Polym. Sci. Polym. Phys. Edn*, **20**, 825–835.
Loutfy, R. O. (1986) *Pure Appl. Chem.*, **58**, 1239–1248.
Machado, A. V., Covas, J. A. & Van Duin, M. (1999) *J. Appl. Polym. Sci.*, **71**, 135–141.
Mackison, R., Brinkworth, S. J., Belchamber, R. M. *et al.* (1992) *Appl. Spectrosc.*, **46**, 1020–1024.
Macosko, C. (1994) *Rheology: Principles, Measurements, and Applications*, New York:, VCH.
Maffezzoli, A., Trivisano, A., Opalicki, M., Mijovic, J. & Kenny, J. M. (1994) *J. Mater. Sci.*, **29**, 800–8.
Mailhot, B. & Gardette, J.-L. (1996) *Vib. Spectrosc.*, **11**, 69–78.
Malinowski, E. R. & Howery, D. G. (1980) *Factor Analysis in Chemistry*, New York: John Wiley.
Mcgovern, S., Royce, B. S. H. & Benziger, J. (1985) *J. Appl. Opt.*, **24**, 1512–1514.
Melby, E. G. & Castro, J. M. (1989) Glass-reinforced thermosetting polyester molding: materials and processing, in Allen, G. (Ed.) *Comprehensive Polymer Science*, Oxford: Pergamon.
Mijovic, J., Bellucci, F. & Nicolais, L. (1995) *J. Electrochem. Soc.*, **142**, 1176–1182.
Mijovic, J. & Wang, H. T. (1986) *J. Appl. Polym. Sci.*, **37**, 2661–2673.
Mikes, F., Baselga, J. & Paz-Abuin, S. (2002a). *Eur. Polym. J.*, **38**, 2393–2404.
Mikes, F., Gonzalez-Benito, F., Serrano, B., Bravo, J., Baselga, J. (2002b) *Polymer*, **43**, 4331–4339.
Miller, C. E. (1991) *Appl. Spectrosc. Rev.*, **26**, 277–339.
Milner, S. (1996) *J. Rheology*, **40**, 303–315.
Morgan, R. J. & Mones, E. T. (1987) *J. Appl. Polym. Sci.*, **33**, 999–1020.

Mori, H., Kono, H., Terano, M., Nosov, A. & Zakharov, V. A. (1999) *Macromol. Rapid Commun.*, **20**, 536–540.
Myrick, M. L., Angel, S. M. & Desiderio, R. (1990) *Appl. Opt.*, **29**, 1333–1344.
Niemela, P. & Suhonen, J. (2001) *Appl. Spectrosc.*, **55**, 1337–1340.
Noll, W. (1968) *Chemistry and Technology of Silicones*, New York: Academic Press.
Donnell, J. H. & O'Sullivan, P. W. (1981) *Polym. Bull.*, **5**, 103–110.
Oh, S. J., Lee, S. K. & Park, S. Y. (2006) *Vib. Spectrosc.*, **42**, 273–277.
Ozpozan, T., Schrader, B. & Keller, S. (1997) *Spectrochim. Acta A*, **53**, 1–7.
Paik, H.-J. & Sung, N.-H. (1994) *Polym. Eng. Sci.*, **34**, 1025–1032.
Palmese, G. & Gillham, J. (1987) *J. Appl. Polym. Sci.*, **34**, 1925–1939.
Parker, H.-Y., Westmoreland, D. G. & Chang, H.-R. (1996) *Macromolecules*, **29**, 5119–5127.
Paul, S. (1989) Crosslinking: chemistry of surface coatings, in Allen, G. (Ed.) *Comprehensive Polymer Science*, Oxford: Pergamon.
Peinado, C., Alonso, A., Salvador, E. F., Baselga, J. & Catalina, F. (2002) *Polymer*, **43**, 5355–5361.
Pekcan, O. & Kaya, D. (2001) *Polymer*, **42**, 7865–7871.
Pekcan, O., Kaya, D. & Erdogan, M. (2001) *Polymer*, **42**, 645–650.
Pekcan, O., Yilmaz, Y. & Okay, O. (1997) *Polymer*, **38**, 1693–1698.
Pelikan, P., Ceppan, M. & Liska, M. (1994) *Applications of Numerical Methods in Molecular Spectroscopy*, Boca Raton, FL: CRC Press.
Pell, R. J., Callis, J. B. & Kowalski, B. R. (1991) *Appl. Spectrosc.*, **45**, 808–818.
Pelletier, M. J. (2003) *Appl. Spectrosc.*, **57**, 20A–39A.
Perrier, S. & Haddleton, D. M. (2003) *In situ* NMR monitoring of living radical polymerization, in Puskas, J., Long, T. E., Storey, R. (Eds.) In situ *Spectroscopy of Monomer and Polymer Synthesis*, New York: Kluwer Academic.
Poisson, N., Lachenal, G. & Sauterau, H. (1996) *Vib. Spectrosc.*, **12**, 237–247.
Powell, G. R., Crosby, P. A., Fernando, G. F. *et al.* (1995) *Proc. SPIE*, **2444**, 386–395.
Powell, G. R., Crosby, P. A., Waters, D. N. *et al.* (1998) *Smart Mater. Structures*, **7**, 557–68.
Prime, R. (1997) Thermosets, in Turi, E. (Ed.) *Thermal Characterization of Polymeric Materials*, 2nd edn, New York: Academic Press.
Prime, R., Burns, J., Karmin, M., Moy, C. & Tu, H. (1988) *J. Coatings Technol.*, **60**, 55–60.
Puskas, J. E., Lanzendorfer, M. G. & Pattern, W. E. (1998) *Polym. Bull.*, **40**, 55–61.
Quinn, A. C., Gemperline, P. J., Baker, B., Zhu, M. & Walker, D. S. (1999) *Chemom. Intell. Lab. Syst.*, **45**, 199–214.
Quirin, J. C. & Torkelson, J. M. (2003) *Polymer*, **44**, 423–432.
Rakicioglu, Y., Schulman, J. M. & Schulman, S. G. (2001) Applications of chemiluminescence in organic analysis, in Garcia-Campana, A. M. & Baeyens, W. R. G. (Eds.) *Chemiluminescence in Analytical Chemistry*, New York: Marcel Dekker.
Ramis, X., Cadenato, A., Morancho, J. & Salla, J. (2003) *Polymer*, **44**, 2067–2079.
Ramis, X. & Salla, J. (1997) *J. Polym. Sci. B, Polym. Phys.*, **35**, 371–388.
Reis, M. M., Araujo, P. H. H., Sayer, C. & Giudici, R. (2004) *Ind. Eng. Chem. Res.*, **43**, 7243–7250.
Remillard, J. T., Jones, J. R., Poindexter, B. D., Helms, J. H. & Weber, W. H. (1998) *Appl. Spectrosc.*, **52**, 1369–1376.
Reshadat, R., Desa, S., Joseph, S. *et al.* (1999) *Appl. Spectrosc.*, **53**, 1412–1418.
Rettig, W. (1986) *Angew. Chem. Int. Edn. Engl.*, **25**, 971–988.
Rey, L., Galy, J., Sauterau, H. *et al.* (2000) *Appl. Spectrosc.*, **54**, 39–43.
Rheometrics-Scientific (1990) *Understanding Rheological Testing – Thermoplastics*, Piscataway, NJ: Rheometrics Inc.
Richardson, M. J. (1989) Thermal analysis, in Allen, G. (Ed.) *Comprehensive Polymer Science*, Oxford: Pergamon.

Rohe, T., Becker, W., Kolle, S., Eisenreich, N. & Eyerer, P. (1999) *Talanta*, **50**, 283–290.

Rohe, T., Becker, W., Krey, A. *et al.* (1998) *J. Near Infrared Spectrosc.*, **6**, 325–332.

Roper, T. M., Guymon, C.A., Hoyle, C.E. (2005) *Rev. Sci. Instrum.*, **76**, 054102-1–054102-8.

Roper, T. M., Lee, T. Y., Guymon, C. A. & Hoyle, C. E. (2005) *Macromolecules*, **38**, 10 109–10 116.

Rouse, P. (1953) *J. Chem. Phys.*, **21**, 1272.

Rozenberg, B. A. (1986) Kinetics, thermodynamics and mechanism of reactions of epoxy oligomers with amines, in Dusek, K. (Ed.) *Advances in Polymer Science 75: Epoxy Resins and Composites: II*, Berlin: Springer-Verlag.

Sahre, K., Schulze, U., Hoffmann, T. *et al.* (2006) *J. Appl. Polym. Sci.*, **101**, 1374–1380.

Sanghera, J. S. & Aggarwal, I. D. (1999) *J. Non-Cryst. Solids*, **256–257**, 6–16.

Sanjana, Z. (1986) *Polym. Eng. Sci.* **26**, 373–379.

Santos, A. F., Silva, F.M., Lenzi, M.K., Pinto, J.C. (2005) *Polym. Plast. Technol. Eng.*, **44**, 1–61.

Sasic, S., Kita, Y., Furukawa, T. *et al.* (2000) *Analyst*, **125**, 2315–2321.

Scarlata, S. F. & Ors, J. A. (1986) *Polym. Commun.*, **27**, 41–42.

Scherzer, T. & Decker, U. (1999) *Vib. Spectrosc.*, **19**, 385–398.

Schmidt, L. E., Leterrier, Y., Vesin, J.-M., Wilhelm, M. & Manson, J.-A. E. (2005) *Macromol. Mater. Eng.*, **290**, 1115–1124.

Schweinsberg, D. P. & George, G. A. (1986) *Corros. Sci.*, **26**, 331–340.

Scott, G. (1993) Oxidation and stabilization of polymers during processing, in Scott, G. (Ed.) *Atmospheric Oxidation and Anti-oxidants*, Amsterdam: Elsevier.

Scott, T. F., Cook, W. D. & Forsythe, J. S. (2002) *Polymer*, **43**, 5839–5845.

Senturia, S. & Sheppard, N. (1986) *Adv. Polym. Sci.*, **80**, 3–43.

Senturia, S., Sheppard, N., Lee, H. & Day, D. (1982) *J. Adhesion*, **15**, 69–90.

Sewell, G. J., Billingham, N. C., Kozielski, K. A. & George, G. A. (2000) *Polymer*, **41**, 2113–2120.

Shaffer, O., Bagheri, R., Qian, J. *et al.* (1995) *J. Appl. Polym. Sci.*, **58**, 465–484.

Sherman, B., Neaffer, R. & Galiatsatos, V. (1996) *Trends Polym. Sci.*, **4**, 72–73.

Shim, M. G. & Wilson, B. C. (1997) *J. Raman Spectrosc.*, **28**, 131–142.

Shimoyama, M., Hayano, M., Matsukawa, K. *et al.* (1998) *J. Polym. Sci. B: Polym. Phys.*, **36**, 1529–1537.

Shimoyama, M., Maeda, H., Matsukawa, K. *et al.* (1997) *Vib. Spectrosc.*, **14**, 253–259.

Simon, G. (1997) *Trends Polym. Sci.*, **5**, 394–400.

Simpson, J. & Bidstrup, S. (1995) *J. Polym. Sci. B: Polym. Phys.*, **33**, 55–62.

Sircar, A. K. (1997) Elastomers, in Turi, E. A. (Ed.) *Thermal Characterization of Polymeric Materials*, San Diego, CA: Academic Press.

Skytt, M.-L., Faernert, G., Jansson, J.-F. & Gedde, U. W. (1996) *Polym. Eng. Sci.*, **36**, 1737–1744.

Sohma, J. (1989a) Mechanochemical degradation, in Allen, G. (Ed.) *Comprehensive Polymer Science*, Oxford: Pergamon.

Sohma, J. (1989b) *Prog. Polym. Sci.*, **14**, 451–596.

Song, J. C. & Neckers, D. C. (1996) *Polym. Eng. Sci.*, **36**, 394–402.

Sourour, S. & Kamal, M. R. (1976) *Thermochim. Acta*, **14**, 41–59.

Spragg, R. A. (1984) *Perkin-Elmer Infrared Bull.*, **100**, 7.

St John, N. A. (1993) Spectroscopic Studies of the Cure and Structure of an Epoxy Amine Network, Thesis, Department of Chemistry, St Lucia, The University of Queensland.

St John, N. A. & George, G. A. (1992) *Polymer*, **33**, 2679–2688.

St John, N. A. & George, G. A. (1994) *Prog. Polym. Sci.*, **19**, 755–795.

St John, N. A., George, G. A., Cole-Clarke, P. A., Mackay, M. E. & Halley, P. J. (1993) *High Perform. Polym.*, **5**, 212–236.

Starnes, W. H. (1981) Mechanistic aspects of the degradation and stabilization of poly(vinyl chloride), in Grassie, N. (Ed.) *Developments in Polymer Degradation – 3*, London: Applied Science.

Stellman, C. M., Aust, J. F. & Myrick, M. L. (1995) *Appl. Spectrosc.*, **49**, 392–394.
Storey, R. F., Donnalley, A. B. & Maggio, T. L. (1998) *Macromolecules*, **31**, 1523–1526.
Storey, R. F., Maggio, T. L. & Brister, L. B. (1999) *ACS Polym. Preprints*, **40**, 964–965.
Strehmel, B., Strehmel, V., Timpe, H.-J. & Urban, K. (1992) *Eur. Polym. J.*, **28**, 525–533.
Strehmel, B., Strehmel, V. & Younes, M. (1999) *J. Polym. Sci. B: Polym. Phys.*, **37**, 1367–1386.
Strehmel, V. & Scherzer, T. (1994) *Eur. Polym. J.*, **30**, 361–368.
Stutz, H. & Mertes, J. (1989) *J. Appl. Polym. Sci.*, **38**, 781–787.
Sung, C. S. P., Chin, I.-J. & Yu, W. C. (1985) *Macromolecules*, **18**, 1510–1512.
Sung, C. S. P., Pyun, E. & Sun, H.-L. (1986) *Macromolecules*, **19**, 2922–2932.
Taramae, N., Hiroguchi, M. & Tanaka, S. (1982) *Bull. Chem. Soc. Japan*, **55**, 2097–2100.
Tonelli, A. E. & Srinivasarao, M. (2001) *Polymers from the Inside Out; An Introduction to Macromolecules*, New York: Wiley-Interscience.
Tredwell, C. J. & Osborne, A. D. (1980) *J. Chem. Soc. Faraday II*, **76**, 1627–1637.
Tuchbreiter, A., Marquardt, J., Kappler, B., Honerkamp, J., Mulhaupt, R. (2004) *Macromol. Symp.*, **213**, 327–333.
Van Assche, G., Van Hemelrijck, A., Rahier, H. & Van Mele, B. (1997) *Thermochim. Acta*, **304/305**, 317–334.
Vatanparast, R., Li, S., Hakala, K. & Lemmetyinen, H. (2001) *J. Appl. Polym. Sci.*, **82**, 2607–2615.
Vatanparast, R., Li, S., Hakala, K., Lemmetyinen, H. (2002) *J. Appl. Polym. Sci.*, **83**, 1773–1780.
Vatanparast, R., Li, S., Lemmetyinen, H. (2001) *J. Appl. Polym. Sci.*, **82**, 2607–2615.
Vedula, S. & Hansen, M. G. (1998) *J. Appl. Polym. Sci.*, **68**, 873–889.
Vieira, R. A. M., Sayer, C., Lima, E. L. & Pinto, J. C. (2001) *Polymer*, **42**, 8901–8906.
Vorobyova, O. & Winnik, M. A. (2001) *J. Polym. Sci. B: Polym. Phys.*, **39**, 2302–2316.
Wang, B., Gong, W., Liu, W. *et al.* (2003) *Polymer*, **44**, 4047–4052.
Wang, F. W., Lowry, R. E. & Cavanagh, R. R. (1986a) *Polymer*, **26**, 1657–1661.
Wang, F. W., Lowry, R. E. & Fanconi, B. M. (1985) *Polym. Mater. Sci. Eng.*, **53**, 180–184.
Wang, F. W., Lowry, R. E. & Fanconi, B. M. (1986b) *Polym. Preprints*, **27**, 306–307.
Wetton, R. (1986) Dynamic mechanical thermal analysis of polymers and related materials, in Dawkins, J. (Ed.) *Developments in Polymer Characterization*, London: Elsevier.
Wetton, R. & Duncan, J. (1993) *Society of Plastic Engineers ANTEC Proceedings ANTEC 1993*, pp. 2991–2995.
Woerdeman, D. L. & Parnas, R. S. (2001) *Appl. Spectrosc.*, **55**, 331–337.
Wróblewski, S., Trzebiatowska, K., Jedrzejewska, B. *et al.* (1999) *J. Chem. Soc., Perkin Trans. 2*, 1909–1917.
Wu, C., Danielson, J. D. S., Callis, J. B., Eaton, M. & Ricker, N. L. (1996) *Process Control and Quality*, **8**, 1–23.
Xanthos, M. (Ed.) (1992) *Reactive Extrusion: Principles and Practice*, Munich:, Hanser-Verlag.
Yamada, B., Azukizawa, M., Yamazoe, H., Hill, D. J. T. & Pomery, P. J. (2000) *Polymer*, **41**, 5611–5618.
Younes, M., Wartewig, S., Lellinger, D., Strehmel, B. & Strehmel, V. (1994) *Polymer*, **35**, 5269–5278.
Young, P. R., Druy, M. A., Stevenson, W. A. & Compton, D. A. C. (1989) *SAMPE J.*, **25**, 11–16.
Yu, W. C. & Sung, C. S. P. (1990) *Macromolecules*, **23**, 386–390.
Yu, W. C. & Sung, C. S. P. (1995) *Macromolecules*, **28**, 2506–2511.
Zhang, X., Garcia-Campana, A. M. & Baeyens, W. R. G. (2001) Application of chemiluminescence in inorganic analysis, in Garcia-Campana, A. M. & Baeyens, W. R. G. (Eds.) *Chemiluminescence in Analytical Chemistry*, New York: Marcel Dekker.
Zweifel, H. (1998) *Stabilization of Polymeric Materials*, Berlin: Springer-Verlag.

4 Chemorheological techniques for reactive polymers

4.1 Introduction

This chapter highlights the importance of chemorheology in determining cure properties of reactive systems. A brief introduction to experimental rheology has been provided in Chapter 3 to provide a baseline knowledge of experimental rheology. In this chapter we examine a description of chemorheology in terms of basic chemorheology, chemoviscosity, gelation and vitrification transitions and ultimate properties. Finally, examples of chemorheological analysis will be discussed. (We will briefly summarize chemorheological data and models in this chapter, but only for reference to chemorheological testing. A more extensive examination of chemorheology and modelling of systems will be presented in Chapter 5.)

4.2 Chemorheology

The definition of chemorheology (in this text) is the study of the deformation properties of reactive polymer systems. Figure 4.1 shows a schematic representation of the structural development during thermoset cure.

Step (a) shows unreacted monomers, and cure proceeds to step (b), at which there is the formation of some branched molecules. By step (c) the cure has progressed to the gel point, such that an infinite network is formed across the whole structure. Further cure can occur to point (d), at which the material becomes fully cured and vitrification is reached.

The essential elements of a chemorheological study are

- fundamental chemorheology
- chemoviscosity profiles
- gelation
- vitrification
- ultimate chemorheological properties
- modelling

We shall focus on modelling in Chapter 5. This chapter will focus on the other chemorheological properties and use examples from the literature to clarify key points.

4.2.1 Fundamental chemorheology

Many chemorheological studies omit what we have termed fundamental chemorheological behaviour, by which we mean the basic tests that can assess the fundamental rheological

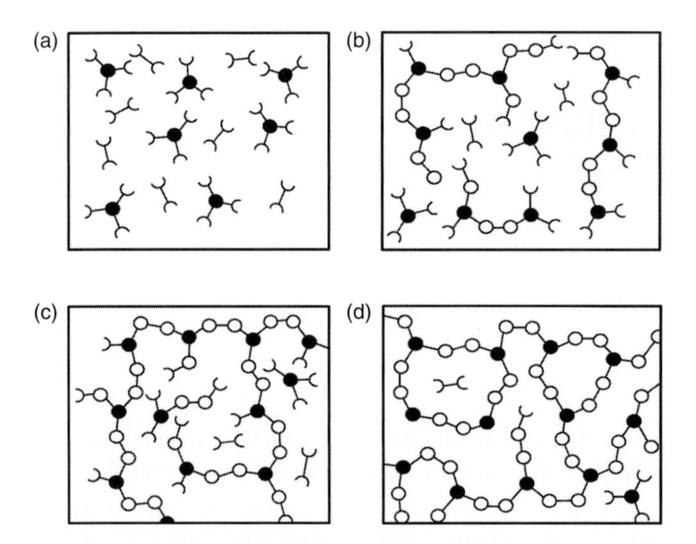

Figure 4.1. A schematic diagram of structural development during thermoset cure from (a) unreacted monomers, (b) formation of initial links, (c) the gel point or a path of covalent links across the sample and (d) the cured crosslinked system with some unreacted groups.

behaviour of the system. We define fundamental rheological behaviour in terms of such characteristics as linear viscoelastic behaviour, wall-slip behaviour and the presence of a yield stress and an understanding of the steady-viscosity–dynamic-viscosity relationship. These are extremely important for full characterization of a fluid, and the fact that many reactive systems are non-linear and exhibit yield stresses and wall slip means that these tests are very important. We shall now examine these fundamental chemorheological tests in more detail.

Linear viscoelastic behaviour

In many studies it is presumed that linear viscoelastic behaviour always occurs, but this is not the case for many reactive systems. Conventional experimental rheology utilizes a dynamic strain sweep, which examines the dynamic rheological response to varied strain amplitudes, at a fixed frequency. If the system shows an effect of strain amplitude on dynamic properties (such as G' or G'') the system is said to be exhibiting a non-linear (viscoelastic) response. If the properties are independent of strain amplitude, then the system is said to be exhibiting linear viscoelastic behaviour. Figure 4.2 shows the response of an industrial epoxy-resin moulding compound (approximately 70 wt.% silica) at 90 °C at strain amplitudes of 0.1% to 10% for frequencies of 1, 10 and 100 rad/s.

Here it can be seen that the response is highly non-linear since the dynamic viscosity is dependent on the strain amplitude. Thus any further testing assuming linear viscoelastic behaviour is not possible for this system under these conditions.

The investigation of non-linear behaviour is an active field for many other polymer systems. For example Payne (1962) interpreted a maximum in G'' (with respect to strain) in the non-linear behaviour of filled suspensions, which has been denoted the Payne effect. Maier and Goritz (1996) and Wilhelm *et al.* (2000) examined the use of Fourier-transform analysis of non-sinusoidal waveforms produced by materials exhibiting non-linear behaviour. This concept of using Fourier-transform rheology to characterize the non-linear

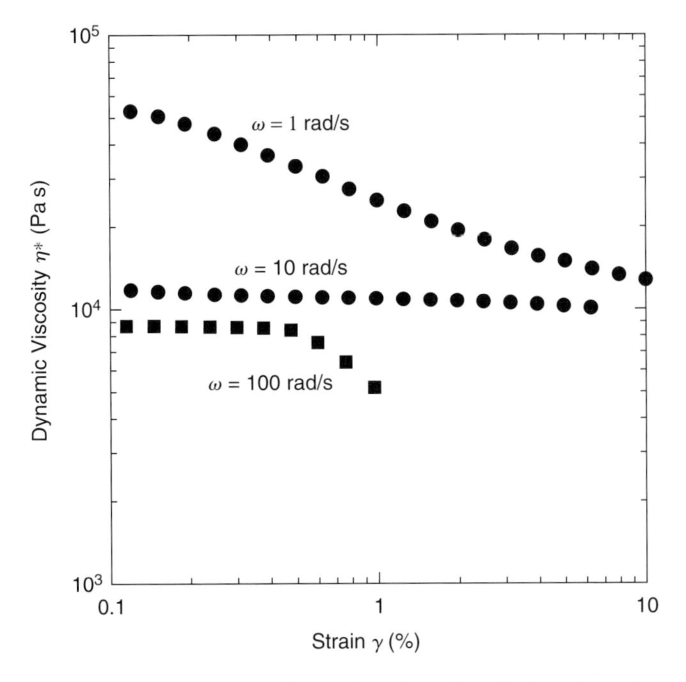

Figure 4.2. Dynamic properties as functions of strain in a dynamic step-strain test to determine the linear viscoelastic region for a highly filled epoxy-novolac moulding sample used for computer-chip encapsulation.

response of materials is related to the work by Giacomin and Dealy (1993) that examined the non-linear response of thermoplastic polymer melts to very large strains. This work focusses on the ability to generate 'effective' G' and G'' data from non-linear responses.

It should be noted that the linear viscoelastic behaviour needs to be examined as a function of temperature and cure level during the reaction in order to describe fully the viscoelastic behaviour of the system.

Yield stress

There are various forms of yield-stress test that can assess the yield stress (σ_Y) of a material, and they are typically separated into indirect and direct methods. Indirect methods monitor properties that may in turn be related to a yield stress, whereas direct methods measure the yield stress directly. Various techniques are described in more detail in Gupta (2000). Here we shall describe two methods that have been used successfully in our laboratory.

An example of an indirect method is the use of a steady-shear-rate sweep test, which examines the steady-shear viscosity (η) and shear stress (σ) as functions of shear rate (γ'). The results, once obtained, may be plotted in terms of steady-shear viscosity (η) versus shear stress (σ). The existence of an infinite-viscosity asymptote at low levels of shear stress is indicative of a yield stress, and the asymptotic stress is defined as the yield stress (σ_Y). If there is no asymptote, then the material is deemed to have no measurable yield stress. This is shown diagrammatically in Figure 4.3.

An example of a direct method is the vane technique (Collyer and Clegg, 1988). Here a vane rheometer is used (see Figure 4.4), in such a way that the vane attachment is used to monitor the shear response of a rotating fluid bath.

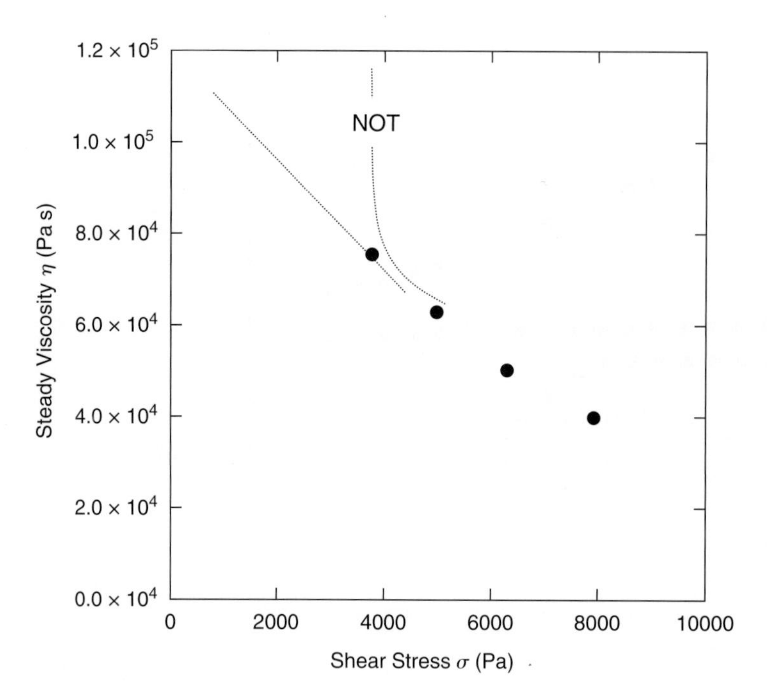

Figure 4.3. Steady-shear viscosity as a function of shear stress to examine the presence of a yield stress for a highly filled epoxy-novolac moulding sample used for computer-chip encapsulation.

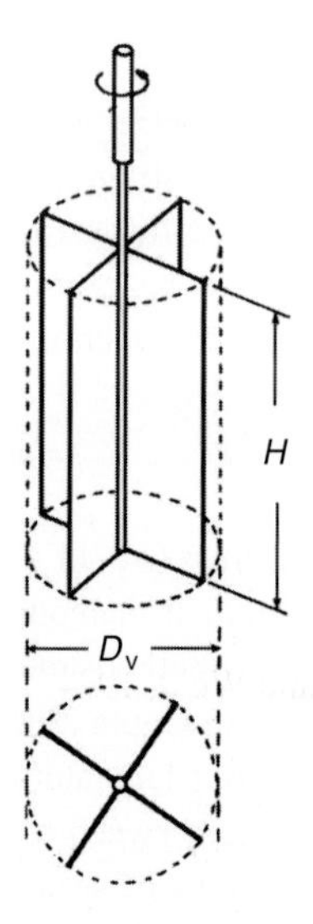

Figure 4.4. The vane-rheometer configuration used for direct yield-stress measurements. Reprinted from Figure 9.9 (Collyer and Clegg, 1988).

The advantage of the vane technique is that material is allowed in between the prongs of the vane and in effect shears the material upon itself (rather than inducing shearing between a mechanical fixture and the material, as would be seen if a standard cup-and-bob or Couette rheometer were used). This eliminates the potential for slippage between a mechanical fixture and the material. A typical response is given in Figure 4.5.

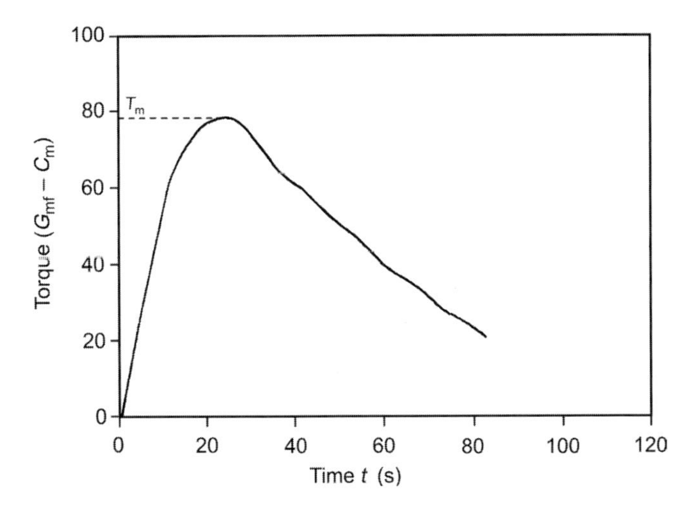

Figure 4.5. Typical response from a Vane rheometer. Reprinted from Figure 9.10 (Collyer and Clegg, 1988).

The yield stress may then be calculated on the basis of a simple force-balance calculation from the following equation;

$$T_m = \pi D_v{}^3/2(H_v/D_v + 1/3)\sigma_y, \tag{4.1}$$

where T_m is the maximum in torque in a steady time test (in which the torque (T) is monitored as a function of time (t) at a fixed shear rate (γ')), D_v is the external diameter of the vane, H_v is the height of the vane and σ_y is the yield stress. Once again, with reactive systems the effects of temperature and cure on the yield stress should be considered in order to obtain a full chemorheological description of the material.

Wall slip

There is also a variety of wall-slip techniques used in rheological analysis (Gupta, 2000). The methods described here are flow visualization, capillary flow and torsional flow.

Flow visualization essentially determines the slip velocity at the wall via particle-tracking software and data capture, and is typically used only for model systems and surfaces (Piau *et al.*, 1995). However, some work on visualization of wall slip in highly filled reactive epoxy-resin moulding compounds has been reported (Manzione and Weld, 1994, Manzione, 1995).

Capillary rheometry and parallel-plate rheometry use the fact that wall slip will manifest itself as a geometry-dependent phenomenon. That is, wall slip will appear as a geometric effect on 'apparent' rheological properties. In the capillary-rheometer technique, slip will manifest itself as an effect of capillary diameter (D) on the shear stress (τ_w). Wall slip in capillary rheology can be calculated from an analysis that involves the following:

(a) determining the wall shear stress ($\tau_w = \Delta P/[4(L/D)]$) versus apparent shear rate ($\gamma'_{app} = 32Q/(\pi D^3)$) curves for a variety of diameters, where ΔP is the pressure drop across the capillary, L is the capillary length, D is the capillary diameter and Q is the flow rate;
(b) plotting the apparent shear rate (γ'_{app}) versus $1/D$ for various values of the shear stress (τ_w); and

(c) calculating the slip velocity from the slope of the curves in (b) or via the expression

$$8V_s = d(\gamma'_{app})/d(1/D), \tag{4.2}$$

which is derived by following the Rabbinowitsch correction (Rabbinowitsch, 1929) and will provide a wall slip $V_s(\tau_w)$ function for the system.

In parallel-plate geometry, slip will appear as an effect of gap height on shear stress for tests conducted with differing gaps between parallel plates. Yoshimura and Prudhomme (1988) showed that wall slip may be obtained from the following expression:

$$V_s = [\gamma'_{RH1} - \gamma'_{RH2}]/[2(1/H_1 - 1/H_2)], \tag{4.3}$$

Where γ'_{RH1} and γ'_{RH2} are the shear rates at the plate edge for plates of gaps H_1 and H_2, respectively.

Figure 4.6 shows that all wall-slip techniques can be used successfully to determine the wall slip of a material over a wide range of shear stresses. Once again, for full characterization of wall slip in reactive systems the effects of temperature and extent of cure on wall slip should be determined.

The steady-viscosity–dynamic-shear-viscosity relationship

The relationship between steady-shear viscosity and dynamic-shear viscosity is also a common fundamental rheological relationship to be examined. The Cox–Merz empirical rule (Cox, 1958) showed for most materials that the steady-shear-viscosity–shear-rate relationship was numerically identical to the dynamic-viscosity–frequency profile, or $\eta(\gamma') = \eta^*(\omega)$. Subsequently, modified Cox–Merz rules have been developed for more complex systems (Gleissle and Hochstein, 2003, Doraiswamy *et al.*, 1991). For example Doriswamy *et al.* (1991) have shown that a modified Cox–Merz relationship holds for filled polymer systems for which $\eta(\gamma') = \eta^*(\gamma\omega)$, where γ is the strain amplitude in dynamic shear.

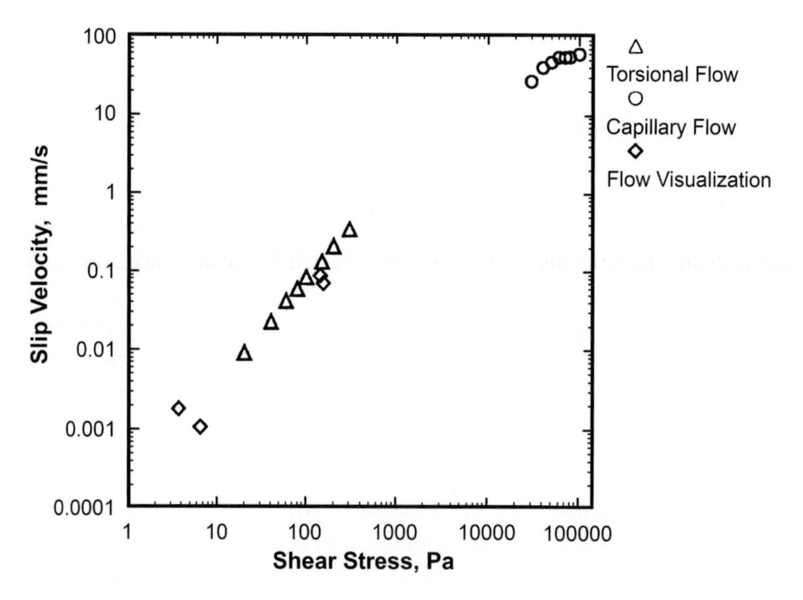

Figure 4.6. Wall-slip tests with a range of rheometer systems. Reprinted from Figure 9.11 (Collyer and Clegg, 1988).

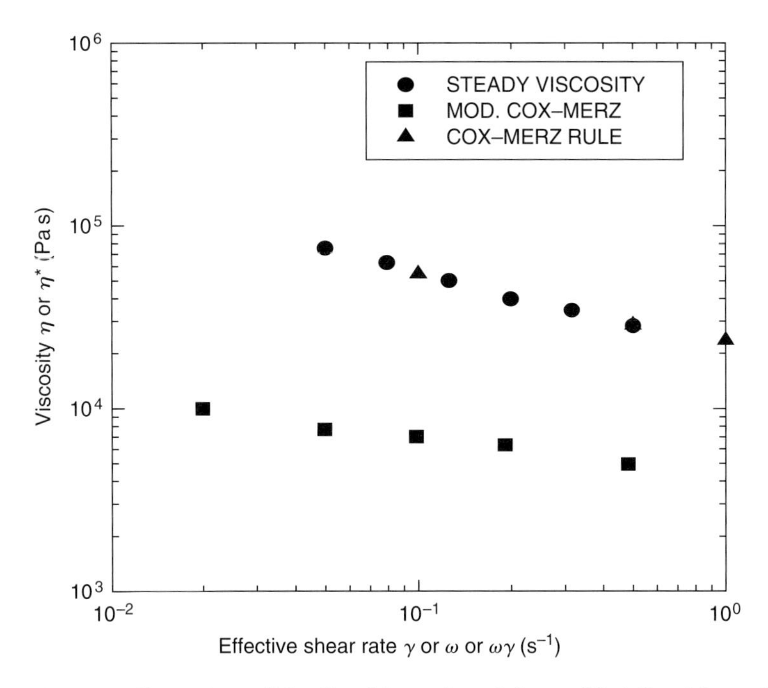

Figure 4.7. Comparison of the Cox–Merz rule and the modified Cox–Merz rule for a highly filled epoxy-novolac moulding sample used for computer-chip encapsulation.

Figure 4.7 shows the steady- and dynamic-viscosity profiles as functions of shear rate for a filled reactive epoxy-resin moulding compound. Here, interestingly, the Cox–Merz rule provides a better correlation than does the modified Cox–Merz rule.

Overall these fundamental chemorheological tests will provide essential characterization of the typically complex reactive systems under investigation. As mentioned previously, to achieve full characterization we are next interested in chemoviscosity profiles.

4.3 Chemoviscosity profiles

4.3.1 Chemoviscosity

The chemoviscosity of thermosetting resins is affected by many variables. In a major review, Ryan 1984) expressed the chemoviscosity (η) as a function of pressure (P), temperature, time, shear rate (γ') and filler properties (F), as shown by the following general equation:

$$\eta = \eta(T, P, \gamma', t, F). \tag{4.4}$$

The effects of each variable on the chemoviscosity are usually examined by separate tests such as of

- cure effects,

$$\eta_{\mathrm{c}} = \eta_{\mathrm{c}}(T, t), \tag{4.5}$$

- shear-rate effects,

$$\eta_{\mathrm{sr}} = \eta_{\mathrm{sr}}(\gamma', T), \tag{4.6}$$

- filler effects,

$$\eta_{\mathrm{f}} = n_{\mathrm{f}}(F), \tag{4.7}$$

or

$$\eta_{\mathrm{f}} = \eta_{\mathrm{c}}(F, T, t), \tag{4.8}$$

or

$$\eta_{\mathrm{f}} = \eta_{\mathrm{sr}}(F, \gamma', t). \tag{4.9}$$

Recent work has included tests that examine the effects of shear rate and cure simultaneously (which will also be discussed below). The models derived from these tests are recombined to provide an overall chemoviscosity model to be used in processing applications. The effect of pressure on chemoviscosity has not been studied extensively; however, system pressure may be relevant to high-pressure injection-moulding and transfer-moulding processes.

The *cure effects* on the chemoviscosity are two-fold; the viscosity will initially decrease due to the increase in thermal effects but will eventually increase due to formation of the crosslinked network via the curing reaction. This is shown schematically for the chemoviscosity of a polyester resin during injection moulding in Figure 4.8.

Initially, during plasticization, the viscosity of the resin decreases due to shear heating and thermal effects (stage I). As the temperature (or time) increases, the curing reaction begins and the decrease in viscosity due to heating is compensated by the increase in viscosity due to the curing process. At the point of minimum viscosity the polyester is injected into the mould (stage II). Finally, the viscosity of the resin increases as the material is transformed into an infusible solid (stage III). The cure effects on viscosity (η_{c}) may be determined by measuring the activation energies of the reaction and of viscous flow.

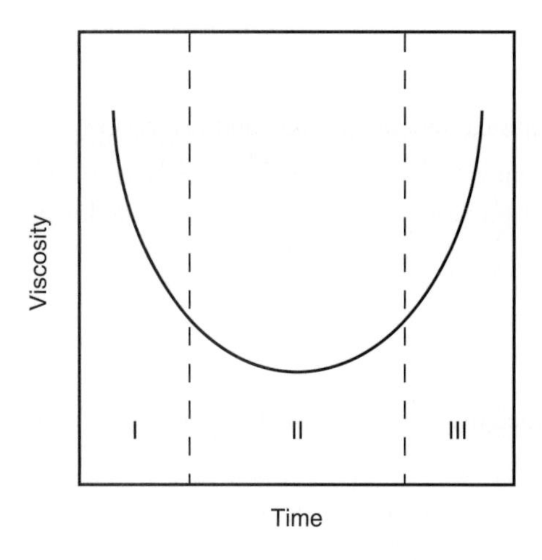

Figure 4.8. The viscosity of a thermoset polymer during processing.

The effects of temperature and time on the chemoviscosity can also be described explicitly in terms of the extent of cure (α) from knowledge of the kinetics of the cure (i.e. (T, t)) and temperature by the following equation:

$$\eta_c = \eta_c(T, \alpha). \tag{4.10}$$

Extensive work in the literature has focussed on the determination of the cure effects on the chemoviscosity of thermosetting resin, and a summary of the effects of cure on various chemorheological models is given in Table 4.1.

These models, which examined the effects of cure on the chemoviscosity, range from simple empirical models (Malkin and Kulichikin, 1991, Lane, 1987, Mussati and Macosko, 1973, Kamal and Sourour, 1973, Roller, 1976, Dusi *et al.*, 1982) (used for epoxy-resin and polyurethane systems), probability-based and molecular models (Vinogradov and Malkin, 1977, Lipshitz and Macosko, 1976, 1977, Mijovic *et al.*, 1996) (used for polyurethane and epoxy-resin systems), gelation models (Malkin and Kulichikin, 1991, Castro and Macosko, 1982, Yang and Suspene, 1991) (used for polyester, epoxy-resin and melamine systems) and Arrhenius models (Martin *et al.*, 1989, Kamal, 1974, Kojima *et al.*, 1986, Tungare *et al.*, 1986, Dusi *et al.*, 1983) (used for many thermosets, including epoxies and polyimides) to detailed models based on free-volume analyses (Hale *et al.*, 1989, Chiou and Letton, 1992) (used for epoxy-resin systems). Further reviews of curing-effect models are given by Roller (1976) and Ryan (1984). Great accuracy in determination of the cure effects on viscosity is required, since they have a large effect on the processing of thermosets, especially during the final curing stage. Accurate data are also needed for determining the parameters in some of the models.

Of course, it should be noted that *cure conversion or kinetic models* themselves should be accurately determined, because they must be used in parallel with cure models of chemoviscosity. There are essentially two forms of kinetic model used to describe thermoset curing reactions, namely empirical and mechanistic models. Empirical models assume an overall reaction order and fit this model to the kinetic data. This type of model provides no information on the kinetic mechanisms of the reaction, and is predominantly used to provide models for industrial samples. Mechanistic models are derived from an analysis of the individual reactions involved during curing, which requires detailed measurements of the concentrations of reactants, intermediates and products. Essentially, mechanistic models are intrinsically more complex than empirical models; however, they are not restricted by compositional changes, as are empirical models. Typical kinetic models used in the analysis of thermosetting chemical reactions are listed in Table 4.2.

These models include simplistic empirical models (Malkin and Kulichikin, 1991, Martin *et al.*, 1989, Ryan, 1973, 1984, Arrelano *et al.*, 1989, Han and Lem, 1984, Hale *et al.*, 1989, Dutta and Ryan, 1979, Lane, 1987) (used for epoxy-resin and polyester systems) and complex mechanistic models derived from chemical analysis and probability theories (Yang and Suspene, 1991, Gupta, 1990, Riccardi and Williams, 1986, Batch and Macosko, 1987, Pannone and Macosko, 1988, Mussati and Macosko, 1973) (used for epoxy-resin, polyester and polyurethane systems). Determination of the most appropriate kinetic model for an application will depend on the type of system and the accuracy and form of results required.

Evaluation of the appropriate kinetic model parameter from kinetic measurement defines the degree of conversion as a function of time and temperature, $\alpha(T, t)$. This relationship is then used together with chemoviscosity and gel effects to define the thermoset process.

The *shear-rate effect* on the viscosity of thermosetting resins, $\eta_{sr}(\gamma)$, is also essential to the determination of the chemoviscosity, η. The exact relationship will depend on the type of

Table 4.1. Kinetic models for reactive systems

Model	Expression	System	Notes
First order	$\frac{\mathrm{d}a}{\mathrm{d}t} = k(1-a)$ $k = A\exp(-E/RT)$	Epoxy resin (DGEBA/DCA) (Dusi *et al.*, 1982)	k, rate constant E_a, activation energy A, coefficient
Second order	$\frac{\mathrm{d}a}{\mathrm{d}t} = k(1-a)^2$ $k = A\exp(-E/RT)$	Epoxy resin (Gonzalez-Romero and Casillas, 1989) (DGEBA/amines) (Dusi *et al.*, 1983)	
*n*th order	$\frac{\mathrm{d}a}{\mathrm{d}t} = k(1-a)^n$ $k = A\exp(-E/RT)$	Polyester (Lee, 1981) Epoxy resin (epoxy novolac) (Martin *et al.*, 1989) Epoxy resin (Knauder *et al.*, 1991)	n, reaction order
Series of *n*th-order reactions	$\frac{\mathrm{d}a}{\mathrm{d}t} = \sum_i g_i \frac{A_i}{\beta}\exp\left(-\frac{E_i}{RT}\right)(1-a_i)^{n_i}$	Tri-epoxy system (TGMDA/novolac/carboxylate/DDS) (Castro and Macosko, 1982)	g_i, factor A_i, coefficient β, heating rate n_i, reaction order
Polynomial	$\frac{\mathrm{d}a}{\mathrm{d}t} = k\exp\left(-\frac{E_a}{RT}\right)(a_0 + a_1 a + a_2 a^2)$	Epoxy resin (epoxy novolac/silica filler) (Ryan and Kamal, 1976)	a_0, a_1, a_2, constants
Autocatalytic, 1	$\frac{\mathrm{d}a}{\mathrm{d}t} = (k_1 + k_2 a^m)(1-a)^n$ $k_1 = k_0\exp[-E_1/(RT)]$ $k_2 = k_0\exp[-E_2/(RT)]$	Polyester (Bidstrup *et al.*, 1986) Epoxy novolac + filler (Mussati and Macosko, 1973) Epoxy resin (DGEBA + amine filler) (Nass and Seferis, 1989) Thermosets (Kamal and Ryan, 1980)	n, m, reaction orders k_1, k_2, rate constants E_1, E_2, activation energies
Autocatalytic, 2	$\frac{\mathrm{d}a}{\mathrm{d}t} = (k_1 + k_2 a)(1-a)(B-a)$	Epoxy resin (TGDDM/DDS) (Kascaval *et al.*, 1993)	B, stoichiometry factor
Mechanistic	$\frac{a}{a_{\mathrm{gel}}} = f(\text{concentration})$	Unsaturated polyester/styrene (Lane *et al.*, 1986) Epoxy resin (TGDDM/DDS) (Sourour and Kamal, 1976)	a_{gel}, conversation at gelation
Self-acceleration	$\frac{\mathrm{d}a}{\mathrm{d}t} = k(1-a)(1+Ca)$	Thermosets (Gonzalez-Romero and Casillas, 1989)	C, constant
Self-inhibition	$\frac{\mathrm{d}a}{\mathrm{d}t} = k(1-a)(1-\varepsilon a)$	Thermosets (Gonzalez-Romero and Casillas, 1989) – diffusion-limited	ε, constant
Combined	$\frac{\mathrm{d}a}{\mathrm{d}t} = k(1-a)(1+Ca)(1-\varepsilon a)$	Thermosets (Gonzalez-Romero and Casillas, 1989)	

Table 4.2. Chemorheological models for the effect of curing

Model	Expression	System	Comments
Macosko	$\frac{1}{t_1} = C\ \exp[-E_a/(RT)]$	Phenolic, epoxy, EPDM rubber (Roller, 1976)	t_1, time to reach viscosity η_1 C, constant E_a, activation energy
First-order, isothermal	$\eta_c = \eta_0\ \exp(\theta t)$	Epoxy resins (Rydes, 1993), Ng and Manas-Zloczower, (1993)	Initial stages of cure Doesn't take into account gel point η_0, initial viscosity θ, constant
First-order, non-isothermal	$\eta_c = \eta(T)\exp(\phi kt)$ $\phi = f(a)$	Epoxy resins (Ng and Manas-Zloczower, 1989)	η (T), viscosity as a function of temperature (T) k, rate constant ϕ, function of conversion (a)
Molecular, 1	$\eta_c = kM^a$ $a = 1$ for $\overline{M} < \overline{M}_c$ $a = 3.5$ for $\overline{M} > \overline{M}_c$	Linear polymers (Han and Lem, 1983a)	k, constant $\overline{M}$, molar mass $\overline{M}_c$, critical molar mass
Molecular, 2	$\eta_c = A \exp\left(\frac{D}{RT}\right)\left(\frac{\overline{M}}{\overline{M}_0}\right)^{\left(\frac{C}{RT}+S\right)}$	Polyurethanes (Han and Lem, 1983b)	A, D, C, S, constants $\overline{M}_0$, initial molar mass
Empirical, 1	$\eta_c = Kt^a$ $a = a_1$ for $t < t_c$ $a = a_2$ for $t_c < t < t_p$ $a = a_3$ for $t_p < t$	Epoxy resins (Gonzalez-Romero and Casillas, 1989)	Homogeneous reaction up to termination of reaction No diffusion limitations or gel effects considered a, K, constants t_c, t_p, characteristic times
Empirical, 2	$\log \eta_c = \log \eta_v + \frac{E_v}{RT} + ka$	Epoxy resin (TGDDM/DDS) (Kascaval *et al.*, 1993)	k, constant E_v, viscous activation energy η_v, initial viscosity
Empirical, 3	$\frac{\eta}{\eta_0} = \left(\frac{1+kt}{1-t/t*}\right)^a$	Polyurethane (Gonzalez-Romero and Casillas, 1989)	Reaction and gel effects with no phase separation. k, constant

Table 4.2. *(cont.)*

Model	Expression	System	Comments
Empirical, 4	$\log\left(\frac{\eta_c}{\eta_m}\right) = \pm a \log\left(\frac{t}{t*}\right) + b \log\left(\frac{1}{1 - t/t*}\right)$ $a = 1$ for $t < t_c$ $a = 3.5$ for $t > t_c$	Polyurethane (Gonzalez-Romero and Casillas, 1989)	Gel effects including phase separation η_m, viscosity of the medium a, b, constants η_m, changes at t_c and t_p $\pm$, change as $t > t_c$
Gel model	$\frac{\eta_c}{\eta_0} = \left(\frac{a*}{a* - a}\right)^{A+Ba}$	Thermosets (Lem and Han, 1983a)	$a*$, conversion at gel point A, B, constants
Microgel, 1	$\frac{\eta}{\eta_0} = \left(1 - \frac{t}{t*}\right)^b$	Melamine formaldehyde/epoxy resin (Gonzalez-Romero and Casillas, 1989)	Non-homogeneous reaction Includes formation of microgel particles. t^*, gel time b, constant
Microgel	$\log \eta_s = A + B\left(\frac{t}{t*}\right)$ $\eta_s = \frac{\eta_c}{\eta_o} - 1$	Polyester/styrene (Lane *et al.*, 1986)	Non-homogeneous reaction Three distinct regions of microgelation, transition and macrogelation. A, B, constants
Arrhenius first-order (isothermal)	$\ln \eta = \ln \eta_v + \frac{E_v}{RT} + t k_k \exp\left(\frac{E_k}{RT}\right)$	Epoxy resin (Han and Lem, 1983c)	η_v, initial viscosity E_v, viscous activation energy k_k, rate constant E_k, kinetic activation energy
(non-isothermal)	$\ln \eta = \ln \eta_v^t + \frac{E_v}{RT} + k_k \int \exp\left(\frac{E_k}{RT}\right) dt$	Epoxy resin (Dusi *et al.*, 1982) (DGEBA/DCA) Polyamide fibre (Kamal and Sourour, 1973) Epoxy resin (Mijovic and Lee, 1989) (DGEBA/DCA)	
Arrhenius *n*th order	$\ln \eta = \ln \eta_v + \frac{E_v}{RT} + k_k \int (1 - a)^n \exp\left(\frac{E_k}{RT}\right) dt$	Epoxy resin (Knauder *et al.*, 1991)	n, reaction order
Modified Arrhenius			
first order	$\ln \eta = \ln \eta_0 + \phi_1 k_k \exp[-E_k/(RT)]t$	Epoxy resin (Lem and Han, 1983c)	ϕ_1, entanglement factor
*n*th order	$\ln \eta = \ln \eta_0 + \frac{\phi}{n - 1}$ $\ln\left[1 + (n - 1) k_k \exp\left(-\frac{E_k}{RT}\right) dt\right]$	Epoxy resin (Lem and Han, 1983c)	

WLF	$\ln \eta_c = \ln \eta_a + \ln \overline{M}_w + \frac{E}{RT} + \frac{C_1(T - T_g)}{C_2 + T - T_g}$ $M_w = M_w(a)$ $T_g = T_g(a)$ $a = a(t)$	Epoxy resins (Lem and Han, 1983b)	η_a, initial viscosity C_1, C_2, functions of a T_g, glass-transition temperature
Modified WLF, 1	$\eta_c = \eta_0 \exp\left(\frac{1}{f_a/B + a_1/[B(T - T_g)]}\right) M_w$ $M_w = M_w(a)$ $T_g = T_g(a)$	Epoxy resin (Epoxy novolac)/silica filler (Mussati and Macosko, 1973)	f_a, a, B, constants T_g, glass-transition temperature
Modified WLF, 2	$\ln\left(\frac{\eta_c(T)}{\eta_c(T_g)}\right) = \frac{C_1(a)[T - T_g(a)]}{C_2(a) + T - T_g(a)}$	Multiple epoxy system (TGMDA + novolac + carboxylate/DDS) (Castro and Macosko, 1982)	C_1, C_2, T_g, functions of a

Table 4.3. Chemorheological models for the effect of shear rate

Model	Form	Systems	Notes
Power law	$\eta_{SR} = A\gamma'^{n-1}$	Epoxy resin (epoxy novolac/silica filler) (Mussati and Macosko, 1973) Epoxy resin (epoxy/diamine) (Riccardi and Vazquez, 1989)	A, constant n, index
Bueche–Harding	$\frac{\eta_{SR}}{\eta_0} = \frac{1}{(1+\lambda\omega)^{3/4}}$	Thermosets (Prime, 1973)	η_0, initial viscosity
Carreau	$\frac{\eta_{SR}}{\eta_0} = (1+(\lambda\omega)^2)^{n-1/2}$ $= (1+(\lambda\gamma')^2)^{n-1/2}$	Thermosets (Prime, 1973)	λ, relaxation time η_0, zero-time viscosity
Cross	$\frac{\eta_{SR}}{\eta_0} = \frac{1}{1+(\lambda\omega)^{1-n}}$ $= \frac{1}{1+(\eta_0\gamma'/\tau^*)^{1-n}}$	Diallyl phthalate (Prime, 1973)	τ^*, stress at transition
Ellis	$\frac{\eta_{SR}}{\eta_0} = \frac{1}{(\eta/\eta_0)\lambda\omega}\frac{1-n}{n}$ $= \frac{1}{1+(\tau/\tau_{1/2})}$	Diallyl phthalate (Prime, 1973)	$\tau_{1/2}$, stress at $\eta = \eta_0/2$
WLF	$\eta_{SR} = \frac{a_t}{A_0 + (A_1\omega a_t)^{A_2}}$, $a_t = \exp\left(-\frac{C_1(T-T_0)}{C_2+T-T_0}\right)$	Filled epoxy resin (Ryan, 1984)	A_0, A_1, A_2, constants C_1, C_2, constants T_0, reference temperature Assumes Cox–Merz rule is valid

system used. For example, Sundstrom and Burkett (1981) and Hartley and Williams (1981) found that polyesters and polyurethane exhibit essentially Newtonian behaviour, whereas epoxies and phenolics show marked shear thinning. More detailed rheological models used to characterize the effect of shear rate on chemoviscosity are displayed in Table 4.3.

The power-law model is the most extensively used shear-rate model for thermosets and has been used for unfilled (Ryan and Kamal, 1976, Kascaval *et al.*, 1993, Riccardi and Vazquez, 1989) and filled (Ryan and Kamal, 1976, Knauder *et al.*, 1991) epoxy-resin systems. Sundstrom and Burkett (1981) showed that there was a good fit of the viscosity of diallyl phthalate to the Cross model. The viscosity of polyesters has been modelled by Yang and Suspene (1991) using a Newtonian model. The WLF model has been used by Pahl and Hesekamp (1993) for a moderately filled epoxy-resin system. Rydes (1993) showed that the viscosity of DMC polyesters followed a power-law relationship at high shear rate.

The *filler effects* on the chemoviscosity of thermosetting resins have not been studied extensively, but are vital to understanding the rheology of filled thermosets. For example, the effects of filler concentration on viscosity can be used in process control to monitor batch-to-batch variations or to provide essential information for research into alternative filler/resin batches. Ng and Manas-Zloczower (1993) examined an epoxy-resin system with silica filler and established that the elastic modulus of the resin can be expressed in terms of

Table 4.4. Chemorheological models combining all effects

Model	Expression	System
Power law/WLF	$\eta = \eta_0 \exp\left(\frac{1}{\frac{f_g}{B} + \frac{\alpha}{B}(T - T_g)}\right) M_w$ $M_w = f(\alpha)$ $T_g = T_g(\alpha)$ $\eta_0 = A\gamma^{n-1}$	Epoxy resin (Epoxy novolac/Silica filler) (Hale *et al.*, 1989)
Power law/WLF/ Conversion	$\eta = A_1 \gamma'^{A_2} \left(\frac{\alpha_g}{\alpha_g - \alpha}\right)^{A_3 + A_4 \alpha} \exp\left(\frac{C_1(T - T_g)}{C_2 + T - T_g}\right)$	Epoxy resin (Castro and Macosko, 1982)
WLF/Arrhenius	$\ln \eta^* = \ln \eta^*(T, \omega) + k_a \int \exp\left(\frac{\Delta E_k}{RT}\right) dt$ $\eta^*(T, \omega) = \frac{a_T(T)}{A_0 + A_1[\omega a_T(T)]^{A_2}}$	Filled epoxy resin (Pahl and Hesekamp, 1993) Assumes Cox–Merz rule is valid
Power law/ Arrhenius/ molecular, 1	$\eta = \mu \left(\frac{\gamma'}{\gamma'_0}\right)^{n-1} \exp[-b(T - T_0) + aP] M_\omega^m \frac{d\gamma}{dr}$ $M_w = f(\alpha)$	Epoxy resin (Epoxy/Diamine) (Riccardi and Vazquez, 1989)
Power law/ Arrhenius/ molecular, 2	$\eta = A\gamma'^B \exp\left(\frac{C}{T}\right) \left(\frac{\alpha_g}{\alpha_g - \alpha}\right)^{D + E\alpha}$	Epoxy resin (Castro and Macosko, 1982, Nguyen, 1993)
Carreau	$\frac{\eta - \eta_a}{\eta_0 - \eta_a} = \left[1 + \left(\frac{\eta_0 \gamma'}{\tau^*}\right)^2\right]^{n-1/2}$ where $-\eta_0 = \mathrm{A} \exp\left(\frac{\mathrm{E}}{\mathrm{T}}\right) \left(\frac{\alpha_g}{\alpha_g - \alpha}\right)^{B + C\alpha}$	Epoxy resin (Peters *et al.*, 1993)
Mould flow	$\ln \eta = A_1 + A_2 \ln \gamma' + A_3 T + A_4 \ln \gamma'^2$ $+ A_5 \ln(\gamma' T) + A_6 T^2 + (A_7 + A_8 X) \ln\left(\frac{\alpha_g}{\alpha_g - \alpha}\right)$	Epoxy resin (Peters *et al.*, 1993)

filler concentration and the modulus of the matrix, and that the gel time decreases with greater filler concentration. Dutta and Ryan (1979) examined the effects of fillers on the kinetics of an epoxy–diamine reaction, for which they found that the type of filler affected the reaction kinetics. They noted that carbon-black fillers increase the reaction rate via pre-exponential terms and that silica fillers affect the reaction rate through activation energies. Han and Lem (1983a, 1983b, 1983c, 1984, Lem and Han, 1983a, 1983b, 1983c) noted the effects of various fillers on the kinetics and chemorheology of polyester resins. They found that increasing the concentration of clay or glass fibre exaggerated the degree of shear thinning at low shear rates (a behaviour similar to that of concentrated suspensions in Newtonian fluids), whereas the addition of $CaCO_3$ at all concentrations produced Newtonian behaviour. Han and Lem (1984) found an interaction among fillers, thickeners and low-profile additives that produced complex rheological behavior. Kalyon and Yilmazer (1990) and Metzner (1985) stated that the effect of fillers on the viscosity of highly filled polymeric suspensions induces wall slip, flow instabilities, yield stresses and dilatancy. Kalyon and Yilmazer (1990) also noted that filler concentration, surface interactions, orientation in the flow field, particle shape, particle diameter and resin properties will influence the effect of fillers on the overall viscosity.

By combining the effects of cure, shear rate and filler, a complete model for the chemoviscosity can be established. Examples of these combined models are shown in Table 4.4.

It should be noted that the effects of fillers may be incorporated into the cure and shear-rate effects. The main forms of combined-effects model consist of WLF, power-law or Carreau shear effects, Arrhenius or WLF thermal effects and molecular, conversion or empirical cure effects. Nguyen (1993) and Peters *et al.* (1993) used a modified Cox–Merz relationship to propose a modified power-law model for highly filled epoxy-resin systems. Nguyen (1993) also questions the validity of the separability of thermal and cure effects in the derivation of combined models.

Together with the chemoviscosity modelling, other variables are important to the processing of thermosets. These include

- $\eta_{\min}$ – the minimum viscosity (as a function of temperature) for injection or transfer moulding, and
- dT/dt – the optimum heating rate to ensure that the minimum-viscosity the and gelation times are optimized.

The measurement of these properties, together with an accurate chemoviscosity model, enables prediction of optimum flow conditions.

4.3.2 Gel effects

Various viscosity models have implicitly included the effects of gelation on the chemoviscosity, and these were reviewed in Table 4.2 incorporating gelation–conversion and glass-transition-temperature effects implicitly in the cure effects on chemoviscosity. Explicit models for the expression of gel time versus temperature and time are sparse, with empirical measurements mainly being used.

Thus, with a knowledge of the effects of gelation and an accurate chemoviscosity model, the optimum flow and cure conditions can be established. The measurement techniques for chemoviscosity and gelation effects will be reviewed next.

4.4 Chemorheological techniques

Testing procedures in the measurement of the chemorheology of thermosetting resins usually incorporate the following techniques:

- reference to international or national standards (these, in some cases, use techniques that mimic true rheological characterization given below)
- measuring the effect of shear rate and temperature on chemoviscosity: $\eta_{sr} = \eta_{sr}(\gamma', T)$
- determining the effect of curing on the chemoviscosity: $\eta_c = \eta_c(T, t) = \eta_c(T, a)$
- measuring the effect of fillers on the chemoviscosity: $\eta_{sr}(F)$ and $\eta_c(F)$
- measuring the effect of temperature and fillers on the gel time: $t_{gel}(T, F)$
- measuring the combined effects of shear rate and curing: $\eta_{src} = \eta_{src}(T, a)$
- Determining optimum process properties: minimum viscosity, $\eta_{\min}$, and optimum heating rate, $(dT/dt)_{opt}$.

Each of these techniques will be discussed in detail below.

Table 4.5. Standard tests for reactive polymers

Standard	Designation	Notes
Flow tests		
Cup flow	BS2782-86	Slow-curing thermosets
	Method 720 B	Measure time to fill
		Set closing speed, force, amount of flash and temperature
Cup flow	ASTM D731-84	Measure time to fill given mould
		Set closing speed, force, amount of flash and various temperatures
Spiral flow	ASTM D3123-83	Measure length of flow along spiral die at end of test
		Set pressure and temperature from transfer press
Impregnated resin flow	ASTM D3795-70 Method 105 D	Measure flow from flash or central piece in plate press
		Set temperature and pressure for epoxies, phenolics and silicones
Roller mixer	ASTM D3795-90	Measure torque over time in roller mixer
		Measure initial maximum, minimum and final maximum torques and corresponding times
		Set temperatures, roller speed and charge volume
Viscosity tests		
Capillary rheometer	ASTM D3835-90	Measure shear effects on viscosity at high rates
		Given sample conditioning (ASTM D618), given pressures or flow rates, determine viscosity via various corrections
		Determine dwell time that does not influence results (due to curing)
Dynamic properties	ASTM D4065-90	Viscoelastic properties
	ASTM D4473-90	Cure effects on viscosity
		Dynamic properties of supported and unsupported resins
		Isothermal and non-isothermal tests
Viscosity of epoxy resins	ASTM D2393-86	Measure pre-cure viscosity of filled and unfilled epoxy liquids
		Viscosity versus spindle speed at desired temperature
Gel-point tests		
Rotating rod	ISO 2535-74	Measure gel time at 50 Pa s of polyesters
		Rotating rod at given speed, temperature and rod dimensions
Hand probe	ASTM D2471-79	Measure gel time and exotherm temperature
		Wooden hand-stirred probe at given temperature and rod dimensions
Glass rod	BS 2782	Measure gel time of phenolics
	(M:835A-80)	Stir resin with glass rod until 'rubbery' at given temperatures
Test tube	(M:835B-80)	Measure gel time of polyesters
		Resin test tube at given temperature and volume
Gel timer	(M:835C-80)	Measure gel time from restricted motion of plunger for epoxies and polyesters
Knife	(M:835D-80)	Vertical reciprocating disc and plunger is stopped when material gels at given temperature
Press	(M:835E-80)	Measure gel time
		Stir sample on hot plate with knife
		Measure gel time for impregnated resin
		Exude material from between plates with wire at given temperature and size of plates
Dynamic properties	ASTM 4065-90	Viscoelastic properties
	ASTM 4473-90	Gel point from crossover tests and time for dynamic viscosity to reach 100 Pa s
		For supported and unsupported resins
		Isothermal and non-isothermal tests

4.4.1 Standards

Standards for the chemoviscosity of reactive fluids are shown in Table 4.5. Clearly these tests mainly are typical of rapid, quality-control measurements; however, in contrast the standard proposed by Rydes (1993) for using capillary and slit-die rheometry is based on true rheological analysis. Alternative rheometers used for reactive systems are summarized in detail in Table 4.6.

4.4.2 Chemoviscosity profiles – shear-rate effects, $\eta_s = \eta_s(\gamma, T)$

The effects of shear rate on the viscosity of thermosets are usually measured on uncatalysed resins or resins that have undergone minimal curing. This is because the effects of shear rate will be most prevalent in the early stages of the process, during transfer or injection of the resin.

The measurement techniques involve either steady shear or dynamic shear. In steady-shear tests the viscosity (η) is measured as a function of the steady-shear rate (γ), over a range of temperatures. These tests are known as *isothermal steady-shear-rate sweeps*. Common shear rates encountered in processing of thermoset resins extend from $1\ s^{-1}$ to $10\,000\ s^{-1}$, and typically two measuring systems are required to cover this range. For example, ASTM 3835/90 outlines a procedure for steady-shear-rate sweeps with a capillary rheometer for the flow of thermoplastic and thermosetting polymers. In this standard, checking for cure effects is carried out by determination of the maximum dwell time of a thermoset. Eley (1983) and Ryan (1973) used a cone-and-plate rheometer for low-shear-rate sweeps in the characterization of thermoset powders and filled epoxy resins, respectively. Ryan (Ryan and Kamal, 1976, Ryan, 1984) measured the steady-shear viscosity of unfilled and filled epoxy resins, using cone-and-plate and capillary rheometry, and in addition measured the effects of shear rate at various degrees of pre-cure.

Dynamic-shear measurements are of the complex viscosity (η^*) as a function of the dynamic oscillation rate (ω), at constant temperature. These tests are defined as *isothermal dynamic frequency sweeps*. Since the dynamic frequency sweeps are conducted at a given amplitude of motion, or strain, it is necessary to ensure that the sweeps are conducted in the region where the response is strain-independent, which is defined as the linear viscoelastic region. This region of strain independence is determined by an *isothermal strain sweep*, which measures the complex viscosity as a function of applied strain at a given frequency. This ensures that a strain at which the dynamic frequency sweep may be conducted in the linear viscoelastic region is selected.

The complex viscosity can be related to the steady-shear viscosity (η) via the empirical Cox–Merz rule, which notes the equivalence of steady-shear and dynamic-shear viscosities at given shearing rates: $\eta(\gamma) = \eta^*(\omega)$. The Cox–Merz rule has been confirmed to apply at low rates by Sundstrom and Burkett (1981) for a diallyl phthalate resin and by Pahl and Hesekamp (1993) for a filled epoxy resin. Malkin and Kulichikin (1991) state that for highly filled polymer systems the validity of the Cox–Merz rule is doubtful due to the strain dependence at very low strains and that the material may partially fracture. However, Doraiswamy *et al.* (1991) discussed a modified Cox–Merz rule for suspensions and yield-stress fluids that equates the steady viscosity with the complex viscosity at a modified shear rate dependent on the strain, $\eta(\gamma) = \eta^*(\gamma_m \omega)$, where γ_m is the maximum strain. This equation has been utilised by Nguyen (1993) and Peters *et al.* (1993) for the chemorheology of highly filled epoxy-resin systems.

Table 4.6. Rheometers used for reactive polymers

Rheometer	Measurement	Advantages/Limitations
Parallel-plate (Martin *et al.*, 1989, Hale *et al.*, 1989, Kojima *et al.*, 1986, Sundstrom and Burkett, 1981, Pahl and Hesekamp, 1993, Ryan and Kamal, 1976, Ng and Manas-Zloczower, 1993, Nguyen, 1993, Peters *et al.*, 1993, Kenny *et al.*, 1991, Choi and Lin, 1991, Halley *et al.*, 1994)	**Isothermal steady-shear sweep** $\eta = \eta(\gamma', T)$, low γ' slip yield stress **Isothermal dynamic shear sweep** $\eta = \eta(\gamma', T)$ low to high γ' from Cox–Merz rule **Isothermal dynamic time test** $\eta = \eta(T, t)$ t_{gel} from G' and G'' crossover **Non-isothermal dynamic temperature ramp** $\eta = \eta(T, t)$ and $T(t)$ t_{gel} from $\tan\delta_{max}$ **Isothermal multiwave test** t_{gel} from $\tan\delta \neq f(\omega)$ **Isothermal relaxation test** t_{gel} from $G(t) \propto t^{-n}$	**Advantages** Most common geometry Small sample size Easy to clean Various plate designs **Limitations** Non-uniform shear rate Low to moderate shear rates
Cone-and-plate (Martin *et al.*, 1989, Ryan, 1984, Eley, 1983, Ryan and Kamal, 1976)	**Isothermal steady-shear sweep** $\eta = \eta(\gamma', T)$ low γ' **Isothermal dynamic-shear sweep** $\eta = \eta(\gamma', T)$ low to high γ' from Cox–Merz rule	**Advantages** Small sample size Uniform shear rate **Limitations** Difficult to measure slip Edge-fracture problem Limited to low shear rate
Eccentric disc (Collyer and Clegg, 1988, Sundstrom and Burkett, 1981)	**Isothermal dynamic-shear sweep** $\eta = (\gamma', T)$ low γ' from Cox–Merz rule	**Advantages** Small sample size **Limitations** Only one deformation mode Sample exposure
Capillary (Ryan, 1984, Ryan and Kamal, 1976, Rydes, 1993, Blyler *et al.*, 1986)	**Isothermal steady-shear sweep** $\eta = \eta(\gamma', T)$ high γ' slip yield stress	**Advantages** Large shear rates **Limitations** Large sample size Cannot monitor time effects Difficult to clean End effects
Slit (Collyer and Clegg, 1988, Han and Wang, 1994)	**Isothermal steady-shear sweep** $\eta = \eta(\gamma', T)$ high γ' slip yield stress	**Advantages** Sample flow Large shear rates **Limitations** Cannot monitor time effects Difficult to clean
Squeezing flow (Tungare *et al.*, 1986)	**Isothermal steady time test** $\eta = \eta(T, t)$ at low γ' **Non-isothermal steady time test** $\eta = \eta(T, t)$ at low γ'	**Advantages** Simple flow **Limitations** Limited time Low shear rates

Table 4.6. (*cont.*)

Rheometer	Measurement	Advantages/Limitations
Sliding-plate (Giacomin *et al.*, 1989)	**Isothermal steady-shear sweep** $\eta = \eta(\gamma', T)$, low γ' to moderate γ'	**Advantages** Less edge failure Direct stress measure **Limitations** End effects Limited travel
Vane (Barnes and Carnali, 1990)	**Isothermal steady-shear sweeps** $\eta = (\gamma', T)$, low γ' yield stress	**Advantages** Direct determination of yield stress No slip effects **Limitations** Low shear rates only Difficult to clean
Magnetic-needle (Chu and Hilfiker, 1989)	**Isothermal steady-shear sweeps** $\eta(\gamma', T)$ low γ'	**Advantages** No moving parts Completely isolated system **Limitations** Low shear rates Fixed-rate tests
Helical-screw (Collyer and Clegg, 1988, Jones *et al.*, 1992)	**Isothermal steady-shear sweeps** $\eta(\gamma', T)$ moderate to high γ' **Isothermal steady time test** $\eta(T, t)$, at low γ'	**Advantages** Closed system Used for suspensions and curing systems **Limitations** Difficult to clean Large sample size Shear rate affects curing viscosity
Dielectric-measurement (Hsieh and Su, 1992, Lane *et al.*, 1986, Lane, 1987, Gorto and Yandrastis, 1989)	**Isothermal frequency** $\eta(T, t)$ from correlation t_{gel} from dielectric parameters **Non-isothermal frequency** $\eta(T, t)$ from correlation t_{gel} from dielectric properties	**Advantages** Small sample On-line measurements **Limitations** No direct measurement of viscosity
Vibrating-paddle (Pethrick, 1993)	**Isothermal damping** $\eta(T, t)$ at low frequency t_{gel} **Non-isothermal damping** $\eta(T, t)$ at low frequency t_{gel}	**Advantages** Small sample size Easy to clean Various geometries **Limitations** Limited frequency range Slip effects
Pulse (Carriere *et al.*, 1992)	**Isothermal impulse** $\eta_0(t)$ at given impulse	**Advantages** Quick test **Limitations** Only zero-shear viscosity
NMR imaging (Jackson, 1992)	**Isothermal curing** $\eta = \eta(T, t)$ from known correlation	**Advantages** Isolated system **Limitations** Requires correlation

Table 4.6. (*cont.*)

Rheometer	Measurement	Advantages/Limitations
Spiral die mould with numerical analysis (Gonzales and Shen, 1992)	**Isothermal tests** $\eta = \eta(T, t, \gamma)$ from inverse problem solution	**Advantages** Good for fast-curing systems **Limitations** Requires complex numerical analysis
Torsional braid analyser (Enns and Gillham, 1976)	**Isothermal damping** t_{gel} (T, t) from damping of a fixed oscillation	**Advantages** Good for fast-curing systems **Limitations** Requires complex numerical analysis

The complex viscosity as a function of frequency, maximum strain and temperature is generally determined with one rheometer. Standard ASTM 4440-84/90 defines the measurement of rheological parameters of polymer samples using dynamic oscillation. This standard reiterates the importance of determining the linear viscoelastic region prior to performing dynamic frequency sweeps.

The presence of apparent wall slip both in steady- and in dynamic-shear sweeps can be expected for highly filled resins. The determination of wall slip has been discussed by Yoshimura (Yoshimura and Prudhomme, 1988, Yoshimura and Prud'homme, 1988) for steady- and dynamic-shear flows. The effects of wall slip have been documented for polymer solutions and suspensions and more recently for polymer melts. An excellent review of the rheology of filled thermoplastics and thermosets is presented by Utracki and Fisa (1982), who investigated the effect of fillers on the rheological properties of polymer suspensions and melts. More specifically Halley (1993) discusses the effects of wall slip on the steady- and dynamic-shear flow of LLDPE. A review of the effects of wall slip on the flow of polymer melts has been given by Petrie and Denn (1976). The understanding of slip effects for thermosets, however, is limited to only a few studies. The effects of wall slip as a function of shear stress at the wall have been modelled for liquid epoxy resins filled with calcium carbonate by Riccardi and Vazquez (1989). In general, the effect of wall slip can be examined by using some of the techniques listed in Table 4.6. Expressions for the true viscosity and the slip velocity can be derived from analysis of data gathered using these techniques. However, the slip velocity can be determined easily in only a few rheometers.

The measurement of yield stress at low shear rates may be necessary for highly filled resins. Doraiswamy *et al.* (1991) developed the modified Cox–Merz rule and a viscosity model for concentrated suspensions and other materials that exhibit yield stresses. Barnes and Carnali (1990) measured yield stress in a Carboxymethylcellulose (CMC) solution and a clay suspension via the use of a vane rheometer, which is treated as a cylindrical bob to monitor steady-shear stress as a function of shear rate. The effects of yield stresses on the rheology of filled polymer systems have been discussed in detail by Metzner (1985) and Malkin and Kulichikin (1991). The appearance of yield stresses in filled thermosets has not been studied extensively. A summary of yield-stress measurements is included in Table 4.6.

The most common measuring systems used for the effects of shear rate on the chemoviscosity of thermosets have been the parallel-plate rheometer and the capillary rheometer; however, the choice of rheometer is dependent on the type of system to be studied. The advantages and limitations of these systems are presented in Table 4.6.

4.4.3 Chemoviscosity profiles – cure effects, $\eta_c = \eta_c(\alpha, T)$

The effects of cure on the viscosity of thermoset resins are monitored on the catalysed resin and are associated with the late injection, packing and curing cycles of a thermoset process. There are essentially two types of test methods, namely isothermal and non-isothermal tests, which are usually performed in dynamic or steady shear.

In one type of isothermal test the sample is dynamically sheared at a low oscillation rate and strain, so as not to disturb the curing network. The strain may be high initially to determine the complex viscosity of the resin, since the viscosity may be quite low and a higher strain will increase the accuracy of the instrument. However, the strain must be reduced once curing has begun, in order to ensure that the test is conducted within the linear viscoelastic region. These tests are defined as *isothermal dynamic time tests* and are used to determine the changes in complex viscosity over time at a fixed frequency and at various temperatures. Lane *et al.* (1986) and Malkin and Kulichikin (1991) measured the changes in complex viscosity for DGEBA/DCA and TGDDM/DDS epoxy-resin systems via isothermal dynamic time tests using parallel-plate rheometry. These studies also incorporated dielectric measurements during the curing in an attempt to find a correlation between complex viscosity and dielectric measurements and to relate these measurements to the degree of cure. Kojima *et al.* (1986) measured the cure effects on the complex viscosity of a polyimide fibre resin through isothermal dynamic time tests. Pahl and Hesekamp (1993) monitored the complex viscosity of filled epoxy resins through the use of these tests. Standard ASTM 4440-84/90 recommends a procedure for determining the curing viscosity of thermoset resins via isothermal dynamic time tests. Pichaud *et al.* (1999) and Maazouz *et al.* (2000) characterized the viscosity as a function of temperature and conversion for an epoxy–amine and an RTM polyester–polyurethane hybrid system (respectively). The authors of both studies describe the viscosity as a function of temperature, extent of reaction and a conversion-dependent critical exponent (i.e. Macosko's model in Table 4.2).

Isothermal tests have also been carried out using steady shear via *isothermal steady time tests* to determine the change in steady-shear viscosity as a function of time at a given shear rate. However, the effects of shear rate on network formation of the resin must be evaluated in order to validate these tests. Mussati and Macosko (1973) measured the curing viscosities of a phenolic resin, an epoxy resin and an EPDM rubber as functions of time at various temperatures. Han and Lem (Han and Lem, 1983a, 1983b, 1983c, 1984, Lem and Han, 1983a, 1983b, 1983c) investigated the effects of curing on the chemoviscosity and normal stresses of filled polyesters.

Non-isothermal dynamic tests defined as *dynamic temperature ramps* are used to measure the changes in complex viscosity during an imposed temperature ramp at a given oscillatory frequency and strain. This ramp may be chosen to be linear or non-linear in order to derive information on viscosity during curing or to model processing conditions. Also, the strain may be changed during the ramp, yet it must be held within the linear viscoelastic region and not interfere with the network formation. Martin *et al.* (1989) measured the complex viscosity of epoxy resins during various linear temperature ramps to evaluate model parameters for the curing model, $\eta_c(T, t)$. Chiou and Letton (1992) investigated the curing of a three-component epoxy-resin system via a non-isothermal dynamic temperature ramp in order to evaluate the complex curing effects on the viscosity. Tajima and Crozier (1983) evaluated the curing models for various epoxy-resin systems using temperature-ramp tests on samples partially reacted to various extents of cure. Kenny *et al.* (1991) and Lane *et al.* 1986) examined the relationship between curing complex viscosity and dielectric properties through temperature-ramp tests. A standard procedure for dynamic temperature ramps tests is given in ASTM 4440-84/90, where ramp rates of 3–5 °C/min are recommended, in order to allow the sample to equilibrate at a given temperature. Although most industrial ramp rates are non-linear, this linear rate is representative of typical curing rates used.

An inherent assumption when using the above dynamic techniques is that the complex viscosity gives a good representation of the steady-shear viscosity during the curing reaction. This has been validated for many systems. However, care should be taken when relating the effects of cure on complex viscosity to the processing viscosity; in other words the Cox–Merz rule or a similar relationship must be validated.

Steady-shear temperature ramp tests are used to monitor the change in steady-shear viscosity at a given shear rate during a linear or non-linear temperature ramp. Tungare (1986) used a squeezing flow viscometer to measure the change in viscosity during a linear temperature ramp of an embedded epoxy resin and developed a model for the curing viscosity using an optimization technique.

The most common geometry used for the determination of the cure viscosity is that of the parallel-plate rheometer (Martin *et al.* 1989, Hale *et al.* 1989, Sundstrom and Burkett, 1981, Kojima *et al.* 1986, Pahl and Hesekamp, 1993, Ryan and Kamal, 1976, Ng and Manas-Zloczower, 1993, Nguyen, 1993, Peters *et al.* 1993, Kenny *et al.* 1991, Choi and Lin, 1991, Halley *et al.* 1994). This rheometer may be operated at low shear rates under isothermal and non-isothermal conditions. Alternative measurements have been used to measure the effects of cure on the viscosity, and these are summarized in Table 4.6.

4.4.4 Filler effects on viscosity: $\eta_{sr}(F)$ and $\eta_c(F)$

The effect of fillers on the viscosity of thermosetting resins is a recent area of study and little has been done to establish testing methods for these effects. Han and Lem (1983a, 1983b, 1983c, 1984, Lem and Han, 1983a, 1983b, 1983c) examined the effects of fillers, thickeners and low-profile additives by investigating the addition of each component to a model polyester. Ng and Manas-Zloczower (1993) used isothermal dynamic time tests to evaluate the effect of filler loading on the elastic modulus of filled epoxy resins. They fitted their results to an empirical model for filled systems. Kalyon and Yilmazer (1990) described isothermal dynamic sweeps and steady sweeps to examine the effects of filler loading, filler size and matrix viscosity for a concentrated solution of ammonium sulfate in a poly(butylacrylonitrile) (PBAN)-based terpolymer on the rheological properties. They noted that increasing the filler concentration increases slip, induces plug flow and causes flow instabilities. Kalyon and Yilmazer (1990) also highlight the non-linear nature of the filled system, even at low strains, and propose a non-linear complex viscosity based on maximum values of the response. Aranguren *et al.* (1992) examined the effects of treated and untreated silica fillers on the uncured rheology of various polydimethylsiloxane (PDMS) resins. They used dynamic-shear tests at low strains to find that the storage modulus (G') increases with increasing filler loading and is greater with less surface deactivation. Kalyon and Yilmazer (1990) and Metzner (1985) described the effect of dilatancy (or shear thickening) for some highly filled suspensions, produced via steady-shear sweeps. Han (1981) noted that the viscosity of the unfilled resin, filler loading, length of fibres, fibre orientation and use of coupling agents all must be considered when measuring the viscosity of such systems.

It is evident that strict control over the composition of a filled thermoset system is vital to establishing the chemoviscosity of the system. Also, the measurement of samples with various filler loadings, filler types, surface treatments and filler dimensions would aid in fully describing the effect of fillers and enable better characterization of filled samples.

4.4.5 Chemoviscosity profiles – combined effects, $\eta_{\mathrm{all}} = \eta_{\mathrm{all}}(\gamma', \alpha, T)$

The work of several authors (Peters *et al.*, 1993, Halley *et al.*, 1994) has demonstrated the use of *non-isothermal dynamic sweep tests* to examine the combined effects of shear rate and curing on the chemoviscosity of a highly filled epoxy resin simultaneously. These tests use a selected temperature ramp with repeated dynamic rate sweeps to investigate the effects on the chemoviscosity. The advantage of these tests is that the effects of shear rate and cure are not separated, which is similar to processing conditions.

Eom *et al.* (2000) used a time–cure–temperature superposition principle upon isothermal dynamic data obtained at various temperatures to predict instantaneous viscoelastic properties during cure.

4.4.6 Process parameters

Optimum process parameters such as the minimum processing viscosity (η_{min}) and the optimum heating rate can be determined by performing a series of non-isothermal dynamic temperature ramps. Using various heating rates, the optimum minimum processing viscosity at the desired heating rate can be established. The optimum heating rate is determined by optimizing the time needed to reach the minimum viscosity and the time over which this minimum viscosity extends. It should be noted that increasing the heating rate by too much can cause large temperature differences in the moulded part. The time at which the viscosity is at or near its minimum value is known as the processing window. Non-isothermal dynamic temperature tests are generally performed on a parallel-plate rheometer at fixed frequencies and strains to determine this window.

This review of chemorheological techniques highlights the numerous possibilities for monitoring chemoviscosity and gel-point measurements. Determination of the appropriate techniques must be based on the following considerations:

- material type and form
- initial and final viscosities
- desired ranges for shear rate, temperature and curing
- gelation properties
- filler characteristics

The selection of the optimum chemorheological techniques can then be used to influence the choice of rheometer for chemorheological measurements. Final selection of the rheometer design must be determined by the considerations given in the previous section, as well as the following:

- cost
- ease of use
- reproducibility of results
- versatility

4.5 Gelation tests

One of the key transitions that also needs to be characterized is gelation. The physical nature and models of gelation have been described in Chapter 2, so here we wish to examine the methods used to determine gelation. The determination of the gel point of thermosets may be monitored by the following methods:

- isothermal dynamic time tests
- isothermal steady time tests
- isothermal multiwave tests
- isothermal step strain-relaxation tests
- dynamic temperature-ramp tests

In *isothermal dynamic time* tests the storage modulus (G') and loss modulus (G'') are monitored at a given frequency (ω) and strain (γ) over time. The gel point, or the initial formation of an infinite network, has been reported in the literature (Winter and Chambon, 1986) as the time at which G' and G'' cross over. However, Winter (1987) notes that the crossover point is valid only for stoichiometrically balanced systems or those with excess hardener. For these systems the gel point is equivalently defined as the point where the loss tangent ($\tan\delta = G''/G'$) reaches unity. Standard ASTM 4473-90 also describes the technique determining for the crossover-point analysis via isothermal dynamic time tests for thermosetting resins. Scanlan and Winter (1991) and Winter *et al.* (1994) expand on Winter's initial work by analysing effects of gelation on the gel constitutive equation (i.e. the constitutive equation at gelation) and its key parameters of gel strength (S) and power law index (n). They note the characterization of network structure via analysis of S and n and the effects of M_w on these parameters. Arbabi and Sahimi (1990), Lairez *et al.* (1992) and Eloundou *et al.* (1996a, 1996b) extend the characterization of gelation to analysis of critical viscoelastic properties near gelation and correlate the results to known molecular theories.

Arrelano *et al.* (1989) evaluated the gel points of various DGEBA-based epoxy-resin systems via the crossover method. They also defined the vitrification point as the maximum in G'' in an isothermal dynamic time test. In all these gel-point measurements the frequency must be chosen such that the relaxation of the network is enabled during data sampling. This is represented as

$$t_{\mathrm{exp}} = \frac{2\pi}{\omega} > t_{\mathrm{relax}}, \tag{4.11}$$

where t_{exp} is the experiment time and t_{relax} is the characteristic relaxation time of the system. Thus, a low enough frequency must be used to enable equilibrium values to be obtained. However, as DeRosa and Winter (1993) explain, the time must not be too large (frequency must not be too low), in order that one does get gel effects during the experiment. Harran and Laudouard (1986) also used isothermal dynamic time tests to measure the gelation and vitrification of DGEBA epoxy-resin systems. The gelation point in their study was, however, measured as the change in slope of G'' with time (due to the slowing of the reaction rate). Harran and Laudouard (1986) also noted that there was a minimum temperature for this inflection-point method that was not related to vitrification. The vitrification point is determined by the maximum in $\tan\delta$ (or G''), as was done by Lairez *et al.* (1992). However, Lange *et al.* (1996) prefer to define vitrification as a transition process rather than a single transition point, and define its initiation as where the dynamic modulus (G^*) becomes non-zero and its end when the dynamic modulus plateaus.

Isothermal steady time tests are used to determine the gel point of a thermoset system as the point at which the shear viscosity tends towards infinity. In these tests the viscosity is measured as a function of time at a constant shear rate. This method has the following major disadvantages. Firstly, the infinite viscosity can never be measured due to equipment limitations and thus the gel time must be obtained by extrapolation. Secondly, shear flow may destroy or delay network formation. Finally, gelation may be confused with vitrification or phase separation since both these processes lead to an infinite viscosity (St John *et al.*, 1993). However, some work by Matejka (1991) and Halley *et al.* (1994) has shown that extrapolation to zero values of reciprocal viscosity or normal stress (i.e. extrapolation to infinite viscosity and normal stress) can be used with some success.

An *isothermal multiwave test* is a recent technique that uses a Fourier-transformation response to a multiple-frequency wave to determine the gel point. This test is used to measure the dynamic response of a fluid to simultaneously applied sine waves of various dynamic frequencies, each with the same maximum strain. The gel point is determined as the point at which the loss tangent becomes independent of frequency. Winter (1987) describes this method as useful for determining the gel point of any resin system. De Rosa and Winter (1993) note that for entangled polymers a frequency window must be chosen such that the system is in the terminal zone. The lower limit of frequency is expressed by Equation (4.11) as previously mentioned, and the upper limit is determined by the crossover of G' and G'' with respect to frequency. Matejka (1991) finds the gel point of an epoxy-resin system via the independence of the loss tangent with respect to frequency, but uses a series of isothermal dynamic rate sweeps. The advantage of the multiwave test is that a series of frequencies can be imposed simultaneously, thereby eliminating the need for performing multiple isothermal dynamic time tests, which may suffer from test-time effects (Lairez *et al.* 1992).

Winter (Winter and Chambon, 1986, Winter, 1987) also described an *isothermal dynamic relaxation test* to measure the gel point. He noted that the gel point coincides with a power-law relationship between the relaxation modulus (G) and relaxation time (t) ($G = St^{-n}$, where S and n are constants. An isothermal step strain test measures the relaxation modulus as a function of time after an instantaneously applied strain. The gel time can be measured as the point at which the profile of the relaxation modulus can be expressed by this power law. This model is equivalent to $\tan\delta$ being independent of frequency.

Dynamic temperature-ramp tests are used to measure the storage and loss modulis of a system as a function of time during a given temperature ramp. The gel time is expressed as the time at which the loss tangent reaches a maximum. The cure time has also been defined by Choi and Lin (1991) as the intersection of the increasing storage modulus and the plateau modulus, determined using a dynamic temperature-ramp test. Nguyen (1993) also used this test to measure the gel time of a filled epoxy-resin system. Peters *et al.* (1993) expressed reservations about the validity of this type of test for highly filled epoxy resins due to frequency dependences. Standard ASTM 4473-90 also describes an experimental procedure using the temperature-ramp test to determine the gel point of thermosets via the maximum in $\tan\delta$.

An extensive review of the use of various gelation tests (G'/G'' crossover, independence of $\tan\delta$ with respect to frequency, infinite viscosity and infinite normal stress, gel filtration and DSC theoretical gel conversion) is given by Halley *et al.* (1994) for unfilled epoxy-resin systems. Here no differences in gelation-time measurements between techniques (outside experimental error) are found for the TGDDM/DDS system studied. Other authors have reviewed other gelation techniques, for example Heise *et al.* (1990) and Huang and Williams (1994) examine gel techniques such as detection of viscosity reaching a pre-set high value, G'/G'' crossover and sol–gel methods. In more recent work Lange *et al.* (2000) have shown that there is good agreement for use of the inflection in heat capacity (C_p), a maximum in the loss tangent ($\tan\delta$) and the expected time for vitrification conversion ($t_{\alpha\text{vit}}$) as methods for assessing vitrification in curing systems. Of course, ultimate glass-transition temperatures can also be obtained via isothermal dynamic data when examining fully cured materials. Interesting work by Hagen *et al.* (1994) compares DMTA and DSC methods for ultimate T_g and notes the effects of frequency on the T_g measurement. Hagen (1994) also notes an effective frequency for DSC measurements at 10 000 Hz.

The effect of fillers on the gel point of thermosets has not been studied extensively. Ng and Manas-Zloczower (1993) used isothermal dynamic time tests to measure the crossover point for a silica-filled epoxy resin. They noted a decrease in gel time with increasing filler loading. Metzner (1985) also noted that the storage modulus and loss modulus increased by different amounts with filler loading. Therefore, the gel-point tests for highly filled systems must involve knowledge of the effect of filler characteristics at various levels.

Commonly, gel-point tests are performed with parallel-plate rheometers. Large parallel plates and strains may be used to gather data prior to gelation, whereas after gelation smaller strains and diameters are generally used. There are other methods discussed in the literature to determine gel-point measurements and these are presented in Tables 4.5 and 4.6.

References

Aranguren, M., Mora, E., Degroot, J. & Macosko, C. (1992) *J. Rheology*, **36**, 1165–1182.

Arbabi, S. & Sahimi, M. (1990) *Phys. Rev. Lett.*, **65**, 725–728.

Arrelano, M., Velasquez, P. & Gonzalez-Romero, V. (1989) *Society of Plastics Engineers Annual Technical Conference ANTEC 1989*, p. 838.

Barnes, H. & Carnali, J. (1990) *J. Rheology*, **34**, 841.

Batch, G. & Macosko, C. (1987) *Society of Plastics Engineers Annual Technical Conference ANTEC 1987*, p. 974.

Bidstrup, W., Sheppard, N. & Senturia, S. (1986) *Polym. Eng. Sci.*, **26**, 359–361.

Blyler, L. Jr, Bair, H., Hubbauer, P. *et al.* (1986) *Polym. Eng. Sci.*, **26**, 1399.

Carriere, C., Bank, D. & Christenson, C. (1992) *Polym. Eng. Sci.*, **32**, 426.

Castro, J. & Macosko, C. (1982) *Polym. Eng. Sci.*, **28**, 250.

Chiou, P. & Letton, A. (1992) *Polymer*, **33**, 3925.

Choi, J. & Lin, C. (1991) *Society of Plastics Engineers Annual Technical Conference ANTEC 1991*, p. 1746.

Chu, B. & Hilfiker, R. (1989) *Rev. Sci. Instrum.*, **60**, 3828.

Collyer, A. & Clegg, D. (1988) *Rheological Measurements*, Essex: Elsevier Applied Science.

Cox, W. (1958) *J. Polym. Sci.*, **28**, 619.

De Rosa, M. & Winter, H. (1993) *Society of Plastics Engineers Annual Technical Conference ANTEC 1993*, p. 2620.

Doraiswamy, D., Mujumdar, A., Tsao, I. *et al.* (1991) *J. Rheology*, **35**, 647–685.

Dusi, M., May, C. & Seferis, J. (1982) *ACS Org. Coat. Appl. Polym. Sci. Proc.*, **47**, 635.

Dusi, M., May, C. & Seferis, J. (1983) *ACS Symp. Series*, **227**, 301–318.

Dutta, A. & Ryan, M. (1979) *J. Appl. Polym. Sci.*, **24**, 635.

Eley, R. (1983) *ACS Symp. Series*, **227**, 281–299.

Eloundou, J., Feve, M., Gerard, J., Harran, D. & Pascault, J. (1996a) *Macromolecules*, **29**, 6907–6916.

Eloundou, J., Gerard, J., D, H. & Jp, P. (1996b) *Macromolecules*, **29**, 6917–6927.

Enns, E. & Gillham, J. (1976) *J. Appl. Polym. Sci.*, **28**, 2562.

Eom, Y., Boogh, L., Michaud, V., Sunderland, P. & Manson, J.-A. (2000) *Polym. Eng. Sci.*, **40**, 1281–1292.

Giacomin, A. & Dealy, J. (1993) Large amplitude oscillatory strain measurements, in Collyer, A. A. (Ed.) *Techniques in Rheological Measurements*, London: Chapman & Hall.

Giacomin, A., Samurkas, T. & Dealy, J. (1989) *Polym. Eng Sci.*, **29**, 499.

Gleissle, W. & Hochstein, B. (2003) *J. Rheology*, **47**, 897–910.

Gonzales, U. & Shen, S. (1992) *Polym. Eng. Sci.*, **32**, 172.

Gonzalez-Romero, V. & Casillas, N. (1989) *Polym. Eng. Sci.*, **29**, 295.

Gorto, J. & Yandrastis, M. (1989) *Polym. Eng. Sci.*, **29**, 278–284.

Gupta, A., Macosko, C. W. (1990) *J. Polym. Sci.B, Polym. Phys.*, **28**, 2585–2606.

Gupta, R. (2000) *Polymer and Composite Rheology*, New York: Marcel Dekker.

Hagen, R., Salmen, L., Lavebratt, H. & Stenberg, B. (1994) *Polym.Testing*, **13**, 113–128.

Hale, A., Garcia, M., Macosko, C. & Manzione, L. (1989) *Society of Plastics Engineers Annual Proceedings ANTEC 1989*, p. 796.

Halley, P. (1993) The Effect of Surfaces and Structures on the Rheology and Processing of LLDPE, Thesis, Brisbane, University of Queensland.

Halley, P., Mackay, M. & George, G. (1994) *High Perform. Polym.*, **6**, 405–414.

Han, C. (1981) *Multiphase Flow in Polymer Processing*, New York: Academic Press.

Han, C. & Lem, K. (1983a) *J. Appl. Polym. Sci.*, **28**, 743.

Han, C. & Lem, K. (1983b) *J. Appl. Polym. Sci.*, **28**, 763.

Han, C. & Lem, K. (1983c) *J. Appl. Polym. Sci.*, **28**, 3155.

Han, C. & Lem, K. (1984) *Polym. Eng. Sci.*, **24**, 473.

Han, C. & Wang, K. (1994) *Society of Plastics Engineers Annual Technical Conference ANTEC 1994*, p. 935.

Harran, D. & Laudouard, A. (1986) *J. Appl. Polym. Sci.*, **32**, 6043.

Hartley, M. & Williams, H. (1981) *Polym. Eng. Sci.*, **21**, 135.

Heise, M., Martin, G. & Gorto, J. (1990) *Polym. Eng. Sci.*, **30**, 83–89.

Hsieh, T. & Su, A. (1992) *J. Appl. Polym. Sci.*, **44**, 165.

Huang, M. & Williams, J. (1994) *Macromolecules*, **27**, 7423–7428.

Jackson, P. (1992) *J. Mater. Sci.*, **27**, 1302.

Kalyon, D. & Yilmazer, U. (1990) Rheological behaviour of highly filled suspensions which exhibit wall slip, in Collyer, A., Utracki, L. A. (Eds.) *Polymer Rheology and Processing*, London: Elsevier.

Kamal, M. (1974) *Polym. Eng. Sci.*, **14**, 231–239.

Kamal, M. & Ryan, M. (1980) *Polym. Eng. Sci.*, **20**, 859.

Kamal, M. & Sourour, S. (1973) *Polym. Eng. Sci.*, **13**, 59–64.

Kascaval, C., Musata, F. & Rosu, D. (1993) *Angew. Makromol. Chem.*, **209**, 157.

Kenny, J., Trivisano, A. & Berglund, L. (1991) *SAMPE J.*, **27**, 39–46.

Knauder, E., Kubla, C. & Poll, D. (1991) *Kunstoffe German Plastics*, **81**, 39.

Kojima, C., Hushower, M. & Morris, V. (1986) *Society of Plastics Engineers Annual Technical Conference ANTEC 1986*, p. 344.

Lairez, D., Adam, M., Emery, J. & Durand, D. (1992) *Macromolecules*, **25**, 286.

Lane, J. (1987) *J. Comp. Mater.*, **21**, 243.

Lane, J., Seferis, J. & Bachmann, M. (1986) *Polym. Engin. Sci.* **26**, 346–353.
Lange, J., Altmann, N., Kelly, C. & Halley, P. (2000) *Polymer*, **41**, 5949–5955.
Lange, J., Manson, J. & A, H. (1996) *Polymer*, **37**, 5859–5868.
Lee, L. (1981) *Polym. Eng. Sci.*, **21**, 483.
Lem, K. & Han, C. (1983a) *J. Appl. Polym. Sci.*, **28**, 779.
Lem, K. & Han, C. (1983b) *J. Appl. Polym. Sci.*, **28**, 3185.
Lem, K. & Han, C. (1983c) *J. Appl. Polym. Sci.*, **28**, 3207.
Lipshitz, S. & Macosko, C. (1976) *Polym. Eng. Sci.*, **16**, 503.
Lipshitz, S. & Macosko, C. (1977) *J. Appl. Polym. Sci.*, **21**, 2029.
Maazouz, A., Dupuy, J. & Seytre, G. (2000) *Polym. Eng. Sci.*, **40**, 690–701.
Maier, P. & Goritz, D. (1996) *Kaut. Gummi Kunststoffe*, **49**, 18–21.
Malkin, A. & Kulichikin, S. (1991) *Adv. Polym. Sci.*, **101**, 218–254.
Manzione, L. (1995) Processing epoxy molding compounds, in *Thermoset Polymer Meeting for J. K Gillham's 60th birthday*, Princeton, MA: Princeton University.
Manzione, L. & Weld, J. (1994) *Society of Plastics Engineers Annual Technical Conference ANTEC 1994*, pp. 1371–1376.
Martin, G., Tungare, A., Fuller, B. & Gorto, J. (1989) *Society of Plastics Engineers Annual Technical Conference ANTEC 1989*, p. 1079.
Matejka, L. (1991) *Polym. Bull.*, **26**, 109.
Metzner, A. (1985) *J. Rheology*, **29**, 739.
Mijovic, J., Andjelic, S., Fitz, B. *et al.* (1996) *J. Polym. Sci., B: Polym. Phys.*, **34**, 379–388.
Mijovic, J. & Lee, C. (1989) *J. Appl. Polym. Sci.*, **38**, 2155.
Mussati, F. & Macosko, C. (1973) *Polym. Eng. Sci.*, **13**, 236.
Nass, K. A. & Seferis, J. C. (1989) *Polym. Eng. Sci.*, **29**, 315.
Ng, H. & Manas-Zloczower, I. (1989) *Polym. Eng. Sci.*, **29**, 1097.
Ng, H. & Manas-Zloczower, I. (1993) *Polym. Eng. Sci.*, **33**, 211–216.
Nguyen, L. (1993) *Proceedings from the 43rd IEEE Electronic Component and Technology Conference*, Buena Vista, FL, pp. 1–10.
Pahl, M. & Hesekamp, D. (1993) *Appl. Rheology*, 70.
Pannone, M. & Macosko, C. (1988) *Polym. Eng. Sci.*, **28**, 660.
Payne, A. (1962) *J. Appl. Polym. Sci.*, **6**, 57–63.
Peters, G., Spoelstra, A., Meuwissen, M., Corbey, R. & Meijer, H. (1993) Rheology and rheomerty for highly filled reactive materials, in Dijksman, J., Nieuwstadt, F. T. M. (Eds.) *Topics in Applied Mechanics*, Dordrecht: Kluwer Academic Publishers.
Pethrick, R. (1993) in Collyer, A. (Ed.) *Techniques of Rheological Measurements*, London: Chapman & Hall.
Petrie, C. & Denn, M. (1976) *AIChE J.*, **22**, 209.
Piau, J., Kissi, N. & Mezghani, A. (1995) *Non-Newtonian Fluid Mechanics*, **59**, 11–30.
Pichaud, S., Duteurtre, X., Fit, A. *et al.* (1999) *Polym. Int.*, **48**, 1205–1218.
Prime, R. (1973) *Polym. Eng. Sci.*, **13**, 365.
Rabbinowitsch, B. (1929) *Z. Physik. Chem. (Leipzig)* A **145**, 1–26.
Riccardi, C. & Vazquez, A. (1989) *Polym. Eng. Sci.*, **29**, 120.
Riccardi, C. & Williams, R. (1986) *Polymer*, **27**, 913.
Roller, M. (1976) *Polym. Eng. Sci.*, **16**, 687.
Ryan, M. (1973) Thesis, Montreal, McGill University.
Ryan, M. (1984) *Polym. Eng. Sci.*, **24**, 698.
Ryan, M. & Kamal, M. (1976) *Proceedings of the VII International Congress on Rheology*, pp. 289–290.

Rydes, M. (1993) *Aspects of the Rheology of Unsaturated Polyester Dough Molding Compounds*, London: University of West London and National Physical Laboratories.
Scanlan, J. & Winter, H. (1991) *Macromolecules*, **24**, 47–54.
Sourour, S. & Kamal, M. (1976) *Thermochim. Acta*, **14**, 41.
St John, N., George, G., Cole-Clarke, P., Mackay, M. & Halley, P. (1993) *High Perform. Polym.*, **5**, 21.
Sundstrom, D. & Burkett, S. (1981) *Polym. Eng. Sci.*, **21**, 1108.
Tajima, Y. & Crozier, D. (1983) *Polym. Eng. & Sci.*, **23**, 186.
Tungare, A., Martin, G. & Gorto, J. (1986) *Society of Plastics Engineers Annual Technical Conference ANTEC 1986*, p. 330.
Utracki, L. & Fisa, B. (1982) *Polym. Composites*, **3**, 193–210.
Vinogradov, G. & Malkin, A. (1977) *Rheology of Polymers*, Moscow: Mir.
Wilhelm, M., Reinheimer, P., Ortseifer, M., Neidhofer, T. & Spiess, H. (2000) *Rheologica Acta*, **39**, 241–246.
Winter, H. (1987) *Polym. Eng. Sci.*, **27**, 1698.
Winter, H. & Chambon, F. (1986) *J. Rheology*, **30**, 367–382.
Winter, H., Izuka, A. & Derosa, M. (1994) *Polym. Gels Networks*, **2**, 239–245.
Yang, Y. & Suspene, L. (1991) *Polym. Eng. & Sci.*, **31**, 321.
Yoshimura, A. & Prudhomme, R. (1988) *J. Rheology*, **32**, 53–67.
Yoshimura, A. & Prud'homme, R. (1988) *J. Rheology*, **32**, 575.

5 Chemorheology and chemorheological modelling

5.1 Introduction

Chapters 3 and 4 presented chemical, physical and chemorheological techniques useful for characterizing various reactive polymer systems. This chapter will now focus on a review of chemorheological analyses for a variety of polymer systems, including detailed experimental findings and chemorheological modeling.

5.2 Chemoviscosity and chemorheological models

Chemorheology is defined as the study of the viscoelastic behaviour of reacting polymer systems. This involves examining the effects on chemoviscosity of chemical reactions (cure conversion, cure kinetics) and processing conditions (temperatures, shear rates), as well as gelation and vitrification. In Chapter 4 we briefly summarized chemoviscosity models that highlight effects of cure ($\eta_c = \eta_c(T, \alpha)$), shear rate ($\eta_{sr} = \eta_{sr}(\gamma, T)$) and filler ($\eta_f = \eta_f(F, T, t)$) in Tables 4.4–4.6. This chapter will examine the development of chemorheology and chemorheological modelling in more detail by examining the chemorheology and chemoviscosity models of unfilled reactive systems, overviewing the effects of fillers on chemoviscosity and then presenting chemoviscosity data and models for filled systems. It is hoped that, by presenting the data and models in more depth, a better understanding of the chemorheology of systems will be obtained.

5.2.1 Neat systems

Chemorheological models for neat (unfilled) curing systems can be grouped into the following categories:

- simple empirical models
- Arrhenius models
- structural and free-volume models
- probability-based and molecular models

Simple empirical models

Malkin and Kulichikin (1991) initially reviewed the rheokinetics of cured polymers and highlighted the first empirical chemorheological models. They showed that for a simple homogeneous reaction with no diffusion limitations or gel effects for reacting epoxy-resin systems the chemoviscosity could be described by

$$\eta = Kt^a, \tag{5.1}$$

with $a = a_1$ for $t < t_c$, $a = a_2$ for $t_c < t < t_p$, and $a = a_3$ for $t_p > t$, where η is the viscosity, K, a_1, a_2 and a_3 are fitting parameters, t_c and t_p are characteristic times and t is time.

For reacting polyurethane systems with gel effects Malkin and Kulichikin (1991) also showed that the chemoviscosity could be described by

$$\eta/\eta_0 = [(1 + kt)/(1 - t/t^*)]^a, \tag{5.2}$$

where η_0 is the initial viscosity, k is a constant and t^* is the gel time.

If one also considers phase separation for a reacting polyurethane system with gel effects, Malkin and Kulichikin (1991) showed that the chemoviscosity could then be described by

$$\ln(\eta/\eta_m) = \pm a \ln(t/t^*) + b \ln[1/(1 - t/t^*)], \tag{5.3}$$

with $a = 1$ for $t < t_c$ and $a = 3.5$ for $t > t_c$, where η_m is the viscosity of the mediums, a and b are constants, t_c is a characteristic time and the sign $\pm$ changes as $t > t_c$.

Lane and Khattack (1987) showed also for an epoxy resin that the chemoviscosity could be described by

$$\ln \eta = \ln \eta_0 + E_v/(RT) + Ka, \tag{5.4}$$

where E_v is the viscous activation energy and a is the cure conversion.

Mussati and Macosko (1973) fitted steady- and dynamic-shear viscosities of phenolics, epoxies and EPDM rubbers to the following model:

$$1/t_1 = C \exp[-E/(RT)], \tag{5.5}$$

where t_1 is the time taken to reach a specific viscosity η_1, C is a constant and E is the activation energy for this reaction up to t_1.

First-order isothermal (Kamal and Sourour, 1973) and non-isothermal (Dusi *et al.*, 1982, 1983) chemoviscosity profiles were presented for epoxy-resin systems in the forms:

$$\eta = \eta_0 \exp(\theta t) \tag{5.6}$$

for isothermal cure, where θ is a constant, and

$$\eta = \eta(T)\exp(\phi kt) \tag{5.7}$$

for non-isothermal cure, where $\eta(T)$ is the viscosity as a function of temperature, ϕ is a constant and k is a rate constant. Both of these equations are for the initial stages of cure and do not consider gel effects.

Castro and Macosko 1982) considered gel effects and provided the following expression for general thermoset systems:

$$\eta/\eta_0 = a^*/(a^* - a)^{A+Ba}, \tag{5.8}$$

where a is the conversion, a^* is the conversion at the gel point and A and B are constants.

Yang and Suspene (1991) and Malkin and Kulichikin (1991) showed that for melamine formaldehyde and polyester–styrene systems prone to microgel formation the following chemoviscosity models were relevant:

$$\eta/\eta_0 = (1 - t/t^*)^b, \tag{5.9}$$

from Yang and Suspene (1991), and

$$\ln(\eta/\eta_0 - 1) = A + B(t/t^*), \tag{5.10}$$

from Malkin and Kulichikin (1991), where A, B and b are constants and t^* is the gel time. Here distinct regions for microgelation, transition and macrogelation are observed and modelled.

Arrhenius models

Dusi *et al.* (1987) examined the chemorheology of Fiberite976 resin (TGDDM/DDS). The chemoviscosity is described by a simple Arrhenius relationship as shown by the following equation:

$$\eta = \eta_\infty \exp[U/(RT) + ka], \tag{5.11}$$

where η is the viscosity, η_∞ and k are constants, U is the activation energy for viscosity, R is the universal gas constant, a is the conversion and T is the temperature.

Martin *et al.* (1989) and Kojima *et al.* (1986) showed the validity of the following Arrhenius first-order model for the chemoviscosity of epoxy-resin and polyamide systems:

$$\ln \eta(t, T) = \ln \eta_0 + E_v/(RT) + K_k \int_0^t \exp[E_k/(RT)]dt, \tag{5.12}$$

where $\eta(t, T)$ is the viscosity as a function of time (t) and temperature (T), η_0 is the zero-time viscosity, K_k is the apparent rate constant, E_v and E_k are the activation energies for viscosity and kinetics, respectively, and R is the gas constant.

A modified Arrenhius first-order reaction is given by Dusi *et al.* (1983),

$$\ln \eta(t, T) = \ln \eta_0 + \phi \exp[-E_k/(RT)]t, \tag{5.13}$$

where ϕ is an entanglement factor.

For an nth-order reaction the Arrhenius equation becomes (Knauder *et al.*, 1991)

$$\ln \eta(t, T) = \ln \eta_0 + E_v/(RT) + K_k \int_0^t (1 - a)^n \exp[E_k/(RT)]dt, \tag{5.14}$$

where n is the reaction order.

A modified Arrenhius nth-order reaction is again given by Dusi *et al.* (1983),

$$\ln \eta(t, T) = \ln \eta_0 + \phi/(n - 1) \ln\{1 + (n - 1) K_k \exp[-(E_k/RT)]t\}, \tag{5.15}$$

where ϕ is an entanglement factor.

Martin *et al.* (1989b) focus on cure of epoxies and cyanate ester resins for lamination of prepregs into multilayered circuit boards. The following dual Arrhenius engineering model was fitted to viscosity profiles:

$$\ln \eta(t, T) = \ln \eta_0 + \int_0^t k \, dt, \tag{5.16}$$

with $\eta_0 = \eta_\infty \exp[E_\eta/(RT(t))]$ and $k = k_x \exp[-E_k/(RT(t))]$, where $\eta(t, T)$ is the viscosity as a function of time (t) and temperature (T), η_0 is the zero-time viscosity, k is the apparent rate constant, η_∞ and k_x are pre-exponential factors, E_η and E_k are activation energies for flow and kinetics, respectively, R is the gas constant and $T(t)$ is the variation of temperature with time.

There is reasonable agreement with the viscosity data except that some deviations occur at long times. Use of this equation with mass- and heat-transfer equations for prepreg production is useful for optimizing processing conditions.

Cheng *et al.* (1994) examine the chemorheology of a DGEBA/2,4,6-*tris*(dimethylaminomethyl)phenol (TDAP) system using the dual Arrhenius model

$$\ln \eta(t,T) = \ln \eta_\infty + \Delta E_\eta/(RT(t)) + \int_0^{t_c} k_{1\infty} \exp[-\Delta E_{k_1}/(RT)]\mathrm{d}t + \int_{t_c}^{t} k_{2\infty} \exp[-\Delta E_{k_2}/(RT)]\mathrm{d}t, \tag{5.17}$$

where η is the viscosity, η_∞ is a constant, E_η is the activation energy of the viscosity, R is the gas constant, T is the temperature (K), $k_{1\infty}$ and $k_{2\infty}$ are pre-exponential factors and E_{k_1} and E_{k_2} are activation energies for reaction rates k_1 (before the cure time, t_c) and k_2 (after t_c), and t is the time.

There is reasonably good agreement with non-isothermal viscosity data (although deviations occur at higher conversions).

Eley (1983) also describes the chemorheology of thermoset coating systems by utilizing the dual Arrhenius model, with acceptable results. Dusi *et al.* (1983) describe the use of an expanded double Arrhenius model for describing the chemorheology. This model is shown as

$$\ln \eta(t,T) = \ln \eta_x + E_\eta/(RT) + [\phi/(n-1)] \ln[1 + (n-1)k_x] \int_0^t \exp[-E_k/(RT)]\mathrm{d}t, \tag{5.18}$$

where n is the order of the reaction and ϕ is related to the degree of chain entanglement.

The results give a good fit to an accelerated (by BF3-MEA) TGDDM/DDS system; however, they do not predict results for the unaccelerated system well at longer times. The introduction of the order of reaction greatly improved fits.

Roller (1986) reviews the rheology of curing thermosets, describing both isothermal (Equation (5.19)) and non-isothermal (Equation (5.20)) models for chemoviscosity:

$$\ln \eta(t) = \ln \eta_\infty + E_\eta/(RT) + tk_\infty \exp[E_k/(RT)], \tag{5.19}$$

$$\ln \eta(t,T) = \ln \eta_\infty + E_\eta/(RT) + \int_0^t k_\infty \exp[E_k/(RT)]\mathrm{d}t. \tag{5.20}$$

Additionally, Roller (1986) discusses an empirical model by Macosko like that in Equation (5.8). These models are also discussed in Tajima and Crozier (1986).

Structural and molecular models

Bidstrup and Macosko (1990) developed a chemorheological model that was based on branching theory for a diglycidyl ether of bisphenol-A/diaminodiphenyl sulfone (DGEBA/DDS) epoxy–amine system. This model was based on a establishing link between rheology and structure from the use of kinetic data and branching theory in order to calculate M_w and other average structural parameters, and correlating these data with chemoviscosity data. Correlations of both steady-shear viscosity and dynamic viscosity to structural properties were assessed, with the importance of using the same samples for both kinetic and rheological tests being highlighted. The model may be described by

$$\eta = f^*F, \tag{5.21}$$

where η is the viscosity, f is a friction factor (such as that derived by branching theory) and F is a structure factor (such as the M_w or M_w^a). For example, Berry and Fox (1968) showed that

$$f = f_0 \exp\{B/[f_g + a_1(T - T_g)]\}, \tag{5.22}$$

where f_0 is the initial friction term, f_g is the friction factor at the glass-transition temperature, T_g, and B and a_1 are constants.

Vinogradov and Malkin (1977) showed that for linear polymer polymerizations

$$\eta = kM_w^a, \tag{5.23}$$

where $a = 1$ for $M_w < M_c$ and $a = 3.5$ for $M_w > M_c$, k is a constant, M_w is the molecular weight (molar mass) and M_c is a critical molar mass.

Richter and Macsoko (1980) describe the chemoviscosity of a reactive polyurethane (4,4′-diphenyl methane diisocyanate, polyester triol and dibutyltin dilaurate catalyst) reasonably well by the following model:

$$\eta = A \exp[D/(RT)](M_w/M_{w0})^{C/(RT)+S}, \tag{5.24}$$

where A, D, C and S are constants, M_w is the molecular weight of the system and M_{w0} is the initial M_w.

Pichaud *et al.* (1999) highlight the chemorheology and dielectrics of the cure of DGEBA with isophorone diamine (IPD). The kinetics are well described by an autocatalytic model and the chemorheology is well described by the Macosko model

$$\eta(T)/\eta_0(T) = (a_{gel}/(a_{gel} - a))^{A+Ba}, \tag{5.25}$$

with $\eta_0(T) = A_\eta \exp[E_\eta/(RT)]$, where a_{gel} is the conversion at gelation, a is the conversion, A and B are parameters, $\eta_0(T)$ is the viscosity of the uncured material at $a = 0$, A_η is the pre-exponential factor, E_η is the activation energy for viscosity and R is the gas constant. Here dielectric data could also be related to kinetics and chemoviscosity via experimental relationships.

Free-volume models

Chiou and Letton (1992) examine the chemorheology of a complex industrial resin, Hercules 3501-6, containing a major epoxide (TGDDM), two minor epoxides (alicyclic diepoxy carboxylate and epoxy cresol novolac), a hardener (diaminodiphenyl sulfone, DDS) and a Lewis-base catalyst (boron trifluroride monoethylamine complex). They found that both the conventional WLF model (Equation (5.26)) (Ferry, 1980) and a modified WLF model (Equation (5.27) were appropriate to describe the chemorheology:

$$\ln\left(\frac{\eta(T)}{\eta(T_g)}\right) = -\frac{C_1(a)(T - T_g(a))}{C_2(a) + T - T_g(a)}, \tag{5.26}$$

$$\ln\left(\frac{\eta(T)}{\eta(T_{g0})}\right) = -\frac{C_1(T - T_g(a))}{C_2(a) + T - T_g(a)}, \tag{5.27}$$

where η is the viscosity, $\eta(T_g)$ is the viscosity at the glass-transition temperature T_g (or a constant), $\eta(T_{g0})$ is the reference viscosity at T_g, C_1 and C_2 are parameters as a function of conversion (a), and T is the temperature.

The modification represents the case in which an iso-free-volume assumption (or $C_1 =$ constant) applies. This work also highlighted the use of three independent nth order reaction rates for understanding the kinetics of the complex system.

Hesekamp and Pahl (1996) describe the chemoviscosity of an epoxy–amine (DGEBA/TMAB) adhesive using a modified WLF model shown as

$$\ln\left(\frac{\eta(T,X)}{\eta_0^*}\right) = -\frac{C_1(T-T_0)}{C_2+T-T_0} - \frac{b_1(T-T_g(X))}{(b_2+T-T_g(X)} + \frac{b_1(T-T_g(X=0))}{b_2+T-T_g(X=0)}, \tag{5.28}$$

where b_1 and b_2 are new parameters. The first term is the temperature dependence, the second term is the crosslinking effect and the final term is for normalization or to cancel out the crosslinking effect in uncured systems. The isothermal and non-isothermal chemoviscosity results are well fitted by this model.

Hou and Bai (1988, 1989) have investigated the chemorheology of Hercules 3501-6 and found good agreement with a modified WLF chemoviscosity model. The equation is summarized as

$$\ln\left(\frac{\eta_T}{\eta_{T_g}}\right) = -\frac{C_1(T-T_g)}{C_2+T-T_g}, \tag{5.29}$$

with $T_g = A(T)a(t) + B(T)$ and $a = a(T, t)$.

Kenny (Kenny *et al.*, 1989a, 1989b, Kenny and Opalicki, 1993) examined the kinetics and chemorheology of epoxy-resin-based composite laminates for which first the effects of degree of cure and temperature on resin viscosity were characterized and then, given thermal profiles from kinetic, mass- and heat-transfer calculations, the viscosity profiles during processing were determined. Here they describe the changes in viscosity of their TGDDM–DDS system by

$$\frac{\eta(T,a)}{\eta(T_0)} = g\left(\frac{M_w(a)}{M_{w0}}\right)^{3.4} \frac{\exp[C_1(T_r-T_{g0})/(C_2+T_r-T_{g0})]}{\exp[C_1(T-T_g(a))/(C_2+T-T_g(a))]}, \tag{5.30}$$

where $\eta(T, a)$ is the viscosity at temperature T and conversion a, $\eta(T_0)$ is the viscosity at reference temperature T_r, g is the ratio of the radius of gyration of a branched chain to that of a linear chain at the same molecular weight (molar mass), $M_w(a)$ is the molecular weight at conversion a, M_{w0} is the molecular weight at the initial T_0, C_1 and C_2 are WLF constants, T_{g0} is the glass-transition temperature at T_0 and $T_g(a)$ is the glass transition temperature at a conversion of a. This equation is shown to work well for the epoxy-resin laminate profiles examined under selected non-isothermal processing conditions.

Shear effects

Simple chemoviscosity models have been used to examine the effects of shear rate on the chemoviscosity $\eta = \eta(\gamma, T)$ as shown in Table 4.3. For example, Sundstrom and Burkett (1981) and Hartley and Williams (1981) found that polyesters and polyurethanes exhibit essentially Newtonian behaviour, whereas epoxies and phenolics show marked shear thinning. The power-law model is the most extensively used shear-rate model for thermosets and has been used for unfilled (Ryan and Kamal, 1976, Kascaval *et al.*, 1993,

Riccardi and Vazquez, 1989) and filled (Ryan and Kamal, 1976, Knauder *et al.*, 1991) epoxy-resin systems. Sundstrom and Burkett (1981) also obtained a good fit of the viscosity of diallylphthalate to the Cross model. The viscosity of polyesters has been modelled by Yang and Suspene (1991) using a Newtonian model. The WLF model has been used by Pahl and Hesekamp (1993) for a moderately filled epoxy-resin system. Rydes (1993) showed that the viscosity of DMC polyesters followed a power-law relationship at high shear rate.

Cure and shear effects

Ryan and Kamal (1976) discuss a combined chemoviscosity model as described in Equation (5.31) for an unfilled liquid epoxy resin (DGEBA/MPDA and Fiberite industrial resin), namely

$$\eta = A\gamma'(n-1)\exp[-E_\eta/(RT)], \tag{5.31}$$

where η is the viscosity, γ' is the shear rate, n is a power-law index, E_η is the activation energy for flow and A is a parameter (depending on the cure).

Macosko (1985) describes the importance of models that combine chemical kinetics, branching theory and composition and processing conditions (temperature, shear rates) in chemorheological models for structure–property relations.

Kamal (1974) summarizes the chemorheology data of various epoxy-resin moulding compounds and describes the relationship in general as

$$\eta = \eta(\gamma', T, a), \tag{5.32}$$

where η is the viscosity, γ' is the shear rate, T is the temperature and a is the conversion.

Han and Lem (1983) reviewed the cure kinetics, chemorheology and properties of curing polyester resins.

5.2.2 Filled systems

It is necessary, before we summarize the effects of fillers on the chemoviscosity of reactive systems, to highlight a review of the effects of fillers on rheology in general. (A more introductory discussion of the rheology of filled systems (suspensions) is given in Chapter 2.)

Rheology of non-reactive filled systems

Common fillers used for polymers are shown in Table 5.1 (Sekutowski, 1996). Commercial coupling agents are shown in Table 5.2 (Mitsuishi and Kawasaki, 1996). Barnes (1989) reviews shear thickening in non-aggregating particles in Newtonian liquids and discusses the parameters that control shear thickening, including the particle-size distribution, shape, volume, viscosity and type of deformation. The increase in viscosity is related to the transition from two-dimensional layered structure to a three-dimensional form. The severity of shear thickening is proportional to the concentration and the maximum packing fraction (which in turn is related to the particle-size distribution and shape). Shear thickening is reduced by reducing particle size (by delaying the onset) or by using a mixture of particle sizes (by increasing the maximum packing fraction).

Malkin (1990) reviews the rheology of filled polymers and highlights the importance of yield stresses, non-Newtonian flow, wall slip and normal stresses in filled polymer flow. Details of the effects of particle shape, concentration and adsorption on these phenomena are discussed.

Table 5.1. Common fillers and their properties

Classification	Density (g/cm^3)	Size (μm)	Hardness (mohs)	Aspect ratio	CAS number
Calcium carbonate	2.7		2.5–3.5	Low	471-34-1
Coarse-ground		10–40			
Fine-ground		1–10			
Precipitated		0.7–2.0			
Silica			6.5–7	Low	7631–86-9
Ground	2.0	1–30			
Precipitated	1.9–2.2	0.02–0.01[a]			
Fumed	2.2	<0.1[a]			
Gel	1.3	0.01–0.02[a]			
Kaolin	2.6		2.5–3	Moderate	1332-58-7
Air-classified		0.2–10			
Water-washed		0.2–10			
Calcined		0.8–3			
Mica	2.8	2–500	2.5–3	High	12001-26-2
Talc	2.8	1–75	1	Moderate	14807-96-6
Wollastonite	2.9	3–10	4.5–5.5	High	14567-51-2
Feldspar/nepheline	2.6	5–15	5.5–6.5	Low	68476-25-5/ 37244-96-5
Barium sulfate			3–4	Low	
Barytes	4.5	1–20			13462-86-7
Blanc fixe	4.4	0.2–5			7727-43-7
Carbon black	1–2.3	0.01–0.1[a]	—	Low	1333-86-4
Alumina trihydrate	2.4	0.4–20	2.5–3.5	Moderate	14762-49-3
Glass fibre	2.5	—	—	High	
Organic			—		
Wood flour	0.7	40–325[b]		Low	
Organic fibre	0.7–1.6	30–100[b]		High	
Nutshell flour	1.4	325[b]		Low	
Starch	—	2–150		High	
Corncob	1.2	—		—	
Rice hull	1	100–125[b]		—	
Beads				Low	
Glass	2.5	5–300	5.5		
Glass, hollow	0.2	1–200			

[a] Silica and carbon black form aggregates 10–100 times the size of the primary particles.
[b] Mesh size.

Aranguren *et al.* (1992) examined the rheology of PDMS filled with untreated and silane-treated fused silica. It was found that the dynamic moduli increase with increasing concentration of silanol groups (as present on the untreated silica) on the surface, due to the increase in interactions with the PDMS. Direct interactions and indirect entanglement interactions are thought to be the types of interactions occurring.

Gupta and Seshadri (1986) discussed calculations of the maximum loading level of filled liquids. They found that a geometric method for calculating the maximum packing fraction works well for polydisperse spherical suspensions.

Table 5.2. Commercial filler coupling agents

Name	Chemical formula	Minimum covered area (m^2/g)	Solubility in water	Solvent for modification
Silane-based				
γ-Chloropropyl trimethoxy silane	$ClC_3H_6Si(OCH_3)_3$	394	None	Water (pH 4.0–4.5)
Vinyl triethoxy silane	$CH_2{=}CHSi(OC_2H_5)_3$	411	None	Water (pH 3.0–3.5)
Vinyl trimethoxy silane	$CH_2{=}CHSi(OCH_3)_3$	526	None	Water (pH 4.0–4.5)
Vinyl *tris*(β-methoxy-ethoxy silane)	$CH_2{=}CHSi(OC_2H_4OCH_3)_3$	279	<5%	Water
γ-Methacryloxy propyl trimethoxy silane	$CH_2{=}CCH_3COOC_3H_6Si(OCH_3)_3$	316	None	Water (pH 3.5–4.0)
β-(3,4-Epoxycylohexyl ethyltrimethoxy silane)	(epoxycyclohexyl)-$C_2H_4Si(OCH_3)_3$	318	None	Water/ethanol – 3 : 1
γ-Glycidoxypropyl trimethoxy silane	$CH_2{-}O{-}CHCH_2OC_3H_6Si(OCH_3)_3$	332	<5%	Water
γ-Mercaptopropyl trimethoxy silane	$HSC_3H_6Si(OCH_3)_3$	399	None	Water (pH 4.5–5.0)
γ-Aminopropyl triethoxy silane	$NH_2C_3H_6Si(OC_2H_5)_3$	354	<5%	Water
N-β-(Aminoethyl)-γ-aminopropyl methoxy silane	$NH_2C_2H_4NHC_3H_6Si(OCH_3)_3$	353	<5%	Water

Name	Chemical formula
Titanate-based	
Isopropyl triisostearoyl titanate	$CH_3{-}CH(CH_3){-}O{-}Ti[O{-}C({=}O){-}C_{17}H_{35}]_3$
Isopropyl tridodecylbenzenesulfonyl titanate	$CH_3{-}CH(CH_3){-}O{-}Ti[O{-}S(O)(O){-}C_6H_4{-}C_{12}H_{25}]_3$
Isopropyl tri(dioctylpyrophosphato)titanate	$CH_3{-}CH(CH_3){-}O{-}Ti[O{-}P(O)(OH){-}O{-}P(O)(O{-}C_8H_{17})_2]_3$
Tetraisopropyl di(diocitylphosphito)titanate	$[CH_3{-}CH(CH_3){-}O]_4{-}Ti[P{-}(O{-}C_8H_{17})_2OH]_2$
Tetraoctyloxtitanium di(ditridecyl phosphite)	$(C_8H_{17}{-}O)_4{-}Ti[P{-}(O{-}C_{13}H_{27})_2OH]_2$
Tetra(2,2 diallyloxymethy1-1 butoxy titanium di(ditridecyl)phosphite	$[C_2H_5{-}C(CH_2{-}O{-}CH_2{-}CH{=}CH_2)_2{-}CH_2{-}O]_4Ti{-}[P{-}(O{-}C_{13}H_{27})_2OH]_2$
Titanium di(dioctylpyrophosphate)oxyacetate titanate	$(O{=}C{-}O{-})(CH_2{-}O{-})Ti[O{-}P(O)(OH){-}O{-}P(O){-}(O{-}C_8H_{17})_2]_2$
Di(dioctylpyrophosphato)ethylene titanate	$(CH_2{-}O{-})(CH_2{-}O{-})Ti[O{-}P(O)(OH){-}O{-}P(O){-}(O{-}C_8H_{17})_2]_2$

Lepez *et al.* (1990) examined the rheology of glass-bead-filled HDPE and PS. They found that a Cross model describes the viscosity–shear-rate relationship, a Quemada model describes the concentration dependence of the viscosity, and a compensation model applies for the temperature dependence of the viscosity. This model is expressed as

$$\eta^*(\omega, \phi, T) = \frac{1}{1 + \{a\exp[E_v/(RT)]\omega\}^P} \frac{1}{(1-\phi/\phi_m)^2 A(\phi=0)\exp[E_v/(RT)]}, \quad (5.33)$$

where η^* is the dynamic viscosity, ω is the frequency, T is the temperature, a is a parameter related to the relaxation time, E_v is the activation energy of flow, P is a shear-thinning parameter, ϕ is the volume fraction of filler, ϕ_m is the maximum volume fraction of filler and $A(\phi = 0)$ is a parameter related to E_v. Thus this model allows characterization of the viscosity of a filled system with only the average maximum packing volume (a function of the size distribution and shape of the filler), the flow activation energy of the thermoplastic, the shear-thinning parameter and an adjustable parameter (related to the activation energy of flow and the relaxation time).

Shang *et al.* (1995) show that the work of adhesion between a silica filler surface and a polymer matrix is directly related to the dynamic viscosity and moduli. Additionally, at lower frequencies there is a greater influence of the work of adhesion. The influence is shown to be described well by an effective increase in interphase thickness due to the increase in the work of adhesion, such that polymer chains are effectively immobilized around the filler, and the friction between the immobilized layer and the polymer then governs the dynamic rheology. It was noted that the immobilized layer could be reduced in extent at higher frequencies.

Maurer (1990) discusses the use of an interlayer model to describe the physical properties of filled composite systems.

Salovey and co-workers (Sun *et al.*, 1992a, 1992b, 1993, Gandhi *et al.*, 1990, Argarwal and Salovey, 1995, Zou *et al.*, 1990, 1992a, 1992b) examined the rheology of model filled polymers in a series of papers. They firstly prepared a range of model filler particles of PS, PMMA and modified particles to be introduced into PS and PMMA matrices. They examine mixing and time dependence, the rheology of monodisperse and crosslinked polymer fillers and the effects of chemical interactions and M_w of modified PMMA and PS particles on the rheology of the filled systems. The series gives an insight into the effects of filler–matrix interphases in the rheology and processing of filled systems.

Jeyaseelan and Giacomin (1995) examine the use of large-amplitude oscillatory-shear (LAOS) rheology of filled polymer melts (HDPE filled with carbon black) and use transient-network theory (which separates filler and polymer entanglement effects) to describe the non-linear flow behaviour.

Watanabe *et al.* (1996) describe the non-linear dynamic rheology of a concentrated spherical silica-particle-filled ethylene glycol/glycerol system. The strain dependence was well described by the BKZ-type constitutive equation, until the shear-thickening regime. The shear thickening was qualitatively described in relation to the structure of the suspended filler particles.

Yilmazer and Kalyon (Yilmazer and Kalyon, 1989, Kalyon and Yilmazer, 1990) described the rheology of highly filled suspensions (ammonium sulfate in poly(butadiene acrylonitrile acrylic acid)). They examined steady and dynamic rheology and found that wall slip and non-linear dynamic rheology are prevalent in these highly filled systems.

Parthasarathy and Klingenberg (1995a, 1995b) examined the steady shear and microstructure of electro-rheological suspensions. They note there is a highly non-linear rheological response and that the non-linearity is related to rearrangement of structure rather than gross deformation of structure.

Gahleitner *et al.* (1994) describe the rheology of talc-filled PP. They note the effects of filler concentration, particle size and dispersion on linear viscoelastic properties.

Utracki and Fisa (1982) and Metzner (1985) review the rheology of (asymmetric) fibre- and flake-filled plastics, noting the importance of the filler–polymer interface, filler–filler interactions, filler concentration and filler-particle properties in determining rheological phenomena such as yield-stress, normal-stress and viscosity profiles (thixotropy and rheopexy, dilatancy and shear thinning).

Quemada (1978a, 1978b) examined the rheology and modelling of concentrated dispersions and described simple viscosity models that incorporate the effects of shear rate and concentration of filler and separate effects of Brownian motion (or aggregation at low shear) and particle orientation and deformation (at high shear). The ratio of structure-build-up and -breakdown rates is an important parameter that is influenced by the ratio of the shear rate to the particle diffusion. A simple form of viscosity relation is given here:

$$\eta_r = 1\Big/\left(1 - \frac{1}{2}K\phi\right)^2, \qquad (5.34)$$

with $K = (k_0 + k_\infty \gamma^P)(1 + \gamma^P)$, where η_r is the reduced viscosity, ϕ is the particle concentration, k_0 is the intrinsic viscosity at $\gamma' = 0$, k_∞ is the intrinsic viscosity at $\gamma' = \infty$, P is a shear-thinning parameter and γ' is the shear rate.

Rucker and Bike (1995) examined the rheological properties of silica-filled PMMA. They showed the existence of a yield stress and a poor fit of viscosity data to existing filler models.

Seymour *et al.* (1982) highlighted the rheology of linear and gelled random styrene acrylate resins filled with carbon black for toner usage. The melt rheology was strongly affected by the initial structure in the resin and by the dispersion of the carbon black.

Van der Werff and de Kruif (1989) examined the scaling of rheological properties of a hard-sphere silica dispersion (sterically stable monodisperse silica in cyclohexane) with particle size, volume fraction and shear rate. The shear-thinning behaviour was found to scale with the Péclet number ($Pe = 6\pi\eta_s a^3\gamma/(k_B T)$, or the ratio of shear time to structure-build-up time, where a is the particle radius, η_s is the viscosity of the solution, γ is the shear rate, k_B is the Boltzmann constant and T is the temperature) and at higher volume fractions the shear thinning transition shifts to lower Pe.

Doraiswamy *et al.* (1991) developed a non-linear rheological model combining elastic, viscous and yielding phenomena for filled polymers. The model predicts a modified Cox–Merz relationship for filled melts:

$$\eta(\gamma') = \eta^*(\gamma_0\omega), \qquad (5.35)$$

where $\eta(\gamma')$ is the steady viscosity as a function of the shear rate (γ') and η^* is the dynamic viscosity as a function of the maximum strain (γ_0) and frequency (ω). The model gives a good prediction of dynamic- and steady-shear properties of a highly silica-filled PE.

Bicerano *et al.* (1999) provide a simplified scaling viscosity model for particle dispersions that states the importance of the shear conditions, the viscosity profile of the dispersing fluid, the particle volume fraction and the morphology of the filler in terms of its aspect ratio, the length of the longest axis and the minimum radius of curvature induced by flexibility.

Computer simulations of concentrated suspensions are also provided by Phillips *et al.* (1988) and Boersma *et al.* (1995), which provide insights into the mechanics of shear-thickening behaviour.

Rheology of reactive filled systems

We are now in a position to examine the chemorheology and modelling of filled reactive systems, by combing our knowledge of cure, shear-rate and filler effects, as described above, with an examination of the current literature on filled reactive systems.

Filled epoxy-resin systems

Dutta and Ryan (1979) examined the effects of fillers (carbon black and silane-surface-treated silica) on the cure of DGEBA/MPDA epoxy–amine systems. They found that the rate constants of the cure reaction are affected by the presence of the fillers in an unusual fashion (a function of temperature and concentration) with respect to concentrations up to 10%. This was postulated to be due to the reactive surface groups on the fillers. The reaction order, however, is not affected.

Kenny *et al.* (1995) examined the chemorheology and processing of short-fibre-reinforced BMC polyesters. They found that the presence of fillers ($CaCO_3$ and glass fibres) increases the reaction time. Also they found that the chemorheology could be described by considering both a shear-thinning model (typical of suspension theory) and the Macosko model (Equation (5.8)).

Kogan *et al.* (1988) examined the chemorheology of silica- and carbon-fibre-filled epoxy-resin systems. They found unusual effects of carbon fibre on the uncured rheology and chemorheology of filled epoxy-resin systems, and related these to the anisotropic nature of the filler shape and the effect of filler surfaces on the kinetics.

McGee (1982) examined the effect of fillers on the processing of filled polyesters and investigated effects on processing in terms of effects on thermal properties, kinetics and chemorheology. McGee noted the important effect of the type of filler on the ratio of the polymerization rate to the heat-transfer rate, denoted k'. The smaller the k' indicated, the shorter the gel time. For the system investigated, McGee noted that the primary effect of the filler was to modify the temperature profiles.

De Miranda *et al.* (1997) showed the effects of silica flour on the cure of DGEBA/DDM epoxy-resin systems. They noted that there were appreciable decreases in reaction rates for systems with higher filler concentrations than 10 wt.% at high temperatures and conversions above 50%.

Ng and Manas-Zloczower (1993) examined the chemorheology of a silica-filled epoxy-resin system (DGEBA/MDA/diaminotoluene). They found that the filler increased the reaction rate, decreased the rheologically determined gel times and increased the ability of the system to dissipate heat.

Peng and Riedl (1994) investigated the chemorheology of phenol formaldehyde resins filled with unmodified and methylolated lignin fillers. They found that both fillers reduced the reaction rate and that the chemorheology profile of filled systems is better represented by the WLF model than by the Arrenhius model.

Shaterzadeh *et al.* (1998) notes the importance of filler distribution as well as interphase in the generalized self-consistent three-phase model used to predict the dynamic moduli of an epoxy/A-glass-bead system.

Tungare *et al.* (1988) examined the chemorheology of an epoxy-resin (FR4) and glass-cloth assembly via squeeze-flow geometry and a Newtonian-flow assumption. The results fitted well to the Arrhenius model and the sensitivity of the Arrhenius model parameters is discussed.

Walberer and McHugh (2000) investigated the use of squeeze-flow rheometry to examine the relaxation and elongation rheological behaviour of very highly filled glass-bead–PDMS and reactive-glass-bead–PVA resins. The rheological data presented could not be collected for such systems using conventional shear rheometers.

Boutlin *et al.* (1992) examined the chemorheology of highly filled epoxy-resin moulding compounds. They found that the crossover method is not a good procedure and that the independence of $\tan\delta$ is a better diagnostic for gelation prediction (as defined in Table 4.5). Also they found that steady-shear results can disturb the gel structure and that dynamic data are more reliable. Furthermore, they found that the Macosko model gives a better prediction of the chemorheological data than does the double Arrhenius model (as defined in Tables 4.2 and 4.4).

Chang *et al.* (1998) examined the chemorheology of a highly filled epoxy-resin moulding compound. They found that a coupled power law and the Macosko model adequately describe the chemorheology of the system. This equation is

$$\eta(a, T, \gamma') = \exp[E_a(a)/(RT)]\gamma'^{n(a)}\eta_0(a), \tag{5.36}$$

where

$$E_a(a) = C_1 + C_2 a + C_3 a^2 + C_4 a^3,$$

$$n(a) = C_5 + C_6 a + C_7 a^2 + C_8 a^3,$$

$$\eta_0(a) = \exp(C_9 + C_{10} a + C_{11} a^2 + C_{12} a^3).$$

The authors also note the application of the modified Cox–Merz rule relating dynamic and steady viscosities, $\eta(\gamma') = \eta^*(\gamma_m\omega)$.

Han *et al.* (1997) examined the chemorheology of a highly filled epoxy-resin moulding compound that is characterized by a modifed slit rheometer. Results show that a modified Cox–Merz rule relating dynamic and steady viscosities is established, $\eta(\gamma') = \eta^*(\gamma_m\omega)$. Also the material was shown to exhibit a yield stress at low shear rates and power-law behaviour at higher shear rates. The temperature dependence of the viscosity is well predicted by a WLF model, and the cure effects are described by the Macosko relation.

Kuroki *et al.* (1999) examined the chemorheology of an epoxy resin with 90 wt.% silica. Gelation was predicted by independence of $\tan\delta$, power-law behaviour in dynamic moduli and GPC. A modified power-law–Arrhenius chemorheology model fitted the data well.

Petti (1994) showed the importance of silica-particle shape and maximum packing volume fractions on the chemoviscosity and spiral flow of highly filled epoxy-resin moulding compounds. Ultrahigh concentrations of silica could be used using the correct filler.

Saeki and Kaneda (1988) examined the apparent mean viscosity, minimum viscosity and apparent gel time of an epoxy-resin moulding compound during the transfer-moulding operation by using an instrumented transfer mould.

Spoelstra *et al.* (1996) examined the chemorheology of a highly filled epoxy-resin moulding compound and showed that the modified Cox–Merz rule (as defined in

Section 4.2.1) gives a good description of the steady-shear–dynamic-shear relationship. They found that the following general relationship is a good representation of the chemoviscosity profile:

$$\eta(a, T, \gamma') = \eta_0(a) \exp[E_a(a)/(RT)](\omega\gamma)^{n(a)}. \quad (5.37)$$

They also found that neither the crossover of dynamic moduli nor the independence of the loss tangent was a good diagnostic to measure the gel point.

Baikerikar and Scranton (2000) examined novel photopolymerizable liquid encapsulants (epoxy novolac–vinyl ester) filled with silica fillers. They found that the desired viscosity reduction occurred due to a blending of large and small particles, and that there was an optimum size distribution. A silane coating was also desirable, in order to reduce viscosity.

Filled polyester systems

Rheological properties of mineral-filled and mineral/glass-fibre-filled unsaturated polyester compounds have been examined in a series of excellent papers by Han and Lem (Han and Lem, 1983a, 1983b, 1983c, 1984, Lem and Han, 1983a, 1983b, 1983c) in which they used expressions for the cure kinetics (auto-catalytic equation) and chemorheology (shear-thickening) profiles. Note that these equations presume minimal cure effects on viscosity profiles during shearing, and, although effects of fillers, additives and filler surface treatments of process flows were examined, no formalization of a process model was attempted. However, in later work Han and Lee (Han and Lee, 1987, Lee and Han, 1987) extended their chemorheological modelling to incorporate cure effects by use of the following WLF-based equation:

$$\ln \eta = \ln \eta_{T_g} - a(T - T_g)/(C_1 + T - T_g), \quad (5.38)$$

where η is the viscosity, η_{T_g} is a constant, a is a free-volume parameter, T is the termperature, T_g is the glass-transition temperature and C_1 is a constant. Note that in this equation T_g is a function of cure conversion (a) and hence must be coupled to the kinetic equation.

Lee *et al.* (1981) highlighted the importance of shear and elongational models for polyester materials. These models obey an Arrhenius temperature relationship; however, it is assumed that there is no curing during flow.

Reactive toughened systems

As can be seen in the literature, a large amount of work was done to toughen epoxy-resin systems by the addition of carboxy-terminated butadiene acrylonitrile (CTBN) rubber materials (Dickie and Newman, 1973, Bascom *et al.*, 1975, Kunz-Douglas *et al.*, 1981, Riew, 1984). This work highlighted the importance of particle size and interfacial adhesion between the rubber and the matrix for good toughening. This toughening was seen as arising from extra mechanisms of stress relief such as crazing (Riew, 1984), particle-cavitation-induced shear banding and shear yielding (Bascom *et al.*, 1975, Pearson and Yee, 1989). Other work has focussed on core–shell particles (or graded rubber particles) for which the inner core and the outer shell of a rubber particle can be tailored to toughen specific epoxy-resin matrices. For example, the effects of shell diameter, shell thickness, crosslink density of core and chemical reactivity of shell surface on the toughness improvements in highly crosslinked epoxies have been examined (Nakamura *et al.*, 1986, 1987). Improvements in toughness for core–shell particles have been reported to be due to cavitation, shear yielding and crack deflection (for larger particles) (Sue, 1991), interfacial effects on solubility and dispersion (Sue *et al.*, 1996) and intermediate/strong interfacial bonding (Nakamura *et al.*, 1986). However, although toughness was improved in CTBN and core–shell rubber-modified epoxies, this was typically accompanied at the cost of losses in modulus, glass-transition temperature and solvent resistance (Sue *et al.*, 1996).

Ratna (2001) examined the cure and phase separation of carboxy-terminated poly(2-ethyl hexyl acrylate) (CTPEHA) liquid modifier in DGEBA/diethyltoluene diamine (DETD). It was found that the CTPEHA liquid rubber causes a delay in the polymerization of the epoxy-resin matrix due to chain extension during pre-reaction and the viscosity effect. The T_g decreases with increasing rubber concentration.

Di Pasquale *et al.* (1997) examined the effects of addition of a thermoplastic amine functionalized poly(arylene ether sulfone) on the cure of a DGEBA/DDS system. The fracture toughness increased with addition of the reactive thermoplastic due to the effects of the thermoplastic as a curing agent, an interfacial bonding agent, reducing the crosslink density and inducing phase separation.

Girard-Reydet *et al.* (1998) examined the reaction-induced phase separation in poly-etherimide (PEI)–DGEBA–DDS or MCDEA systems via *in situ* SAXS, light-scattering and TEM methods. The phase separation was dependent on the initial PEI concentration and the ratio of phase separation to polymerization rates. For concentrations near the critical concentration, the system was put into the unstable region and phase separation proceeded by spinodal demixing. This was independent of the polymerization rate. For off-critical concentrations, the homogeneous solution de-mixed slowly via a nucleation- and- growth mechanism, and the cure temperature effects vitrification of the thermoplastic-rich phase and hence controls morphologies.

Ho *et al.* (1996) examined polyol or polysiloxane thermoplastic polyurethanes (TPUs) as modifiers in cresol–formaldehye novolac epoxy resins cured with phenolic novolac resin for computer-chip encapsulation. A stable sea–island dispersion of TPU particles was achieved by the epoxy ring-opening with isocyanate groups of the urethane prepolymer to form an oxazolidone. The flexural modulus was reduced by addition of TPU and also the T_g was increased due to the rigid oxazolidone structure. Mayadunne *et al.* (1999) extended this work to a series of phenol- and naphthol-based aralkyl epoxy resins.

Kim and Kim (1994, 1995) examined the cure and vitrification of ATBN-modified DGEBA/TETA epoxy resins. The presence of the ATBN rubber had only a small effect on kinetics up to 20 wt.%. Reaction injection-moulding experiments and simulations were also conducted on this system.

Kim *et al.* (1993) examined the cure of a polyethersulfone–dicyandiamide–TGDDM system and established a time–temperature-transformation cure diagram for the system. During the early stage of curing the system was in the single-phase regime, and after an induction period phase separation occurred by spinodal decomposition and the domain spacing increased until gelation. Phase separation was characterized by the time variation of light scattering. Gelation and vitrification were monitored via rheological methods.

Kim *et al.* (1999) examined the morphology and cure of semi-IPN epoxy resin or dicyanate–polyimide/polysulfone–carbon-fibre films. Polyimide or polysulfone films were inserted into the curing epoxy–dicyanate monomers to form semi-IPNs with sea–island morphology at the thermoset–thermoplastic interface. The final carbon-fibre-thermoplastic-dicyanate films had fracture toughness three to five times higher than that of unmodified carbon-fibre-dicyanate composites.

Martinez *et al.* (2000) examined the chemorheology and phase separation of polysulfone-modified DGEBA–DDM epoxy-resin mixtures. They found a delay in polymerization due to dilution and viscosity effects. The final morphologies are controlled by curing temperature due to the effect of temperature on phase separation. The fracture toughness increased with increasing immiscibility and was at its maximum for a bicontinuous morphology.

Poncet *et al.* (1999) monitored frequency-dependent dielectric measurements to examine the phase-separation process in poly(2,6-dimethyl-1,4-phenylene ether) (PPE) in a DGEBA–MCDEA resin. Dielectric measurements measured the build up in T_g both in the PPE-rich continuous phase and in the epoxy-rich occluded phases for 30–60-wt.% PPE mixtures. In the 30% PPE mixture, the rate of reaction of the thermoset phase is equivalent to that of the neat system due to two opposing effects, namely a slower reaction rate due to dilution and a low level of conversion at vitrification due to the presence of high-T_g PPE. In the 60-wt.% mixture the dilution effect of the PPE has a large effect of decreasing the reaction rate. The continuous thermoplastic-rich phase vitrifies first, followed by the thermoset occluded phase. The final morphology (size of occluded particles and composition of continuous phase) is affected by kinetics, diffusion and viscosity during phase separation.

Woo and Hseih (1998) examined the design of micro-epoxy particles with a thermoplastic polymer shell. As cure progressed, the phenoxy component (polyhydroxyl ether of bisphenol-A) was expelled from the epoxy-resin (TGDDM or DGEBA–DDS) phase during phase separation. At later stages of cure, etherification occurred for high temperatures of cure in the TGDDM–DDS–phenoxy systems, between the pendant hydroxyl group of the phenoxy shell and the residual epoxide of the crosslinked epoxy core particles. This led to a chemical link between the thermoplastic and the epoxy-resin. This phenomenon was not seen for DGEBA–DDS–phenoxy systems, possibly due to the lack of residual epoxide groups.

Bonnet *et al.* (1999a, 1999b) examined the reaction kinetics and rheological behaviour of PEI–DGEBA–MCDEA systems through phase transitions. Initially systems are homogeneous, with no effects of PEI on the kinetics, other than dilution. At a conversion from 0.2 to 0.4, phase separation and a sudden increase in reaction rate occur. This is due to the formation of an epoxy–amine-rich phase at phase separation. The chemorheological behaviour was highly dependent on the concentration of PEI: for low concentrations (lower that 10%–15%) phase separation induced a decrease in viscosity; for concentrations close to the phase-inversion composition (20%) a bicontinous dependence of viscosity on frequency was observed; and at PEI concentrations above 30 wt.% phase separation led to an increase in viscosity. Chemorheological modelling based on miscible polymer blends, filled polymer and chemorheological theories gave good agreement with experiments.

Kim and Char (2000) examine the rheology of a PES–DGEBA–DDM system during isothermal cure. They found a fluctuation in viscosity at phase separation that could be simulated by a two-phase suspension model that incorporated chemoviscosity effects.

Tillie *et al.* (1998) examined the effect of the fibre/matrix interface on the cure of glass-fibre-filled epoxy-resin systems. They found that the introduction of a lower-T_g interphase based on hydroxylated PDMS oligomers allowed an increase in toughness without a reduction in modulus or T_g. This was due to a modification of the stress field under load due to the elastomeric interphase.

Pearson and Lee (1991) examined the effects of particle-size and particle-distribution effects on rubber-toughened epoxy resins. They examined a variety of CTBN liquid rubbers and a methacrylated butadiene styrene core–shell particle in a DGEBA–piperidine system. They found that the toughening mechanism for small particles was internal cavitation of the rubber particles and subsequent formation of shear bands. Subtle toughening mechanisms such as crack deflection and particle bridging were noted for large particle sizes. No synergistic effects were seen for mixtures of particle sizes.

Qian *et al.* (1997) examined the fracture toughness of structural core–shell particles in an epoxy-resin matrix. They focussed on varying the shell composition of polybutadiene-co-styrene (PBS)-core–polymethylmethacrylate (PMMA)-shell particles by incorporating acrylonitrile (AN) comonomer into the PMMA shell at various AN : MMA ratios and by crosslinking of the shell. The degree of particle dispersion was controlled precisely by the AN content and by crosslinking. Microclustered dispersion provided higher toughness than did a uniform particle distribution.

Wilkinson *et al.* (1993) examined the effects of thermoplastic additives on the cure of commercial bismaleimide. A range of thermoplastics based on polyether sulfones, polyether phosphine oxide and polyimides were used in commercial Matrimid bismaleimide resin. Incorporation of a low-M_w polyethersulfone thermoplastic brought about an increase in fracture toughness without increasing viscosity. Reactive end groups contributed greatly to the improvement in fracture toughness by improving linkages between continuous and discrete phases, especially for phase-inverted structures. Improvements in fracture toughness in carbon-fibre prepregs with the toughened bismaleimides were also shown.

Ving-Tung *et al.* (1996) examined the rheological monitoring of phase separation on a high-T_g thermoplastic in a TGDDM/aromatic diamine epoxy resin. The chemorheological behaviour is affected by the vitrification of the thermoplastic-rich phase, gelation of the epoxy-resin-rich phase and the morphological effects of the occluded phase. It is shown that complex viscosity is a sensitive measure of the onset of phase separation and also is sensitive to the type of phase separation that occurs (that is related to the concentration of the thermoplastic phase). Three types of morphology based on the three types of phase separation are found, namely nucleation and growth for 5–10 wt.% of the thermoplastic phase (the epoxy-resin-rich phase is the continuous phase), spinodal decomposition for 12–20 of the wt.% thermoplastic phase (co-continuous phases) and nucleation and growth with phase inversion for above 20 wt.% of the thermoplastic phase (the epoxy-resin phase is the occluded phase); and the chemoviscosity profiles are correlated with these phases. The effects of shear strain have an effect on final morphologies.

Vlassopoulos *et al.* (1998) examined the gelation of three epoxy-rubber thermoset blends (based on TGDDM/DDS/(acrylonitrile/butadiene rubber/methacrylic acid copolymer) of the same chemistry but different pre-cure treatments. The pre-treatments used heat and catalysts to promote epoxy–carboxyl reactions, and there was some evidence of a decrease in gelation time and an effect on pre-gel rheology with these treatments.

More recently a large body of research has focussed on thermoplastic-modified thermoset matrices (Pearson and Yee, 1992, Ohnaga *et al.*, 1994, Hay *et al.*, 1996, Girard-Reydet *et al.*, 1997, Hodgkin *et al.*, 1998). Typically, high-performance high-temperature-resistant thermoplastics such as polysulfones, polyphenylene ethers and polyimides are used. Although a good improvement in toughness can be achieved in these systems without loss in terms of other thermomechanical properties, the viscosity of the system increases (Hodgkin *et al.*, 1998) and thus the processibility decreases. Therefore, there exists a need to develop a system that improves the toughness without deleterious effects on other properties such as strength, thermal and solvent resistance and processibility.

There has been much work on characterizing the phase separation and network development for thermoplastic-modified epoxy-resin systems. Initial work (Yamanaka and Inoue,

1989, Kim *et al.*, 1993) focussed on understanding the development of morphologies in rubber (CTBN)- and thermoplastic (polyethersulfone)-modified epoxy resins by monitoring the onset of phase separation, network gelation, fixation of phase-separated structure, end of phase separation and network vitrification. This morphology development was characterized via light scattering and rheological techniques, and was reported in terms of a modified time–temperature-transformation (TTT) diagram for a range of cure temperatures (Kim *et al.*, 1993). Additional work focussed on the use of frequency-dependent dielectric measurements to monitor *in situ* the phase separation and network development of polyether-modified epoxy resin (Poncet *et al.*, 1999). This study showed that phase separation decreased epoxy reaction rates at high levels of modifier due to dilution and vitrification effects, and that the final morphology is strongly influenced by the interdependence of cure kinetics, viscosity and phase separation. Recent work (Girard-Reydet *et al.*, 1998) has shown that important factors in controlling phase separation in polyetherimide-modified epoxy resins are the proximity of the initial modifier concentration to the critical concentration and the ratio of the phase-separation rate to the polymerization rate. For materials near the critical concentration (as determined by mixing theory for reactive blends (Clarke *et al.*, 1995, Inoue, 1995) and based on critical miscibility) phase separation occurs unstably via spinodal demixing, whereas for off-critical concentrations phase separation proceeded via a stable nucleation–growth mechanism. Additionally both cure and post-cure temperatures controlled the extent of phase separation due to vitrification halting the evolution of morphologies. These models relate the free energy of mixing of a polymer blend to volume fractions and degrees of polymerization of the various phases and an interaction parameter based on enthalpic and entropic terms. Further excellent studies (Bonnet *et al.*, 1999a, 1999b) have shown the effects of the phase separation on the cure kinetics and rheology of polyetherimide-modified DGEBA–MCDEA epoxy resins, highlighting again the interdependence of phase separation and curing epoxy properties. This work has shown that both the rate of epoxy–amine reaction and the viscosity can be modelled both in epoxy-resin and thermoplastic-rich phases via extension of Couchman (1978) (kinetic) and Macosko–Miller (Macosko and Miller, 1976) (viscosity) models with an understanding of phase behaviour.

Epoxy–hyperbranched-polymer systems

Recently the development of dendritic and hyperbranched polymers (HBPs) has attracted much attention (Tomalia, 1985, Newkome *et al.*, 1985, Webster, 1991, Chu and Hawker, 1993, Wooley *et al.*, 1994, Feast and Stanton, 1995, Malmstrom *et al.*, 1995, Kim, 1998). The key features of the macromolecular architecture of dendrimers and HBPs are given in Section 1.2, and their synthesis by stepwise polymerization is discussed in Section 1.2.1. Dendrimers and HBPs are globular macromolecules that have a highly branched structure with multiple reactive chain ends (shell), which converge to a central focal point (core); see Figure 5.1, where I is the core, II is the structure and III is the shell.

Although the terms dendrimer and HBP are often interchanged, they are different and can have quite different properties (Frechet *et al.*, 1996). Strictly dendrimers have a precise end-group multiplicity and functionality, but are very tedious to synthesize, whereas HBPs mimic (but do not achieve) regular dendritic growth, but are easily synthesized. Many HBPs based on polystyrene, polyesters, polyethers, polyphenylene, polyamides and various engineering polymers have been synthesized (Kim, 1998). Key attributes of HBPs are their high potential for crosslinking and blending (due to the large number of reactive end groups)

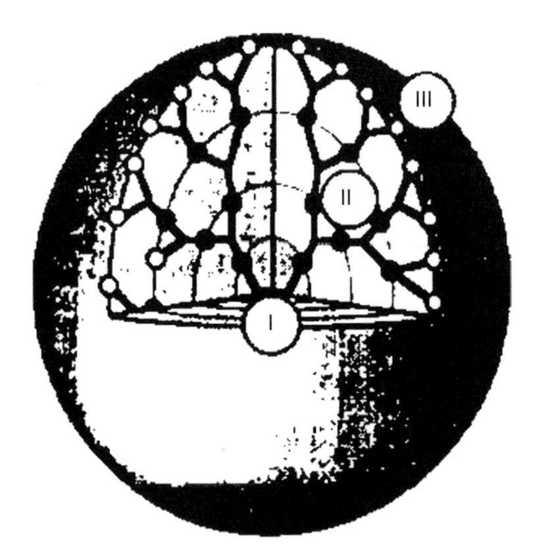

Figure 5.1. The structure of a hyperbranched polymer.

and their low viscosities (lower than those of linear polymers of equivalent molar mass (Wooley *et al.*, 1994) due to the elimination of intermolecular entanglements). It is these properties that are of interest regarding the use of HBPs as tougheners in epoxy-resin systems.

There have been very few reports on epoxy-hyperbranched toughened polymers, but what has been done has shown both the enormous potential and the conflicting nature of results on these systems. Recent work (Boogh *et al.*, 1999) has shown that epoxy-functionalized HBPs can greatly improve toughness (a 250% increase in K_{1c}) without affecting Young's modulus, the glass-transition temperature and processing viscosity of a model DGEBF/isophorondiamine epoxy-resin matrix at very low levels of HBP (5 wt.%). In this study it was noted that the shell chemistry and polarity affected the initial miscibility and kinetics of phase separation, whilst subsequent cure conditions and phase separation controlled processibility and structure. Unfortunately, details of the shell chemistry and cure profiles were not disclosed, thus detailed relationships between phase separation and shell chemistry/cure conditions could not be determined. However, other recent work (Wu *et al.*, 1999) has shown on the contrary that very slight improvements in toughness were obtained for hydroxy- and acetyl-terminated HBPs added to DGEBA/DDS epoxy-resin systems in comparison with the system that contained a linear analogue of the HBP. However, this paper (Wu *et al.*, 1999) does note that the use of an epoxy-terminated end group may promote better interfacial adhesion and thus better toughness. It was also noted that better control of phase separation (via varying cure conditions) may be necessary in order to optimize the final structure and properties.

It is postulated that the role of the HBP in toughening will be to act similarily to a core–shell particle; that is, the core of the HBP will act to cavitate and promote shear yielding, and the shell will be able to be tailored to control aggregation and interactivity with the epoxy-resin matrix. Increases in core M_w should promote cavitation, and shell-chemistry functionalization should increase dissolution and reactivity with the epoxy resin. However, unlike with the core–shell particles, the inherently greater number of shell sites and low viscosity of the HBP will enable the toughening to occur without deleterious effects on other properties.

Epoxy–nanocomposite systems

Epoxy–nanocomposite systems are relatively novel systems examined first by Pinavaia and Gianellis (Wang and Pinnavaia, 1994, Messersmith and Giannelis, 1994). A recent review of epoxy nanocomposites is provided by Becker and Simon (2005). In terms of chemorheology of epoxy–nanocomposite systems, there are relatively few papers, however.

Le Pluart *et al.* (2004) and Becker *et al.* (2003a) showed the effects of nanoclays on dynamic rheology and steady viscosity prior to cure and post-cure. There was a clear increase in low-frequency elastic modulus and low shear viscosity before cure, and a clear increase in fully cured mechanical properties post-cure with increasing clay content.

Becker *et al.* (2003b) examined the influence of an organically modified clay on the curing behavior of three epoxy-resin systems widely used in the aerospace industry and the effects of various structures and functionalities. Diglycidyl ether of bisphenol-A (DGEBA), triglycidyl *p*-amino phenol (TGAP) and tetraglycidyl diamino diphenylmethane (TGDDM) were mixed with an octadecylammonium-ion-modified organoclay and cured with diethyltoluene diamine (DETDA). As the nanofiller concentration increased, the gelation time decreased. It was found that this was due to the fact that the rate of reaction is increased by the addition of the organoclay.

Dean *et al.* (2005) examined epoxy layer silicate nanocomposites and showed that the interlayer spacing increased with the temperature of cure, resulting in intercalated morphologies with varying degrees of interlayer expansion, depending on the cure temperature used. Chemorheological studies of the curing process indicate that inter-gallery diffusion and catalysis occur before curing and are essential for exfoliation, before the morphology is frozen in by gelation and vitrification.

Sun *et al.* (2006) examined the use of novel silica nanofillers in underfill for flip-chip applications, and showed that pre-cure rheology and post-cure values of T_g are effected by nanosilica surface treatment.

Hadjistamov (1999) examined the effect of nanoscale silica on the rheology of silicone oil and uncured epoxy-resin (araldite) systems. Shear thickening and yield-stress-like behaviour were observed and found to be due to a build-up of network structure associated with the nanocomposite phase.

5.2.3 Reactive-extrusion systems and elastomer/rubber-processing systems

Reactive-extrusion and elastomer/rubber-processing systems are intimately linked with their processing (e.g. reactive extrusion), so chemorheological models for these systems will be discussed alongside their process modelling in Chapter 6.

5.3 Chemorheological models and process modelling

Chemorheological models are an integral part of process modelling of reactive polymer systems, as shown below. Integral parts of a flow-modelling procedure are the following:

(1) Physical model
  - physical properties
  - initial conditions and boundary conditions
  - physical flow geometry

(2) Kinetic model
 – kinetic data
(3) Chemorheological model
 – chemoviscosity data
 – gelation data
(4) Governing equations
 – mass balance
 – force/momentum balance
 – energy balance
(5) Mathematical solver process, such as finite-difference or finite-element analyses. When the kinetic and chemorheological models are input into this procedure, process modelling is enabled. Chapter 6, in which process modelling of various reactive systems is investigated, will examine this further in more detail.

References

Aranguren, M., Mora, E., Degroot, J. & Macosko, C. (1992) *J. Rheology*, **36**, 1165–1182.

Argarwal, S. & Salovey, R. (1995) *Polym. Eng. Sci.*, **35**, 1241–1251.

Baikerikar, K. & Scranton, A. (2000) *Polym. Composites*, **21**, 297–304.

Barnes, H. (1989) *J. Rheology*, **33**, 329–366.

Bascom, W., Cottingham, R., Jones, R. & Peyser, P. (1975) *J. Appl. Polym. Sci.*, **19**, 2545.

Becker, O., Cheng, Y., Varley, R. & Simon, G. (2003b) *Macromolecules*, **36**, 1616–1625.

Becker, O. & Simon, G. (2005) *Adv. Polym. Sci.*, **179**, 29–82.

Becker, O., Simon, G., Varley, R. & Halley, P. (2003a) *Polym. Eng. Sci.*, **43**, 850–862.

Berry, G. & Fox, T. (1968) *Adv. Polym. Sci.*, **5**, 261.

Bicerano, J., Douglas, J. & Brune, D. (1999) *J. Mater. Sci. – Rev. Macromol. Chem. Phys.* C, **39**, 561–642.

Bidstrup, S. & Macosko, C. (1990) *J. Polym. Sci. B: Polym. Phys.*, **28**, 691–709.

Boersma, W., Laven, J. & Stein, H. (1995) *J. Rheology*, **39**, 841–860.

Bonnet, A., Pascault, J., Sautereau, H. & Camberlin, Y. (1999b) *Macromolecules*, **32**, 8524–8530.

Bonnet, A., Pascault, J., Sautereau, H., Taha, M. & Camberlin, Y. (1999a) *Macromolecules*, **32**, 8517–8523.

Boogh, L., Pettersson, B. & Manson, J. (1999) *Polymer*, **40**, 2249.

Boutin, L., Ajji, A. & Choplin, L. (1992) *Proceedings of the XIth International Congress on Rheology*, Belgium, pp. 327–329.

Castro, J. & Macosko, C. (1982) *AIChE J.*, **28**, 251–260.

Chang, R., Lin, Y., Lin, F., Yang, W. & Hsu, C. (1998) *Society of Plastic Engineers ANTEC Proceedings ANTEC 1998*, pp. 1397–1401.

Cheng, K., Chiu, W., Hsieh, K. & Ma, C. (1994) *J. Mater. Sci.*, **29**, 721–727.

Chiou, P. & Letton, A. (1992) *Polymer*, **33**, 3925.

Chu, F. & Hawker, C. (1993) *Polym. Bull.*, **30**, 265.

Clarke, N., McGleish, T. & Jenkins, S. (1995) *Macromolecules*, **28**, 4650.

Couchman, P. (1978) *Macromolecules*, **11**, 117–119.

Dean, D., Walker, R., Theodore, M., Hampton, E. & Nyairo, E. (2005) *Polymer*, **46**, 3014–3021.

De Miranda, M., Tomedi, C., Bica, C. & Samios, D. (1997) *Polymer*, **38**, 1017–1020.

Di Pasquale, G., Motto, O., Recca, A. *et al.* (1997) *Polymer*, **38**, 4345–4348.

Dickie, R. & Newman, S. (1973) *J. Appl. Polym. Sci.*, **17**, 65.

Doraiswamy, D., Mujumdar, A., Tsao, I. *et al.* (1991) *J. Rheology*, **35**, 647–685.

Dusi, M., Lee, W., Ciriscoli, P. & Springer, G. (1987) *J. Comp. Mater.*, **21**, 243–261.

Dusi, M., May, C. & Seferis, J. (1982) *ACS Org. Coat. Appl. Polym. Sci. Proc.*, **47**, 635.
Dusi, M., May, C. & Seferis, J. (1983) *ACS Symp. Series*, **227**, 301–318.
Dutta, A. & Ryan, M. (1979) *J. Appl. Polym. Sci.*, **24**, 635.
Eley, R. (1983) *ACS Symp. Series*, **227**, 281–299.
Feast, W. & Stanton, N. (1995) *J. Mater. Chem.*, **5**, 405.
Ferry, J. (1980) *Viscoelasticity of Polymers*, New York: Wiley.
Frechet, J., Hawker, C., Gitsov, I. & Leon, J. (1996) *J. Macromol. Sci. – Pure Appl. Chem.* A, **33**, 1399.
Gahleitner, M., Bernreitner, K. & Neissl, W. (1994) *J. Appl. Polym. Sci.*, **53**, 283–289.
Gandhi, K., Park, M., Sun, L. *et al.* (1990) *J. Polym. Sci. B: Polym Phys.*, **28**, 2707–2714.
Girard-Reydet, E., Sautereau, H., Pascualt, J. P. *et al.* (1998) *Polymer*, **39**, 2269–2280.
Girard-Reydet, E., Vicard, V., Pascault, J. & Sautereau, H. (1997) *J. Appl. Polym. Sci.*, **65**, 2433.
Gupta, R. & Seshadri, S. (1986) *J. Rheology*, **30**, 503–508.
Hadjistamov, D. (1999) *Appl. Rheology*, Sep–Oct, 212–218.
Han, C. & Lee, D. (1987) *J. Appl. Polym. Sci.*, **33**, 2859–2876.
Han, C. & Lem, K. (1983a) *J. Appl. Polym. Sci.*, **28**, 743.
Han, C. & Lem, K. (1983b) *J. Appl. Polym. Sci.*, **28**, 763.
Han, C. & Lem, K. (1983c) *J. Appl. Polym. Sci.*, **28**, 3155.
Han, C. & Lem, K. (1984) *Polym. Eng. and Sci.*, **24**, 473.
Han, S., Wang, K., Hieber, C. & Cohen, C. (1997) *J. Rheology*, **41**, 177–195.
Hartley, M. & Williams, H. (1981) *Polym. Eng. Sci.*, **21**, 135.
Hay, J., Woodfine, B. & Davies, M. (1996) *High Perform. Polym.*, **8**, 35.
Hesekamp, D. & Pahl, M. (1996) *Rheologica Acta*, **35**, 321–328.
Ho, T., Wang, J. & Wang, C. (1996) *J. Appl. Polym. Sci.*, **60**, 1097–1107.
Hodgkin, J., Simon, G. & Varley, R. (1998) *Polym. Adv. Technol.*, **9**, 3.
Hou, T. & Bai, J. (1988) *Chemoviscosity Modelling for Thermoset Resins*, Nasa contractor report 181 718.
Hou, T. & Bai, J. (1989) *Chemoviscosity Modelling for Thermoset Resins*, Nasa contractor report 181 807.
Inoue, T. (1995) *Prog. Polym. Sci.*, **20**, 119.
Jeyaseelan, R. & Giacomin, A. (1995) *Polym. Gels Networks*, **3**, 117–133.
Kalyon, D. & Yilmazer, U. (1990) In Collyer, A., Utracki L. A. (Eds.) *Polymer Rheology and Processing*, London: Elsevier.
Kamal, M. (1974) *Polym. Eng. Sci.*, **14**, 231–239.
Kamal, M. & Sourour, S. (1973) *Polym. Eng. Sci.*, **13**, 59–64.
Kascaval, C., Musata, F. & Rosu, D. (1993) *Angew. Makromol. Chem.*, **209**, 157.
Kenny, J., Apicella, A. & Nicolais, L. (1989a) *Polym. Eng. Sci.*, **29**, 973–983.
Kenny, J., Apicella, A. & Nicolais, L. (1989b) *Makromol. Chem. – Macromol. Symp.*, **25**, 45–54.
Kenny, J. & Opalicki, M. (1993) *Makromol. Chem. – Macromol. Symp.*, **68**, 41–56.
Kenny, J., Opalicki, M. & Molina, G. (1995) *Society of Plastics Engineers Annual Proceedings ANTEC 1995*, pp. 2782–2789.
Kim, B., Chiba, T. & Inoue, T. (1993) *Polymer*, **34**, 2809–2815.
Kim, D. & Kim, S. (1994) *Polym. Eng. Sci.*, **34**, 625–631.
Kim, D. & Kim, S. (1995) *Polym. Eng. Sci.*, **35**, 564–576.
Kim, H. & Char, K. (2000) *Kor-Aust Rheol. J.*, **12**, 77–81.
Kim, Y. (1998) *J. Polym. Sci. A: Polym. Chem.*, **36**, 1685.
Kim, Y., Min, H. & Kim, S. (1999) *Polymer Processing Society Meeting*, Den Bosch (proceedings on CD).
Knauder, E., Kubla, C. & Poll, D. (1991) *Kunstoffe German Plastics*, **81**, 39.
Kogan, E., Kutseba, S. & Kulichikhin, V. (1988) *Khimicheskie Volokna* **3**, 36–37.

Kojima, C., Hushower, M. & Morris, V. (1986) *Society of Plastics Engineers Annual Technical Conference ANTEC 1986*, p. 344.
Kunz-Douglas, S., Beaumont, P. & Ashby, M. (1981) *J. Mater. Sci.*, **16**, 2657.
Kuroki, M., Takahashi, H. & Ishimuro, Y. (1999) *Nihon Reoroji Gakkaishi*, **27**, 235–241.
Lane, J. & Khattack, R. (1987) *Society of Plastics Engineers Annual Technical Conference ANTEC 1987*, p. 982.
Lee, D. & Han, C. (1987) *Polym. Eng. Sci.*, **27**, 955–963.
Lee, L., Marker, L. & Griffih, R. (1981) *Polym. Composites*, **2**, 209.
Lem, K. & Han, C. (1983a) *J. Appl. Polym. Sci.*, **28**, 779.
Lem, K. & Han, C. (1983b) *J. Appl. Polym. Sci.*, **28**, 3185.
Lem, K. & Han, C. (1983c) *J. Appl. Polym. Sci.*, **28**, 3207.
Lepez, O., Chopin, L. & Tanguy, P. (1990) *Polym. Eng. Sci.*, **30**, 821–828.
Le Pluart, L., Duchet, J., Sautereau, H., Halley, P. & Gerard, J. (2004) *Appl. Clay Sci.*, **25**, 207–219.
Macosko, C. (1985) *Brit. Polym. J.*, **17**, 239–245.
Macosko, C. & Miller, D. (1976) *Macromolecules*, **9**, 199–206.
Malkin, A. (1990) *Adv. Polym. Sci.*, **96**, 70–97.
Malkin, A. & Kulichikin, S. (1991) *Adv. Polym. Sci.*, **101**, 218–254.
Malmstrom, E., Johansson, M. & Hult, A. (1995) *Macromolecules*, **28**, 1698.
Martin, G., Tungare, A., Fuller, B. & Gorto, J. (1989) *Society of Plastics Engineers Annual Technical Conference ANTEC 1989*, p. 1079.
Martin, G., Tungare, A. & Gorto, J. (1989b) *Polym. Eng. Sci.*, **29**, 1279–1285.
Martinez, I., Martin, M., Eceiza, A., Oyanguren, P. & Mondragon, I. (2000) *Polymer*, **41**, 1027–1035.
Maurer, F. (1990) Interphase effects on viscoelastic properties of polymer composites, in Ishida, H. E. (Ed.) *Controlled Interphases in Composite Materials*, Amsterdam: Elsevier.
Mayadunne, R. T. A., Rizzardo, E. *et al.* (1999) *Macromolecules*, **32**, 6977–6980.
McGee, S. (1982) *Polym. Eng. Sci.*, **22**, 484–491.
Messersmith, P. & Giannelis, E. (1994) *Chem. Mater.*, **6**, 1719–1725.
Metzner, A. (1985) *J. Rheology*, **29**, 739.
Mitsuishi, M. & Kawasaki, H. (1996) Fillers surface modification, in *CRC Handbook on Polymer Science*, Boca Raton, FL: CRC Press.
Mussati, F. & Macosko, C. (1973) *Polym. Eng. Sci.*, **13**, 236.
Nakamura, Y., Tabata, H. & Suzuki, H. (1986) *J. Appl. Polym. Sci.*, **32**, 4865.
Nakamura, Y., Tabata, H. & Suzuki, H. (1987) *J. Appl. Polym. Sci.*, **33**, 885.
Newkome, G., Yao, Z., Baker, G. & Gupta, V. (1985) *J. Org. Chem.*, **50**, 2003.
Ng, H. & Manas-Zloczower, I. (1993) *Polym. Eng. Sci.*, **33**, 211–216.
Ohnaga, T., Chen, W. & Inoue, T. (1994) *Polymer*, **35**, 3774.
Pahl, M. & Hesekamp, D. (1993) *Appl. Rheology*, 70.
Parthasarathy, M. & Klingenberg, D. (1995a) *Rheologica Acta*, **34**, 417–429.
Parthasarathy, M. & Klingenberg, D. (1995b) *Rheologica Acta*, **34**, 430–439.
Pearson, R. & Lee, A. (1991) *J. Mater. Sci.*, **26**, 3828–3844.
Pearson, R. & Yee, A. (1989) *J. Mater. Sci.*, **24**, 2571.
Pearson, R. & Yee, A. (1992) *Polymer*, **34**, 1379.
Peng, W. & Riedl, B. (1994) *Polymer*, **35**, 1280–1286.
Petti, M. (1994) *7th International SAMPE Electronics Conference*, pp. 526–540.
Phillips, R., Brady, J. & Bossis, G. (1988) *Phys. Fluids*, **31**, 3462–3472.
Pichaud, S., Duteurtre, X., Fit, A. *et al.* (1999) *Polym. Int.*, **48**, 1205–1218.
Poncet, S., Boiteux, G., Pascault, J. *et al.* (1999) *Polymer*, **40**, 6811–6820.
Qian, J., Pearson, R., Dimone, V., Shaffer, O. & El-Aasser, M. (1997) *Polymer*, **38**, 21 30.
Quemada, D. (1978a) *Rheologica Acta*, **17**, 632–642.

Quemada, D. (1978b) *Rheologica Acta*, **17**, 643–653.
Ratna, D. (2001) *Polymer*, **42**, 4209–4218.
Riccardi, C. & Vazquez, A. (1989) *Polym. Eng. Sci.*, **9**, 120.
Richter, E. & Macosko, C. (1980) *Polym. Eng. Sci.*, **20**, 921–924.
Riew, C. & Gillham, J. (Eds.) (1984) *Rubber Modified Thermosets*, New York: American Chemical Society/Wiley.
Roller, M. (1986) *Polym. Eng. Sci.*, **26**, 432–440.
Rucker, D. & Bike, S. (1995) *Mater. Res. Soc. Symp. Proc.*, **385**, 173–178.
Ryan, M. & Kamal, M. (1976) *Proceedings of the VII International Congress on Rheology*.
Rydes, M. (1993) *Aspects of the Rheology of Unsaturated Polyester Dough Molding Compounds*, London: University of West London and National Physical Laboratories.
Saeki, J. & Kaneda, A. (1988) *Kobunshi Ronbunshu*, **45**, 691–697.
Sekutowski, D. (1996) Fillers and reinforcing agents, in *CRC Handbook on Polymer Science*, Boca Raton, FL: CRC Press.
Seymour, M., Karis, T. & Marshall, G. (1982) *Polym. Preprints*, **29**, 272–273.
Shang, S., Williams, J. & Soderholm, K. (1995) *J. Mater. Sci.*, **30**, 4323–4334.
Shaterzadeh, M., Gauthier, C., Gerard, J., Mai, C. & Perez, J. (1998) *Polym. Composites*, **19**, 655–666.
Spoelstra, A., Peters, G. & Meijer, H. (1996) *Polym. Eng. Sci.*, **36**, 2153–2162.
Sue, H. (1991) *Polym. Eng. Sci.*, **31**, 275.
Sue, H., Garcia-Meitin, E. & Pickelman, D. (1996) Fracture behaviour of rubber modified high performance epoxies, in Arends, C. E. (Ed.) *Polymer Toughening*, New York: Marcel Dekker.
Sun, L., Aklonis, J. & Salovey, R. (1993) *Polym. Eng. Sci.*, **33**, 1308–1319.
Sun, L., Park, M., Aklonis, J. & Salovey, R. (1992a) *Polym. Eng. Sci.*, **32**, 1418–1425.
Sun, L., Park, M., Salovey, R. & Aklonis, J. (1992b) *Polym. Eng. Sci.*, **32**, 777–786.
Sun, Y., Zhang, Z. & Wong, C. (2006) *IEEE Trans. Components and Packaging Technol.*, **29**, 190–197.
Sundstrom, D. & Burkett, S. (1981) *Polym. Eng. & Sci.*, **21**, 1108.
Tajima, Y. & Crozier, D. (1986) *Polym. Eng. Sci.*, **26**, 427–431.
Tillie, M., Lam, T. & Gerard, J. (1998) *Comp. Sci. Technol.*, **58**, 659–663.
Tomalia, D. (1985) *Polymer J.*, **17**, 117.
Tungare, A., Martin, G. & Gorto, J. (1988) *Polym. Eng. Sci.*, **28**, 1071–1075.
Utracki, L. & Fisa, B. (1982) *Polym. Composites*, **3**, 193–210.
Van Der Werff, J. & Dekruif, C. (1989) *J. Rheology*, **33**, 421–454.
Ving-tung, C., Boiteux G., Lachenal, G. & Chabert, B. (1996) *Polym. Composites*, **17**, 761–769.
Vinogradov, G. & Malkin, A. (1977) *Rheology of Polymers*, Moscow: Mir.
Vlassopoulos, D., Chira, I., Loppinet, B. & McGrail, P. (1998) *Rheologica Acta*, **37**, 614–623.
Walberer, J. & Mchugh, A. (2000) *J. Rheology*, **44**, 743–757.
Wang, M. & Pinnavaia, T. (1994) *Chem. Mater.* **6**, 468–474.
Watanabe, H., Yao, M., Yamagishi, A. *et al.* (1996) *Rheologica Acta*, **35**, 433–445.
Webster, O. (1991) *Science*, **251**, 887.
Wilkinson, S., Ward, T. & Mcgrath, J. (1993) *Polymer*, **34**, 870–884.
Woo, E. & Hseih, H. (1998) *Polymer*, **39**, 7–13.
Wooley, K., Frechet, J. & Hawker, C. (1994) *Polymer*, **35**, 4489.
Wu, H., Xu, J., Liu, Y. & Heiden, P. (1999) *J. Appl. Polym. Sci.*, **72**, 151.
Yamanaka, K. & Inoue, T. (1989) *Polymer*, **30**, 662.
Yang, Y. & Suspene, L. (1991) *Polym. Eng. Sci.*, **31**, 321.
Yilmazer, U. & Kalyon, D. (1989) *J. Rheology*, **33**, 1197–1212.
Zou, D., Derlich, V., Gandhi, K. *et al.* (1990) *J. Polym. Sci. A, Polym. Chem.*, **28**, 1909–1921.
Zou, D., Ma, S., Guan, R. *et al.* (1992a) *J. Polym. Sci. A, Polym. Chem.*, **30**, 137–144.
Zou, D., Sun, L., Aklonis, J. & Salovey, R. (1992b) *J. Polym. Sci. A, Polym. Chem.*, **30**, 1463–1475.

6 Industrial technologies, chemorheological modelling and process modelling for processing reactive polymers

6.1 Introduction

The aim of this chapter is to describe a range of industrial processing technologies for reactive polymer systems, and specifically to

- characterise the process and highlight important processing-quality-control tests, process variables and typical systems used,
- highlight applications of chemorheology in the process and
- examine the use of chemorheology in modelling of the production process.

In this way we will be bringing the concepts and understanding from all subsequent chapters into practical processing applications in order to aid acquisition of deeper understanding of these processes.

6.2 Casting

6.2.1 Process diagram and description

Casting is a relatively simple process (Figure 6.1) involving the pouring of a thermosetting liquid into a mould, where the liquid hardens into a solid, dimensionally-stable shape.

Examples of products include rod stocks, spheres, gears, bushings and complex moulded items. In casting applications structural properties such as hardness, toughness, dimensional stability and machinability are of most interest.

6.2.2 Quality-control tests and important process variables

Important process variables include

cost
viscosity
the reaction exotherm
shrinkage
pot life

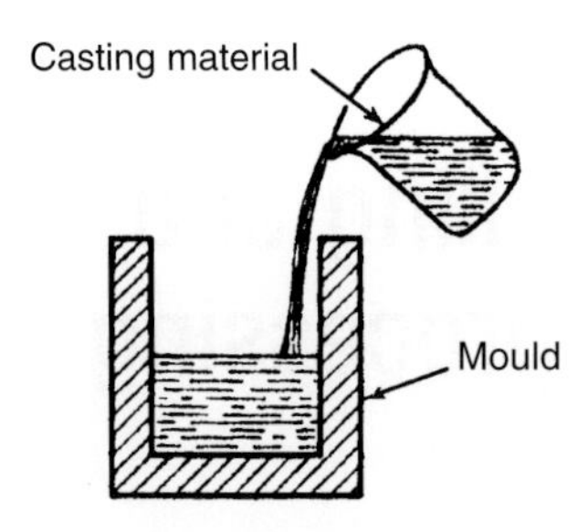

Figure 6.1. A schematic diagram of the casting process. Figure 8.1 (Lee and Neville, 1957) reproduced with permission of the McGraw-Hill Companies.

Cost is typically reduced by minimizing the amount of resin used (either via incorporation of fillers or modifiers, or by foaming). Viscosity is critical for casting operations, in which requirements concerning ease of processibility and large loadings of fillers need to be optimized. Exotherms typically pose processing problems for large casting masses since the cure reaction evolves >100 kJ/mol (Section 3.2.2), which may be reduced by incorporation of fillers, modifiers and the correct curing agent at an optimum concentration. Shrinkage, although not typically a problem with epoxy-resin systems, may be reduced by addition of a filler or modifier. Pot life (or the lifetime of resin in storage prior to use) can be extended by addition of fillers, but this will also increase cure times or involve higher cure temperatures, and thus must be optimized.

6.2.3 Typical systems

Typically epoxy resins are used for castings due to their low cost, low shrinkage, ease of rapid curing and good dimensional stability. Cast-foamed epoxies are also used in the aerospace industry, where light weight and high strength are required.

6.2.4 Chemorheological and process modelling

Mitani and Hamada (2005) achieved a three-dimensional flow simulation of epoxy-resin casting by using a controlled-volume finite-element method that builds on localized mass and energy balances. The simulation used experimental reaction kinetics and chemorheological input data from DSC and dynamic rheometry, respectively. The Kamal equation (as highlighted in Table 4.2 and shown in Equation (6.1a) below) was utilized to model the kinetics. By comparison with Equation (3.14) of Section 3.2.6, this is an adaptation of the general empirical equation for determination of the extent of reaction (conversion) from DSC heat flow during isothermal cure, which provides a measure of the instantaneous reaction rates. When combined with the power-law/Mackosko chemorheological model (as discussed in Table 4.4, extending from Equations (5.8) and (5.25) and shown in Equation (6.2)) this may be used to model the chemorheology profiles:

$$d\alpha/dt = (k_1 + k_2\alpha^{a_1})(1-\alpha)^{a_2}, \qquad (6.1a)$$

$$k_1 = A_1 \exp[-E_1/(RT)], \qquad (6.1b)$$

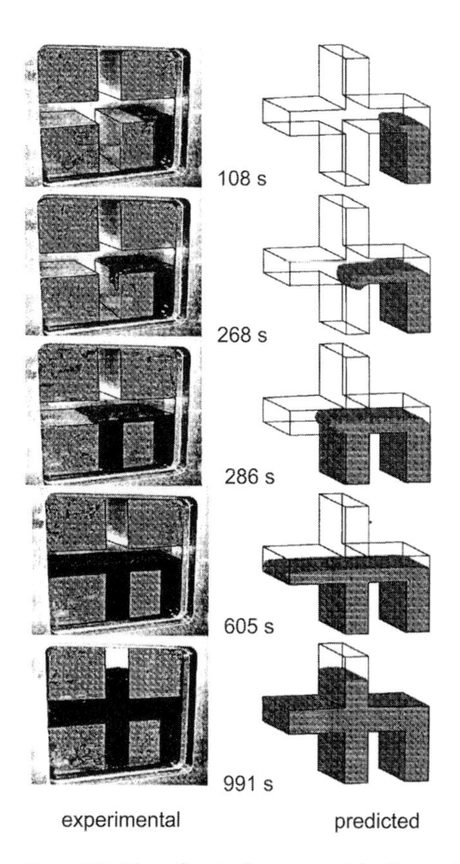

Figure 6.2. Flow-front advancement in the casting process – experimental and model prediction. Reproduced from Mitani and Hamada (2005). Copyright (2005). Reprinted with permission of John Wiley and Sons, Inc. Figure 13.

$$k_2 = A_2 \exp[-E_2/(RT)], \tag{6.1c}$$

$$\eta = B\gamma'^{-n} \exp(T_b/T)\left(\frac{a_{gel}}{a_{gel} - a}\right)^{\delta_1+\delta_2 a} \tag{6.2}$$

where a_1, a_2, E_1, E_2, A_1, A_2, B, n, T_b, a_{gel}, δ_1 and δ_2 are material constants, k_1 and k_2 are reaction rate constants, η is the viscosity, γ' is the shear rate, T is the temperature and a is the conversion. These empirical models predicted the kinetic and chemorheological data well over the processing conditions to be examined (Mitani and Hamada, 2005), but are not based on reaction mechanisms and are valid only insofar as they provide a fit to the data and have a predictive capability.

The model predicts flow-front advancement and temperature profiles during non-isothermal and reactive casting flow, giving good agreement with experimental results. Specifically, this is shown in Figure 6.2, where the flow-front advancement predicted by the simulation agrees well with the experimental verification.

Similar agreement between simulated and experimental temperature profiles is also seen in Mitani and Hamada (2005).

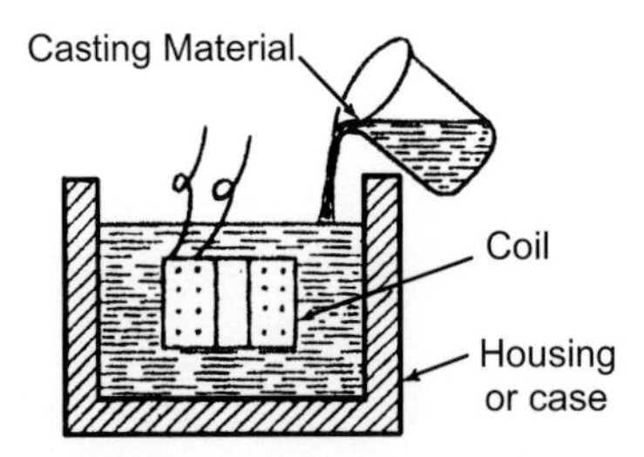

Figure 6.3. A schematic diagram of the potting process. Figure 8.1 (Lee and Neville, 1957) reproduced with permission of the McGraw-Hill Companies.

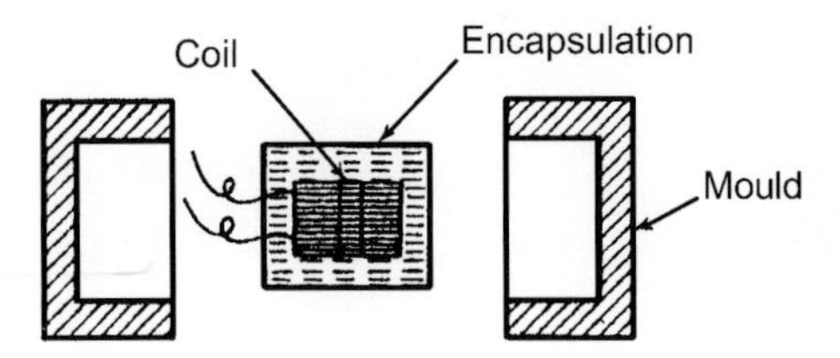

Figure 6.4. A schematic diagram of the encapsulation process. Figure 8.1 (Lee and Neville, 1957) reproduced with permission of the McGraw-Hill Companies.

Isotalo *et al.* (2004) recently added a three-dimensional structural-analysis model to a reactive-processing model for casting, to extend the predictive power of their model to cover filling, curing and post-cure operations.

6.3 Potting, encapsulation, sealing and foaming

6.3.1 Process diagram and description

Potting, encapsulation, sealing and foaming all involve the protection of components by the application of an external thermosetting resin. Potting is shown in Figure 6.3, where a housing or container is filled with a low-viscosity thermoset resin, which embeds and impregnates the internal coil or product.

Encapsulation is described in Figure 6.4, where an item is encapsulated with a shielding material via the application of external moulds.

Sealing as shown in Figure 6.5 involves sealing off a portion of a device to protect it from the external environment.

Foaming is a variant of potting and encapsulation that involves using chemical or physical blowing agents to produce foamed potting or encapsulation products to provide lightweight, rigid structures with good insulating properties.

In potting, encapsulation, sealing and foaming the most important properties are insulation properties, moisture resistance, impact resistance, chemical resistance and electrical resistance.

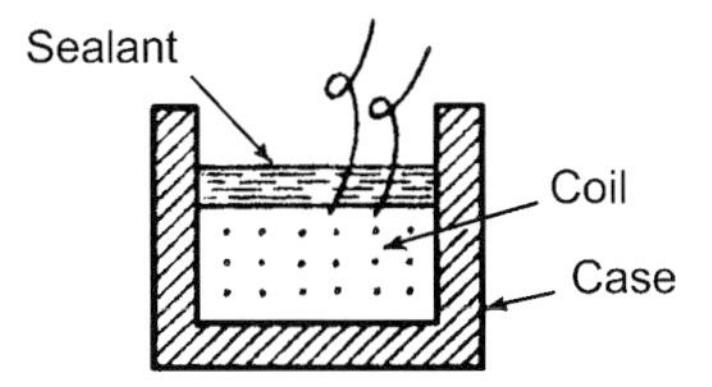

Figure 6.5. A schematic diagram of the sealing process. Figure 8.1 (Lee and Neville, 1957) reproduced with permission of the McGraw-Hill Companies.

6.3.2 Quality-control tests and important process variables

Important process variables include

cost
viscosity
the reaction exotherm
shrinkage
pot life
the coefficient of thermal expansion

Cost is typically reduced by minimizing the amount of resin used (either via incorporation of fillers or modifiers, or by foaming). Low viscosity is critical for potting operations for which a high degree of impregnation is required, and may be achieved via higher processing temperatures and addition of diluents. Exotherms typically pose processing problems for temperature-sensitive components that are to be potted, encapsulated or coated, and may be reduced by incorporation of fillers, modifiers and the correct curing agent. Shrinkage, although not typically a problem with epoxy-resin systems, may be reduced by addition of a filler or modifier. Shrinkage of epoxy-resin systems may, however, be sufficient to damage delicate components (Lee and Neville, 1957). Pot life (or the lifetime of resin from initial mixing to gelation) can be extended by addition of fillers, but this will also increase cure times or involve higher cure temperatures, and thus must be optimized. Typically DSC studies and rheological measurements are used to optimize pot life for each application. The coefficient of thermal expansion of unfilled epoxies is generally higher than those of most inserts used in potting and sealing applications, and is usually reduced via filler addition to match the insert values.

6.3.3 Typical systems

Owing to their barrier resistance, durability and low viscosities, epoxy resins are widely used in the potting industry in applications such as electrical coils for transformers and electrical motors. Epoxy pottings are able to withstand extreme environments and have shown resistance to fungal growth and environmental resistance in arctic, desert and tropical conditions (Lee and Neville, 1957).

Lightweight epoxy-resin foams have found applications as sandwich laminates in the aerospace industry, and can be produced by using chemical foaming agents (which liberate gas as a reaction by-product) or physical foaming agents (use of which involves including fillers containing a gas such as nitrogen).

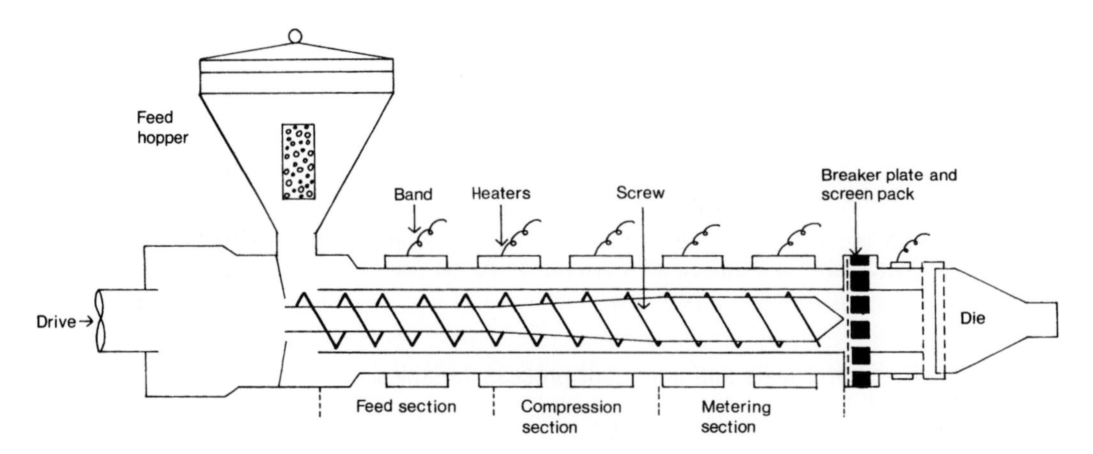

Figure 6.6. A schematic diagram of thermoset extrusion.

6.3.4 Chemorheological and process modelling

Bunyawanichakul *et al.* (2005) recently noted that potting and encapsulation products are still designed mainly using test results, and the lack of an efficient numerical model remains a problem for understanding potting and sandwich processing. However, they did develop a three-dimensional numerical simulation for predicting failure modes and product properties for potting processes. The lack of process modelling for potting and encapsulation processes was also noted in a paper by Murthy and Prasad (2006), which focusses on the importance of good-quality practices (rather than modelling) for potting and encapsulation processes in electronic assembly.

6.4 Thermoset extrusion

6.4.1 Extrusion

Process diagram and description

Extrusion involves mixing and forcing molten material via a rotating screw, which is heated by external barrel heaters and via friction, through a die to impart final profile shape. The screw (shown in Figure 6.6) fits snugly (with sufficient clearance to allow rotation yet minimize back flow) into a cylindrical barrel.

The polymer is fed into the feed hopper, which regulates flow into the screw. As the screw rotates, feed is drawn into the feed section and flows in the feed zone in the helical flow channel between the rotating screw and the barrel wall. Typically this channel depth decreases in the compression zone, which promotes increasing pressure in the extruder to promote mixing and expel air voids. In the final metering zone the channel depth is constant again, and the material is homogenized and supplied to the die zone at constant temperature and pressure to ensure consistent production. The die zone imparts the cross-section to the profile, screens out impurities and distributes the material evenly across the cross-section.

An interesting extension of thermoset extrusion is the use of reactive thermosetting solvents to process intractable engineering polymers. Many engineering polymers have very high glass-transition temperatures (200–250 °C) and thus require high processing temperatures (250–300 °C); however, typically these processing temperatures are equivalent to the thermal or oxidative degradation temperatures and thus processing is prohibitive. New work has focussed on mixing these engineering polymers with reactive solvents that reduce the processing temperature, undergo cure and phase separation, or phase invert after processing such that the reactive (thermoset) solvent is the dispersed phase of the engineering polymer.

Nelissen *et al.* (1992) examined the use of styrene monomer as a reactive solvent for poly (phenylene ether), finding good mixing even without extrusion, and it used to prepare high-T_g foams. Pearson and Yee (1993) and Bucknall and Partridge (1986) have shown that epoxy resins may also be classed as excellent reactive solvents for most of the common high-performance plastics, such as poly(ether sulfone), poly(ether ether ketone), poly(ether imide) and poly(phenylene ether). Venderbosch *et al.* (1994) highlighted effective mixing of PPE–epoxy-resin mixtures in static mixers and twin-screw extruders and use of the epoxy system as a reactive solvent (during processing) and phase-inversion aid (during curing) to produce continuous-phase PPE systems with good mechanical properties. This work was extended to PPE/epoxy/carbon-fibre composites (Venderbosch *et al.*, 1995) in static extrusion mixing and composite moulding processes, in which the epoxy regin also facilitated excellent wetting into the composite. Schut *et al.* (2003) examined the controlled extrusion of reaction-induced phase separation (RIPS) to process syndiotactic polystyrene/ epoxy-resin mixtures that are either thermoset- or thermoplastic-rich.

Quality-control tests and important process variables

Important process variables include

hopper feed control and rates
screw configuration
screw speed
screw temperature profile
die shape and profile

Typical systems

Typical systems are unsaturated polyesters, usually filled with mineral fillers and/or fibres. Newer systems such as thermoset–thermoplastics using reactive solvents for processing high-performance engineering polymers are being researched.

Process modelling

There have not been many papers directly focussing on extrusion of thermosetting polymers, since extrusion is usually used since a primary mixing procedure or a conveying stage to other processes such as thermoset injection moulding and other secondary shaping processes. However, there have been papers focussing on modelling the extrusion of thermoset–thermoplastic materials. As mentioned above, one of the key drivers of this work is the desire to utilize the thermoset phase as a reactive solvent to allow processing of intractable high-performance polymers and also to control phase separation and possibly phase inversion in order to achieve controlled thermoset–thermoplastic morphologies.

Riccardi *et al.* (1996) focussed on the modelling of phase separation in reactive thermoset–solvent high-performance-polymer mixtures by characterizing the interaction parameter as a function of conversion in a thermodynamic model. The reaction-induced phase-separation process (Section 1.3.2) is caused by the increase in the average molar mass of the polymer, superimposed on possible variations of the Flory–Huggins interaction parameter, χ, with conversion α. This secondary effect may favour mixing or de-mixing, depending on whether the interaction parameter decreases or increases, respectively, with conversion.

Girard-Reydet *et al.* (1998) highlight the use of small-angle X-ray scattering (SAXS), light transmission (LT) and light scattering (LS) to model phase structures in thermoset–thermoplastic phase-separation processes, and characterize the phase separation in terms of nucleation and growth or spinodal decomposition.

6.4.2 Pultrusion

Process diagram and description of process

Pultrusion is a continuous process involving pulling a collection of fibres on a creel system in the form of a roving, tow, mat or fabric through a resin bath (for impregnation) and then through a heated die to cure the resin and impart a constant cross-section to the product, Figure 6.7.

The die entrance is typically characterized by a tapered shape to remove the excess resin and make the ingress of material easier. A water cooling channel is placed in the first part of the die, to prevent premature solidification. The heat for polymerization is provided by heating platens placed on the top and bottom surfaces of the die. Outside the die, the cured composite material is pulled by a continuous pulling system (caterpillar or reciprocating pullers) and then travelling cut-off saws are programmed to cut the product into desired lengths. Since the predominant orientation direction is longitudinal, typical products are strong and stiff in tension and bending, but relatively poor in transverse properties (unless mats or fabrics with transverse reinforcement are employed).

Quality-control tests and important process variables

The important process-control parameters in pultrusion manufacturing are the pull speed, fibre volume fraction, viscosity, temperature settings for die heating zones, and the preform plate area ratio (compaction ratio). Key process variables include die pressure and temperature.

Typical systems

Pultrusion-grade resin matrices are available in a variety of systems such as polyester, epoxy, vinyl ester and phenolic resins. Unsaturated polyester resins are most commonly used because of the low heat input required with faster gelation than with other resin systems, although more recently carbon/epoxy-resin and fibreglass/epoxy-resin composite systems have found numerous applications in low-weight, high-strength aerospace and industrial applications. Almost all types of reinforcing materials can be used, including rovings, tows, mats, cloth and any hybrid of these. The most widely used reinforcing material is glass fibre, e.g. E-glass and S-glass fibres.

To produce pultruded products with consistent and high quality, it is important to tailor and control the pultrusion process. To achieve a uniform degree of cure in the cross-section of a product, control of the temperature profile inside the pultrusion die is an essential aspect. Also, to achieve consistent fibre wetting, controlled flow and pressure build-up must

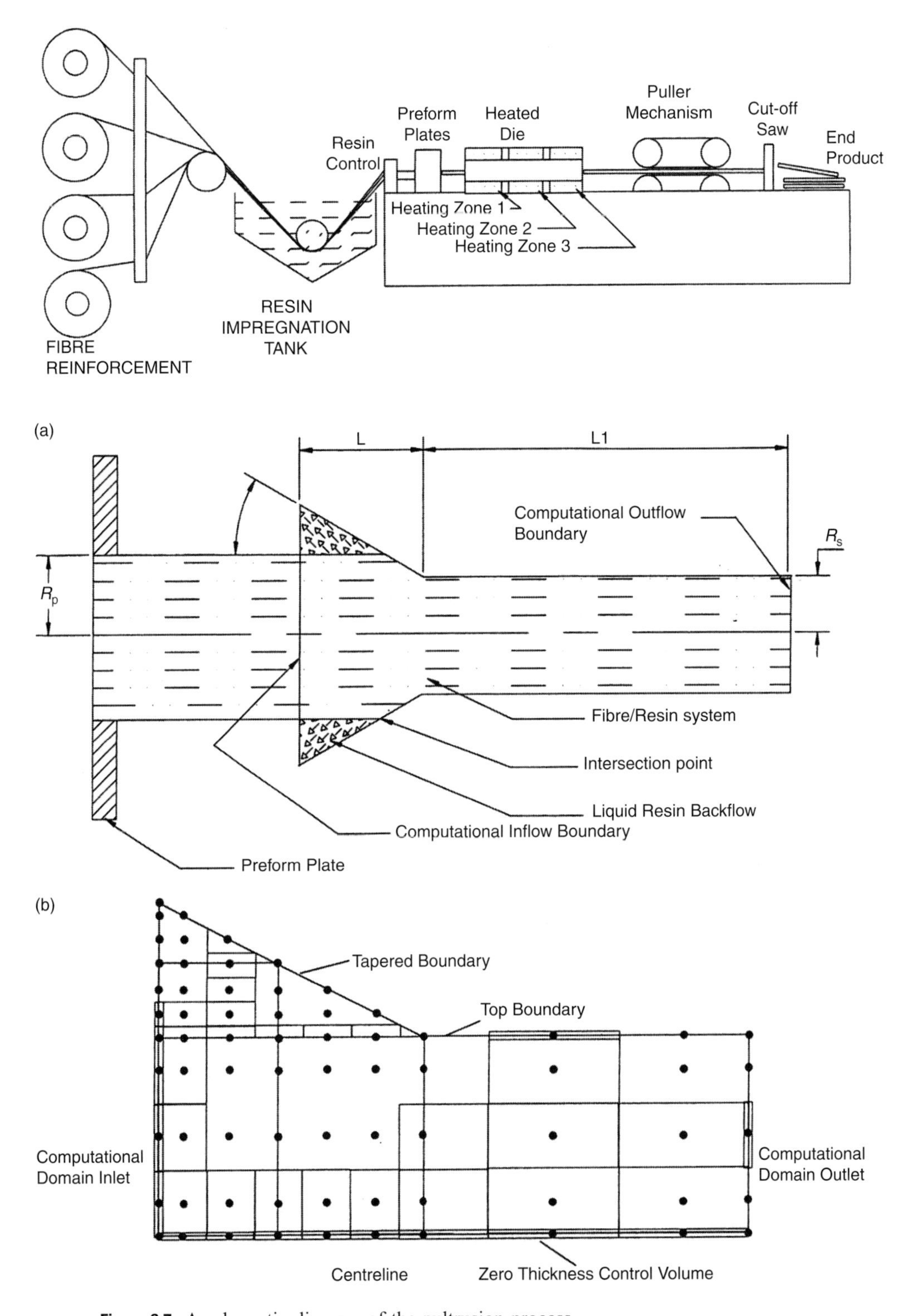

Figure 6.7. A schematic diagram of the pultrusion process.

be achieved in the die. It is therefore important to develop process models to simulate these key variables and the pultrusion process, and to optimize the process.

Chemorheological and process modelling

Cure modelling of polyester pultrusion systems was carried out by Ng and Manas Zloczower (1989). A mechanistic model that couples free-radical polymerization and diffusion control (Section 1.2.3) was used for the cure kinetics and is shown here:

$$\mathrm{d}a/\mathrm{d}t = 2fA_\mathrm{p}\exp[-E_\mathrm{p}/(RT)](I_0 - I)(1 - a)(1 - a/a_\mathrm{f})^n, \tag{6.3}$$

where a is the cure conversion, a_f is the final conversion, n is a pseudo reaction order, I_0 is the initial concentration of initiator, I is the instantaneous concentration of initiator, f is the initiator efficiency, A_p is the pre-exponential factor, E_p is the activation energy for the propagation reaction, R is the gas constant and T is the temperature. By combining this model with an energy balance of the pultrusion process, temperature and conversion profiles during pultrusion are predicted.

Initial modelling on predicting velocity profiles in pultrusion dies was carried out by Gorthala *et al.* (1994). Here a two-dimensional mathematical model in cylindrical co-ordinates with a control-volume-based finite-difference method was developed for resin flow, cure and heat transfer associated with the pultrusion process. Raper *et al.* (1999) and Gadam *et al.* (2000) highlight process models for pultrusion built on mass balances and Darcy's law for flow through porous media, to predict velocity and pressure fields in pultrusion dies. The pressure rise in the die inlet contributes to a major extent by enhancing fibre wet-out and suppressing void formation in the manufactured composite. In this work, permeability, fibre porosity and the chemoviscosity of the reactive resin must be coupled together to predict pressure and velocity profiles. Interestingly, Gadam *et al.* (2000) use an empirical chemoviscosity equation incorporating effects of temperature and cure level:

$$\eta(x) = \eta_\infty[E/(RT(x)) + ka(x)], \tag{6.4}$$

where $\eta(x)$ is the viscosity as a function of distance (x) in the die, η_∞, E, R and k are constants, T is the temperature and a is the cure level. In this model increases in the pull speed, fibre volume fraction, viscosity and preform plate area ratio all resulted in increased pressure rises, as expected from experimental results.

Three-dimensional analyses of heat transfer and cure in pultrusion of epoxy-resin composites have been examined by Chachad *et al.* (1995, 1996) and Liu *et al.* (2000). Carlone *et al.* (2006) review finite-difference and finite-element process models used for predicting heat transfer and cure in pultrusion. In this work they recommend the following empirical nth-order cure model for predicting cure kinetics of epoxy-resin composites, which is then coupled to the system's energy balance to predict thermal properties and cure conversion:

$$\mathrm{d}a/\mathrm{d}t = k_0\exp[-E/(RT)](1 - a)^n, \tag{6.5}$$

where k_0, E and n are constants, a is the cure conversion and T is the temperature. This model assumes simple nth-order kinetics and does not incorporate factors such as autocatalysis and similar reaction-dependent rates that are well known for epoxy resins from both thermal (Section 3.2.6) and spectroscopic (Sections 3.3.5 and 3.3.6) analysis. (Note that in this analysis it is also presumed that the resin fully wets the fibres and does not flow during the process – namely that is there is no chemoviscosity term in the analysis.)

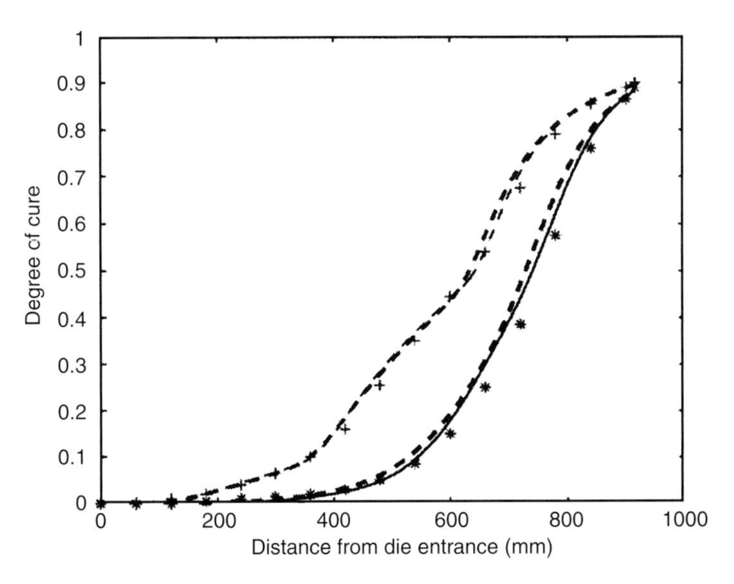

Figure 6.8. Experimental and model predictions of cure profiles in pultrusion.

This simplified kinetic model has also been used in other pultrusion modelling studies by Liu *et al.* (2000) and Valliappan *et al.* (1996).

Carlone *et al.* (2006) examined using finite-difference and finite-element discretization methods to model the process with respect to cure and thermal properties, and found very good agreement between these models and experimental data from the literature as found in Figure 6.8.

On closer inspection, relatively more accurate values for the temperature peak and mean degree of cure have been found using the finite-element model, which has provided an overestimated value of the standard degree of curc. The finite-difference model seems to provide underestimated values for all parameters considered; however, both analyses produce good agreement with experimental values.

6.5 Reactive extrusion

6.5.1 Process diagram and description

There are various processes that can be conducted via reactive extrusion (Figure 6.9):

polymerization
grafting
compatibilization and
controlled rheology

polymerization during reactive extrusion involves bulk polymerization of monomers into polymers to produce the fabricated article continuously from the extruder. All types of polymerization (radical polymerization, ionic polymerization, ring-opening polymerization and stepwise polymerization) have been attempted for a range of

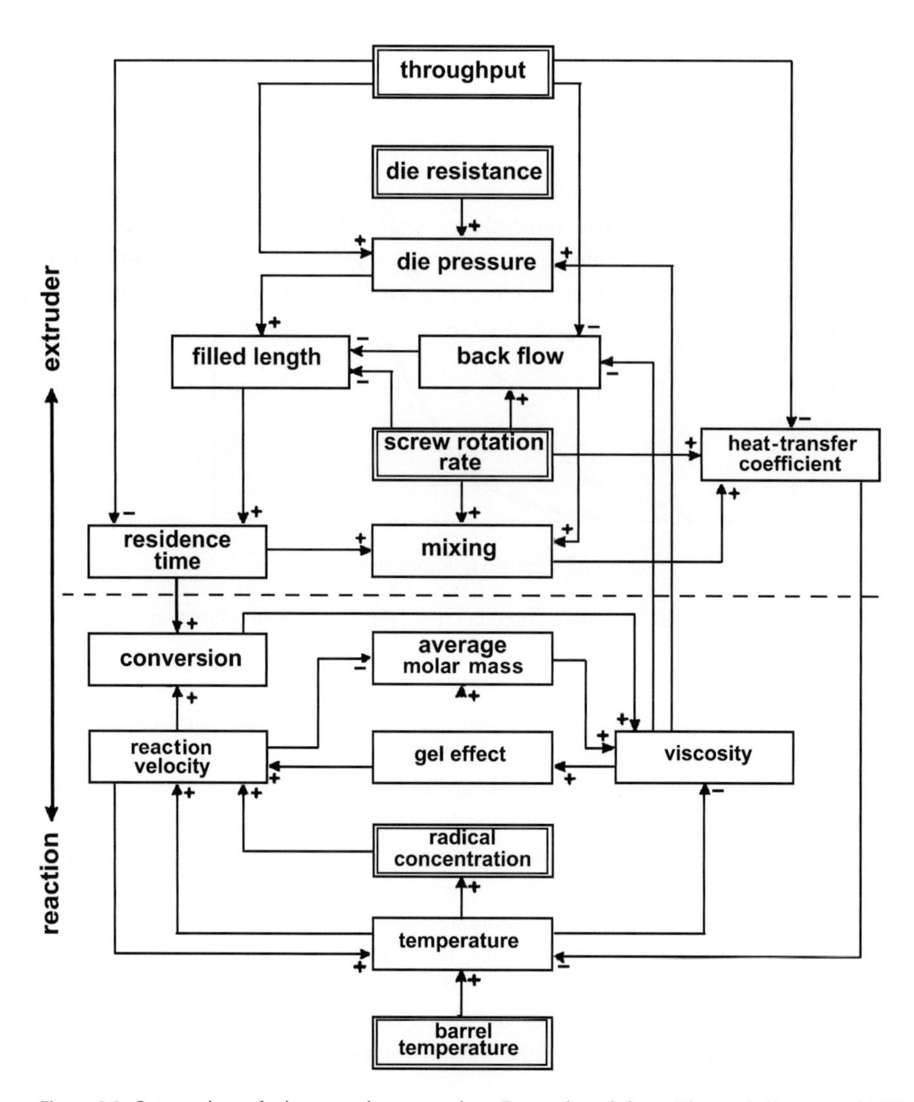

Figure 6.9. Interactions during reactive extrusion. Reproduced from Figure 3 (Janssen, 1998). Copyright (1994). Reprinted with permission of John Wiley and Sons, Inc.

polymer systems (such as methacrylates, polycarbonates, polyhydroxyalkanoates, styrene, lactam and lactide). The individual polymerization and copolymerization reactions, and their chemistry and kinetics, have been discussed in Chapter 1 in Sections 1.2.1 and 1.2.2 (stepwise and ring-opening polymerization) and 1.2.3 and 1.2.4 (ionic and free-radical polymerization). The attractiveness of polymerization during extrusion is that it allows high throughputs and control of molar mass; however, limitations to do with short residence times (i.e. about 5 minutes) and with the complex control needed for kinetics, heat transfer and mass transfer need to be addressed in each production process.

Polymer grafting involves introducing functional groups (typically via free-radical reactions) in to a matrix polymer. The types of molecular architecture achievable by grafting and the chemical processes were described in Section 1.2.4. Polymer grafting via extrusion has the advantages of there being no use of solvents, continuous processing and rapid preparation, and limitations arising from the high temperatures required, intimate mixing and side reactions of degradation, crosslinking and discolouration. The factors affecting polymer grafting are the mixing efficiency in the extruder, free-radical concentration, backbone-chain chemistry, extrusion processing conditions (geometry, pressure, temperature profiles, residence times) and by-product removal. Systems can be developed from monomer–polymer grafting or polymer–polymer grafting. Typical monomers include styrene and acrylates, and the final grafted polymers are useful for grafted block copolymers, compatibilizers, adhesion promoters, surface modifiers, nucleating agents, coupling agents between fillers and matrices and dispersing agents.

Reactive compatibilization processes in reactive extrusion generally involve reduction of interfacial tension, developing fine and stable morphological structure, stable reactions across phase boundaries and reactions that occur rapidly enough during the extrusion residence time. Typical systems are polymer alloys and blends that have morphologies optimized via reactive compatibilization with added fillers, interfacial agents, emulsifiers, coupling agents or block copolymers.

Controlled-rheology reactive extrusion typically involves developing high-melt-flow-rate (low-viscosity) polymers through controlled-molar-mass breakdown or degradation during extrusion. The chemistry of degradation reactions and their control are discussed in Section 1.4. This controlled-viscosity degradation involves addition of free-radical-generating agents (hydroperoxides or peroxides) or degradation catalysts, temperature-profile control and shear control. In practice masterbatching has been used in production to facilitate mixing of the peroxide (typically with an absorbant) with the polymer at high levels, and this masterbatch is then subsequently added to the base polymer. Of course, to maintain control over the viscosity drop, typically scavengers (such as nitroxyl radicals) are also added to control degradation effects. (The range of stabilizer strategies that can be used is given in Section 1.4.) Chain-branching chemicals have also been used to produce long-chain branched polyolefins. The most abundantly used systems for controlled-rheology reactive extrusion are high-melt-flow polypropylene polymers and their blends (Moad, 1999b).

6.5.2 Quality-control tests and important process variables

Important material and process variables for reactive extrusion have been summarized by Janssen (1998) and are

conversion times and reaction rates
screw-rotation rate
filled length
barrel-temperature profile
screw profile
back flow
throughput rate
die resistancc
residence time and mixing

6.5.3 Typical systems

A large range of polymers can be obtained by polymerization in reactive extrusion, and the techniques involved and polymers produced include the following (Fink, 2005):

radical polymerization, poly(styrene), poly(butyl methacrylate);
ring-opening polymerization, poly(lactide);
anionic polymerization, poly(styrene), styrene–butadiene copolymer, polyamide-12;
metathesis polymerization, poly(octenylene);
grafting; poly(ε-caprolactone)-grafted starch, starch and poly(acrylamide), poly(propylene) wood-flour composites, poly(styrene)-grafted starch.

In reactive polymerization it is important to control extrusion conditions (extruder profiles, temperature profiles, residence times) as well as having a good understanding of the polymerization kinetics, thermodynamics and properties (i.e. reaction rates, conversion times, heats of polymerization, ceiling termperatures of the polymerization reactions, viscosity (M_w) changes). All of the processes discussed in Section 1.2 are brought to bear in the environment of the extruder.

Moad (1999a, 1999b) reviewed the chemistry of polyolefin graft polymerization. The following classifications were used:

free-radical-induced grafting of unsaturated monomers
end-functional polyolefins via the ene reaction
hydrosilation
carbene insertion
transformation of pending functional groups (transesterifcation, alcoholysis)

Moad also notes that the most common grafting modifications made to polyolefins are via maleic anhydride, maleate esters, styrene, maleimides, acrylates and their esters, and vinyl silanes. Other polymer systems (Fink, 2005) that undergo grafting are polystyrene/maleic anhydride (useful for PA6/PS blends), PVC/butylmethacrylate (for improved processibility), PET/nadic anhydride, starch/vinyl acetate and starch/methyl acrylate (for improved water resistance).

Reactive compatibilization of blends and filled systems is common, and Table 6.1 highlights typical systems and their compatibilizers.

Here there are many compatibilized systems, which can be loosely grouped into the following categories:

compatibilization by additives (such as inorganic fillers, modified fillers, modified nanolays, modified rubber particles)
block copolymers
coupling agents (such as peroxides, diamines, epoxies, anhydrides)
ionomeric compatibilizers (as in PEVOH and PETG blends)
end-functional polymers (as produced via the iniferter (initiator-transfer-termination) method, reversible chain termination and atom-transfer radical polymerization]

Common peroxide degradants and stabilizers used in controlled-rheology reactive extrusion are shown in Table 17.2 and 17.8 of Fink (2005).

The predominant system for controlle-rheology reactive extrusion is polypropylene (PP) with systems such as ultrahigh-melt-flow polypropylene, high-impact polypropylene blends

Table 6.1. Reactive compatabilization

Blend	Compatibilizer
LDPE/PA6	Diethylsuccinate, glycidal methacrylate, ethylene–acrylic acid copolymer, polyethylenc-g-malcic acid
LLDPE/PET	Ethylene–propylene copolymer-g-metacrynol carbamate
LDPE/starch	Poly(ethylene-co-glycidal methacrylate)
HDPE/HIPS	Styrene/ethylene butylene/styrene block copolymer
HDPE/PA12	Maleic anhydride
PP/PS	Styrene–butadiene triblock copolymer, styrene–ethylene–propylene copolymer
PP/PA6	ε-caprolactam and maleic anhydride grafted poly(propylene)
PA6/PPO	Styrene–maleic anhydride copolymer
PBT/EVA	Maleic anhydride
PC/PVDF	Copolymer of methyl methacrylate and acrylic acid
PE/wood flour	Maleated and acrylic acid grafted polyethylene
PET/PA6	Acylic modified polyolefin ionomer

(with PP homopolymer, ethylene–propylene rubber, and ethylene-rich ethylene–propylene copolymer phases) and long-chain branched polypropylene (in which branching reactions are increased over β-scission) having been examined.

6.5.4 Chemorheological and process modelling

Janssen (1998) examines the interactions during reactive extrusion and summarizes them in the following interaction diagram

Here clear interactions between (process) extruder parameters and (material) reaction parameters are shown. Janssen (1998) also described three instabilities in reactive extrusion, namely thermal instabilities (runaway thermal conditions due to reaction effects and thermal inhomogeneities), hydrodynamic instabilities (unstable flow conditions at intermediate conversions) and chemical instabilities (auto-catalytic gel effects, inhibiting-reaction ceiling temperatures and phase-separation effects under conditions of high shearing). De Graff *et al.* (1997) develops a model to predict the residence-time distribution during reactive extrusion of grafting polystyrene onto starch polymers. In this mode the residence-time distribution during twin-screw extrusion is modelled by combining the mass balance along the extruder and die with the reaction kinetics of the grafting reaction. Verification experiments were performed with residence-time-distribution tests and gave good agreement with the model.

Maier and Lambla (1995) model the conversion of a nonylphenyl ethoxylate (NP8) onto a pre-maleated ethylene–propylene rubber (EPR-ma) along the reactive extruder by using a kinetics model derived from batch experiments and a residence-time-distribution model. Mathematically this model is given by the equation:

$$C_c = \int_0^\infty C_c(t)E(t)\,dt, \tag{6.6}$$

where C_c is the average conversion, $C_c(t)$ is the conversion at time t (from batch kinetics) and $E(t)$ is the residence-time distribution (from mass and force balances on the extruder). Chen *et al.* (1996) used the same process model as Equation (6.6) to model the conversion

Table 6.2. Reactive extrusion systems

Paper	System	Model	Outputs
Zagal *et al.* (2005)	Methyl methacrylate	Polymerization model (second-order thermal self-initiation) coupled with flow and energy equations	Pressure, temperature, residence time, filling ratio, number-average, and weight-average molar masses (M_n and M_w) along the extruder length can be calculated
Berzin *et al.* (2006), Vergnes and Berzin (2004), Vergnes *et al.* (1998)	Controlled degradation of PP	One-dimensional global model of polymer flow in co-rotating, self-wiping, twin-screw extruders, coupled with degradation reaction kinetics	Temperature, pressure, shear rate, viscosity, residence time
Dhavalikar and Xanthos (2003, 2004)	Chemical modification of polyethylene terephthalate (PET)	Kinetic model and chemorheological model coupled with qualitative process model	Effects of stoichiometry, temperature, rate of shear, chemical composition and molar mass (M_w) on torque changes, residual carboxyl content and insoluble content.
Zhu *et al.* (2005)	Polymerization of caprolactone	Review of one- and three-dimensional process models coupled with kinetics and chemorheological models	Predictions of mixing mechanism, heat generation and heat loss are better from three-dimensional model
Choulak *et al.* (2004)	Polymerization of caprolactone	A one-dimensional dynamic model of a twin-screw reactive extrusion	Predicts pressure, filling ratio, temperature, molar conversion profiles and residence-time distributions under various operating conditions
Kim and White (2004)	Polymerization of poly-ε-caprolactone	Engineering model linking residence-time-model, shear-effects model and reaction conversion during extrusion	Reaction conversion and M_w from extruder via correlating specific mechanical energy and M_w
Semsarzadeh *et al.* (2004), Puaux *et al.* (2006)	Poly(urethane-isocyanurate) (PUIR)	Coupled kinetic model with axial dispersion residence-time-distribution model.	Conversion and residence-time distributions are predicted
Roy and Lawal (2004)	Polypropylene and polystyrene reactive-extrusion systems	Two-dimensional isothermal single-screw process model combining kinetics	Concentration profiles at various axial locations in the extruder are presented for various

Table 6.2. (*cont.*)

Paper	System	Model	Outputs
		(first order) and rheology (power law) models	values of power-law index, flow rate and the homogeneous and heterogeneous reaction rate parameters
Oliveira *et al.* (2003)	Degradation of polypropylene	Kinetic modeling of PP degradation in reactive extrusion	Final degree of degradation

profile during extrusion via reaction- and residence-time-distribution modelling for monoesterification of a styrene–maleic anhydride polymer with an alcohol.

More recent work has extended the modelling of reactive extrusion processes to full-scale process modelling by the combination of kinetic, chemorheological and flow/energy-balance modelling. This work is summarized in Table 6.2, highlighting the system used, the model description and the expected model outputs. Full references are provided for readers with further interest in these models.

6.6 Moulding processes

6.6.1 Open-mould processes

Process diagram and description

Open-mould processes are thermoset moulding processes with open moulds in the male or female form. In this process the mould is made from a pattern and the reinforced thermoset is applied to the mould to reproduce the shape and finish of the mould. Typical examples would be boat hulls made from fibreglass-reinforced gel coats. Figure 6.10 shows the basic contact-moulding process.

Here the mould surface is polished and treated with mould-release agent to promote release of the finished part from the mould. Then a thermoset resin (usually termed a gel coat and typically a thermosetting polyester resin with pigments) is applied via brush or spray to the mould surface, typically to thicknesses of about 0.5–1.0 mm. This coating process controls surface features of the final product. When the gel coating layer has set, a layer of fine glass surfacing tissue is applied, followed by layers of chopped strand mat (CSM). Additional layers of gel coat resin and woven fibre mat can be applied sequentially, to ensure that the mat is wetted out and impregnated by the resin and allowed to gel (but not cure fully) before the next layer is applied. When the final layer is applied the whole laminate is fully cured (typically at room temperature).

Quality-control tests and important process variables

Important material and process variables include

gel-coat pot life
layup time

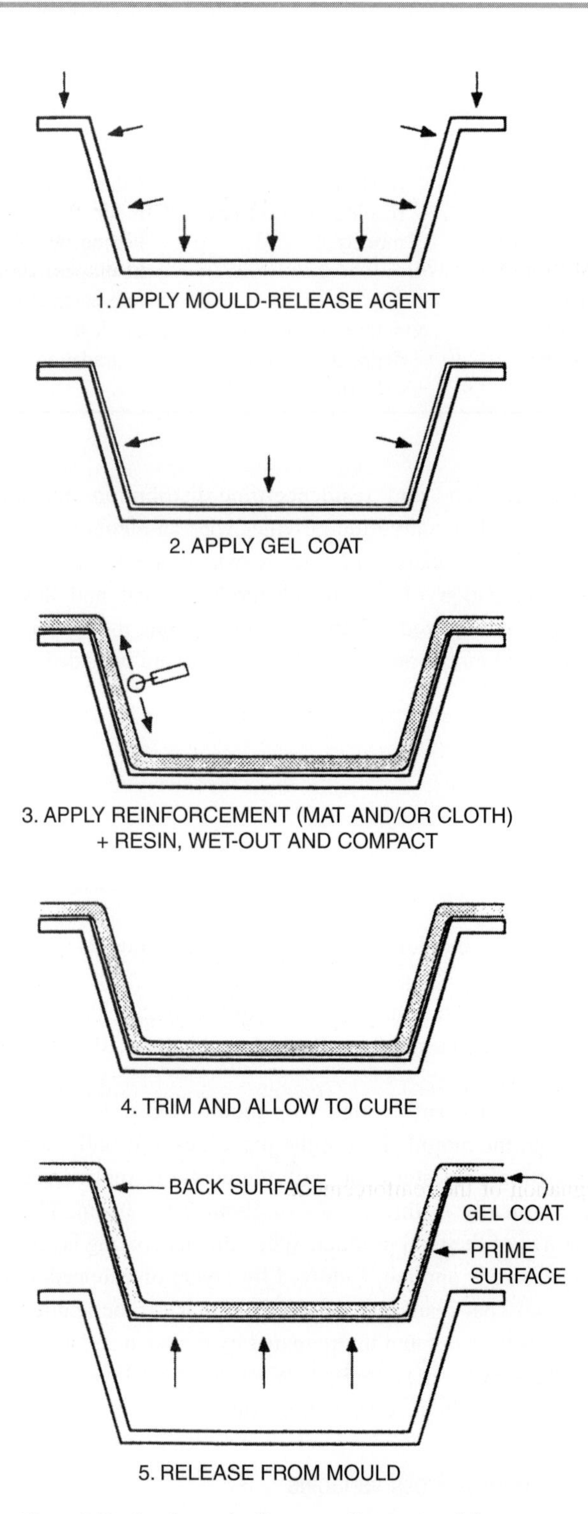

Figure 6.10. A schematic diagram of open moulding.

the gel-coating process
cure time
mould-cleaning time

The process is very labour intensive, but survives because it is adaptable to many mould designs and is useful for making relatively large mouldings (since there is no size limitation), which would be cost-prohibitive with other processes (such as injection moulding). Advances in processing include

pre-impregnating fibre reinforcement mats with resin
the sprayup process of spraying gel coat with chopped fibres
preheating of moulds
infrared heaters
vacuum-assisted consolidation of layers

Typical systems

Typical systems include room-temperature-curable polyester resins, epoxy resins and vinyl ester resins, and glass-fibre and mat reinforcement.

Chemorheological and process modelling

Open-mould processes have traditionally developed through trial-and-error processing and have not been modelled formally, unlike the similar case of process modelling in the metal industry, as reviewed by Thomas (2002).

6.6.2 Resin-transfer moulding

Process diagram and description

In resin-transfer moulding (RTM) there is typically a two-part mould, which forms a cavity corresponding to the final product shape. In the process, as shown in Figure 6.11, reinforcement, typically being a woven roving, cloth or mat, is placed (dry) in the female cavity, and then the male half is placed into position.

The liquid resin is then injected into the mould, typically from the bottom mould, assisted by pressure and vacuum to aid in providing an even resin distribution and achieving the removal of voids. A key processing consideration here is that the resin flow should match the wetting out and impregnation of the reinforcement.

Quality-control tests and important process variables

Important material and processing variables include

the resin's pot life
the resin's chemoviscosity profile
the vacuum applied
the resin's flow rate
mould temperatures and heating

Typical systems (thermoset, filled thermoset)

Typically vinyl ester, epoxy-resin and polyester systems with cure promoters are used at room temperature or under heated conditions.

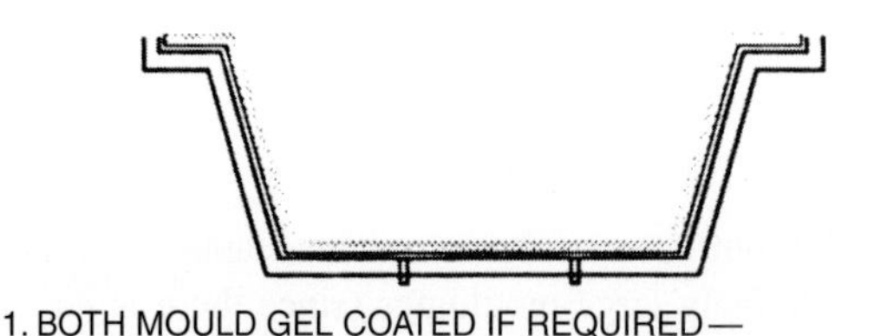

1. BOTH MOULD GEL COATED IF REQUIRED—
REINFORCEMENT PLACED DRY IN FEMALE MOULD

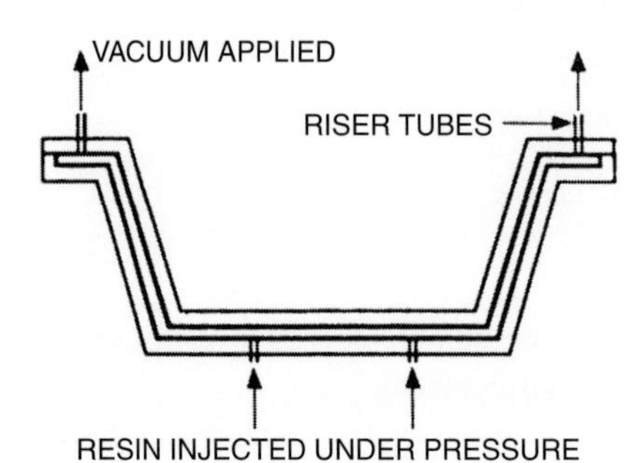

RISER TUBES

2. MOULD HALVES CLOSED—RESIN INJECTED AT BOTTOM—
VACUUM APPLIED AT TOP—RISER TUBES PINCHED OFF
AS RESIN RISES AND IS ALLOWED TO CURE—
THEN MOULD OPENED AND PART RELEASED

Figure 6.11. A schematic diagram of resin-transfer moulding.

Chemorheological and process modelling

Cure kinetics and chemorheological models are key for characterizing and modelling RTM processes. For example, Lobo (1992) utilized the following kinetic and chemorheological models for a vinyl ester RTM resin. The kinetic model is given by

$$\mathrm{d}a/\mathrm{d}t = k(1-a)^n, \tag{6.7}$$

$$k = A\exp[-E_a/(RT)], \tag{6.8}$$

where a is the conversion, n is the reaction order, k is the reaction rate, A is the pre-exponential factor, E_a is the activation energy and T is the absolute temperature. The chemorheological model (on extending from Macosko's equations, (5.8) and (5.25)) is given by

$$\eta = \frac{B\exp(T_b/T)}{1+(\eta_0\gamma'/\tau^*)^{1-n}}[a_g/(a_g-a)]^{C_1+C_2a}, \tag{6.9}$$

where η is the viscosity, T is the absolute temperature, a is the conversion, a_g is the conversion at gelation, and C_1, C_2, η_0, τ^*, B and T_b are constants. One can see that these are phenomenological models based on nth-order kinetics and a combined Arrenhius-shear-thinning–Macosko-cure chemorheological model (as originally presented in Table 4.4). Often the Kamal cure model (refer to Equation (6.10)) combining nth-order and auto-catalytic kinetics is utilized for RTM resins (Kamal and Sourour, 1973):

$$\mathrm{d}a/\mathrm{d}t = (k_1+k_2a^m)(1-a)^n, \tag{6.10}$$

$$k_1 = A_1 \exp[-B_1/(RT)], \tag{6.11}$$

$$k_2 = A_2 \exp[-B_2/(RT)], \tag{6.12}$$

where a is the conversion, T is the temperature, R, A_1, A_2, B_1 and B_2 are constants, and m and n are reaction orders.

Process modelling of RTM polyesters was examined by Kenny *et al.* (1990), who identified the following (auto-catalytic) cure kinetic and (empirical Castro–Macosko (Castro and Macosko 1982) model, as highlighted in (Equation 5.8)) chemorheological models for use in their process simulation:

$$\mathrm{d}a/\mathrm{d}t = k(a_\mathrm{f} - a)^n a^m, \tag{6.13}$$

$$\eta = K_\eta \exp[E_\eta/(RT)][a_\mathrm{f}/(a_\mathrm{f} - a)]^{A+Ba}, \tag{6.14}$$

where a is the cure conversion, t is time, k is a constant, a_f is the final conversion, m and n are constants, K_η is the viscosity constant, E_η is the activation energy for viscosity, R is the gas constant, T is temperature, and A and B are constants. In this work these models are incorporated into a flow model to predict degree-of-cure profiles and experimental temperature profiles in various moulds.

Of course, in RTM process modelling one must combine the above kinetic and chemoviscosity models into mass, momentum and energy balances within a flow simulation. Specifically, the momentum balance must combine any flows induced by pressure and any flows into porous media (as characterized by Darcy's law). Simple one-dimensional RTM flow modelling and two- and three-dimensional RTM simulations have been summarized by Rudd *et al.* (1997) and show the importance of kinetic, rheological and permeability coefficients to the simulation of pressure-profile and flow-front predictions.

6.6.3 Compression, SMC, DMC and BMC moulding

Process diagram and description

Compression, SMC (sheet moulding compound), DMC (dough moulding compound) and BMC (bulk moulding compound) moulding differ from RTM, in that the feed material typically contains the reinforcement, rather than having the reinforcement mat placed inside the mould. A simple process diagram is shown in Figure 6.12.

The terminology is related to the feedstock material, in that SMCs are typically in the form of sheets 5–6 mm thick, DMCs are softer 20–50-mm pelletized feedstocks and BMCs are more meterable 20–50-mm pelletized feedstocks. Compression moulding differs from SMC, DMC and BMC moulding in that charge is not allowed to spill out of the mould, and that the mould holds constant (typically higher) pressure. The feed materials typically consist of pre-blended thermosetting resin, reinforcement and filler. The matched metal dies are opened and a predetermined charge of material is placed into position. The moulds are closed under pressure, which forces the charge throughout the mould cavity, and the charge cures in the heated moulds. The moulds are opened and the cured product is ejected. This moulding-cycle time is of the order of minutes.

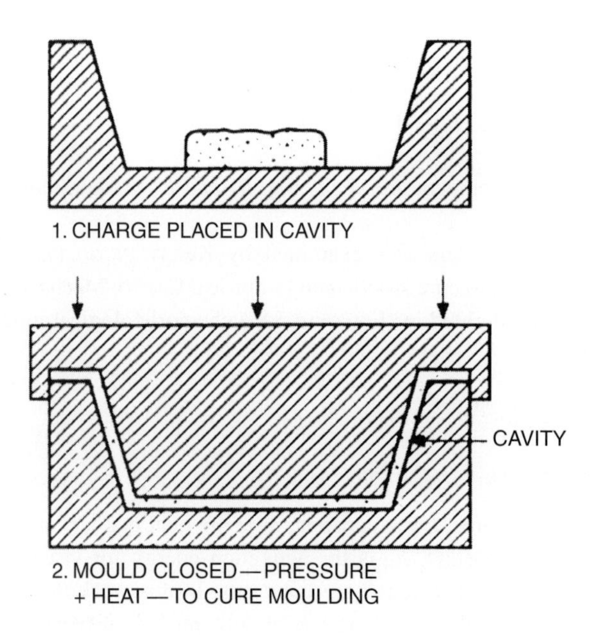

Figure 6.12. A schematic diagram of compression moulding.

Quality-control tests and important process variables

Important material and process variables are

- the charge distribution in the mould
- the charge's pot life
- moulding time
- moulding pressure profile
- moulding temperature profile
- cycle time

Typical systems

Typical systems are polyester and phenolic resins filled with glass fibres and mineral fillers.

Chemorheological and process modelling

The kinematics of flow in compression moulding of SMCs has been examined by Barone and Caulk (1985), who used flow visualization with alternating coloured sheets in compression moulding to find that SMCs deform in uniform extension within individual layers with slip at the mould wall. Additionally, at lower compression speeds there is interlayer flow.

Kau and Hagerman (1986) experimentally mapped pressure and temperature profiles for SMC compression moulding and noted the presence of two clear flow and cure regions, together with a correlation with lubrication flow.

Rheological properties of mineral-filled and mineral/glass-fibre-filled unsaturated polyester DMCs have been presented by Gandhi and Burns (1976), who found that a simple power law was useful to characterize the chemorheology of both systems (note that limited cure effects during compressional flow are assumed). This work was extended by a series of excellent papers by Han and Lem (Han and Lem, 1983a, 1983b, 1983c, Lem and Han, 1983a, 1983b, 1983c), who used the following expressions for the cure kinetics (autocatalytic equation) and chemorheology (shear-thickening) profiles:

$$\mathrm{d}a/\mathrm{d}t = (k_1 + k_2 a^m)(1 - a)^n, \tag{6.15}$$

$$N_1 = A\sigma^{n_1}, \tag{6.16}$$

where a is the cure conversion, k_1 and k_2 are rate constants, m and n are rate orders, N_1 is the normal-stress difference, A and n_1 are power-law contants and σ is the shear stress. Note that these equations presume minimal cure effects on viscosity profiles during shearing, and, although effects of fillers, additives and filler surface treatments of process flows were examined, no formalization of a process model was made. However, in later work Han and Lee (Han and Lee, 1987, Lee and Han, 1987) extended their chemorheological modelling (as first presented in Equation (5.29)) to incorporate cure effects by writing the following WLF-based equation:

$$\ln \eta = \ln \eta_{T_g} - a(T - T_g)/(C_1 + T - T_g), \tag{6.17}$$

where η is the viscosity, η_{T_g} is a constant, a is a free-volume parameter, T is the temperature, T_g is the glass-transition temperature and C_1 is a constant. Note that in this equation T_g is a function of cure conversion (a) and hence must be coupled to the kinetic equation (like in Equation (6.15)).

Lee *et al.* (1981) highlighted the importance of shear and elongational flows for SMC polyesters in flow modelling. They described the following rheological models:

$$\eta_e = 3\eta/[(1 + \lambda\varepsilon)(1 - 2\lambda\varepsilon)] \tag{6.18}$$

and

$$(\eta - \eta_\infty)(\eta_0 - \eta_\infty) = [1 + (\lambda\gamma)^2]^{(n-1)/2}, \tag{6.19}$$

where η_e is the elongation viscosity, ε is the elongation rate, λ is a relaxation time, η is the shear viscosity, η_∞ is the infinite-shear viscosity, η_0 is the zero-shear-rate viscosity, γ is the shear rate and n is a power-law index. It is also assumed that η_0, λ and n obey an Arrhenius relationship with temperature; however, it is assumed that there is no cure during flow. In this work the predominant flow is shown to be biaxial elongational flow when lubricated moulds are used. Non-isothermal and isothermal flows are modelled.

6.6.4 Transfer moulding

Process diagram and description

Transfer moulding is a variant of compression moulding in which the injection of the resin is controlled by a transfer ram, as shown in Figure 6.13.

The process involves a mould cavity and a transfer cavity. The filled charge is initially placed in the transfer cavity, and then heated until it has softened. Then pressure is applied to the ram, causing the charge to flow through the transfer port to the mould cavity. Excess charge is employed to allow a hold pressure to be applied to the mould cavity and to accommodate sample shrinkage during cure. The pressure is maintained during cure and released when the material has gelled.

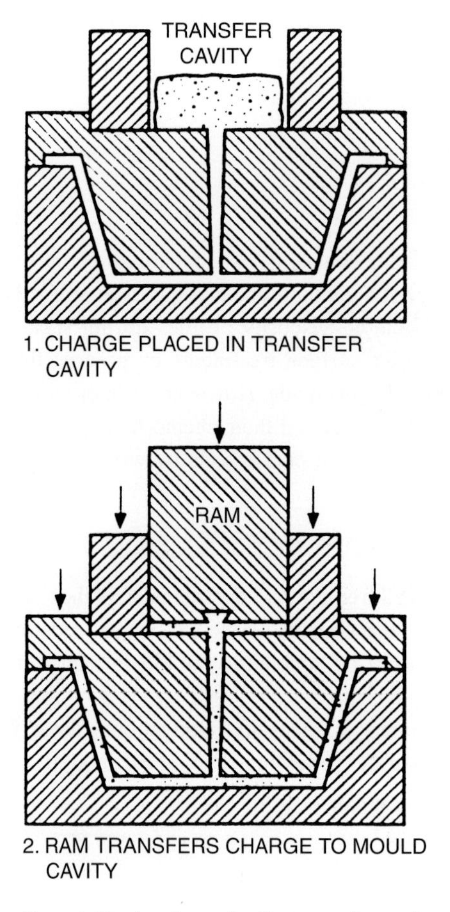

Figure 6.13. A schematic diagram of transfer moulding.

Quality-control tests and important process variables

Important material and process variables are

the resin's cure kinetics
the resin's chemovsicosity
the resin's gel time
transfer-ram injection speed
temperature profiles of transfer pot and mould
hold (cure) time in the mould

Typical systems (thermoset, filled thermoset)

Typical systems are epoxy resins (typically epoxy-novolac systems) with silica fillers, hardeners, catalyst and rubber modifiers used in integrated chip packaging. This process is well illustrated by Figure 6.14.

Here the transfer-moulding process where by the epoxy–silica charge is transferred to the mould to encapsulate the integrated circuit is shown. Note that in this case the fragile integrated-circuit pattern is placed inside the heated mould (typically at about 170–180 °C)

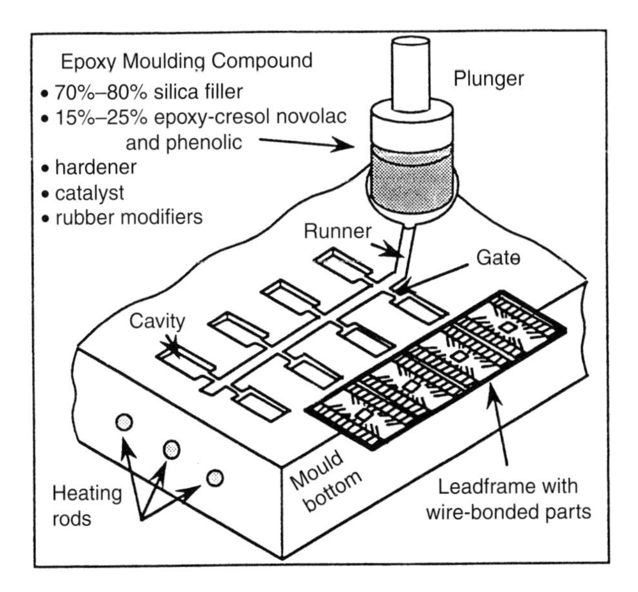

Figure 6.14. Transfer moulding of computer-chip packaging systems Figure 1 from (Nguyen, 1993) © 1993 IEEE.

prior to transfer of the filled epoxy-resin encapsulant. The filled epoxy-resin is heated (to about 80–90 °C) before being injected into the mould. This injection is conducted in such a way that the viscosity is at a minimum value (e.g. the temperature is high enough to melt the filled resin and reduce the viscosity due to thermal effects, but not high enough to induce high viscosities from cure effects) such that minimal damage is done to the integrated circuitry.

Chemorheological and process modelling

Manzione *et al.* (1988) presented an empirical model for transfer moulding of filled epoxy-resin systems for integrated-circuit encapsulation. The following chemorheological model (combining temperature, shear and cure effects) was used to aid flow-balancing calculations:

$$\eta = \eta_\infty \exp[E_\eta/(RT)]\gamma^{1-n}[a_{\text{gel}}/(a_{\text{gel}} - a)], \tag{6.20}$$

where η is the viscosity, η_∞ is a constant, E_η is the activation energy of flow, R is the gas constant, T is the temperature, γ is the shear rate, n is a power-law index, a_{gel} is the gel conversion and a is the conversion. This has also been highlighted in Table 4.4 and extends from the general form of the Ryan model in Equation (5.31).

Nguyen *et al.* (1992, 1993) highlighted a full process model for transfer moulding of highly filled epoxies in integrated-circuit encapsulation. This model used the following kinetic and chemorheological models:

$$\mathrm{d}a/\mathrm{d}t = (k_1 + k_2 a^{m_1})(1 - a)^{m_2}, \tag{6.21}$$

with $k_1 = A_1 \exp(-E_1/T)$ and $k_2 = A_2 \exp(-E_2/T)$, where a is the conversion, k_1 and k_2 are rate constants, m_1 and m_2 are reaction orders, A_1 and A_2 are reaction rate constants, and E_1 and E_2 are reaction rate energies, and

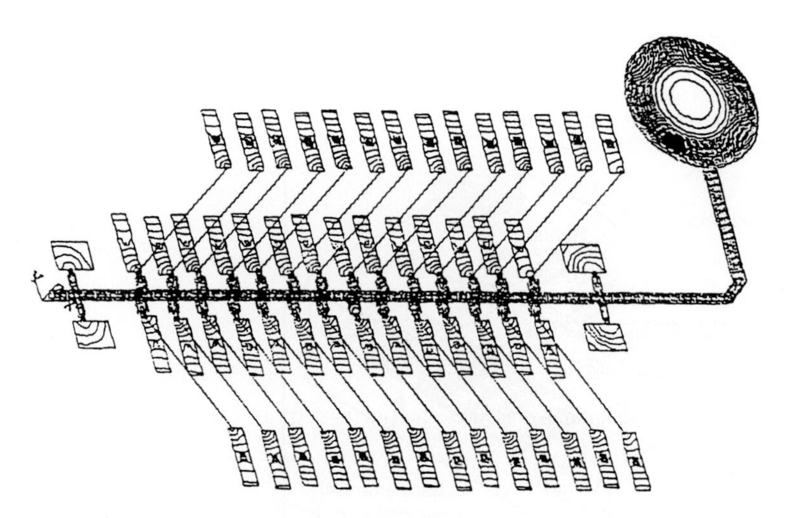

Figure 6.15. Modelling transfer moulding of computer-chip packaging systems. Adapted Figure 7 from (Nguyen, 1993) © 1993 IEEE.

$$\eta(T,\gamma,a) = \eta_0(T)/[I + \eta_0(T)\gamma/\tau^*]^{1-n}[a_{\text{gel}}/(a_{\text{gel}} - a)]^{C_1+C_2a}, \tag{6.22}$$

where $\eta(T, \gamma, a)$ is the viscosity as a function of temperature (T), shear rate (γ) and cure conversion (a), $\eta_0(T)$ is the zero-shear viscosity (which is a (usually Arrhenius-type) function of temperature), τ^* is a critical shear stress, a_{gel} is the gel conversion, a is the conversion and C_1 and C_2 are constants.

These semi-empirical kinetic and chemorheological models are combined with a finite-element flow model (incorporating mass, momentum and energy balances) to predict flow and process problems such as wire sweep (breakage of the intricate integrated-circuit wiring in the transfer mould during processing due to high flow viscosities). A typical process simulation of filling a multiple-cavity mould is shown in Figure 6.15

Turng and Wang (1993) also highlighted the simulation of microelectronics encapsulation using epoxy-resin moulding compounds using a semi-empirical kinetic model (as previously used in Equation (6.21)) and the following chemorheological model:

$$\eta = B\exp(T_b/T)[a_{\text{gel}}/(a_{\text{gel}} - a)]^{C_1+C_2a}, \tag{6.23}$$

where η is the viscosity, B and T_b are constants, a_{gel} is the gel conversion, a is the conversion and C_1 and C_2 are constants. Similarity to the Macosko model (Equations (5.8) and (5.25)) is evident here.

6.6.5 Reaction injection moulding

Process diagram and description

Reaction injection moulding (RIM) is a low-pressure process that allows two reactive streams, A and B say, to meet and react to form a cured polymer in the mould. Figure 6.16 shows the process in a simplified form.

Here two liquid streams, typically streams of a polyol and isocyanate for polyurethane RIM, are mixed in metered (usually stoichiometric) proportions as they are injected into the mould cavity. There are well-controlled, separate delivery systems to ensure that both

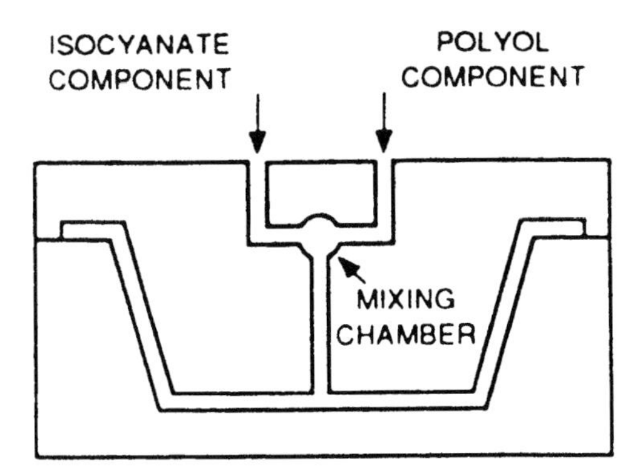

Figure 6.16. A schematic diagram of reaction injection moulding (RIM).

streams are delivered, in correct proportions, to the mould. The runner system is also designed to ensure that the two components impinge and mix in a series of turbulent mixing chambers prior to flow into the mould cavity. The materials then cure rapidly at room temperature

Quality-control tests and important process variables

Important process and material variables include

the resin's chemistry and cure kinetics
the resin's chemoviscosity
premixing-chamber design
mould temperature profile

Typical systems (thermoset, filled thermoset)

Typical systems are two-part polyurethane systems that may produce rigid, rubbery, foamed or filled products. Typical fillers include chopped fibres and mineral fillers. Composite systems may also be produced by having pre-placed reinforcements in the mould, in which case the technique is known as reinforced reactive injection moulding (RRIM).

Chemorheological and process modelling

Haagh *et al.* (1996) modelled the filling stage of a RIM process. They used the following equations to describe the cure kinetics and chemorheology:

$$\mathrm{d}a/\mathrm{d}t = k_1 \exp[-E_\mathrm{a}/(RT)](1-a)^n \tag{6.24}$$

and

$$\eta = \begin{cases} \eta_0 a_T b_T [1 + (\lambda\gamma')^2]^{(n_1-1)/2} & \text{for } a < 1\% \text{ (Carreau model)} \\ \eta_0 a_T b_T{}^{n_1-1}\gamma' & \text{for } a > 1\% \text{ (power-law model),} \end{cases} \tag{6.25}$$

where

$$a_T = \exp\left(\frac{-c_1(T-T_0)}{c_2+T-T_0}\right) \tag{6.25a}$$

(a)

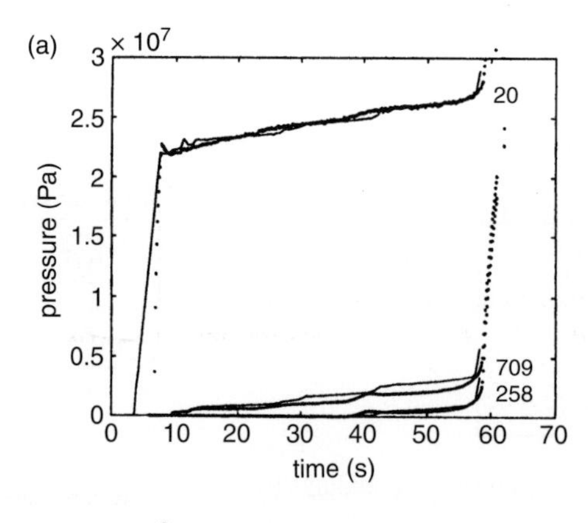

(b)

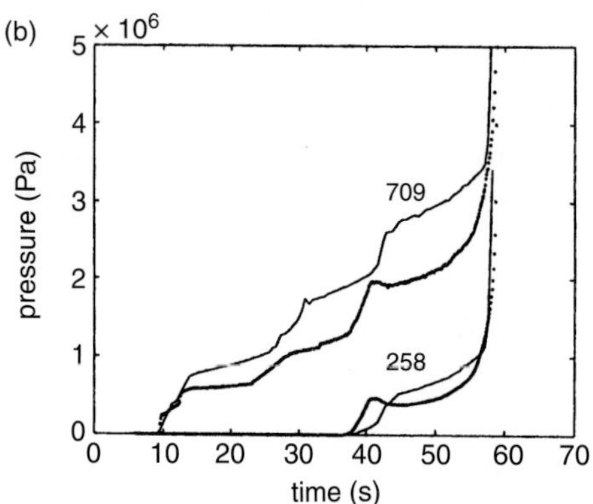

Figure 6.17. Agreement between simulation and experiment for RIM mould filling of an insulator.

$$B_T = \exp(a_T + b_T), \tag{6.25b}$$

$$\eta_0 = \eta_0(\alpha) \quad \text{and} \quad n_1 = n_1(\alpha), \tag{6.25c}$$

where α is the conversion, t is the time, E_a is the activation energy of cure, R is the gas constant, T is the temperature, n is the reaction order, η is the viscosity, η_0 is the zero-shear viscosity, a_T and b_T are thermal parameters, λ is a relaxation time, γ' is the shear rate, n_1 is the power-law coefficient and C_1, C_2 and b are constants. These kinetic and chemorheological models are used in a flow-filling simulation to predict flow-front flow, velocity profiles and pressure profiles for a reactive EVA copolymer. Here very good agreement with simulation and experiment for mould filling of an insulator is found, and shown in terms of pressure profiles in Figure 6.17.

Macosko (1989) examined process modelling for RIM processes. Interestingly, the kinetic models examined here are pseudo-mechanistic, like the following for polyurethane RIM processing:

$$\begin{aligned} -\mathrm{d}[\mathrm{NCO}]/\mathrm{d}t = {} & A_1 \exp(-E_1/T)[\mathrm{C}]^a[\mathrm{NCO}]^b[\mathrm{H}]^c \\ & + A_2 \exp(-E_2/T)[\mathrm{H}]^d, \end{aligned} \tag{6.26}$$

where [NCO] is the isocyanate concentration, [C] is the catalyst concentration, [H] is the hydrogen concentration, t is the time, A_1, E_1, A_2 and E_2 are energy parameters, and a, b, c

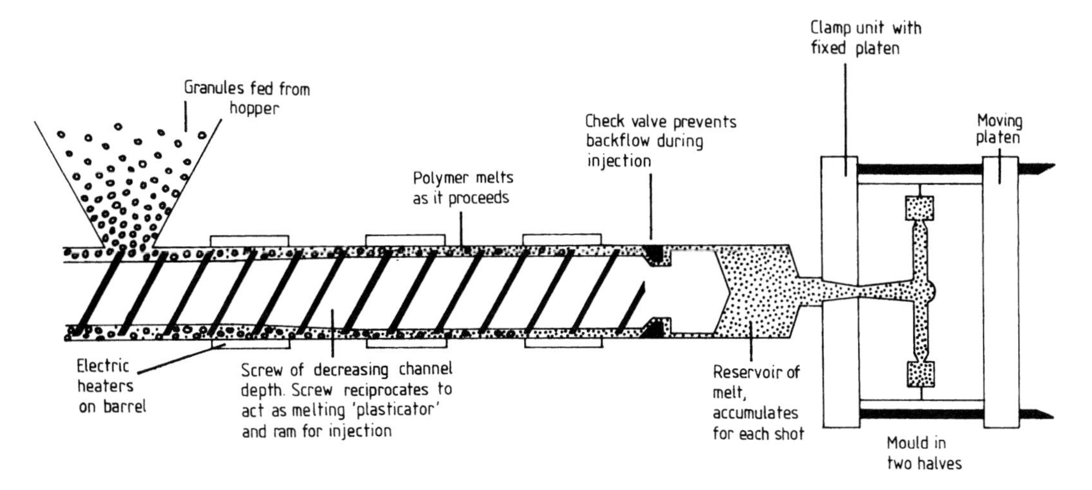

Figure 6.18. A schematic diagram of thermoset injection-moulding.

and d are constants. Macosko (1989) highlights the use of the following general chemorheological models in a RIM mould-filling simulation:

$$\eta = \eta_0 a_g / (a_g - a)^{C_1 + C_2 a} \tag{6.27}$$

or

$$\eta = \eta_0 M_w / M_{w_0}^a, \tag{6.28}$$

with $\eta_0 = A_n \exp[E_\eta/(RT)]$, where η is the viscosity, A_η and E_η are parameters, R is the gas constant, T is the temperature, M_w is the molecular weight (molar mass) at conversion a, M_{w_0} is the original molecular weight, a_g is the gel conversion and a is the conversion.

6.6.6 Thermoset injection moulding

Process diagram and description

Thermoset injection moulding is a high-pressure process typically used for DMC or BMC polyester materials, and is shown in Figure 6.18.

The BMC materials are fed from a hopper into the plasticization section of a relatively low-temperature (to minimize cure) single-screw extruder, where the material is mixed, metered and sheared into a homogenous mixture and collected in an injection cavity is the end of the screw. Once this cavity has been filled the entire screw is hydraulically plunged forwards and the material in the extruder cavity is injected into the mould under high pressures (typically 15 000–30 000 psi). The mould is at high temperature to facilitate curing of the part in the mould.

Quality-control tests and important process variables

Important material and process variables are

the resin's cure kinetics
the resin's chemoviscosity profile
the resin's gel properties

the screw element's profile
the screw's temperature profile
the screw processing conditions
injection speed
the mould's temperature profile
the mould cure time

Typical systems (thermoset, filled thermoset)

Typical systems include BMC and DMC polyester systems, which are usually filled with long (>1 mm) glass fibres.

Chemorheological and process modelling

Gibson and Williamson (1985b, 1985a) examined the shear and extensional flow of BMC materials in injection moulding. Extensional and shear viscosities are found to be important in injection moulding, and here they use the following power-law relationships:

$$\eta = B\gamma' n - 1, \tag{6.29}$$

$$\lambda = A\varepsilon^{n-1}, \tag{6.30}$$

where η is the shear viscosity, λ is the elongation viscosity and n is the power-law index. Note that A and B are constants for isothermal flow or Arrhenius functions (i.e. $A = A_0 \exp[-H/(RT)]$ and $B = B_0 \exp[-H/(RT)]$) for non-isothermal flow. Interestingly here, even though thermal and shear rates are considered, no effect of cure is considered.

Kamal and Ryan (1980) presented a review of early kinetic and chemorheological models of thermoset materials for injection moulding. They highlight the following model:

$$\mathrm{d}a/\mathrm{d}t = (k_1 + k_2 a^m)(1 - a)^n, \tag{6.31}$$

$$\eta = \eta(T, \gamma', a), \tag{6.32}$$

where a is the cure conversion, k_1 and k_2 are constants, m and n are reaction orders, η is the chemoviscosity and $\eta(T, \gamma, a)$ is a generic expression for η as a function of temperature T, shear rate γ' and conversion a. Examples of simple flow filling models for injection moulding were also presented.

Knauder *et al.* (1991) highlighted the following kinetic and chemorheolgical model for injection-moulded epoxy resins:

$$\mathrm{d}a/\mathrm{d}t = k_0 \exp[-E_a/(RT)](1 - a)^n, \tag{6.33}$$

$$\ln \eta = \ln \eta_\infty + E_{a1}/(RT) + k_\infty \int_0^{t^*} (1 - a)^n \exp[-E_{a2}/(RT)]\mathrm{d}t, \tag{6.34}$$

where a is the conversion, t is the time, t^* is the end-of-reaction time, k_0 is the rate constant, E_a is the activation time, R is the gas constant, T is the temperature, n is the reaction order, η is the viscosity and η_∞, E_{a1}, E_{a2} and k_∞ are constants. Here the reaction is of nth order and the chemoviscosity profile combines an Arrhenius model with a cure model (but no shear rate effects are considered) as has been discussed in Section 5.2.1. The model provides local and time-dependent temperature profiles, shear rates and shear-stress and pressure profiles in the mould.

Shen (1990) reviews the flow modelling of injection moulding of thermosets. The kinetic model was the Kamal model (as shown above in Equation (6.31)) and the chemoviscosity model is given by a modified WLF model for low shear rates (as shown by Equation (6.35)) and a modified Cross model for high shear (as shown by Equation (6.36)), namely

$$\ln(\eta/\eta_g) = -C_1(T - T_g)/(C_2 + T - T_g), \tag{6.35}$$

with $T_g = T_g(a)$, $C_1 = C_1(T)$ and $C_2 = C_2(T)$, where η is the viscosity, a is the conversion, T is the temperature, η_g is the viscosity at T_g, T_g is the glass-transition temperature at a, and C_1 and C_2 are functions of temperature; or

$$\eta(a, \gamma', T) = \eta_0/[1 + (\eta_0\gamma'/\tau^*)^{1-n}], \tag{6.36}$$

with $n = n(a)$ and $\tau^* = \tau^*(a)$, where η is the viscosity, a is the conversion, T is the temperature, η_0 is the zero shear viscosity, γ' is the shear rate, τ^* is a yield-stress term and n is the Cross-model index.

6.6.7 Press moulding (prepreg)

Process diagram and description

Press moulding is used for making flat panels from prepreg (pre-impregnated) sheets. The laminate is compressed between a pair of matching heated platens at relatively low pressure to a fixed thickness. Excess resin is allowed to run off from the platens to enable laminates of correct fibre content to be produced. The prepreg is typically covered with a sheet of porous release film and bleeder (absorbent) cloths to absorb excess resin. Application of pressure is typically carried out at gelation due to the likelihood of problems if it is applied earlier (excess resin loss) or later (no resin flow through the prepreg). Owing to obvious wastage problems this process is cost-effective only with high-value applications.

Quality-control tests and important process variables

Important material and process variables include

the resin's cure chemistry
the resin's chemoviscosity
mould pressure and timing

Typical systems (thermoset, filled thermoset)

Typical systems include polyester/glass-fibre sheeting.

Chemorheological and process modelling

Hou (1986) presented a resin-flow model for long-fibre-reinforced-composite prepreg lamination processing. The model includes the following chemorheological model:

$$\eta = \eta_0 \exp(kt), \tag{6.37}$$

with

$$\eta_0 = \eta_\infty \exp[E_\eta/(RT)], \tag{6.38}$$

$$k = k_\infty \exp[-E_k/(RT)], \tag{6.39}$$

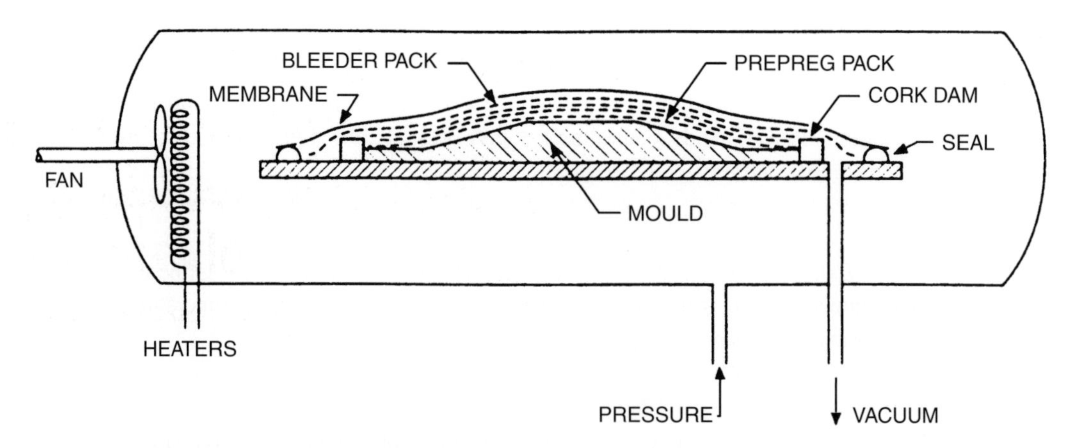

Figure 6.19. A schematic diagram of the autoclave process.

where η is the viscosity, η_0 is the zero-time viscosity, k is the viscosity rate constant, E_η is the activation energy for viscosity and E_k is the activation energy for viscous rate rise. The similarity to the Arrhenius models in Section 5.2.1 is evident.

Overall the model combines mass and force balances incorporating the chemorheological model (above), considerations for viscous flow and flow through fibre bundles, to simulate loss of resin during prepreging, the fibre volume fraction, the pressure distribution, temperature effects and part dimensions. Verification experiments show that the model predicts experimental results to within 8%.

6.6.8 Autoclave moulding (prepreg)

Process diagram and description

Autoclave moulding is typically used in the aerospace industry for the production of high-value composites from prepregs. The laminate, which is covered on both sides by a fine polyester cloth peel-ply (for enhancing the surface effect), is built up on the mould surface. The top surface of the laminate is covered by a porous release film and bleeder cloth. The whole assembly is then covered with a non-porous membrane, which is sealed to the mould, and then placed inside an autoclave as shown in Figure 6.19.

A vacuum is then drawn inside the cover membrane (to remove volatiles and porosity) while the pressure and temperature inside the autoclave are separately controlled (to provide even control of pressure across the surface, and thermal control of cure). Optimization of the application of pressure and vacuum will prevent the formation of dry laminates (which typically occurs when pressure is applied too early and low-viscosity resin is forced out) and porous laminates (when pressure is applied too late with high-viscosity cured resins). In general, consistent mouldings of high quality can be produced, but the process is slow and capital-intensive.

Quality-control tests and important process variables

Important material variables are

resin cure
the resin's chemoviscosity
the resin's gel point

autoclave pressure and temperature profiles
the vacuum applied
cure time

Typical systems (thermoset, filled thermoset)

Typical systems include epoxy carbon-fibre and epoxy glass-fibre prepregs for computer and aerospace applications.

Chemorheological and process modelling

Blest *et al.* (1999) examine the modelling and simulation of resin flow, heat transfer and curing of multilayer composite laminates during autoclave processing. An empirical cure equation is used:

$$\mathrm{d}a/\mathrm{d}t = \begin{cases} (C_1 + C_2 a)(1 - a)(0.47 - a) & \text{for } a < 0.3 \\ C_3(1 - a) \quad \text{for } a > 0.3, \end{cases} \tag{6.40}$$

where a is the conversion, t is time and C_1, C_2 and C_3 are constants. The resin's viscosity is modelled via the expression

$$\eta = \eta_\infty \exp[U/(RT) + \chi a], \tag{6.41}$$

where η is the viscosity, η_∞ is a constant, R is the gas constant, T is the temperature, χ is a temperature-dependent parameter and a is the conversion. However, for the purposes of the flow model it is found that the resin's viscosity is relatively constant during initial flow (since filling is faster than cure), enabling a simpler combined-flow model to be used. Temperature and cure profiles are well predicted using this flow model. Castro (1992) highlights wider implications for simpler flow-modelling techniques that decouple filling from reaction stages for thermoset flow modelling because in practical flow processes filling is much faster than cure reactions.

6.7 Rubber mixing and processing

Rubbers or elastomers are materials that are vulcanized or irreversibly cured through chemical crosslinking to a lightly crosslinked network polymer (see Sections 1.1.4 and 1.4.1). Vulcanization is typically carried out with heating and a curing agent such as sulfur. The elastomeric systems used (and discussed in the following section) range from natural to synthetic filled systems.

Elastomers can be prepared and processed using a variety of mixing and processing equipment, similarly to thermoplastic materials, and these will be addressed in this section.

6.7.1 Rubber mixing processes

Process diagram and description

Rubber mixing processes are used to produce an elastomeric compound with ingredients sufficiently incorporated, dispersed, distributed and plasticized that it will be able to be processed and will cure consistently in subsequent secondary processing.

Internal batch mixers are batch systems that have both a lower-shear homogenizing region and a localized high-shearing section. Refer to the schematic disgram in Figure 6.20.

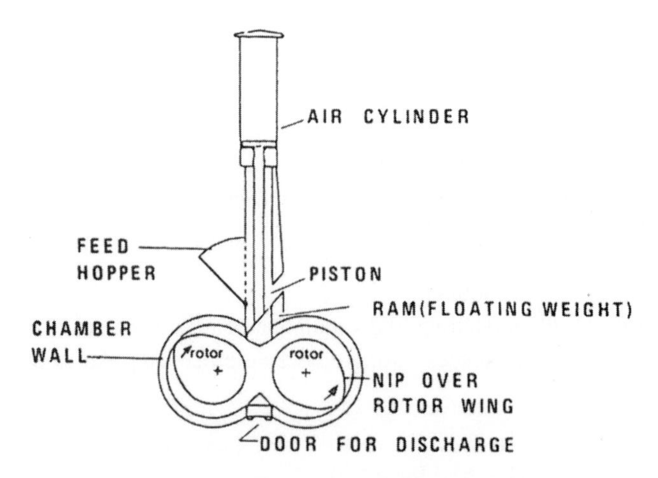

Figure 6.20. A schematic diagram of an internal rubber mixer. Reprinted from Figure 1 (Cherimisinoff, 1986). Copyright (1986), with permission from Elsevier.

There are two basic designs of rotors in internal mixers, non-intermeshing (Banbury type) and intermeshing (intermix type). Intermeshing rotors generally provide more consistent heat transfer, and thus are better for long mixing of heat-sensitive materials, although in general rotor design has proven to be relatively unimportant to mixing efficiency, given the large elongation forces seen in mixers, irrespective of rotor design (Cherimisinoff, 1986).

Mixing mills are open two-roller mills consisting of two parallel horizontal rollers rotating in opposite directions, that pull the ingredients through the gap (or nip) between the rollers. The back roller usually rotates faster than the front roller, and most of the work is done on the slower front roller. Rollers can be heated centrally.

Mixing extruders are automated-feed internal mixers in which the feed is conveyed from the feed hopper into the mixing section (note that the extrusion section is principally used for conveying, not mixing). The mixing section is similar to that of an internal mixer where material is intensively sheared between the rotor and the wall.

Quality-control tests and important process variables

Key process and material variables are

- viscosity (or viscosity reduction)
- power consumption (which concerns key events such as wetting, dispersion and plasticization)
- dispersion and distribution
- scorch stability (maximum permissible heat history)
- cure rate

Typical systems (thermoset, thermoset-filled)

Elastomers include natural rubber (polyisoprene), synthetic polyisoprene, styrene–butadiene rubbers, butyl rubber (isobutylene–isoprene), polybutadiene, ethylene–propylene–diene (EPDM), neoprene (polychloroprene), acrylonitrile–butadiene rubbers, polysulfide rubbers, polyurethane rubbers, crosslinked polyethylene rubber and polynorbornene rubbers. Typically in elastomer mixing the elastomer is mixed with other additives such as carbon black, fillers, oils/plasticizers and accelerators/antioxidants.

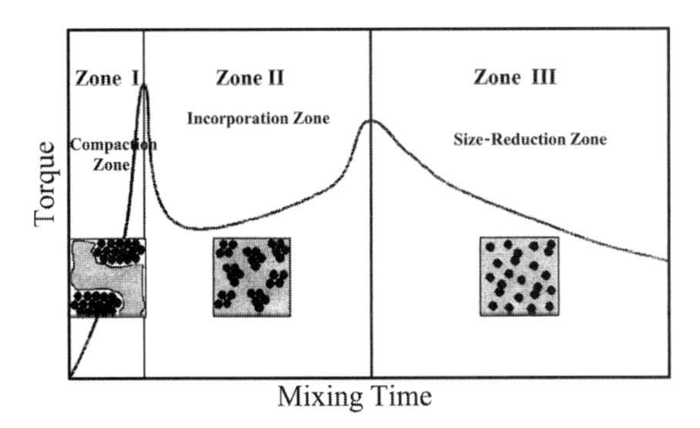

Figure 6.21. Mixing-time zones in rubber mixing. Figure 1 (Campanelli *et al.*, 2004). Reprinted with permission of John Wiley and Sons, Inc.

Process modelling

Campanelli *et al.* (2004) discuss a model of rubber mixing in an internal mixer based on kinetic, thermodynamic and rheological equations that is used to determine the extent of dispersion, batch temperature and relative batch viscosity over time. The chemoviscosity model is complex in that a model must be developed for the compaction zone, the incorporation zone and the size-reduction zone, as defined by the mixing-time zones in Figure 6.21.

For example a combined chemoviscosity model, combining the effects of temperature and fillers as shown in the following equation is useful for the incorporation zone:

$$\eta = \eta_{\text{ref}} \exp[-A(T - T_{\text{ref}})/(B + T - T_{\text{ref}})](1 - A_1\phi)^{B_1}, \tag{6.42}$$

where η is the viscosity, η_{ref} is a reference viscosity at temperature T_{ref}, T is temperature, A, B, A_1 and B_1 are constants and ϕ is the filler concentration. The following equation is useful for the size-reduction zone:

$$\begin{aligned} \eta &= \eta_{\text{i}}(1 + 2.5\phi_{\text{f}} + 6.2\phi_{\text{f}}^2)(1 + 2.5\phi_{\text{p}} + 6.2\phi_{\text{p}}^2), \\ \phi_{\text{f}} &= \phi_{\text{f0}} \exp(-k_{\text{f}}t) + C_1, \\ \phi_{\text{p}} &= \phi_{\text{p0}} \exp(-k_{\text{p}}t) + C_2, \end{aligned} \tag{6.43}$$

where η is the viscosity, η_{i} is the initial batch viscosity, ϕ_{f} is the fragment volume fraction, ϕ_{p} is the parent-filler volume fraction, ϕ_{f0}, ϕ_{p0}, C_1 and C_2 are constants and k_{f} and k_{p} are breakdown rate constants. Useful predictive model correlations to experimental mixing temperatures and torques are provided for rubber–carbon-black mixtures.

Schwartz (2001) and Haberstroh and Linhart (2004) note that rubber mixing processes are inherently extremely complex and use artificial neural networks to evaluate rheometric properties and process models for mixing of rubber compounds.

6.7.2 Rubber processing

Process diagram and description

Rubber extrusion is used to convert feed into a continuous finished product such as rods, tubes or profiles. Feed material is introduced into a hopper and conveyed forwards by a rotating screw (which is driven by an electric motor). This is similar to the process shown in Figure 6.18 for thermoset extrusion.

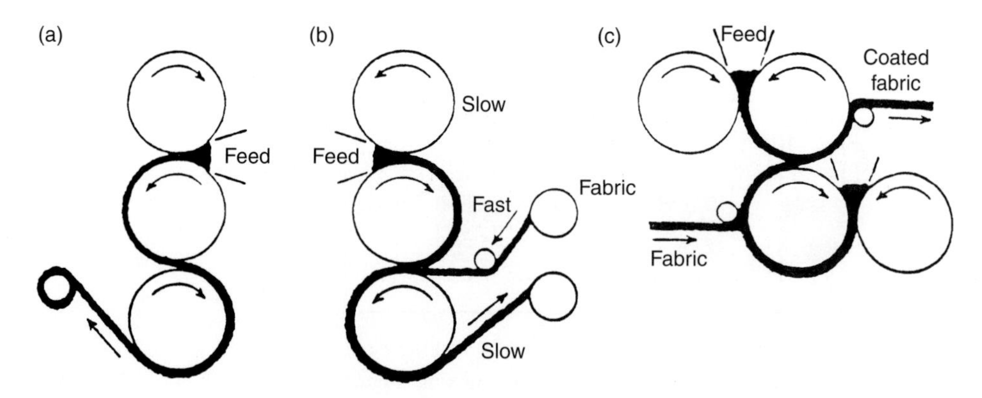

Figure 6.22. Common calendaring operations. From Figure 12 (Cherimisinoff, 1986). Copyright (1986), with permission from Elsevier.

The external barrel is heated, which, together with frictional shear heating, softens the feed material. The material has a reduced viscosity due to the process and is quite deformable by the end of the screw where the die imparts the final shape to the profile. The screw has three sections – the feed section (which conveys feed material and helps expel voidage gases), the transition section (which heats and mixes the materials) and the compression section (which homogenizes the material and builds up the pressure necessary to convey the material consistently through the die).

Rubber calendering involves two rollers rotating in opposite directions. The gap between the rollers is the nip, through which the material is processed to form sheets or laminates. Common calendering operations include sheeting, fractioning onto a fabric and double coating (which are shown in Figure 6.22).

Sheeting is essentially the forming of sheet products via two- or three-roller calendering and a controlled take-up reel. Frictioning involves wiping an elastomer into a textile substrate in a three-roller calendar system that operates at uneven roller speeds to promote wiping. The coating operation is similar to fractioning except that the rollers are controlled at even speeds and the elastomer is merely pressed against a substrate.

Rubber moulding processes are similar to thermoplastic and thermoset moulding processes, except that generally the elastomer has a relatively high initial viscosity. Fortunately, most elastomers have a shear-thinning viscosity profile with increasing shear rate, so they can be processed at high shear rates. Compression moulding is done by pressing elastomeric feed between two metal plates at shear rates in the range 1–10 s^{-1} (Figure 6.23). Since this is a relatively low shear rate, the viscosities are high and processing times are longer.

Transfer moulding is a variation of compression moulding, where by a rubber preform is prepared, heated in the transfer chamber and injected via a plunger into a mould; refer to Figures 6.24(a) and (b).

Injection moulding is completed by conveying the feed materials via a single-screw extruder into a shot volume and then injecting the shot volume into the mould cavity (Figure 6.25).

Shear-rate ranges for injection moulding are 1000–10 000 s^{-1}, thus processing viscosities are low and cycle times are reduced. Of course, higher pressures and more complex control systems are encountered in injection-moulding processes, constituting a more continuous processing method (continuous feed system) than compression or transfer moulding.

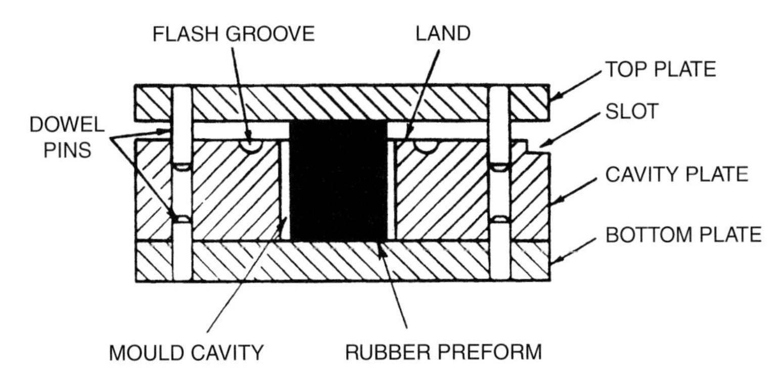

Figure 6.23. A schematic diagram of rubber compression moulding. Adapted from Figure 17 (Cherimisinoff, 1986). Copyright (1986), with permission from Elsevier.

Quality-control tests and important process variables

Important material and process variables for extrusion are

viscosity (or Mooney viscosity)
output rate
extrusion temperature profile
scorch stability (maximum permissible heat history)
extrudate temperatures

Process variables important to calendering include

roller speeds
nip settings
temperature of rollers
fabric properties
final thickness of coatings

Important processing and material variables for rubber moulding processes are

the rubber's viscosity profile
extrusion or plunger speeds
extrusion or plunger thermal profiles
moulding temperatures and time

Typical systems

Typical elastomer systems have been mentioned in the previous section, Section 6.7.1.

Process modelling

Isayev and Wan (1998) and Deng and Isayev (1991) examined the simulation of a natural-rubber injection-moulding process. A shear-rate- and temperature-dependent viscosity function based on the modified Cross model was fitted to experimentally measured viscosity data, and a non-isothermal vulcanization model with non-isothermal induction time was fitted to the non-isothermal curing kinetic data obtained by differential scanning calorimetry (DSC). These data were utilized in the simulation of the cavity-filling and curing stages to predict pressure traces during filling and curing conversions.

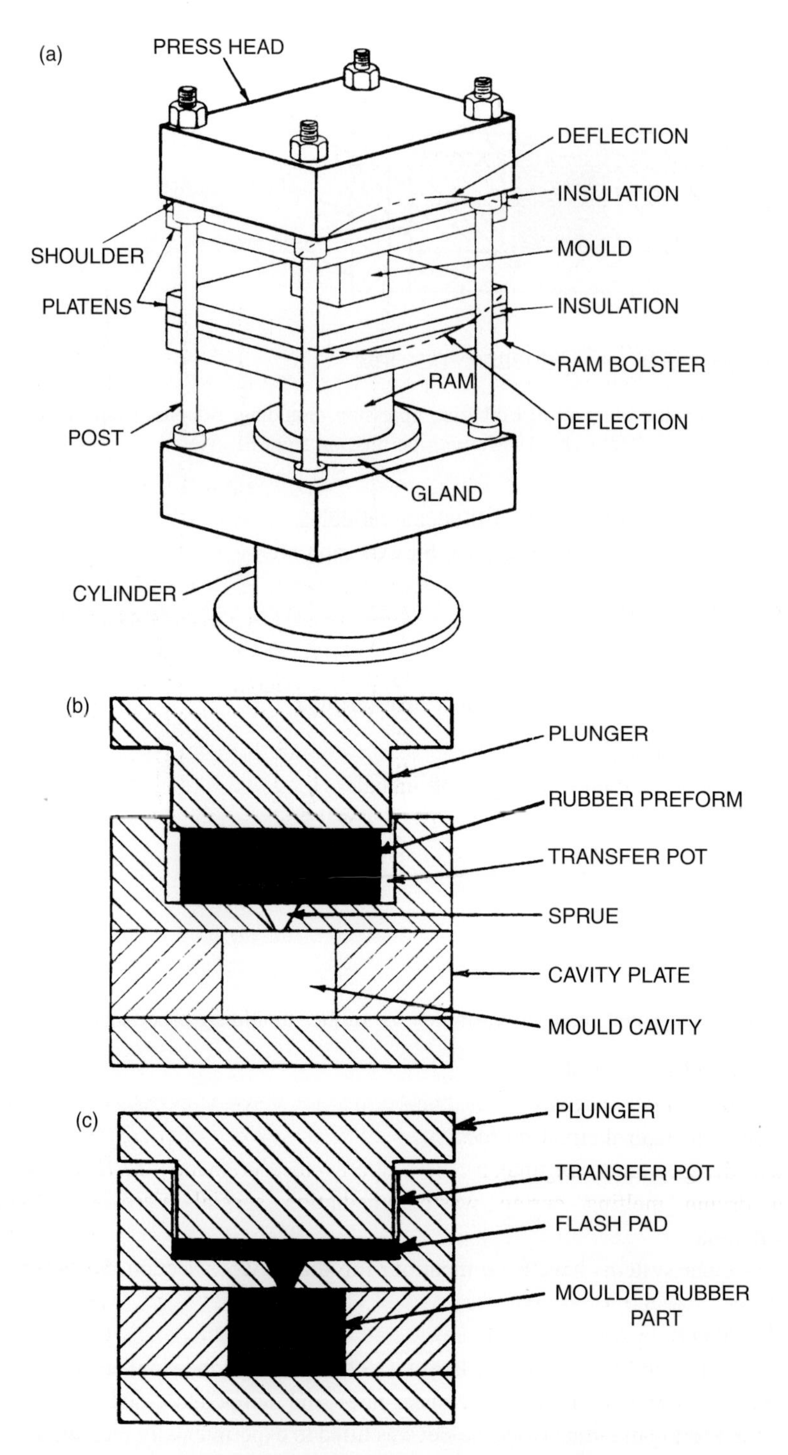

Figure 6.24. A schematic diagram of rubber transfer moulding. Adapted from Figures 19 and 20 (Cherimisinoff, 1986). Copyright (1986), with permission from Elsevier.

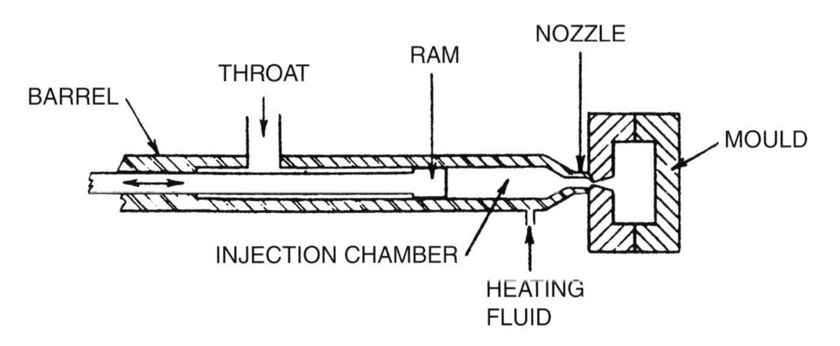

Figure 6.25. A schematic diagram of rubber injection moulding. Adopted from Figure 22 (Cherimisinoff, 1986). Copyright (1986), with permission from Elsevier.

Wan and Isayev (1996) examined a hybrid approach of control-volume finite-element and finite-difference modelling of injection moulding of rubber compounds. The effect of vulcanization on viscosity and yield stress during cavity filling is reported. On comparing two versions of the modified Cross viscosity models – with and without the effect of cure – the use of a viscosity model that accounts for the cure was found to improve the accuracy of the cavity-pressure-prediction models. When the modified Cross model was further extended to include the yield stress and was implemented in the simulation program a significant improvement in the prediction of cavity pressure was obtained in the case of low injection speed.

Haberstroh *et al.* (2002) modelled the injection moulding of liquid silicone rubber via the use of curing kinetics, flow models and the pressure–volume–temperature behaviour.

Nichetti (2003, 2004) examined the use of a cure model to predict mechanical properties of isothermally vulcanized moulded items.

6.8 High-energy processing

6.8.1 Microwave processing

Process diagram and description

Clark and Sutton (1996) reviewed the theory of microwave–material interactions, diagnostics, process control, sterilization, medical treatments, analytical characterization, equipment design, modelling, dielectric-property measurements, materials synthesis, sintering, drying, melting, curing, waste remediation, special microwave effects and process scale-up for microwave processing. A significant increase in research into microwave processing of ceramics and polymers began in the 1980s, most probably as the result of two factors: (a) many laboratories were equipped with inexpensive home microwave ovens (operating at 2.45 GHz) that could easily be modified for high-temperature research; and (b) researchers were reporting successes and unusual effects with microwave processing. As a consequence of the first factor, most of the studies have been, and still are being, performed at 2.45 GHz.

Microwaves (0.3–300 GHz) lie between radiowave (RF) and infrared (IR) frequencies in the electromagnetic radiation spectrum. Microwaves can be reflected, absorbed and/or transmitted by materials. Microwaves can interact with materials through either polarization or conduction processes. Polarization involves the short-range displacement of charge through the formation and rotation of electric dipoles, whereas conduction requires the long-range (compared with rotation) transport of charge (Clark and Sutton, 1996). During interaction, energy in the form of heat is generated in the material primarily through absorption. At room temperature, many ceramics and polymers do not absorb appreciably at 2.45 GHz. However, their absorption can be increased by increasing the temperature (e.g. once a material has been heated to a critical temperature, T_c, microwave absorption becomes sufficient to cause self-heating), adding microwave-absorbing additives (e.g. SiC, carbon, binders), altering the microstructure or defect structure, changing their form (e.g. bulk versus powder), or changing the frequency of the incident radiation (Clark and Sutton, 1996).

Microwave processing has been used to process thermoset polymers and polymer composites, including polyesters, polyurethanes polyimides and epoxies, and in most studies it has been concluded that the curing speed is faster using microwave energy (Clark and Sutton, 1996). The effects of continuous and pulsed microwave irradiation on the polymerization rate and final properties also have been studied, and it was demonstrated that, for certain epoxies, a pulsed microwave cure resulted in improvements in mechanical properties, better temperature uniformity and a faster polymerization rate (Thuillier and Jullien, 1989).

Wei *et al.* (1995) showed for an epoxy–amine system that

- increased reaction rates were observed in the microwave cure as compared with thermal cure at the same temperature;
- the primary-amine–epoxy reactions are more dominant in the microwave cure than in the thermal cure relative to the secondary-amine–epoxy and etherification reactions; and
- there is a difference in the crosslinked network structure between the microwave and thermally cured DGEBA/DDS.

Microwave processing has also been suggested to be able to aid pultrusion and autoclave processing (Clark and Sutton, 1996).

Quality-control tests and important process variables

Important material and process variables include

- microwave intensity
- microwave pulse
- material kinetics and gelation characteristics
- material additives and structure
- operating temperature

Typical systems (thermoset, thermoset-filled)

As previously mentioned many studies have been performed on microwave curing of thermosets, including polyesters, polyurethanes polyimides and epoxies.

Process modelling

Hedreul *et al.* (1998) examined a model of the cure kinetics of a thermally and microwave-cured rubber-modified epoxy-resin formulation. The phenomenological cure kinetic model used was

$$da/dt = (k_1 + k_2 a^m)(1 - a)^n, \tag{6.44}$$

$$k_1 = A_1 \exp(-E_1/T),$$

$$k_2 = A_2 \exp(-E_2/T),$$

where α is the cure conversion, t is the time, k_1 and k_2 are cure parameters, A_1, A_2, E_1 and E_2 are constants, m and n are reaction orders and T is temperature. Interestingly, the cure kinetics from thermal and microwave processing were similar.

Liu *et al.* (2004, 2005) examined a three-dimensional non-linear coupled auto-catalytic cure kinetic model and transient-heat-transfer model solved by finite-element methods to simulate the microwave cure process for underfill materials. Temperature and conversion inside the underfill during a microwave cure process were evaluated by solving the non-linear anisotropic heat-conduction equation including internal heat generation produced by exothermic chemical reactions.

Zhao *et al.* (2001) examined a model for computing the electromagnetic field and power-density distribution in a cavity, and their effects on cure of a thin epoxy-resin layer during a novel microwave rapid-prototyping process.

6.8.2 Ultraviolet processing

Process diagram and description

One of the most efficient methods for rapid generation of highly crosslinked polymers is by exposing multifunctional monomers to UV radiation in the presence of a photoinitiator. A liquid resin can be transformed within a fraction of a second into a solid polymer that is very resistant to chemical, heat and mechanical treatments. A review of UV curing of polymers is presented by Decker (1989) .

Ultroviolet curing is a process that transforms a multifunctional monomer and a photoinitiator (since most monomers do not produce initiating species with sufficiently high yield upon exposure, it is often necessary to introduce a photosensitive initiator that will make the polymerization start at illumination) into a crosslinked macromolecule by a chain reaction initiated by reactive species generated by UV irradiation. A unique advantage of UV initiation, compared with thermal polymerization, is that very high concentrations of radical species can be attained simply by increasing the light intensity, and the depth of cure can be finely controlled through the photoinitiator concentration which directly affects the light absorbance (Decker, 1994).

Quality-control tests and important process variables

Important material and process variables for UV processing include

UV-irradiation time,
UV-irradiation power
curing temperature
the material's cure kinetics and
the material's chemoviscosity

Typical systems (thermoplastic, thermoset)

Typical systems include unsaturated polyesters (relatively slow cure used in the wood-finishing industry), thiolene systems (which are useful for coatings, adhesives and sealants), acrylate and methacrylate systems (which find wide uses), epoxides (microcircuits) and vinyl ether systems.

Process modelling

Ultraviolet processing has been restricted to lab-scale studies; however, many authors have examined kinetic cure modelling of these systems during the development of new UV-cure materials.

Lee and Cho (2005) developed a cure model for UV nano-imprinting. It was shown via fibre-optic FTIR techniques (as developed by Decker and Moussa (1990)) that the degree of cure had an exponential relation to the UV-irradiation time, power and temperature.

Cho and Hong (2005) used photodifferential scanning calorimetry to investigate the photocuring kinetics of UV-initiated cationic photopolymerization of 1,4-cyclohexane dimethanol divinyl ether (CHVE) monomer with and without a photosensitizer, 2,4-diethylthioxanthone (DETX) in the presence of a diaryliodonium-salt photoinitiator. Two kinetic parameters, the rate constant (k) and the order of the initiation reaction (m), were determined for the CHVE system using an auto-catalytic kinetics model as shown in the following equation:

$$\mathrm{d}a/\mathrm{d}t = ka^m(1-a)^n, \tag{6.45}$$

where a is the cure conversion, k is the cure constant, and m and n are reaction orders. The photosensitized system gave much higher k and m values than did the non-photosensitized one (due to the photosensitization effect).

Abadie *et al.* (2002) showed that the following auto-catalytic kinetic model was useful for modelling the UV cure of epoxy-resin systems:

$$\mathrm{d}a/\mathrm{d}t = (k_1 + k_2a^m)(1-a)^n, \tag{6.46}$$

where a is the cure conversion, k_1 and k_2 are cure constants (and Arrhenius exponential functions of temperature), and m and n are reaction orders.

6.8.3 Gamma-irradiation processing

Process diagram and description

Gamma (γ) irradiation first found major utilization in microbiocidal applications, owing to its ability to inactivate many microorganisms. The application to polymer reactions, particularly grafting, requires knowledge of the free-radical chain reactions (Section 1.2.3) as well as degradation reactions (Section 1.4) that can lead to loss of mechanical properties of the substrate. Gamma radiation is made up of pulses of electromagnetic energy (γ-ray photons) from a radioactive source such as cobalt-60. Reactors have common features such as shielding, a radioactive source and an efficient mechanism for inserting and removing the material to be irradiated by the source.

Quality-control tests and important process variables

Important material and process variables for γ-processing include

γ-irradiation time
γ-irradiation power

curing temperature
the material's cure kinetics and
the material's chemoviscosity

Typical systems (thermoplastic, thermoset)

Applications of γ-irradiation as a cold-sterilization method for medical devices have concerned plastic and rubber catheters, drainage bags, gauze and surgical gloves (Clegg and Collyer, 1991) due to its effective penetration into the materials and through packaging. Examples of materials, applications and possible side effects for materials used in γ-irradiation of medical devices are shown in Table 6.3.

Process modelling

Little process modelling has been developed due to the low industrial usage of γ-irradiation processing for reactive polymers. However, many researchers are examining cure models for promising materials.

Alessi *et al.* (2005) examined the γ-irradiation and electron-beam processing of an epoxy-resin system in the presence of a photoinitiator. They showed that increasing the irradiation dose frequency and photoinitiator concentration greatly increases the temperature reached by the samples. The increase in temperature of the system during irradiation is related to the balance between the heat-evolution rate, due to both polymerization reaction and radiation absorption, and the rate of heat release towards the environment. High dose rates and high photoinitiator concentrations increase the reaction rate and the difference between the heat produced and the heat released to the environment (Alessi *et al.* (2005).

Nho *et al.* (2004) also examined γ-irradiation and electron-beam irradiation of epoxy-resin systems. They showed that the gel fraction of two epoxy resins increased markedly with increasing dose, and that the gel fraction of the epoxy resin irradiated by γ-irradiation was higher than that of the epoxy resin irradiated by electron-beam irradiation (possibly due to the greater depth of cure with γ-irradiation).The degree of curing of epoxy resins irradiated in nitrogen was higher than that of epoxy resins irradiated in air (due to inhibition of cure by oxygen) for both electronic-beam and γ-ray initiation.

Palmas *et al.* (2001) highlighted the γ-irradiation curing and thermal ageing of EPDM synthetic elastomers studied by both ^{13}C high-resolution and ^{1}H wideline solid-state NMR.

6.8.4 Electron-beam-irradiation processing

Process diagram and description of process

Electron accelerators are utilized in industrial processes such as crosslinking of polymers, curing rubbers, curing paints and adhesives, sterilization of medical devices and pasteurization of foodstuffs.

The interaction of high-energy electrons and materials depends on the kinetic energy of the electron and the composition of the irradiated materials, so effective operation is conditioned by this interaction and knowledge of the effective penetration depth of the electrons into the material. This penetration depth is the depth at which the absorbed dose has fallen to an arbitrary level (ideally similar to that of the unirradiated surface). An electron processing facility consists of

Table 6.3. Examples of irradiated polymers

Polymer	Application examples	Processing	Stability up to 25 kGy	Stability >25 kGy	Comments
ABS[a]	Closure-piercing devices, roller clamps	Injection moulding	Good	To 1000 kGy	Rigid, most often translucent/ opaque
Acrylics	Luer connectors, injection sites	Injection moulding	Good	To 1000 kGy	Clear with slight yellowness after irradiation
Cellulosics	Ventilation filters	Various	Good	To 200 kGy	Repeated irradiation embrittles
Fluoroplastics	Tubing	Extrusion	Formulation-specific		PTFE unstable, to FEB stable at 25 kGy and higher
Polyamides (nylon)	Fluid filters	Extrusion	Good	To 500 kGy	May harden on irradiation
PE	Protector caps	Injection moulding	Good	To 1000 kGy	Translucent/opaque, soft/ waxy fell
PP	Hypodermic syringes	Injection moulding	Formulation-specific		Embrittles over time
Polystyrene	Syringe plungers	Injection moulding	Good	To 10000 kGy	Rigid
PVC	Tubing, drip chambers	Injection moulding, extrusion	Formulation-specific		Discolours
SAN[b]	Hypodermic syringes	Injection moulding	Good	To 5000 kGy	Rigid, clear

Adapted from Clegg and Collyer (1991, Table 6.4).

[a] Acrylonitrile–butadiene–styrene.

[b] Polystyrene–acrylonitrile.

an accelerator (which is rated for the material's thickness and density, and takes into account optimizing of size, cost, dose distribution and efficiency)
a product conveyor (for continuous, batch or bulk liquid/powder processing)
a radiation shield (designed with knowledge of the total radiation flux)
a ventilation system (for removal of ozone)
a safety system (for high voltage, pressures and radiation)
a control system

The advantages of electron-beam processing include that it can be carried out at room temperature and on the final product part, the fact that there is no need for catalysts or solvents and flexible control of radiation dose and application. Limitations are the high capital cost and the limited penetration depth (in comparison with higher-energy γ-radiation).

Quality-control tests and important process variables

Important material and process variables for electron-beam processing include

irradiation time
irradiation power
curing temperature
the material's cure kinetics and
the material's chemoviscosity

Typical systems

An example system is PVC and polyethylene wire and cabling irradiated to improve stress-cracking resistance, abrasion resistance, high-temperature properties and flame retardance, via controlled electron-beam crosslinking (Loan, 1977). Additionally electron-beam crosslinking is utilized to impart memory into a polymer system, such as crosslinked PE materials for heat-shrinkable films and pipe applications (Baird, 1977). The control of electron-beam processing has advanced the quality of cell size and shape of PE foams via control of crosslink distribution (Paterson, 1984).

Process modelling

Cleland *et al.* (2003) reviewed the application of electron beams with various materials; however, little process modelling has been developed due to the low industrial usage of electron-beam processing for reactive polymers. However, many researchers are examining cure models for promising materials.

Behr *et al.* (2006) showed the benefits of electron-beam curing of glass-fibre-reinforced-composite veneer specimens for dental applications. Under electron-beam cure the reconstructions became stiffer and resisted higher load.

Chen *et al.* (2006) highlight a novel calorimetric method to examine cure during electron-beam processing of epoxy resins via an energy-balance model and use of a novel experimental calorimeter.

Das *et al.* (2005) examined the electron-beam processing of hydrogenated acrylonitrile–butadiene rubber sheets. The gel content and dynamic storage modulus increased with the radiation dose.

Klosterman *et al.* (2003) examined the electron-beam curing of a 'dual-curing' (the monomer contains both acrylate and acetylene reactive groups) liquid-crystal (LC) resin. It was shown that electron-beam curing in the LC phase allows tighter molecular packing and more efficient polymerization, resulting in higher crosslink densities.

6.9 Novel processing

6.9.1 Rapid prototyping and manufacturing

The development of rapid prototyping and rapid manufacturing is closely aligned with the application of computers in manufacturing such as in computer-aided design (CAD), computer-numerical-control (CNC) machining and computer-aided manufacturing (CAM). To differentiate terms, rapid prototyping is when prototypes for visual and design purposes only are produced via a rapid-protyping technique, whereas rapid manufacturing is the rapid production of parts that are functional as products in themselves. Rapid prototyping and manufacturing has also been known as additive fabrication, three-dimensional printing, solid freeform fabrication and layered manufacturing.

There are many classifications of rapid-prototyping techniques, but perhaps the simplest is the broad classification into liquid-based and solid-based techniques. Liquid-based techniques are focussed on systems that are initially in the liquid state and are controllably cured into a final product shape. Solid-based techniques involve an initial solid preform or powdered sample that is melted, fused or sintered together into a final product.

Since this book is focussed on chemorheological processes, we will focus on rapid prototyping and manufacturing processes that include reactive processing (rather than the rapid-prototyping techniques involving only physical transformations, i.e. sintering of metals, ceramics or thermoplastics).

Stereolithography

Process diagram and description

Stereolithography builds plastic parts layer by layer by passing a laser beam (guided by a three-dimensional image) on the surface of a pool of liquid photopolymer. This photopolymer quickly solidifies where the laser beam strikes the surface of the liquid. Once one layer is completely traced, the part is lowered a small distance into the pool, allowing new liquid to cover the completed layer, and then a second layer is traced on top of the first. The layers bond and eventually form a complete, three-dimensional object after many such layers have been formed. A schematic diagram of this process is shown in Figure 6.26.

From a material viewpoint, the process may be described as follows (Flach and Chartoff, 1995a): the photosensitive resin is a mixture of a base resin (typically a multifunctional acrylate oligomer), a crosslinker (monomeric multifunctional acrylate), a diluent (monomeric acrylate), a photoinitiator and various additives (surfactants, stabilizers, etc.). The photoinitiator absorbs the UV light and undergoes a reaction leading to the formation of free radicals. These free radicals initiate the polymerization reaction as described in Section 1.2.3 and Table 1.5. As polymerization proceeds, the viscosity of the mixture in the direct vicinity of the laser increases until gelation, then vitrification occurs (as described in Section 1.2.4 for networks from addition polymerization) and the material can be considered solid. The chemical reaction is quite fast, with a high degree of conversion to a crosslinked network being reached within 20 ms of exposure.

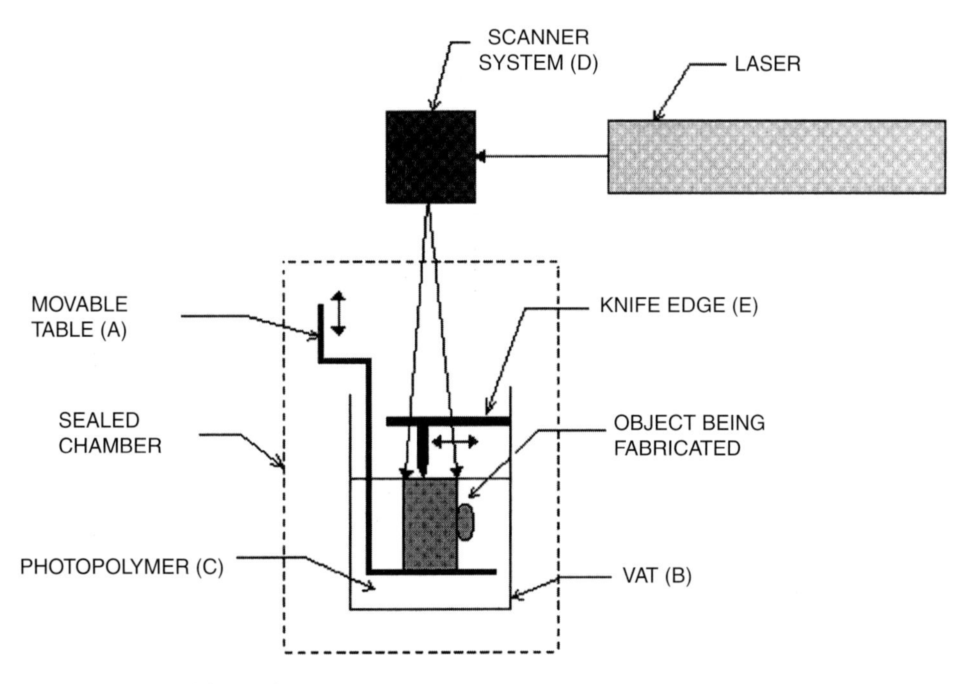

Figure 6.26. Stereolithography.

Advantages of stereolithography are the excellent dimensional accuracy, excellent surface finish, variable vat size and ability to use a range of materials. The disadvantages are the need for handling of liquid photopolymers and the requirement of post-processing (removal of supports) and post-curing in some cases.

Quality-control tests and important process variables

Process variables that control the functionality of the part include

physical and chemical properties of the resin
speed and resolution of the laser optical scanning system
the laser power, wavelength and spot size
the post-curing process

The stereolithography system is controlled by the interconnected CAD system, computer-controlled optical scanning system, control system for the platform and levelling wiper or blade for the recoating process (in between laser curing steps).

Typical systems

Photopolymer systems are photocurable resins incorporating reactive liquid monomers, photoinitiators, chemical modifiers and fillers. Typically stereolithography utilizes UV radiation, so UV-curable systems are used. Free-radical-photopolymerizable acrylate systems were originally used; however, newer cationic epoxy-resin and vinyl ether systems (based on iodinium- or sulfonium-salt cationic initiators) are now being utilized.

An interesting extension of this system is the Soliform system, which focusses on developing rapid-manufacturing moulds via stereolithography for injection moulds, vacuum moulds and cast moulds, via the use of acrylate–urethane resins of inherently higher flexural modulus (Chua *et al.*, 2003).

Process modelling

Very little chemorheological or process modelling has been completed for stereolithography, and most operation and optimization has been done on a trial-and-error basis. However, there are a few key papers that aid the modelling of this process. Flach and Chartoff (1995a, 1995b) developed a numerical model for laser-induced photopolymerization that simulates important aspects of stereolithography. The model considers irradiation, chemical reaction and heat transfer in a small zone of material exposed to a stationary UV laser source, and generates outputs including spatial and temporal variations in the conversion, depletion of photoinitiator and local variations of temperature in and around the region irradiated by the laser light. Here the kinetic model for conversion of monomer is given by

$$\mathrm{d}[\mathrm{M}]/\mathrm{d}t = -k_\mathrm{p}[\mathrm{M}](\phi I_\alpha/k_\mathrm{t})^{0.5}, \tag{6.47}$$

where [M] is the monomer concentration, k_p is the propagation rate constant, k_t is the termination rate constant, I_α is the absorbed light and ϕ is the quantum yield for initiation. This is equivalent to the standard free-radical polymerization of Equation (1.51), in Section 1.2.3. In this model there is assumed to be no material flow, and the temperature rise is due to heat of polymerization alone. Results provided insights into effects of laser dwell time, depth penetration of cure and the overall uniformity of polymer formed during the stereolithography process. Bartolo (2006) more recently presented an extension to this work in a phenomenological numerical model for the curing processes of thermosetting resins for stereolithography. The cure model was developed for light- and/or thermally initiated systems and incorporates inputs such as initiator concentration, temperature and light intensity in order to predict output variables such as the diffusion effects after vitrification, the phenomenon of unimolecular termination and shrinkage effects. The model was experimentally verified.

Tang *et al.* (2004) presented a cure model that captures transient, thermal and chemical effects that are ignored in typical threshold-based models. This new model incorporates as inputs photoinitiation rates, reaction rates, diffusion and temperature distributions, and is able to determine the spatial and temporal distributions of monomer and polymer concentrations, molar masses, crosslink densities and degree of cure.

Solid ground curing (SGC)

Process diagram and description

Solid ground curing (SGC) is also a photolithographic technique; however, unlike stereolithography, which uses a single laser point for curing, SGC requires a mask during curing. In the SGC process a photomask (item 1 of Figure 6.27) is prepared (by transfer of CAD data to a mask generator via an ionographic process). A thin layer of photopolymer is spread on the base surface (item 2), and then the photomask is placed above the base and aligned (item 3) with a collimated UV lamp. The UV light is turned on (item 4) and the part of the resin exposed through the mask is hardened, and unsolidified resin is removed (item 5). Following this, melted wax is spread, solidified (cooled – item 6) and milled to a smooth and precise layer height (item 7). Layer final curing is then achieved via processing under a further UV lamp (item 8). This process is then repeated in a layer-by-layer process. The wax interlayers may be removed at the end of the process via dissolution or heating.

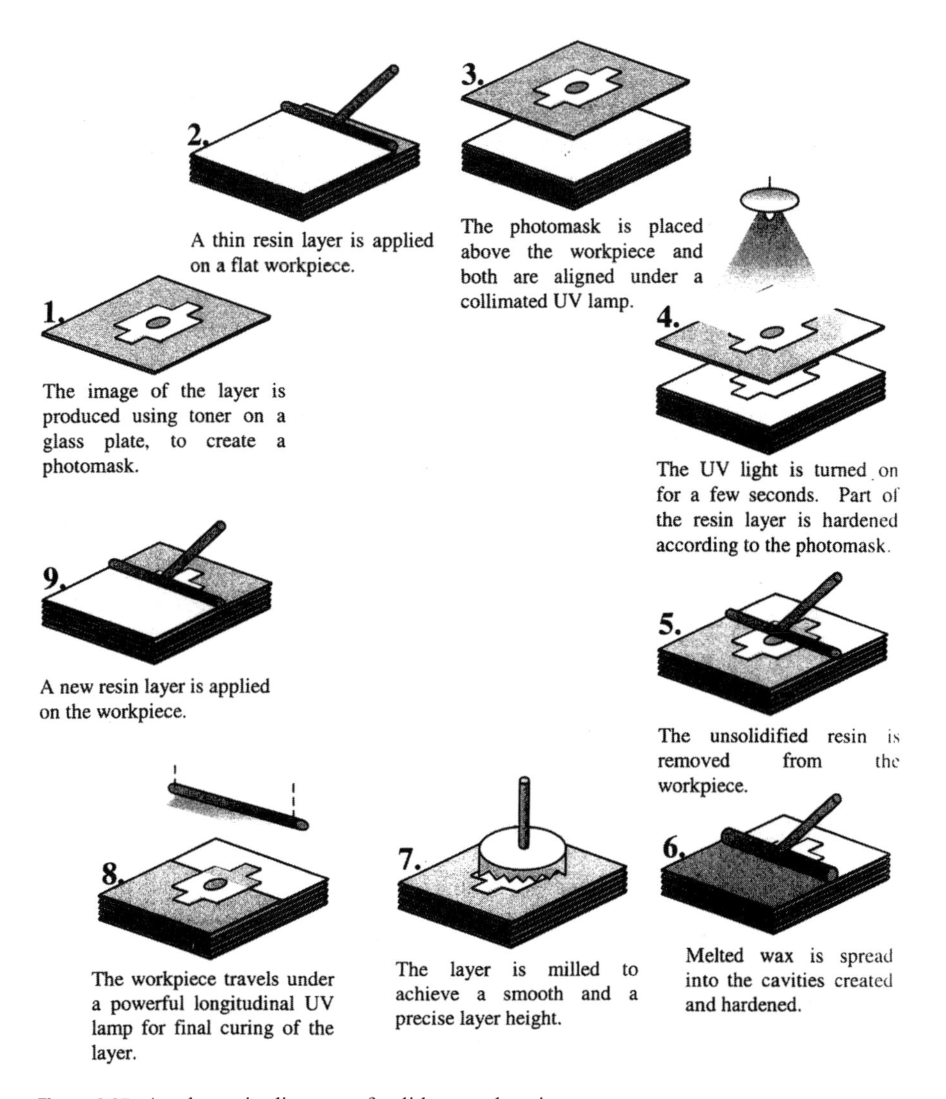

Figure 6.27. A schematic diagram of solid ground curing.

Advantages of this process include instantaneous curing of the whole of the cross-section, minimization of shrinkage and internal stresses (due to the full cure of every layer) the absence of any requirement for final curing and the fact that there is no need for a support structure. Since wax supports the structure. The disadvantages are that wax-cleaning processes are required and the capital cost is high.

Quality-control tests and important process variables

The SGC system incorporates a CAD system, resin systems for the photocurable resin, the soluble wax support and ionographic toner for creating photomasks, a vacuum wiping system, a UV curing system and a dewaxing machine. Important process variables are

physical and chemical properties of the photocurable resin
UV curing conditions

wax application and cooling conditions
layer milling

Typical systems (thermoplastic, thermoset)

Typical photocurable resins for photolithography, such as acrylates and epoxy-resin systems, are employed.

Process modelling

Very little chemorheological modelling has been completed for SGC processing. Gu *et al.* (2001) examined the effects of the SGC processing parameters on dimensional accuracy and mechanical properties. Four key outputs were defined, namely x- and y-dimensional accuracy, impact strength and the elastic modulus of prototypes. Six input processing parameters (layer thickness, first exposure time, second exposure time, tray speed, cool-wax temperature and cooling time) that had possible effects on the four outputs were identified. Because of the interactions among the input parameters and the limited number of experiments, neural networks were used to establish the relationships between the key inputs and outputs. Key results were the following.

Layer thickness and initial exposure time were key parameters; they were highly interactive and increasing them increased the modulus. The second exposure time and cooling time also affect the modulus. Decreasing the cooling temperature increased the modulus.

Layer thickness, initial exposure time and tray speed affected the, x-direction dimensional accuracy (the other parameters have negligible effect on the x-dimensional accuracy).

Tray speed and initial exposure time have the largest effect on the y-direction dimensional accuracy (the other parameters have negligible effect on the y-dimensional accuracy).

Tray speed, cooling temperature and cooling time have the greatest effect on the impact strength (the other parameters have negligible effect on the impact strength).

6.9.2 Microlithography

Process diagram and description

In general there are two lithographic processes used in computer-circuit fabrication – photolithography and radiation (X-ray, laser, electron-beam and deep-UV (248- and 193-nm wavelength)) lithography. The principal difference is the radiation source and wavelength, which in turn define the feature size which can be achieved.

Circuit or mask fabrication has the following procedures.

1. Resist deposition: deposition of the resist (material) film from dilute-solution spinning and baking onto a substrate.
2. Pattern exposure: a pattern is generated by a mask or controlled radiation-beam steering, where the exposed material is degraded, chemically modified or crosslinked.
3. Pattern development: the pattern is developed by solvent interaction to form a positive (irradiated material is removed) or negative (non-irradiated material is removed) and then baked.

4. Transfer of pattern to substrate: exposure of the patterned substrate to an etchant or gas plasma allows features to be transferred to the substrate.
5. Removal of resist: the resist is removed by oxygen ashing without degradation by product formation.

Key properties of the material (resist) to be used in microlithography are that

the resist must be able to change form in response to the radiation
the resist must adhere well to the substrate and not swell or blister when in contact with the solvent
the resist must withstand plasma treatment and ion bombardment
the resist must produce a uniform coating

In a chip-fabrication process, the substrate may undergo two or three deposition processes, 20 patterning processes, 10 etching procedures and further ion implantation and metallization procedures.

Quality-control tests and important process variables

Important process and material variables include

resist spinning
pattern exposure radiation
pattern development: process
transfer of pattern to substrate
removal of resist

Typical systems

Typical resists include cyclized polyisoprene with a photosensitive crosslinking agent (ex bisazide) used in many negative photoresists, novolac resins with diazoquinone sensitizers and imidazole catalysts for positive photoresists, poly(oxystyrenes) with photosensitizers for UV resists, polysilanes for UV and X-ray resists, and polymethacrylates and methacrylate–styrenes for electron-beam resists (Clegg and Collyer, 1991). Also note the more recent use of novolac/diazonaphthoquinone photoresists for mid-UV resists for DRAM memory chips and chemically amplified photoacid-catalysed hydroxystyrene and acrylic resists for deep-UV lithography (Choudhury, 1997).

Process modelling

Cole *et al.* (2001) reviewed physical models for most stages of the microlithography process. For the curing stage, they state that there has been no recent development of cure models since the work of Dill (1975). Dill (1975) presented models that incorporate both light absorption and photopolymer cure kinetics modelling, as shown by the following equations:

$$a = A[\mathrm{M}](z,t) + B, \tag{6.48}$$

$$\mathrm{d}[\mathrm{M}]/\mathrm{d}t = -I(z,t)[\mathrm{M}](z,t)C, \tag{6.49}$$

where a is the absorption coefficient, A, B and C are constants, [M] is the concentration of photosensitive material at position z and time t, and I is the energy at position z and time t. One can see clearly that the absorption and cure are coupled.

Some work has been done to model microlithography processing (including broader material-property considerations). Baek *et al.* (2004) highlight the importance of the presence of a uniform liquid layer between the last objective lens and the photoresist, in optical projection lithography. They also conclude that the refractive index of the liquid must be known to and maintained within a few parts per million for optimal processing. Hu and Tsai Ky (2006) highlight thermal modelling of microlithography to model the warpage of wafers during processing, using temperature measurements and first-principles thermal modelling via a simple heat balance.

6.10 Real-time monitoring

The industrial application of the sensor technologies described earlier (Chapter 3) is the basis for real-time monitoring of processing of reactive polymers. The critical parameters to be monitored may be seen from an examination of the chemorheological process-control equations described in the previous sections:

temperature (particularly the reaction exotherm)
viscosity (particularly gelation if a crosslinking system is concerned)
conversion (i.e. the extent of chemical reaction)
pressure (e.g. back pressure due to reaction by-products; premature gel formation etc.)

The changes in these parameters within the liquid polymer can be compared with the values predicted by the particular chemorheological equation to determine whether the process is operating within specifications. The particular challenge in process operations compared with laboratory and model studies is the decision regarding where to place the sensors and how to use the information as a decision-making tool.

One of the first consolidated approaches to autoclave curing of thermosetting and thermoplastic composite materials was that of Ciriscioli and Springer (1990). They showed that the evolution of an expert system with sensors (for viscosity, pressure and temperature as well as dimensional change) is the logical extension of the analytical model approach discussed in this chapter. The analytical model, being based on heat- and mass-transfer equations is, of course, an advance on the purely empirical approach based on experience of product quality without any real understanding of the underlying chemical and physical changes occurring during material processing. The expert system requires a set of rules that are developed from the analytical model since they are able to predict the changes to reaction time for each change of one of the processing parameters. In its most sophisticated embodiment, a system with a chemorheological model and appropriate sensors can adjust the process cycle to accommodate changes in the reactivity of starting materials as well as changes in temperature and pressure that may be routinely monitored (Ciriscioli and Springer, 1990).

6.10.1 Sensors for real-time process monitoring

The use of sensors for temperature and pressure under processing conditions is routine, and such sensors are standard features of commercial systems, but the measurement of rheological properties, such as viscosity, and chemical properties, such as extent of reaction, requires specialized sensors that must be able to withstand the electrically noisy and aggressive environment of the production line.

Sensors for chemical conversion (extent of reaction)

Hardware sensors for the on-line monitoring of polymerization such as in batch reactors have been reviewed (Kammona *et al.* 1999). In Section 3.4 the use of ruggedized conversion sensors for fibre-optic near-infrared (NIR) spectroscopy during extrusion was described. In many ways, the requirements are simpler than for control of a batch reaction (as in a polymer-synthesis autoclave) since

(i) there are no solvents, so the spectrum is simpler; and
(ii) the reactive extruder works with systems at high viscosity, so the concentration of the species being analysed is higher.

However, there is a particular requirement that governs the sensor:

(iii) the residence time in a reactive extruder is short (of the order of minutes), so a good SNR is required for quantitative analysis of the species (which is necessary for computing the conversion).

Fischer *et al.* (1997, 2006) have described an approach using combined NIR, Raman and ultrasound sensors with a melt-at-die interface for reactive extrusion. The system is shown in Figure 6.28, and it is noted that the combination of multiple sensors offers complementary information and allows optimization and choice of the best method for each system being reacted.

For example, NIR spectra are temperature-sensitive, so control to $\pm 2\,^{\circ}\mathrm{C}$ is required. Ultrasonics using two probes at 5 MHz offers the possibility of a robust and less expensive method, but it is still under development and there are few commercially available systems. Chemometrics (Section 3.5) offers the best method of data analysis, and, by using PLS with three different data pre-treatments, it was possible to obtain real-time values of the composition of a fire-retardant to better than 0.1% accuracy (Fischer *et al.*, 2006).

As is noted in the following sections, dielectric sensors have usually been regarded as providing local viscosity data (Senturia and Sheppard, 1986), but detailed studies over a wide frequency range have shown that the absolute value of the static dielectric constant (ε_0) may also provide data on conversion comparable to those obtained by DSC (Casalini *et al.*, 1997). However, to achieve this it is noted that ε_0 must be measured to high accuracy (a relative error of $\sim 10^{-3}$) for this to be a viable technique.

Sensors for viscosity

The changes in dielectric properties during curing of thermosets have been well characterized (Kranbuehl, 1986, Pethrick and Hayward, 2002), and one of the first attempts at real-time process control was through the use of dielectric sensors for monitoring viscosity (Senturia and Sheppard, 1986). As discussed in Section 3.6.5, microdielectrometry using small embedded chips provides a continuous measurement of changes in capacitance and dielectric loss so that these provide an analogue of viscosity. While the precise chemical species responsible for the change in electrical properties

Figure 6.28. A reactive extruder, showing the metal-at-die interface housing the real-time sensor probes for NIR, Raman and ultrasound (US) sensing of conversion and component concentration (Fischer *et al.*, 2006). Photograph provided by professor D. Fischer, Leibniz Institute of Polymer Research, Dresden.

might not be known for a particular commercial resin system, the system may be calibrated using independent rheological measurements so that real-time viscosity monitoring is possible. The continuous development of the technique has resulted in a planar wafer-thin sensor that is able both to monitor and to control processing of high-performance composites of epoxy resins and polyimides (Kranbuehl *et al.*, 1997). Of particular interest is the comparison of the differences between using a standard cure cycle for a high-temperature polyimide, PMR-15 (as provided by the manufacturer), and the use of an intelligent system with criteria for changes in the cure temperature set by the achievement of target dielectric properties (such as the change in slope of the complex permittivity (ε'') with cure time) as the decision point for temperature change. This resulted in either extension or reduction in duration of various parts of the cure cycle due to subtle differences in composition among batches of prepreg (Kranbuehl *et al.*, 1997). This example also illustrates the fact that dielectric sensors are providing not just viscosity information but rather data on chemical changes as well as physical processes such as sensor wet-out as the resin flows. As in all systems that are analogues rather than absolute methods, the system must be trained in order to determine the optimum parameter for control of the temperature–time process cycle in the autoclave.

The use of these sensors is not restricted to high-performance composites, but can be extended to on-line monitoring of resin-transfer moulding (McIlhagger *et al.*, 2000) and glass–polyester prepreg composite cure (Kim and Lee, 2002).

Sensors for temperature, pressure and strain

While most measurements of temperature in industrial processing simply involve the use of thermocouples or platinum resistance thermometers, there are some applications involving microwave fields or chemical environments for which this might not be possible. In that case a fibre-optic sensor is the instrument of choice. These sensors are often also sensitive to strain, so they may be used to obtain a measure of pressure and local deformation. The operating principle is discussed in the following section.

6.10.2 Real-time monitoring using fibre-optics

The advantage of optical methods for industrial process control is that they are not subject to electrical interference and have a high bandwidth for information transfer. The theory of fibre-optics and examples of prototype and laboratory-based systems were described earlier in Section 3.4. The use of NIR combined with chemometrics as described in Section 6.10.1 is an example that requires fibre-optics and has the advantage that the components may be made of quartz or even glass and still operate successfully. It is thus possible to treat the optical fibres as disposable items. For example, in autoclave processing of composites, it is possible to leave the fibre embedded in the part and use the optical fibre for subsequent assessment of the condition of the material.

Fernando and Degamber (2006) recently surveyed the field of process monitoring of composites using optical-fibre sensors. Their survey included both the spectroscopic methods providing absolute conversion data, which have been discussed in detail previously (NIR, Raman, ATR-IR), and the fluorescence and physical optical techniques providing analogues of viscosity and conversion, rather than a direct measurement of the concentration of reacting species. Figure 6.29 shows the adaptation of an industrial autoclave for processing of advanced fibre composites to enable fibre-optics to be used to monitor the cure process of the resin (Fernando and Degamber, 2006).

In principle any of the fluorescence, UV–visible, chemiluminescence or other viscosity-dependent phenomena discussed in detail in Section 3.3.8 could be used with these probes. It is emphasized that with the techniques which rely on an absolute measurement of emitted or transmitted light intensity one should include some internal reference material that is invariant over the full cure cycle in order to allow for drift in the total amount of light reaching the detector due to changes in refractive index or colour of the sample. Other artefacts include the formation of volatiles and bubbles in the optical path, and detachment of the fibre from the resin as cure occurs (due to differential shrinkage), so resulting in light scattering as well as changes in the fibre itself due to transmission losses from bending. Very often the sensitivity of the optical fibre has formed the basis for the cure-monitoring methodology, such as in the use of speckle interferometry to measure the change in refractive index as the resin cures (Zhang *et al.*, 1999).

The temperature dependences of many optical properties of the resin, e.g. fluorescence and Raman scattering (the ratio of Stokes to anti-Stokes intensities), provide an opportunity to use this as a way of monitoring temperature by comparison with a known standard material. Other systems are based on the properties of the fibre itself or a deliberately added dopant rather than the resin being probed. Table 6.4 shows the commercially available temperature probes that are based on optical phenomena and the use of fibre-optics (Fernando and Degamber, 2006).

Table 6.4. Commercially available temperature probes that are based on optical phenomena and fibre-optics

System	Sensor construction	Number of channels and measurements made
Luxtron[a]	A phosphor attached to the end of a quartz fibre	Four, 8 or 12 channels; temperature probe
Metricor[b]	A cavity resonator consisting of a layer of material whose refractive index varies with temperature	Four channels; temperature, pressure, refractive-index probes
Ipitek	A temperature-sensitive semiconductor platelet attached to the end of a quartz probe	Four to 28 channels; temperature probe
Takaoka	A GaAs crystal assembled onto a quartz fibre inserted in an air-filled glass tube	Modular (1–24 channels); temperature probe
FISO	A Fabry–Pérot cavity between two reflectors in a capillary	Sixteen channels; temperature, pressure, strain probes

From Fernando and Degamber (2006).
[a] Now LumaSense Technologies.
[b] Now Photonetics.

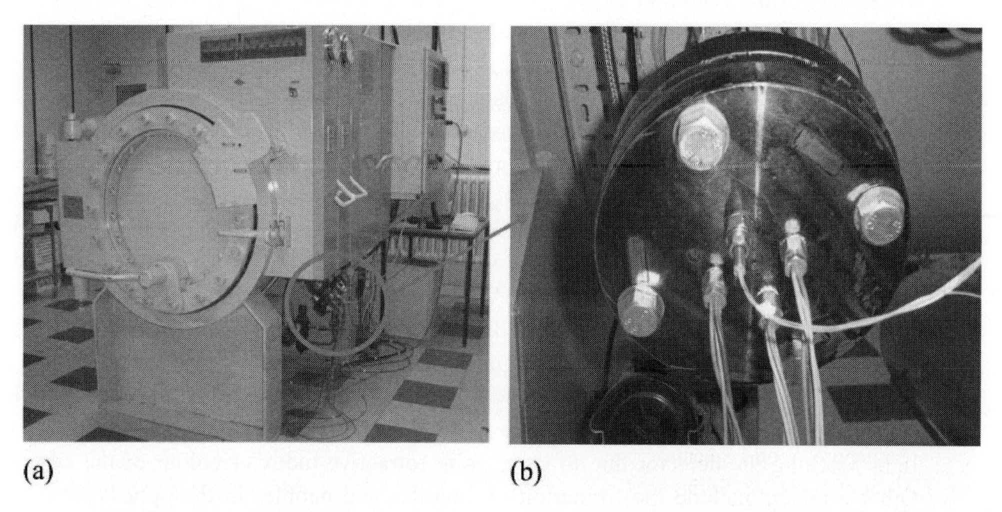

(a) (b)

Figure 6.29. (a) A photograph of an autoclave custom-modified to enable the accommodation of optical and electrical sensor systems for cure monitoring of advanced reinforced composites via contact and non-contact IR spectroscopy and measurements of residual strain and temperature. (b) Details of (a) showing the input and output ports for electrical and optical sensors (Fernando and Degamber, 2006). Reproduced with permission of Maney Pub. Co. Copyright (2006).

Those based on the fluorescence lifetime, rather than intensity (e.g. the Ipitek system), allow one to avoid problems of light loss and other factors that could affect calibration, as discussed above. Assessment of tilted Bragg gratings and long-period gratings on optical fibres has shown them to be a probe of cure of the resin as well as being both temperature- and strain-sensitive (Buggy *et al.*, 2007). The complexity of the response of these and fibre-optics based Fabry–Pérot interferometers to strain, temperature and refractive index makes it necessary to employ combinations of sensors if measurements of all of these properties are required separately.

References

Abadie, M., Chia, N. & Boey, F. (2002) *J. Appl. Polym. Sci.*, **86**, 1587–1591.

Alessi, S., Calderaro, E., Fuochi, P., Lavalle, A., Corda, U., Dispenza, C. & Spadaro, G. (2005) *Nucl. Instrum. Methods Phys. Res. B*, **236**, 55–60.

Baek, S., Cole, D., Rothschild, M., Switkes, M., Yeung, M. & Barouch, E. (2004) *J. Microlithography Microfabrication Microsystems*, **3**, 52–60.

Baird, W. (1977) *Radiat. Phys. Chem.*, **9**, 225.

Barone, M. & Caulk, D. (1985) *Polym. Composites*, **6**, 105.

Bartolo, P. (2006) *CIRP Annals – Manufacturing Technol.*, **55**, 221–225.

Behr, M., Rosentritt, M., Dummler, F. & Handel, G. (2006) *J. Oral Rehabilitation*, **33**, 447–451.

Berzin, F., Vergnes, B., Canevarolo, S., Machado, A. & Covas, J. (2006) *J. Appl. Polym, Sci.*, **99**, 2082–2090.

Blest, D., Duffy, B., McKee, S. & Zulkifle, A. (1999) *Composites A*, **30**, 1289–1309.

Bucknall, C. & Partridge, I. (1986) *Polym. Eng. Sci.*, **26**.

Buggy, S. J., Chehura, E., James, S. W. & Tatam, R. P. (2007) *J. Opt. A: Pure Appl. Opt.*, **9**, S60–S65.

Bunyawanichakul, P., Castanie, B. & Barrau, J. (2005) *Appl. Composite Mater.*, **12**, 177–191.

Campanelli, J., Gurer, C., Rose, T. & Varner, J. (2004) *Polym. Eng. Sci.*, **44**, 1247–1257.

Carlone, P., Palazzo, G. & Pasquino, R. (2006) *Math. Computer Modelling*, **44**, 701–709.

Casalini, R., Corezzi, S., Livi, A., Levita, G. & Rolla, P. A. (1997) *J. Appl. Polym. Sci.*, **65**, 17–25.

Castro, J. (1992) *Polym. Eng. Sci.*, **32**, 715.

Castro, J. & Macosko, C. (1982) *AIChE J.*, **28**, 251–260.

Chachad, Y., Roux, J. & Vaughan, J. (1995) *J. Reinforced Plastics and Composites*, **14**, 495–512.

Chachad, Y., Roux, J., Vaughan, J. & Arafa, E. (1996) *Composites*, **27**, 201–210.

Chen, J., Johnston, A., Petrescue, L. & Hojjati, M. (2006) *Radiation Phys. Chem.*, **75**, 336–349.

Chen, L., Hu, G. & Lindt, J. (1996) *Int. Polym. Processing* **XI**, 329.

Cherimisinoff, N. (1986) V9-Unit operations in rubber processing, in *Encyclopedia of Fluid Mechanics*, Houston, TX: Gulf Publishing Co.

Cho, J. & Hong, J. (2005) *J. Appl. Polym. Sci.*, **97**, 1345–1351.

Choudhury, R. (1997) *Handbook of Microlithography, Micromachining and Microfabrication*, Washington: SPIE.

Choulak, S., Couenne, F., Le Gorrec, Y. *et al.* (2004) *Indust. Eng. Chem. Res.*, **43**, 7373–7382.

Chua, C., Leong, F. & Lim, C. (2003) *Rapid Prototyping – Principles and Applications*, Singapore: World Scientific.

Ciriscioli, P. R. & Springer, G. S. (1990) *Smart Autoclave Cure of Composites*, Lancaster, PA: Technomic.

Clark, D. & Sutton, W. (1996) *Ann. Rev. Mater. Sci.*, **26**, 299–331.

Clegg, D. & Collyer, A. (1991) *Irradiation Effects on polymers*, London: Elsevier.

Cleland, M., Parks, L. & Cheng, S. (2003) *Nucl. Instrum. Methods Phys. Res. B*, **208**, 66–73.

Cole, D., Barouch, E. & Conrad, E. (2001) *Proc. IEEE*, **89**, 1194–1213.

Das, P., Ganguly, A. & Banerji, M. (2005) *J. Appl. Polym. Sci.*, **97**, 648–651.

Decker, C. (1989) Rubber processing, in Cheremisinoff, N. E. (Ed.) *Handbook of Polymer Science and Technology*, New York: Marcel Dekker.

Decker, C.(1994) *Acta Polymica*, **45**, 333–347.

Decker, C. & Moussa, K. (1990) *J. Coatings Technol.*, **62**, 55–61.

Degraff, R., Rohde, M. & Janssen, L. (1997) *Chem. Eng. Sci.*, **52**, 4345.

Deng, J. & Isayev, A. (1991) *Rubber Chem. Technolo.*, **64**, 296–324.

Dhavalikar, R. & Xanthos, M. (2003) *J. Appl. Polym. Sci.*, **87**, 643–652.

Dhavalikar, R. & Xanthos, M. (2004) *Polym. Eng. Sci.*, **44**, 474–486.
Dill, F. (1975) *IEEE Trans. Electron Devices*, **22**, 440–444.
Fernando, G. F. & Degamber, B. (2006) *Int. Mater. Rev.*, **51**, 65–106.
Fink, J. (2005) *Reactive Polymers; Fundamentals and Applications*, New York: William Andrews Publishing.
Fischer, D., Bayer, T., Eichhorn, K.-J. & Otto, M. (1997) *Fresenius J Anal Chem*, **359**, 74–77.
Fischer, D., Sahre, K., Abdelrhim, M., *et al.* (2006) *C. R. Chimie*, **9**, 1419–1424.
Flach, L. & Chartoff, R. (1995a) *Polym. Eng. Sci.*, **35**, 483–492.
Flach, L. & Chartoff, R. (1995b) *Polym. Eng. Sci.*, **35**, 493–502.
Gadam, S., Roux, J., Mccarty, T. & Vaughan, J. (2000) *Composites Sci. Technol.*, **60**, 945–958.
Gandhi, K. & Burns, R. (1976) *Trans. Soc. Rheology*, **20**, 489–502.
Gibson, A. & Williamson, G. (1985a) *Polym. Eng. Sci.*, **25**, 968.
Gibson, A. & Williamson, G. (1985b) *Polym. Eng. Sci.* , **25**, 980.
Girard-Reydet, E., Sautereau, H., Pascualt, J. P. *et al.* (1998) *Polymer*, **39**, 2269–2280.
Gorthala, R., Roux, J. & Vaughan, J. (1994) *Composite Mater.*, **28**, 486–506.
Gu, P., Zhang, X., Zeng, Y. & Ferguson, B. (2001) *J. Manufacturing Systems*, **20**, 250–263.
Haagh, G., Peters, G. & Meijer, H. (1996) *Polym. Eng. Sci.*, **36**, 2579.
Haberstroh, E. & Linhart, C. (2004) *J. Polymer Eng.*, **24**, 325–341.
Haberstroh, E., Michaeli, W. & Henze, E. (2002) *J. Reinforced Plastics and Composites*, **21**, 461–471.
Han, C. & Lee, D. (1987) *J. Appl. Polym. Sci.*, **33**, 2859–2876.
Han, C. & Lem, K. (1983a) *J. Appl. Polym. Sci.*, **28**, 743.
Han, C. & Lem, K. (1983b) *J. Appl. Polym. Sci.*, **28**, 763.
Han, C. & Lem, K. (1983c) *J. Appl. Polym. Sci.*, **28**, 3155.
Hedreul, C., Galy, J., Dupuy, J., Delmotte, M. & More, C. (1998) *J. App. Polym. Sci.*, **68**, 543–552.
Hou, T. (1986) *Society of Plastics Engineers ANTEC Conference, ANTEC 1986*, p. 1300.
Hu, N. & Tsai Ky, T. A. (2006) *Meas. Sci. Technol.*, **17**, 2233–2240.
Isayev, A. & Wan, M. (1998) *Rubber Chem. Technol.*, **71**, 1059–1072.
Isotalo, P., Bednarowski, D. & Nowak, T. (2004) *Int. J. Mater. Product Technol.*, **20**, 239–253.
Janssen, L. (1998) *Polym. Eng. Sci.*, **38**, 2010.
Kamal, M. & Ryan, M. (1980) *Polym. Eng. Sci.*, **20**, 859.
Kamal, M. & Sourour, S. (1973) *Polym. Eng. Sci.*, **13**, 59–64.
Kammona, O., Chatzi, E. G. & Kiparissides, C. (1999) *J. M.S. – Rev. Macromol. Chem. Phys.*, **C39**, 57–134.
Kau, H. & Hagerman, E. (1986) *Society of Plastics Engineers ANTEC Conference Proceedings, ANTEC 1986*, p. 1345.
Kenny, J., Maffezzoli, A. & Nicolais, L. (1990) *Composites Sci. Technol.*, **38**, 339–358.
Kim, B. & White, J. (2004) *J. Appl. Polym. Sci.*, **94**, 1007–1017
Kim, H. G. & Lee, D. G. (2002) *Composite Structures*, **57**, 91–99.
Klosterman, D., Chartoff, R., Tong, T. & Galaska, M. (2003) *Thermochimi. Acta*, **396**, 199–210.
Knauder, E., Kubla, C. & Poll, D. (1991) *Kunstoffe German Plastics*, **81**, 39.
Kranbuehl, D., Hood, D., Wang, Y. *et al.* (1997) *Polym. Adv. Technol.*, **8**, 93–99.
Kranbuehl, D. E. (1986) Electrical methods of characterising cure processes of polymers, in Pritchard, G. (Ed.) *Developments in Reinforced Plastics – 5*, Essex: Elsevier Applied Science.
Lee, D. & Han, C. (1987) *Polym. Eng. Sci.*, **27**, 955–963.
Lee, H. & Neville, K. (1957) *Epoxy Resins. Their Applications and Technology*, New York: McGraw-Hill.
Lee, J. & Cho, D. (2005) *J. Nanosci. Nanotechnol.*, **5**, 1637–1642.
Lee, L., Marker, L. & Griffih, R. (1981) *Polym. Composites*, **2**, 209.

Lem, K. & Han, C. (1983a) *J. Appl. Polym. Sci.*, **28**, 779.
Lem, K. & Han, C. (1983b) *J. Appl. Polym. Sci.*, **28**, 3185.
Lem, K. & Han, C. (1983c) *J. Appl. Polym. Sci.*, **28**, 3207.
Liu, L., Yi, S., Ong, L. & Chian, K. (2004) *Thin Solid Films*, **462**, 436–445.
Liu, L., Yi, S., Ong, L. *et al.* (2005) *IEEE Trans. Electronics Packaging Manufacturing*, **28**, 355–363.
Liu, X., Crouch, I. & Lam, Y. (2000) *Composites Sci. Technol*, **60**, 857–864.
Loan, L. (1977) *Radiat. Phys. Chem.*, **9**, 253.
Lobo, H. (1992) AC Technology Polymer Laboratories Report 1998–492.
Macosko, C. (1989) *Fundamentals of Reaction Injection Molding*, New York: Hanser.
Maier, C. & Lambla, M. (1995) *Polym. Eng. Sci.*, **35**, 1197.
Manzione, L., Osinski, J., Poelzing, G., Crouthamel, D. & Thierfelder, W. (1988) *Society of Plastic Engineers ANTEC Proceedings, ANTEC 1988*, p. 454.
McIlhagger, A., Brown, D. & Hill, B. (2000) *Composites A*, **31**, 1373–1381.
Mitani, T. & Hamada, H. (2005) *Polym. Eng. Sci.*, **45**, 364–374.
Moad, G. (1999a) *Prog. Polym. Sci.*, **24**, 1527–1528.
Moad, G. (1999b) *Prog. Polym. Sci.*, **24**, 81–142.
Murthy, P. & Prasad, R. (2006) *Defence Sci. J.*, **56**, 81–86.
Nelissen, L., Meijer, E. & Lemstra, P. (1992) *Polymer*, **33**, 3734.
Ng, H. & Manas Zloczower, I. (1989) *Polym. Eng. Sci.*, **29**, 1097.
Nguyen, L. (1993) *Proceedings from the 43rd IEEE Electronic Component and Technology Conference*, Buena Vista, FL.
Nguyen, L., Danker, A., Santhiran, N. & Shervin, C. (1992) *ASME Winter Annual Meeting*, Anaheim, CA.
Nho, Y., Kang, P. & Park, J. (2004) *Radiation Phys. Chem.*, **71**, 243–246.
Nichetti, D. (2003) *J. Polym. Eng.*, **23** 399–412.
Nichetti, D. (2004) *European Polym. J.*, **40**, 2401–2405.
Oliveira, J., Biscaia, E. & Pinto, J. (2003) *Macromol. Theor. Simulations*, **12**, 696–704.
Palmas, P., Le Campion, L., Bourgeoisat, C. & Martel, L. (2001) *Polymer*, **42**, 7675–7683.
Paterson, A. (1984) *Proceedings of Radiation Processing for Plastics and Rubbers*, London: London Plastics and Rubber Institute.
Pearson, R. & Yee, F. (1993) *J. Appl. Polym. Sci.*, **48**, 1051.
Pethrick, R. A. & Hayward, D. (2002) *Prog. Polym. Sci.*, **27**, 1983–2017.
Puaux, J., Cassagnau, P., Bozga, G. & Nagy, I. (2006) *Chem. Eng. Processing.*, **45**, 481–487.
Raper, K., Roux, J., Mccarty, T. & Vaughan, J. (1999) *Composites*, **30**, 1123–1132.
Riccardi, C., Borrajo, J., Williams, R. *et al.* (1996) *J. Polym. Sci. B: Polym. Phys.*, **34**, 349–356.
Roy, S., & Lawal, A., (2004) J. Reinforced Plastics Composites, **23**, 685–706.
Rudd, C., Long, A., Kendall, K. & Mangin, C. (1997) *Liquid Molding Technologies*, Cambridge: Woodhead Publishing.
Schut, J., Stamm, M., Dumon, M., Galy, J. & Gerard, J. (2003) *Macromol. Symp.*, **202**, 25–35.
Schwartz, G. (2001) *Rubber Chem. Technol.*, **74**, 116–123.
Semsarzadeh, M., Navarchian, A. & Morshedian, J. (2004) *Adv. Polym. Technol.*, **23**, 239–255.
Senturia, S. D. & Sheppard, N. F. (1986) Dielectric analysis of thermoset cure, in Dusek, K. (Ed.) *Advances in Polymer Science 80: Epoxy Resins and Composites IV*, Berlin: Springer-Verlag.
Shen, S. (1990) *Int. J. Numerical Methods Eng.*, **30**, 1633–1647.
Tang, Y., Henderson, C., Muzzy, J. & Rosen, D. (2004) *Int. J. Mater. Product Technol.*, **21**, 255–272.
Thomas, B. (2002) *Metall. Mater. Trans. B*, **33**, 795–812.
Thuillier, F. & Jullien, H. (1989) *Makromol. Chem.: Makromol. Symp.*, **25**, 63.

Turng, L. & Wang, V. (1993) *J. Reinforced Plastics and Composites*, **12**, 506.
Valliappan, M., Roux, J., Vaughan, J. & Arafat, E. (1996) *Composites B*, **27**, 1–9.
Venderbosch, R., Meijer, H. E. H. & Lemstra, P. J. (1994) *Polymer*, **35**, 4349.
Venderbosch, R., Meijer, H. & Lemstra, P. (1995) *Polymer*, **36**, 1167.
Vergnes, B. & Berzin, F. (2004) *Plastics Rubber Composites*, **33**, 409–415.
Vergnes, B. Della Valle, G. & Delamare, L. (1998) *Polym. Eng. Sci.*, **38**, 1781–1792.
Wan, M. & Isayev, A. (1996) *Rubber Chem. Technol.*, **69**, 294–312.
Wei, J., Hawle, M. & Demeuse, M. (1995) *Polym. Eng. Sci.*, **35**, 461–66.
Zagal, A., Vivaldo-Lima, E. & Manero, O. (2005) *Indust. Eng. Chem. Res.*, **44** 9805–9817.
Zhang, B., Wang, D., Du, S. & Song, Y. (1999) *Smart Mater. Structures*, **8**, 515–518.
Zhao, H., Turner, I., Yarlagadda, P. & Berg, K. (2001) *Int. J. Adv. Manufacturing Technol.*, **17**, 916–927.
Zhu, L., Narh, K. & Hyun, K. (2005) *Adv. Polym. Technol.*, **24**, 183–193.

Glossary of commonly used terms

A	pre-exponential factor for Arrhenius relation
ABS	poly(acrylonitrile-co-butadiene-co-styrene)
AFM	atomic-force microscopy
AIBN	2,2′-azobis-(isobutyronitrile)
ATBN	amine-terminated butadiene-acrylonitrile copolymer
ATR	attenuated total reflectance spectroscopy
ATRP	atom transfer radical polymerization
BHT	butylated hydroxy toluene (antioxidant)
C	Characteristic ratio
c^*	the overlap concentration
c^{++}	the critical packing concentration
c	concentration
C_p	heat capacity
CTBN	carboxy-terminated butadiene–acrylonitrile copolymer
CTPEHA	carboxy-terminated poly(2-ethyl hexyl acrylate)
CTT	conversion–temperature–time diagram
DB	degree of branching
DCP	dicumyl peroxide
DDS	4,4′-diamino-diphenyl sulfone
DETDA	diethyltoluene diamine
D_f	fractal dimension
DGEBA	diglycidyl ether of bisphenol-A
DGEBF	diglycidyl ether of bisphenol-F
DMA	dynamic mechanical analyser
DMA	dynamic mechanical analysis
DMC	dough moulding compound
DMTA	dynamic mechanical thermal analyser
DP	degree of polymerization
DSC	differential scanning calorimetry,
DTA	differential thermal analysis
DVB	divinyl benzene
E_a	activation energy for Arrhenius relation
EGDMA	ethylene glycol dimethacrylate
EPDM	ethylene-propylene-diene monomer (elastomer)
EPR	ethylene-propylene rubber
ESR	electron spin resonance spectroscopy
E_v	quantized energy

EVA	poly(ethylene-co-vinyl acetate)
E_x	activation energy of reaction x (e.g. propagation (p))
f	free-radical escape efficiency
f	functionality
FSD	Fourier self-deconvolution
FT-NIR	Fourier-transform near-infrared spectroscopy
G	Gibbs free energy
G'	storage (solid-like) modulus
G''	loss (viscous-like) modulus
gel T_g	the temperature at which gelation and vitrification coincide
GPC	gel permeation chromatography (a.k.a. SEC)
H	enthalpy
HBP	hyperbranched polymer
HDPE	high-density poly(ethylene)
HIPS	high-impact polystyrene
HPLC	high-performance liquid chromatography
I_{CL}	the intensity of emission of chemiluminescence
IPN	interpenetrating network
iso-T_g-TTT diagram	TTT diagram with lines of constant T_g
k	kinetic rate coefficient
k_B	Boltzmann constant
K	equilibrium constant
LAOS	large-amplitude oscillatory shear
LCST	lower critical solution temperature
LDPE	low-density poly(ethylene)
LLDPE	linear low-density poly(ethylene)
LS	light scattering
M	the ratio of viscosities of the droplet to the viscosity of the solution
M_0 (g/mol)	molar mass of the monomeric repeat unit
MA	maleic anhydride
MALDI-MS	matrix-assisted laser desorption ionisation mass spectrometry
MCDEA	4,4′-methylene *bis*[3-chloro 2,6-diethylaniline]
M_w	weight-average molecular weight
M_n	number-average molecular weight
M_v	viscosity-average molecular weight
M_z	*z*-average molecular weight
MIR	mid-infrared spectroscopy
MPDA	*meta*-phenylene diamine
MY721®	Ciba TGDDM resin
N	normal stress
n_0	refractive index
NIR	near-infrared spectroscopy
NMR	nuclear magnetic resonance spectroscopy
^{13}C NMR	carbon-13 nuclear magnetic resonance spectroscopy
OIT	oxidation induction time
p	degree of polymerization/extent of reaction
p	fraction of bonds

PALS	positron-annihilation spectroscopy
PAN	poly(acrylonitrile)
PB	poly(butadiene)
PBAN	poly(butyl acrylonitrile)
PBT	poly(butylene terephthalate)
PC	poly(carbonate)
p_c	percolation threshold
PCA	principal-component analysis
PCR	principal-component regression
PDI = M_w/M_n	polydispersity index
PE	poly(ethylene)
PEI	poly(ether imide)
PEK	poly(ether ketone)
PES	poly(ether sulfone)
PET	poly(ethylene terephthalate)
PHB	poly(hydroxy butyrate)
PLA	poly(lactic acid)
PLS	partial least squares
PMMA	poly(methyl methacrylate)
POOH	polymer hydroperoxide
PP	poly(propylene)
PPE	poly(2,6-dimethyl-1,4-phenylene ether)
PRESS	predicted residual sum of squares
PS	poly(styrene)
PU	poly(urethane)
PVAc	poly(vinyl acetate)
PVC	poly(vinyl chloride)
R_{rms}	root mean separation of polymer ends
R	gas constant
RAFT	reversible addition–fragmentation chain-transfer polymerization
REX	reactive extrusion
RTM	resin transfer moulding
R_g	radius of gyration
RIM	reaction injection moulding
R_o	actual chain end to end distance
r_p	rate of polymerization
S	entropy
SAN	poly(styrene-co-acrylonitrile)
SANS	small-angle neutron scattering
SAXS	small-angle X-ray scattering
SEBS	poly(styrene-co-ethylene-b-poly(butene-co-styrene))
SEC	size-exclusion chromatography (a.k.a. GPC)
SEM	scanning electron microscopy
SLA	stereolithography
SLS	selective laser sintering
SMCR	self-modelling curve resolution
SNR	signal-to-noise ratio

T	temperature
tan δ	loss tangent
TBA	torsional braid analyser
TETA	triethylene tetramine
T_c	crystallization temperature
T_c	polymerization ceiling temperature
T_c	isothermal curing temperature
TEM	transmission electron microscopy
T_g	glass-transition temperature
T_{g0}	glass-transition temperature of the initial uncured system
$T_{g\infty}$	maximum T_g of the cured system
TGA	thermo-gravimetric analysis
TGAP	triglycidyl *p*-amino phenol
TDAP	2,4,6-*tris*(dimethylaminomethyl)phenol
TGDDM	tetraglycidyl diaminodiphenyl methane
t_{gel}	gelation time
T_m	melting temperature
TMA	thermal mechanical analyser
TMA	thermo-mechanical analysis
TMAB	trimethylene glycol di-*p*-aminobenzoate
TPU	thermoplastic poly(urethane)
T_{rxn}	reaction temperature
TTT	time–temperature–transformation diagram
UCST	upper critical solution temperature
UHMWPE	ultra-high-molecular-weight poly(ethylene)
Ult T_g	the ultimate glass-transition temperature of the fully cured material
V_f	polymer free volume
V_s	specific volume
V_s	wall-slip velocity
WAXD	wide-angle X-ray diffraction
WLF	Williams, Landel and Ferry equation
XRD	X-ray diffraction

δ	solubility parameter
δ^2	cohesive energy density
κ	compressibility
ρ,	density
τ_{CL}	the lifetime of decay in chemiluminescence
τ_F	fluorescence lifetime
Φ_{CL}	chemiluminescence quantum yield
χ	the Flory–Huggins interaction parameter
a	coefficient of thermal expansion
a	cure conversion
a_c	critical conversion
a_{gel}	cure conversion at gelation
a	chain expansion co-efficient
ε''	dielectric loss

ε'	dielectric permittivity
ϕ	volume fraction of particles
ϕ_m	maximum packing volume fraction
γ	shear strain
γ'	steady shear rate
γ'_e	rate of elongation
$[\eta]$	intrinsic viscosity
η^*	dynamic or complex viscosity in oscillatory flow
η_e	elongation viscosity
η_r	η/η_s = the reduced viscosity
$\eta_{r\infty}$	the high-shear-rate viscosity
η_{ro}	the low-shear-rate viscosity
η	viscosity or chemoviscosity
η_{min}	minimum viscosity in thermoset processing
$\eta_c(T, t) = \eta_c(T, a)$	Temperature, time and conversion dependent viscosity during cure
$\eta_{sr}\ (\gamma', T)$	shear rate and temperature dependency of chemoviscosity
λ	relaxation time
θ condition	temperature at which chain expansion coefficient =1
σ	stress
σ_r	the reduced shear stress
σ_Y	yield stress
ω	the dynamic shear rate (or frequency)
ν	frequency of the fundamental vibration

Index